Bird's Basic Engineering Mathematics

Why is knowledge of mathematics important in engineering?

A career in any engineering or scientific field will require both basic and advanced mathematics. Without mathematics to determine principles, calculate dimensions and limits, explore variations, prove concepts, and so on, there would be no mobile telephones, televisions, stereo systems, video games, microwave ovens, computers or virtually anything electronic. There would be no bridges, tunnels, roads, skyscrapers, automobiles, ships, planes, rockets or most things mechanical. There would be no metals beyond the common ones, such as iron and copper, no plastics, no synthetics. In fact, society would most certainly be less advanced without the use of mathematics throughout the centuries and into the future.

Electrical engineers require mathematics to design, develop, test, or supervise the manufacturing and installation of electrical equipment, components or systems for commercial, industrial, military or scientific use.

Mechanical engineers require mathematics to perform engineering duties in planning and designing tools, engines, machines and other mechanically functioning equipment; they oversee installation, operation, maintenance and repair of such equipment as centralised heat, gas, water and steam systems.

Aerospace engineers require mathematics to perform a variety of engineering work in designing, constructing and testing aircraft, missiles, and spacecraft; they conduct basic and applied research to evaluate adaptability of materials and equipment to aircraft design and manufacture and recommend improvements in testing equipment and techniques.

Nuclear engineers require mathematics to conduct research on nuclear engineering problems or apply principles and theory of nuclear science to problems concerned with release, control and utilisation of nuclear energy and nuclear waste disposal.

Petroleum engineers require mathematics to devise methods to improve oil and gas well production and determine the need for new or modified tool designs; they oversee drilling and offer technical advice to achieve economical and satisfactory progress.

Industrial engineers require mathematics to design, develop, test and evaluate integrated systems for managing industrial production processes, including human work factors, quality control, inventory control, logistics and material flow, cost analysis and production coordination.

Environmental engineers require mathematics to design, plan, or perform engineering duties in the prevention, control and remediation of environmental health hazards, using various engineering disciplines; their work may include waste treatment, site remediation or pollution control technology.

Civil engineers require mathematics in all levels in civil engineering – structural engineering, hydraulics and geotechnical engineering are all fields that employ mathematical tools such as differential equations, tensor analysis, field theory, numerical methods and operations research.

Knowledge of mathematics is therefore needed by each of the engineering disciplines listed above.

It is intended that this text – *Bird's Basic Engineering Mathematics* – will provide a step by step approach to learning all the early, fundamental mathematics needed for your future engineering studies.

Now in its eighth edition, *Bird's Basic Engineering Mathematics* has helped thousands of students to succeed in their exams. Mathematical theories are explained in a straightforward manner, supported by practical engineering examples and applications to ensure that readers can relate theory to practice. Some 1,000 engineering situations/problems have been 'flagged-up' to help demonstrate that engineering cannot be fully understood without a good knowledge of mathematics.

The extensive and thorough coverage makes this a great text for introductory level engineering courses – such as for aeronautical, construction, electrical, electronic, mechanical, manufacturing engineering and vehicle technology – including for BTEC First, National and Diploma syllabuses, City & Guilds Technician Certificate and Diploma syllabuses, and even for GCSE revision.

Its companion website provides extra materials for students and lecturers, including full solutions for all 1,700 further questions, lists of essential formulae, multiple choice tests, and illustrations, as well as full solutions to revision tests for course instructors.

John Bird, BSc (Hons), CEng, CMath, CSci, FIMA, FIET, FCollT, is the former Head of Applied Electronics in the Faculty of Technology at Highbury College, Portsmouth, UK. More recently, he has combined freelance lecturing at the University of Portsmouth, with Examiner responsibilities for Advanced Mathematics with City and Guilds and examining for the International Baccalaureate Organisation. He has over 45 years' experience of successfully teaching, lecturing, instructing, training, educating and planning trainee engineers study programmes. He is the author of 146 textbooks on engineering, science and mathematical subjects, with worldwide sales of over one million copies. He is a chartered engineer, a chartered mathematician, a chartered scientist and a Fellow of three professional institutions. He has recently retired from lecturing at the Defence College of Marine Engineering in the Defence College of Technical Training at H.M.S. Sultan, Gosport, Hampshire, UK, one of the largest engineering training establishments in Europe.

To Sue

Bird's Basic Engineering Mathematics

Eighth Edition

John Bird

Routledge
Taylor & Francis Group

LONDON AND NEW YORK

Eighth edition published 2021
by Routledge
2 Park Square, Milton Park, Abingdon, Oxon, OX14 4RN

and by Routledge
52 Vanderbilt Avenue, New York, NY 10017

Routledge is an imprint of the Taylor & Francis Group, an informa business

© 2021 John Bird

First edition published by Newnes 1999
Seventh edition published by Routledge 2017

British Library Cataloguing-in-Publication Data
A catalogue record for this book is available from the British Library

Library of Congress Cataloging-in-Publication Data
A catalog record has been requested for this book

ISBN: 9780367643706 (hbk)
ISBN: 9780367643676 (pbk)
ISBN: 9781003124214 (ebk)

Typeset in Times by
Cenveo Publisher Services

Visit the companion website: www.routledge.com/cw/bird

Contents

Preface

Bird's Basic Engineering Mathematics 8th Edition introduces and then consolidates basic mathematical principles and promotes awareness of mathematical concepts for students needing a broad base for further vocational studies. In this eighth edition, examples and problems where *engineering applications* occur have been 'flagged up', new multiple-choice questions have been added to each chapter, the text has been added to and simplified, together with other minor modifications.

The text covers:

(i) **Basic mathematics** for a wide range of introductory/access/foundation mathematics courses

(ii) Mathematics contents of courses on **Engineering Principles**

(iii) '**Mathematics for Engineering Technicians**' for BTEC First NQF Level 2; *chapters 1 to 12, 16 to 18, 21, 22, 24, and 26 to 28 are needed for this module.*

(iv) The mandatory '**Mathematics for Technicians**' for BTEC National Certificate and National Diploma in Engineering, NQF Level 3; *chapters 7 to 10, 14 to 17, 19, 21 to 24, 26 to 28, 32, 33, 35 and 36 are needed for this module. In addition, chapters 1 to 6, 11 and 12 are helpful revision for this module.*

(v) **GCSE revision**, and for similar mathematics courses in English-speaking countries worldwide.

Bird's Basic Engineering Mathematics 8th Edition provides a lead into *Bird's Engineering Mathematics 9th Edition.*

Each topic considered in the text is presented in a way that assumes in the reader little previous knowledge of that topic.

Theory is introduced in each chapter by an outline of essential theory, definitions, formulae, laws and procedures. However, these are kept to a minimum, for problem solving is extensively used to establish and exemplify the theory. It is intended that readers will gain real understanding through seeing problems solved and then solving similar problems themselves.

This textbook contains over **800 worked problems**, followed by some **1700 further problems** (all with answers - at the end of the book). The further problems are contained within **201 Practice Exercises**; each Practice Exercise follows on directly from the relevant section of work. Fully worked solutions to all 1700 problems have been made freely available to all via the website www.routledge.com/cw/bird – see below. **427 line diagrams** enhance the understanding of the theory. Where at all possible the problems mirror potential practical situations found in engineering and science. In fact, some **1000 engineering situations/problems** have been 'flagged-up' to help demonstrate that engineering cannot be fully understood without a good knowledge of mathematics.

At regular intervals throughout the text are **15 Revision Tests** to check understanding. For example, Revision Test 1 covers material contained in chapters 1 and 2, Revision Test 2 covers the material contained in chapters 3 to 5, and so on. These Revision Tests do not have answers given since it is envisaged that lecturers/instructors could set the Tests for students to attempt as part of their course structure. Lecturers/instructors may obtain solutions to the Revision Tests in an **Instructor's Manual** available online at www.routledge.com/cw/bird – see below.

At the end of the book a list of relevant **formulae** contained within the text is included for convenience of reference.

'**Learning by Example**' is at the heart of *Bird's Basic Engineering Mathematics 8th Edition.*

JOHN BIRD
formerly Royal Naval Defence College of Marine and Air Engineering, HMS Sultan, University of Portsmouth and Highbury College, Portsmouth

Free Web downloads at www.routledge.com/cw/bird

For students

1. **Full solutions** to the 1700 questions contained in the 201 Practice Exercises

2. **List of essential formulae**

3. **Famous engineers/scientists** – From time to time in the text, 18 famous mathematicians/engineers are referred to and emphasised with an asterisk*. Background information on each of these is available via the website. Mathematicians/engineers involved are: **Boyle, Celsius, Charles, Descartes, Faraday, Henry, Hertz, Hooke, Kirchhoff, Leibniz, Morland, Napier, Newton, Ohm, Pascal, Pythagoras, Simpson and Young.**

For instructors/lecturers

1. **Full solutions** to the 1700 questions contained in the 201 Practice Exercises

2. **Full solutions** and marking scheme to each of the **15 Revision Tests**

3. **Revision Tests** – available to run off to be given to students

4. **List of essential formulae**

5. **Illustrations** – all 427 available on Power-Point

6. **Famous engineers/scientists** – 18 are mentioned in the text, as listed previously.

Acknowledgements

The publisher wishes to thank CASIO Electronic Co. Ltd, London for permission to reproduce the image of the Casio fx-991ES calculator on page 27.

The publisher also wishes to thank the AA Media Ltd for permission to reproduce the map of Portsmouth on page 166.

Chapter 1

Basic arithmetic

Why it is important to understand: Basic arithmetic

Being numerate, i.e. having an ability to add, subtract, multiply and divide whole numbers with some confidence, goes a long way towards helping you become competent at mathematics. Of course electronic calculators are a marvellous aid to the quite complicated calculations often required in engineering; however, having a feel for numbers 'in our head' can be invaluable when estimating. Do not spend too much time on this chapter because we deal with the calculator later; however, try to have some idea how to do quick calculations in the absence of a calculator. You will feel more confident in dealing with numbers and calculations if you can do this.

At the end of this chapter you should be able to:

- understand positive and negative integers
- add and subtract integers
- multiply and divide two integers
- multiply numbers up to 12×12 by rote
- determine the highest common factor from a set of numbers
- determine the lowest common multiple from a set of numbers
- appreciate the order of operation when evaluating expressions
- understand the use of brackets in expressions
- evaluate expressions containing $+, -, \times$, and brackets

1.1 Introduction

Whole numbers

Whole Numbers are simply the numbers **0, 1, 2, 3, 4, 5,** ...

Counting numbers

Counting Numbers are whole numbers, but **without the zero**, i.e. **1, 2, 3, 4, 5,** ...

Natural numbers

Natural Numbers can mean either counting numbers or whole numbers.

Integers

Integers are like whole numbers, but they **also include negative numbers**.

Examples of integers include $\ldots -5, -4, -3, -2, -1, 0, 1, 2, 3, 4, 5, \ldots$

Arithmetic operators

The four basic arithmetic operators are add $(+)$, subtract $(-)$, multiply $(\times)$ and divide $(\div)$.

It is assumed that adding, subtracting, multiplying and dividing reasonably small numbers can be achieved without a calculator. However, if revision of this area is needed then some worked problems are included in the following sections.

When **unlike signs** occur together in a calculation, the overall sign is **negative**. For example,

$$3 + (-4) = 3 + -4 = 3 - 4 = -1$$

and
$$(+5) \times (-2) = -10$$

Like signs together give an overall **positive sign**. For example,
$$3 - (-4) = 3 - -4 = 3 + 4 = 7$$
and
$$(-6) \times (-4) = +24$$

Prime numbers

A prime number can be divided, without a remainder, only by itself and by 1. For example, 17 can be divided only by 17 and by 1. Other examples of prime numbers are 2, 3, 5, 7, 11, 13, 19 and 23.

1.2 Revision of addition and subtraction

You can probably already add two or more numbers together and subtract one number from another. However, if you need revision then the following worked problems should be helpful.

Problem 1. Determine $735 + 167$

```
H T U
  7 3 5
+ 1 6 7
-------
  9 0 2
-------
  1 1
```

(i) $5 + 7 = 12$. Place 2 in units (U) column. Carry 1 in the tens (T) column.

(ii) $3 + 6 + 1$ (carried) $= 10$. Place the 0 in the tens column. Carry the 1 in the hundreds (H) column.

(iii) $7 + 1 + 1$ (carried) $= 9$. Place the 9 in the hundreds column.

Hence, $\mathbf{735 + 167 = 902}$

Problem 2. Determine $632 - 369$

```
H T U
  6 3 2
- 3 6 9
-------
  2 6 3
-------
```

(i) $2 - 9$ is not possible; therefore change one ten into ten units (leaving 2 in the tens column). In the units column, this gives us $12 - 9 = 3$

(ii) Place 3 in the units column.

(iii) $2 - 6$ is not possible; therefore change one hundred into ten tens (leaving 5 in the hundreds column). In the tens column, this gives us $12 - 6 = 6$

(iv) Place the 6 in the tens column.

(v) $5 - 3 = 2$

(vi) Place the 2 in the hundreds column.

Hence, $\mathbf{632 - 369 = 263}$

Problem 3. Add $27, -74, 81$ and -19

This problem is written as $27 - 74 + 81 - 19$.

Adding the positive integers:	27
	81
Sum of positive integers is	108
Adding the negative integers:	74
	19
Sum of negative integers is	93
Taking the sum of the negative integers from the sum of the positive integers gives	108
	−93
	15

Thus, $\mathbf{27 - 74 + 81 - 19 = 15}$

Problem 4. Subtract -74 from 377

This problem is written as $377 - -74$. Like signs together give an overall positive sign, hence

$$377 - -74 = 377 + 74$$

```
  3 7 7
+   7 4
-------
  4 5 1
-------
```

Thus, $\mathbf{377 - -74 = 451}$

Problem 5. Subtract 243 from 126

The problem is $126 - 243$. When the second number is larger than the first, take the smaller number from the larger and make the result negative. Thus,

$$126 - 243 = -(243 - 126)$$

```
  2 4 3
- 1 2 6
-------
  1 1 7
-------
```

Thus, $\mathbf{126 - 243 = -117}$

Problem 6.   Subtract 318 from -269

The problem is $-269 - 318$. The sum of the negative integers is

$$
\begin{array}{r}
2\,6\,9 \\
+\,3\,1\,8 \\
\hline
5\,8\,7
\end{array}
$$

Thus, $-269 - 318 = -587$

Now try the following Practice Exercise

Practice Exercise 1   Further problems on addition and subtraction (answers on page 442)

In Problems 1–15, determine the values of the expressions given, without using a calculator.

1. $67\,\text{kg} - 82\,\text{kg} + 34\,\text{kg}$

2. $73\,\text{m} - 57\,\text{m}$

3. $851\,\text{mm} - 372\,\text{mm}$

4. $124 - 273 + 481 - 398$

5. £927 − £114 + £182 − £183 − £247

6. $647 - 872$

7. $2417 - 487 + 2424 - 1778 - 4712$

8. $-38419 - 2177 + 2440 - 799 + 2834$

9. £2715 − £18250 + £11471 − £1509 + £113274

10. $47 + (-74) - (-23)$

11. $813 - (-674)$

12. $3151 - (-2763)$

13. $4872\,\text{g} - 4683\,\text{g}$

14. $-23148 - 47724$

15. \$53774 − \$38441

16. Calculate the diameter d and dimensions A and B for the template shown in Fig. 1.1. All dimensions are in millimetres.

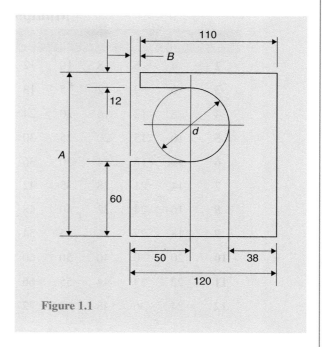

Figure 1.1

1.3   Revision of multiplication and division

You can probably already multiply two numbers together and divide one number by another. However, if you need a revision then the following worked problems should be helpful.

Problem 7.   Determine 86×7

$$
\begin{array}{r}
\text{H T U} \\
8\ 6 \\
\times \quad 7 \\
\hline
6\ 0\ 2 \\
\hline
4
\end{array}
$$

(i) $7 \times 6 = 42$. Place the 2 in the units (U) column and 'carry' the 4 into the tens (T) column.

(ii) $7 \times 8 = 56; 56 + 4\,(\text{carried}) = 60$. Place the 0 in the tens column and the 6 in the hundreds (H) column.

Hence, $86 \times 7 = 602$

A good grasp of **multiplication tables** is needed when multiplying such numbers; a reminder of the multiplication table up to 12×12 is shown below. Confidence with handling numbers will be greatly improved if this table is memorised.

Multiplication table

×	2	3	4	5	6	7	8	9	10	11	12
2	4	6	8	10	12	14	16	18	20	22	24
3	6	9	12	15	18	21	24	27	30	33	36
4	8	12	16	20	24	28	32	36	40	44	48
5	10	15	20	25	30	35	40	45	50	55	60
6	12	18	24	30	36	42	48	54	60	66	72
7	14	21	28	35	42	49	56	63	70	77	84
8	16	24	32	40	48	56	64	72	80	88	96
9	18	27	36	45	54	63	72	81	90	99	108
10	20	30	40	50	60	70	80	90	100	110	120
11	22	33	44	55	66	77	88	99	110	121	132
12	24	36	48	60	72	84	96	108	120	132	144

Problem 8.   Determine 764×38

$$
\begin{array}{r}
764 \\
\times\ \ 38 \\
\hline
6112 \\
22920 \\
\hline
29032 \\
\hline
\end{array}
$$

(i) $8 \times 4 = 32$. Place the 2 in the units column and carry 3 into the tens column.

(ii) $8 \times 6 = 48; 48 + 3$ (carried) $= 51$. Place the 1 in the tens column and carry the 5 into the hundreds column.

(iii) $8 \times 7 = 56; 56 + 5$ (carried) $= 61$. Place 1 in the hundreds column and 6 in the thousands column.

(iv)   Place 0 in the units column under the 2

(v) $3 \times 4 = 12$. Place the 2 in the tens column and carry 1 into the hundreds column.

(vi) $3 \times 6 = 18; 18 + 1$ (carried) $= 19$. Place the 9 in the hundreds column and carry the 1 into the thousands column.

(vii) $3 \times 7 = 21; 21 + 1$ (carried) $= 22$. Place 2 in the thousands column and 2 in the ten thousands column.

(viii) $6112 + 22920 = 29032$

Hence, **$764 \times 38 = 29032$**

Again, knowing multiplication tables is rather important when multiplying such numbers.

It is appreciated, of course, that such a multiplication can, and probably will, be performed using a **calculator**. However, there are times when a calculator may not be available and it is then useful to be able to calculate the 'long way'.

Problem 9.   Multiply 178 by -46

When the numbers have different signs, the result will be negative. (With this in mind, the problem can now be solved by multiplying 178 by 46.) Following the procedure of Problem 8 gives

$$
\begin{array}{r}
178 \\
\times\ \ 46 \\
\hline
1068 \\
7120 \\
\hline
8188 \\
\hline
\end{array}
$$

Thus, $178 \times 46 = 8188$ and **$178 \times (-46) = -8188$**

Problem 10.   Determine $1834 \div 7$

$$
\begin{array}{r}
262 \\
7\overline{)1834}
\end{array}
$$

(i)   7 into 18 goes 2, remainder 4. Place the 2 above the 8 of 1834 and carry the 4 remainder to the next digit on the right, making it 43

(ii) 7 into 43 goes 6, remainder 1. Place the 6 above the 3 of 1834 and carry the 1 remainder to the next digit on the right, making it 14

(iii) 7 into 14 goes 2, remainder 0. Place 2 above the 4 of 1834

Hence, $1834 \div 7 = 1834/7 = \dfrac{1834}{7} = \mathbf{262}$

The method shown is called **short division**.

Problem 11.   Determine $5796 \div 12$

$$
\begin{array}{r}
483 \\
12\overline{)5796} \\
48 \\
\overline{99} \\
96 \\
\overline{36} \\
36 \\
\overline{00}
\end{array}
$$

(i)   12 into 5 won't go. 12 into 57 goes 4; place 4 above the 7 of 5796

(ii) $4 \times 12 = 48$; place the 48 below the 57 of 5796

(iii) $57 - 48 = 9$

(iv)   Bring down the 9 of 5796 to give 99

(v)   12 into 99 goes 8; place 8 above the 9 of 5796

(vi) $8 \times 12 = 96$; place 96 below the 99

(vii) $99 - 96 = 3$

(viii) Bring down the 6 of 5796 to give 36

(ix)   12 into 36 goes 3 exactly.

(x)   Place the 3 above the final 6

(xi) $3 \times 12 = 36$; Place the 36 below the 36

(xii) $36 - 36 = 0$

Hence, $5796 \div 12 = 5796/12 = \dfrac{5796}{12} = \mathbf{483}$

The method shown is called **long division**.

Now try the following Practice Exercise

Practice Exercise 2   Further problems on multiplication and division (answers on page 442)

Determine the values of the expressions given in Problems 1 to 9, without using a calculator.

1. (a) 78×6 (b) 124×7

2. (a) £261×7 (b) £462×9

3. (a) $783\,kg \times 11$ (b) $73\,kg \times 8$

4. (a) $27\,mm \times 13$ (b) $77\,mm \times 12$

5. (a) 448×23 (b) $143 \times (-31)$

6. (a) $288\,m \div 6$ (b) $979\,m \div 11$

7. (a) $\dfrac{1813}{7}$ (b) $\dfrac{896}{16}$

8. (a) $\dfrac{21424}{13}$ (b) $15900 \div -15$

9. (a) $\dfrac{88737}{11}$ (b) $46858 \div 14$

10. A screw has a mass of 15 grams. Calculate, in kilograms, the mass of 1200 such screws ($1\,kg = 1000\,g$).

11. A builder needs to clear a site of bricks and top soil. The total weight to be removed is 696 tonnes. Trucks can carry a maximum load of 24 tonnes. Determine the number of truck loads needed to clear the site.

12. A machine can produce 400 springs in a day. Calculate the number of springs that can be produced using 7 machines in a 5-day working week.

1.4   Highest common factors and lowest common multiples

When two or more numbers are multiplied together, the individual numbers are called **factors**. Thus, a factor is a number which divides into another number exactly. The **highest common factor (HCF)** is the largest number which divides into two or more numbers exactly.
For example, consider the numbers 12 and 15
The factors of 12 are 1, 2, 3, 4, 6 and 12 (i.e. all the numbers that divide into 12).

The factors of 15 are 1, 3, 5 and 15 (i.e. all the numbers that divide into 15).

1 and 3 are the only **common factors**; i.e. numbers which are factors of **both** 12 and 15

Hence, **the HCF of 12 and 15 is 3** since 3 is the highest number which divides into **both** 12 and 15

A **multiple** is a number which contains another number an exact number of times. The smallest number which is exactly divisible by each of two or more numbers is called the **lowest common multiple (LCM)**.

For example, the multiples of 12 are 12, 24, 36, 48, 60, 72, ... and the multiples of 15 are 15, 30, 45, 60, 75, ...

60 is a common multiple (i.e. a multiple of **both** 12 and 15) and there are no lower common multiples.

Hence, **the LCM of 12 and 15 is 60** since 60 is the lowest number that both 12 and 15 divide into.

Here are some further problems involving the determination of HCFs and LCMs.

> **Problem 12.** Determine the HCF of the numbers 12, 30 and 42

Probably the simplest way of determining an HCF is to express each number in terms of its lowest factors. This is achieved by repeatedly dividing by the prime numbers 2, 3, 5, 7, 11, 13, ... (where possible) in turn. Thus,

$$12 = 2 \times 2 \times 3$$
$$30 = 2 \quad \times 3 \times 5$$
$$42 = 2 \quad \times 3 \times 7$$

The factors which are common to each of the numbers are 2 in column 1 and 3 in column 3, shown by the broken lines. Hence, **the HCF is 2 × 3**; i.e. **6**. That is, 6 is the largest number which will divide into 12, 30 and 42.

> **Problem 13.** Determine the HCF of the numbers 30, 105, 210 and 1155

Using the method shown in Problem 12:

$$30 = 2 \times 3 \times 5$$
$$105 = \quad 3 \times 5 \times 7$$
$$210 = 2 \times 3 \times 5 \times 7$$
$$1155 = \quad 3 \times 5 \times 7 \times 11$$

The factors which are common to each of the numbers are 3 in column 2 and 5 in column 3. Hence, **the HCF is 3 × 5 = 15**

> **Problem 14.** Determine the LCM of the numbers 12, 42 and 90

The LCM is obtained by finding the lowest factors of each of the numbers, as shown in Problems 12 and 13 above, and then selecting the largest group of any of the factors present. Thus,

$$12 = 2 \times 2 \times 3$$
$$42 = 2 \quad \times 3 \quad \times 7$$
$$90 = 2 \quad \times 3 \times 3 \times 5$$

The largest group of any of the factors present is shown by the broken lines and is 2 × 2 in 12, 3 × 3 in 90, 5 in 90 and 7 in 42

Hence, **the LCM is 2 × 2 × 3 × 3 × 5 × 7 = 1260** and is the smallest number which 12, 42 and 90 will all divide into exactly.

> **Problem 15.** Determine the LCM of the numbers 150, 210, 735 and 1365

Using the method shown in Problem 14 above:

$$150 = 2 \times 3 \times 5 \times 5$$
$$210 = 2 \times 3 \times 5 \quad \times 7$$
$$735 = \quad 3 \times 5 \quad \times 7 \times 7$$
$$1365 = \quad 3 \times 5 \quad \times 7 \quad \times 13$$

Hence, **the LCM is 2 × 3 × 5 × 5 × 7 × 7 × 13 = 95550**

Now try the following Practice Exercise

> **Practice Exercise 3 Further problems on highest common factors and lowest common multiples (answers on page 442)**
>
> Find (a) the HCF and (b) the LCM of the following groups of numbers.
>
> 1. 8, 12
> 2. 60, 72
> 3. 50, 70
> 4. 270, 900
> 5. 6, 10, 14
> 6. 12, 30, 45

7. 10, 15, 70, 105   8. 90, 105, 300
9. 196, 210, 462, 910   10. 196, 350, 770

1.5 Order of operation and brackets

Order of operation

Sometimes addition, subtraction, multiplication, division, powers and brackets may all be involved in a calculation. For example,

$$5 - 3 \times 4 + 24 \div (3 + 5) - 3^2$$

This is an extreme example but will demonstrate the order that is necessary when evaluating.

When we read, we read from left to right. However, with mathematics there is a definite order of precedence which we need to adhere to. The order is as follows:

Brackets
Order (or p**O**wer)
Division
Multiplication
Addition
Subtraction

Notice that the first letters of each word spell **BODMAS**, a handy aide-mémoire. **O**rder means p**O**wer. For example, $4^2 = 4 \times 4 = 16$
$5 - 3 \times 4 + 24 \div (3 + 5) - 3^2$ is evaluated as follows:

$5 - 3 \times 4 + 24 \div (3 + 5) - 3^2$

$= 5 - 3 \times 4 + 24 \div 8 - 3^2$ (**B**racket is removed and $3 + 5$ replaced with 8)

$= 5 - 3 \times 4 + 24 \div 8 - 9$ (**O**rder means p**O**wer; in this case, $3^2 = 3 \times 3 = 9$)

$= 5 - 3 \times 4 + 3 - 9$ (**D**ivision: $24 \div 8 = 3$)

$= 5 - 12 + 3 - 9$ (**M**ultiplication: $-3 \times 4 = -12$)

$= 8 - 12 - 9$ (**A**ddition: $5 + 3 = 8$)

$= -13$ (**S**ubtraction: $8 - 12 - 9 = -13$)

In practice, **it does not matter if multiplication is performed before division or if subtraction is performed before addition**. What is important is that **the process of multiplication and division must be completed before addition and subtraction**.

Brackets and operators

The basic laws governing the **use of brackets and operators** are shown by the following examples.

(a) $2 + 3 = 3 + 2$; i.e. the order of numbers when adding does not matter.

(b) $2 \times 3 = 3 \times 2$; i.e. the order of numbers when multiplying does not matter.

(c) $2 + (3 + 4) = (2 + 3) + 4$; i.e. the use of brackets when adding does not affect the result.

(d) $2 \times (3 \times 4) = (2 \times 3) \times 4$; i.e. the use of brackets when multiplying does not affect the result.

(e) $2 \times (3 + 4) = 2(3 + 4) = 2 \times 3 + 2 \times 4$; i.e. a number placed outside of a bracket indicates that the whole contents of the bracket must be multiplied by that number.

(f) $(2 + 3)(4 + 5) = (5)(9) = 5 \times 9 = 45$; i.e. adjacent brackets indicate multiplication.

(g) $2[3 + (4 \times 5)] = 2[3 + 20] = 2 \times 23 = 46$; i.e. when an expression contains inner and outer brackets, **the inner brackets are removed first**.

Here are some further problems in which BODMAS needs to be used.

Problem 16. Find the value of $6 + 4 \div (5 - 3)$

The order of precedence of operations is remembered by the word BODMAS. Thus,

$$6 + 4 \div (5 - 3) = 6 + 4 \div 2 \quad \text{(Brackets)}$$
$$= 6 + 2 \quad \text{(Division)}$$
$$= \mathbf{8} \quad \text{(Addition)}$$

Problem 17. Determine the value of $13 - 2 \times 3 + 14 \div (2 + 5)$

$13 - 2 \times 3 + 14 \div (2 + 5) = 13 - 2 \times 3 + 14 \div 7$ (B)
$= 13 - 2 \times 3 + 2$ (D)
$= 13 - 6 + 2$ (M)
$= 15 - 6$ (A)
$= \mathbf{9}$ (S)

Problem 18. Evaluate
$16 \div (2 + 6) + 18[3 + (4 \times 6) - 21]$

$16 \div (2 + 6) + 18[3 + (4 \times 6) - 21]$

$= 16 \div (2 + 6) + 18[3 + 24 - 21]$ (B: inner bracket is determined first)

$= 16 \div 8 + 18 \times 6$ (B)

$= 2 + 18 \times 6$ (D)

$= 2 + 108$ (M)

$= \mathbf{110}$ (A)

Note that a number outside of a bracket multiplies all that is inside the brackets. In this case,

$18[3 + 24 - 21] = 18[6]$, which means $18 \times 6 = 108$

Problem 19. Find the value of
$23 - 4(2 \times 7) + \dfrac{(144 \div 4)}{(14 - 8)}$

$23 - 4(2 \times 7) + \dfrac{(144 \div 4)}{(14 - 8)} = 23 - 4 \times 14 + \dfrac{36}{6}$ (B)

$= 23 - 4 \times 14 + 6$ (D)

$= 23 - 56 + 6$ (M)

$= 29 - 56$ (A)

$= \mathbf{-27}$ (S)

Problem 20. Evaluate
$$\frac{3 + \sqrt{(5^2 - 3^2)} + 2^3}{1 + (4 \times 6) \div (3 \times 4)} + \frac{15 \div 3 + 2 \times 7 - 1}{3 \times \sqrt{4} + 8 - 3^2 + 1}$$

$$\frac{3 + \sqrt{(5^2 - 3^2)} + 2^3}{1 + (4 \times 6) \div (3 \times 4)} + \frac{15 \div 3 + 2 \times 7 - 1}{3 \times \sqrt{4} + 8 - 3^2 + 1}$$

$$= \frac{3 + 4 + 8}{1 + 24 \div 12} + \frac{15 \div 3 + 2 \times 7 - 1}{3 \times 2 + 8 - 9 + 1}$$

$$= \frac{3 + 4 + 8}{1 + 2} + \frac{5 + 2 \times 7 - 1}{3 \times 2 + 8 - 9 + 1}$$

$$= \frac{15}{3} + \frac{5 + 14 - 1}{6 + 8 - 9 + 1}$$

$$= 5 + \frac{18}{6}$$

$$= 5 + 3 = \mathbf{8}$$

Now try the following Practice Exercise

Practice Exercise 4   Further problems on order of precedence and brackets (answers on page 442)

Evaluate the following expressions.

1. $14 + 3 \times 15$

2. $17 - 12 \div 4$

3. $86 + 24 \div (14 - 2)$

4. $7(23 - 18) \div (12 - 5)$

5. $63 - 8(14 \div 2) + 26$

6. $\dfrac{40}{5} - 42 \div 6 + (3 \times 7)$

7. $\dfrac{(50 - 14)}{3} + 7(16 - 7) - 7$

8. $\dfrac{(7 - 3)(1 - 6)}{4(11 - 6) \div (3 - 5)}$

9. $\dfrac{(3 + 9 \times 6) \div 3 - 2 \div 2}{3 \times 6 + (4 - 9) - 3^2 + 5}$

10. $\dfrac{(4 \times 3^2 + 24) \div 5 + 9 \times 3}{2 \times 3^2 - 15 \div 3} +$
$\dfrac{2 + 27 \div 3 + 12 \div 2 - 3^2}{5 + (13 - 2 \times 5) - 4}$

11. $\dfrac{1 + \sqrt{25} + 3 \times 2 - 8 \div 2}{3 \times 4 - \sqrt{(3^2 + 4^2)} + 1} -$
$\dfrac{(4 \times 2 + 7 \times 2) \div 11}{\sqrt{9} + 12 \div 2 - 2^3}$

Practice Exercise 5   Multiple-choice questions on basic arithmetic (answers on page 442)

Each question has only one correct answer

1. $(-5) - (-2) + (-3)$ is equal to:
    (a) -4 (b) 0 (c) -6 (d) -10

2.   Which of the following numbers is not an integer?
    (a) 0 (b) 2 (c) $\dfrac{1}{4}$ (d) -3

3. $6 \times (-2) - 18 \div 2 - 5$ is equal to:
    (a) -18 (b) -26 (c) 16 (d) -6

4. Which of the following is not a prime number?
    (a) 8 (b) 7 (c) 2 (d) 11

5. $15 - 3 \times 2 + 16 \div 2 + 6$ is equal to:
    (a) 23 (b) 26 (c) 38 (d) 11

6. Which prime numbers lies between 19 and 28?
    (a) 20 (b) 23 (c) 25 (d) 27

7. The lowest common multiple of 15 and 18 is:
    (a) 90 (b) 180 (c) 270 (d) 360

8. $45 + 30 \div (21 - 6) - 2 \times 5 + 1$ is equal to:
    (a) -4 (b) 35 (c) -7 (d) 38

9. The highest common factor of 54 and 60 is:
    (a) 2 (b) 3 (c) 6 (d) 12

10. $18 \div 2 + 4 - 10[4 + (5 \times 3) - 21]$ is equal to:
    (a) 11 (b) 33 (c) 23 (d) -11

11. The value of $2 + 2 - 2 \times 2 \div 2$ is:
    (a) 0 (b) 10 (c) 8 (d) 2

12. The H.C.F. of 8, 9 and 25 is:
    (a) 8 (b) 9 (c) 25 (d) 1

13. $(-5)^2 \times 3$ is equal to:
    (a) -75 (b) 15 (c) 75 (d) -15

14. The value of $3(27 - 19) \div \dfrac{(4+2)}{3} + (-1)$ is:
    (a) 10 (b) 24 (c) 13 (d) 11

15. The ratio between the L.C.M and H.C.F. of 5, 15 and 20 is:
    (a) 9:1 (b) 12:1 (c) 11:1 (d) 4:3

For fully worked solutions to each of the problems in Practice Exercises 1 to 4 in this chapter, go to the website:
www.routledge.com/cw/bird

Chapter 2

Fractions

Why it is important to understand: Fractions

Engineers use fractions all the time, examples including stress to strain ratios in mechanical engineering, chemical concentration ratios and reaction rates, and ratios in electrical equations to solve for current and voltage. Fractions are also used everywhere in science, from radioactive decay rates to statistical analysis. Calculators are able to handle calculations with fractions. However, there will be times when a quick calculation involving addition, subtraction, multiplication and division of fractions is needed. Again, do not spend too much time on this chapter because we deal with the calculator later; however, try to have some idea how to do quick calculations in the absence of a calculator. You will feel more confident to deal with fractions and calculations if you can do this.

At the end of this chapter you should be able to:

- understand the terminology numerator, denominator, proper and improper fractions and mixed numbers
- add and subtract fractions
- multiply and divide two fractions
- appreciate the order of operation when evaluating expressions involving fractions

---

2.1 Introduction

A mark of 9 out of 14 in an examination may be written as $\frac{9}{14}$ or 9/14. $\frac{9}{14}$ is an example of a fraction. The number above the line, i.e. 9, is called the **numerator**. The number below the line, i.e. 14, is called the **denominator**.

When the value of the numerator is less than the value of the denominator, the fraction is called a **proper fraction**. $\frac{9}{14}$ is an example of a proper fraction.

When the value of the numerator is greater than the value of the denominator, the fraction is called an **improper fraction**. $\frac{5}{2}$ is an example of an improper fraction.

A **mixed number** is a combination of a whole number and a fraction. $2\frac{1}{2}$ is an example of a mixed number. In fact, $\frac{5}{2} = 2\frac{1}{2}$

There are a number of everyday examples in which fractions are readily referred to. For example, three people equally sharing a bar of chocolate would have $\frac{1}{3}$ each. A supermarket advertises $\frac{1}{5}$ off a six-pack of beer; if the beer normally costs £2 then it will now cost £1.60. $\frac{3}{4}$ of the employees of a company are women; if the company has 48 employees, then 36 are women.

Calculators are able to handle calculations with fractions. However, to understand a little more about fractions we will in this chapter show how to add, subtract, multiply and divide with fractions without the use of a calculator.

> **Problem 1.** Change the following improper fractions into mixed numbers:
>
> $$\text{(a) } \frac{9}{2} \quad \text{(b) } \frac{13}{4} \quad \text{(c) } \frac{28}{5}$$

(a) $\frac{9}{2}$ means 9 halves and $\frac{9}{2} = 9 \div 2$, and $9 \div 2 = 4$ and 1 half, i.e.

$$\frac{9}{2} = 4\frac{1}{2}$$

(b) $\frac{13}{4}$ means 13 quarters and $\frac{13}{4} = 13 \div 4$, and $13 \div 4 = 3$ and 1 quarter, i.e.

$$\frac{13}{4} = 3\frac{1}{4}$$

(c) $\frac{28}{5}$ means 28 fifths and $\frac{28}{5} = 28 \div 5$, and $28 \div 5 = 5$ and 3 fifths, i.e.

$$\frac{28}{5} = 5\frac{3}{5}$$

> **Problem 2.** Change the following mixed numbers into improper fractions:
>
> $$\text{(a) } 5\frac{3}{4} \quad \text{(b) } 1\frac{7}{9} \quad \text{(c) } 2\frac{3}{7}$$

(a) $5\frac{3}{4}$ means $5 + \frac{3}{4}$. 5 contains $5 \times 4 = 20$ quarters. Thus, $5\frac{3}{4}$ contains $20 + 3 = 23$ quarters, i.e.

$$5\frac{3}{4} = \frac{23}{4}$$

The quick way to change $5\frac{3}{4}$ into an improper fraction is $\frac{4 \times 5 + 3}{4} = \frac{23}{4}$

(b) $1\frac{7}{9} = \frac{9 \times 1 + 7}{9} = \frac{16}{9}$

(c) $2\frac{3}{7} = \frac{7 \times 2 + 3}{7} = \frac{17}{7}$

> **Problem 3.** In a school there are 180 students of which 72 are girls. Express this as a fraction in its simplest form

The fraction of girls is $\frac{72}{180}$

Dividing both the numerator and denominator by the lowest prime number, i.e. 2, gives

$$\frac{72}{180} = \frac{36}{90}$$

Dividing both the numerator and denominator again by 2 gives

$$\frac{72}{180} = \frac{36}{90} = \frac{18}{45}$$

2 will not divide into both 18 and 45, so dividing both the numerator and denominator by the next prime number, i.e. 3, gives

$$\frac{72}{180} = \frac{36}{90} = \frac{18}{45} = \frac{6}{15}$$

Dividing both the numerator and denominator again by 3 gives

$$\frac{72}{180} = \frac{36}{90} = \frac{18}{45} = \frac{6}{15} = \frac{2}{5}$$

So $\frac{72}{180} = \frac{2}{5}$ **in its simplest form**.

Thus, $\frac{2}{5}$ **of the students are girls**.

2.2 Adding and subtracting fractions

When the denominators of two (or more) fractions to be added are the same, the fractions can be added 'on sight'.

For example, $\frac{2}{9} + \frac{5}{9} = \frac{7}{9}$ and $\frac{3}{8} + \frac{1}{8} = \frac{4}{8}$

In the latter example, dividing both the 4 and the 8 by 4 gives $\frac{4}{8} = \frac{1}{2}$, which is the simplified answer. This is called **cancelling**.

Addition and subtraction of fractions is demonstrated in the following worked examples.

> **Problem 4.** Simplify $\frac{1}{3} + \frac{1}{2}$

(i) Make the denominators the same for each fraction. The lowest number that both denominators divide into is called the **lowest common multiple** or **LCM** (see Chapter 1, page 6). In this example, the LCM of 3 and 2 is 6

(ii) 3 divides into 6 twice. Multiplying both numerator and denominator of $\frac{1}{3}$ by 2 gives

$$\frac{1}{3} = \frac{2}{6}$$

(iii) 2 divides into 6, 3 times. Multiplying both numerator and denominator of $\frac{1}{2}$ by 3 gives

$$\frac{1}{2} = \frac{3}{6}$$

(iv) Hence,

$$\frac{1}{3} + \frac{1}{2} = \frac{2}{6} + \frac{3}{6} = \mathbf{\frac{5}{6}}$$

Problem 5. Simplify $\frac{3}{4} - \frac{7}{16}$

(i) Make the denominators the same for each fraction. The lowest common multiple (LCM) of 4 and 16 is 16

(ii) 4 divides into 16, 4 times. Multiplying both numerator and denominator of $\frac{3}{4}$ by 4 gives

$$\frac{3}{4} = \frac{12}{16}$$

(iii) $\frac{7}{16}$ already has a denominator of 16

(iv) Hence,

$$\frac{3}{4} - \frac{7}{16} = \frac{12}{16} - \frac{7}{16} = \mathbf{\frac{5}{16}}$$

Problem 6. Simplify $4\frac{2}{3} - 1\frac{1}{6}$

$4\frac{2}{3} - 1\frac{1}{6}$ is the same as $\left(4\frac{2}{3}\right) - \left(1\frac{1}{6}\right)$ which is the same as $\left(4 + \frac{2}{3}\right) - \left(1 + \frac{1}{6}\right)$ which is the same as

$4 + \frac{2}{3} - 1 - \frac{1}{6}$ which is the same as $3 + \frac{2}{3} - \frac{1}{6}$ which is the same as $3 + \frac{4}{6} - \frac{1}{6} = 3 + \frac{3}{6} = 3 + \frac{1}{2}$

Thus, $4\frac{2}{3} - 1\frac{1}{6} = \mathbf{3\frac{1}{2}}$

Problem 7. Evaluate $7\frac{1}{8} - 5\frac{3}{7}$

$$7\frac{1}{8} - 5\frac{3}{7} = \left(7 + \frac{1}{8}\right) - \left(5 + \frac{3}{7}\right) = 7 + \frac{1}{8} - 5 - \frac{3}{7}$$

$$= 2 + \frac{1}{8} - \frac{3}{7} = 2 + \frac{7 \times 1 - 8 \times 3}{56}$$

$$= 2 + \frac{7 - 24}{56} = 2 + \frac{-17}{56} = 2 - \frac{17}{56}$$

$$= \frac{112}{56} - \frac{17}{56} = \frac{112 - 17}{56} = \frac{95}{56} = \mathbf{1\frac{39}{56}}$$

Problem 8. Determine the value of

$$4\frac{5}{8} - 3\frac{1}{4} + 1\frac{2}{5}$$

$$4\frac{5}{8} - 3\frac{1}{4} + 1\frac{2}{5} = (4 - 3 + 1) + \left(\frac{5}{8} - \frac{1}{4} + \frac{2}{5}\right)$$

$$= 2 + \frac{5 \times 5 - 10 \times 1 + 8 \times 2}{40}$$

$$= 2 + \frac{25 - 10 + 16}{40}$$

$$= 2 + \frac{31}{40} = \mathbf{2\frac{31}{40}}$$

⚑ **Problem 9.** A small concert hall is used for an engineering conference. $\frac{2}{5}$ of the delegates were sat in the stalls, $\frac{4}{15}$ were sat in the upper circle and the remaining 210 were sat in the circle. Calculate the number of delegates at the conference.

The fraction of delegates in the stalls and upper circle

$$= \frac{2}{5} + \frac{4}{15} = \frac{6 + 4}{15} = \frac{10}{15} = \frac{2}{3}$$

Therefore, $1 - \frac{2}{3}$ i.e. $\frac{1}{3}$ of the delegates are sat in the circle.

This means that 210 delegates is $\frac{1}{3}$ of the total.

Hence, **the number of delegates at the conference is** $3 \times 210 = \mathbf{630}$ **delegates**

Now try the following Practice Exercise

1. Change the improper fraction $\frac{15}{7}$ into a mixed number.

2. Change the improper fraction $\frac{37}{5}$ into a mixed number.

3. Change the mixed number $2\frac{4}{9}$ into an improper fraction.

4. Change the mixed number $8\frac{7}{8}$ into an improper fraction.

5. A box contains 165 paper clips. 60 clips are removed from the box. Express this as a fraction in its simplest form.

6. Order the following fractions from the smallest to the largest.

$$\frac{4}{9}, \frac{5}{8}, \frac{3}{7}, \frac{1}{2}, \frac{3}{5}$$

7. A training college has 375 students of which 120 are girls. Express this as a fraction in its simplest form.

Evaluate, in fraction form, the expressions given in Problems 8 to 20.

8. $\frac{1}{3} + \frac{2}{5}$ 	9. $\frac{5}{6} - \frac{4}{15}$

10. $\frac{1}{2} + \frac{2}{5}$ 	11. $\frac{7}{16} - \frac{1}{4}$

12. $\frac{2}{7} + \frac{3}{11}$ 	13. $\frac{2}{9} - \frac{1}{7} + \frac{2}{3}$

14. $3\frac{2}{5} - 2\frac{1}{3}$ 	15. $\frac{7}{27} - \frac{2}{3} + \frac{5}{9}$

16. $5\frac{3}{13} + 3\frac{3}{4}$ 	17. $4\frac{5}{8} - 3\frac{2}{5}$

18. $10\frac{3}{7} - 8\frac{2}{3}$ 	19. $3\frac{1}{4} - 4\frac{4}{5} + 1\frac{5}{6}$

20. $5\frac{3}{4} - 1\frac{2}{5} - 3\frac{1}{2}$

21. The movement ratio, M, of a differential pulley is given by the formula: $M = \dfrac{2R}{R - r}$ where R and r are the radii of the larger and smaller portions of the stepped pulley. Find the movement ratio of such a pulley block having diameters of 140 mm and 120 mm.

2.3 Multiplication and division of fractions

Multiplication

To multiply two or more fractions together, the numerators are first multiplied to give a single number and this becomes the new numerator of the combined fraction. The denominators are then multiplied together to give the new denominator of the combined fraction.

For example, $\quad \dfrac{2}{3} \times \dfrac{4}{7} = \dfrac{2 \times 4}{3 \times 7} = \mathbf{\dfrac{8}{21}}$

Problem 10. Simplify $7 \times \dfrac{2}{5}$

$$7 \times \frac{2}{5} = \frac{7}{1} \times \frac{2}{5} = \frac{7 \times 2}{1 \times 5} = \frac{14}{5} = \mathbf{2\frac{4}{5}}$$

Problem 11. Find the value of $\dfrac{3}{7} \times \dfrac{14}{15}$

Dividing numerator and denominator by 3 gives

$$\frac{3}{7} \times \frac{14}{15} = \frac{1}{7} \times \frac{14}{5} = \frac{1 \times 14}{7 \times 5}$$

Dividing numerator and denominator by 7 gives

$$\frac{1 \times 14}{7 \times 5} = \frac{1 \times 2}{1 \times 5} = \mathbf{\frac{2}{5}}$$

This process of dividing both the numerator and denominator of a fraction by the same factor(s) is called **cancelling**.

Problem 12. Simplify $\dfrac{3}{5} \times \dfrac{4}{9}$

$$\frac{3}{5} \times \frac{4}{9} = \frac{1}{5} \times \frac{4}{3} \text{ by cancelling}$$
$$= \frac{4}{15}$$

Problem 13. Evaluate $1\dfrac{3}{5} \times 2\dfrac{1}{3} \times 3\dfrac{3}{7}$

Mixed numbers **must** be expressed as improper fractions before multiplication can be performed. Thus,

$$1\frac{3}{5} \times 2\frac{1}{3} \times 3\frac{3}{7} = \left(\frac{5}{5}+\frac{3}{5}\right) \times \left(\frac{6}{3}+\frac{1}{3}\right) \times \left(\frac{21}{7}+\frac{3}{7}\right)$$
$$= \frac{8}{5} \times \frac{7}{3} \times \frac{24}{7} = \frac{8 \times 1 \times 8}{5 \times 1 \times 1} = \frac{64}{5}$$
$$= 12\frac{4}{5}$$

Problem 14. Simplify $3\dfrac{1}{5} \times 1\dfrac{2}{3} \times 2\dfrac{3}{4}$

The mixed numbers need to be changed to improper fractions before multiplication can be performed.

$$3\frac{1}{5} \times 1\frac{2}{3} \times 2\frac{3}{4} = \frac{16}{5} \times \frac{5}{3} \times \frac{11}{4}$$
$$= \frac{4}{1} \times \frac{1}{3} \times \frac{11}{1} \text{ by cancelling}$$
$$= \frac{4 \times 1 \times 11}{1 \times 3 \times 1} = \frac{44}{3} = 14\frac{2}{3}$$

Division

The simple rule for division is **change the division sign into a multiplication sign and invert the second fraction**.

For example, $\dfrac{2}{3} \div \dfrac{3}{4} = \dfrac{2}{3} \times \dfrac{4}{3} = \dfrac{8}{9}$

Problem 15. Simplify $\dfrac{3}{7} \div \dfrac{8}{21}$

$$\frac{3}{7} \div \frac{8}{21} = \frac{3}{7} \times \frac{21}{8} = \frac{3}{1} \times \frac{3}{8} \text{ by cancelling}$$
$$= \frac{3 \times 3}{1 \times 8} = \frac{9}{8} = 1\frac{1}{8}$$

Problem 16. Find the value of $5\dfrac{3}{5} \div 7\dfrac{1}{3}$

The mixed numbers must be expressed as improper fractions. Thus,

$$5\frac{3}{5} \div 7\frac{1}{3} = \frac{28}{5} \div \frac{22}{3} = \frac{28}{5} \times \frac{3}{22} = \frac{14}{5} \times \frac{3}{11} = \frac{42}{55}$$

Problem 17. Simplify $3\dfrac{2}{3} \times 1\dfrac{3}{4} \div 2\dfrac{3}{4}$

Mixed numbers must be expressed as improper fractions before multiplication and division can be performed:

$$3\frac{2}{3} \times 1\frac{3}{4} \div 2\frac{3}{4} = \frac{11}{3} \times \frac{7}{4} \div \frac{11}{4} = \frac{11}{3} \times \frac{7}{4} \times \frac{4}{11}$$
$$= \frac{1 \times 7 \times 1}{3 \times 1 \times 1} \text{ by cancelling}$$
$$= \frac{7}{3} = 2\frac{1}{3}$$

Now try the following Practice Exercise

Practice Exercise 7   Multiplying and dividing fractions (answers on page 443)

Evaluate the following.

1. $\dfrac{2}{5} \times \dfrac{4}{7}$ 2. $5 \times \dfrac{4}{9}$

3. $\dfrac{3}{4} \times \dfrac{8}{11}$ 4. $\dfrac{3}{4} \times \dfrac{5}{9}$

5. $\dfrac{17}{35} \times \dfrac{15}{68}$ 6. $\dfrac{3}{5} \times \dfrac{7}{9} \times 1\dfrac{2}{7}$

7. $\dfrac{13}{17} \times 4\dfrac{7}{11} \times 3\dfrac{4}{39}$ 8. $\dfrac{1}{4} \times \dfrac{3}{11} \times 1\dfrac{5}{39}$

9. $\dfrac{2}{9} \div \dfrac{4}{27}$ 10. $\dfrac{3}{8} \div \dfrac{45}{64}$

11. $\dfrac{3}{8} \div \dfrac{5}{32}$ 12. $\dfrac{3}{4} \div 1\dfrac{4}{5}$

13. $2\dfrac{1}{4} \times 1\dfrac{2}{3}$ 14. $1\dfrac{1}{3} \div 2\dfrac{5}{9}$

15. $2\dfrac{4}{5} \div \dfrac{7}{10}$ 16. $2\dfrac{3}{4} \div 3\dfrac{2}{3}$

17. $\dfrac{1}{9} \times \dfrac{3}{4} \times 1\dfrac{1}{3}$ 18. $3\dfrac{1}{4} \times 1\dfrac{3}{5} \div \dfrac{2}{5}$

19. A ship's crew numbers 105, of which $\frac{1}{7}$ are women. Of the men, $\frac{1}{6}$ are officers. How many male officers are on board?

20. If a storage tank is holding 450 litres when it is three-quarters full, how much will it contain when it is two-thirds full?

21. Three people, P, Q and R, contribute to a fund. P provides 3/5 of the total, Q provides 2/3 of the remainder and R provides £8. Determine (a) the total of the fund and (b) the contributions of P and Q.

22. A tank contains 24,000 litres of oil. Initially, $\frac{7}{10}$ of the contents are removed, then $\frac{3}{5}$ of the remainder is removed. How much oil is left in the tank?

2.4 Order of operation with fractions

As stated in Chapter 1, sometimes addition, subtraction, multiplication, division, powers and brackets can all be involved in a calculation. A definite order of precedence must be adhered to. The order is:

Brackets

Order (or p**O**wer)

Division

Multiplication

Addition

Subtraction

This is demonstrated in the following worked problems.

Problem 18. Simplify $\frac{7}{20} - \frac{3}{8} \times \frac{4}{5}$

$$\frac{7}{20} - \frac{3}{8} \times \frac{4}{5} = \frac{7}{20} - \frac{3 \times 1}{2 \times 5} \text{ by cancelling}$$

$$= \frac{7}{20} - \frac{3}{10} \qquad (M)$$

$$= \frac{7}{20} - \frac{6}{20}$$

$$= \frac{1}{20} \qquad (S)$$

Problem 19. Simplify $\frac{1}{4} - 2\frac{1}{5} \times \frac{5}{8} + \frac{9}{10}$

$$\frac{1}{4} - 2\frac{1}{5} \times \frac{5}{8} + \frac{9}{10} = \frac{1}{4} - \frac{11}{5} \times \frac{5}{8} + \frac{9}{10}$$

$$= \frac{1}{4} - \frac{11}{1} \times \frac{1}{8} + \frac{9}{10} \text{ by cancelling}$$

$$= \frac{1}{4} - \frac{11}{8} + \frac{9}{10} \qquad (M)$$

$$= \frac{1 \times 10}{4 \times 10} - \frac{11 \times 5}{8 \times 5} + \frac{9 \times 4}{10 \times 4}$$

(since the LCM of 4, 8 and 10 is 40)

$$= \frac{10}{40} - \frac{55}{40} + \frac{36}{40}$$

$$= \frac{10 - 55 + 36}{40} \qquad (A/S)$$

$$= -\frac{9}{40}$$

Problem 20. Simplify

$$2\frac{1}{2} - \left(\frac{2}{5} + \frac{3}{4}\right) \div \left(\frac{5}{8} \times \frac{2}{3}\right)$$

$$2\frac{1}{2} - \left(\frac{2}{5} + \frac{3}{4}\right) \div \left(\frac{5}{8} \times \frac{2}{3}\right)$$

$$= \frac{5}{2} - \left(\frac{2 \times 4}{5 \times 4} + \frac{3 \times 5}{4 \times 5}\right) \div \left(\frac{5}{8} \times \frac{2}{3}\right) \quad (B)$$

$$= \frac{5}{2} - \left(\frac{8}{20} + \frac{15}{20}\right) \div \left(\frac{5}{8} \times \frac{2}{3}\right) \quad (B)$$

$$= \frac{5}{2} - \frac{23}{20} \div \left(\frac{5}{4} \times \frac{1}{3}\right) \text{ by cancelling} \quad (B)$$

$$= \frac{5}{2} - \frac{23}{20} \div \frac{5}{12} \quad (B)$$

$$= \frac{5}{2} - \frac{23}{20} \times \frac{12}{5} \quad (D)$$

$$= \frac{5}{2} - \frac{23}{5} \times \frac{3}{5} \text{ by cancelling}$$

$$= \frac{5}{2} - \frac{69}{25} \quad (M)$$

$$= \frac{5 \times 25}{2 \times 25} - \frac{69 \times 2}{25 \times 2} \quad (S)$$

$$= \frac{125}{50} - \frac{138}{50} \quad (S)$$

$$= -\frac{13}{50}$$

Problem 21. Evaluate

$$\frac{1}{3} \text{ of } \left(5\frac{1}{2} - 3\frac{3}{4}\right) + 3\frac{1}{5} \div \frac{4}{5} - \frac{1}{2}$$

$$\frac{1}{3} \text{ of } \left(5\frac{1}{2} - 3\frac{3}{4}\right) + 3\frac{1}{5} \div \frac{4}{5} - \frac{1}{2}$$

$$= \frac{1}{3} \text{ of } 1\frac{3}{4} + 3\frac{1}{5} \div \frac{4}{5} - \frac{1}{2} \qquad \text{(B)}$$

$$= \frac{1}{3} \times \frac{7}{4} + \frac{16}{5} \div \frac{4}{5} - \frac{1}{2} \qquad \text{(O)}$$

(Note that the 'of' is replaced with a multiplication sign.)

$$= \frac{1}{3} \times \frac{7}{4} + \frac{16}{5} \times \frac{5}{4} - \frac{1}{2} \qquad \text{(D)}$$

$$= \frac{1}{3} \times \frac{7}{4} + \frac{4}{1} \times \frac{1}{1} - \frac{1}{2} \text{ by cancelling}$$

$$= \frac{7}{12} + \frac{4}{1} - \frac{1}{2} \qquad \text{(M)}$$

$$= \frac{7}{12} + \frac{48}{12} - \frac{6}{12} \qquad \text{(A/S)}$$

$$= \frac{49}{12}$$

$$= \mathbf{4\frac{1}{12}}$$

Now try the following Practice Exercise

Practice Exercise 8   Order of operation with fractions (answers on page 443)

Evaluate the following.

1. $2\frac{1}{2} - \frac{3}{5} \times \frac{20}{27}$ 2. $\frac{1}{3} - \frac{3}{4} \times \frac{16}{27}$

3. $\frac{1}{2} + \frac{3}{5} \div \frac{9}{15} - \frac{1}{3}$ 4. $\frac{1}{5} + 2\frac{2}{3} \div \frac{5}{9} - \frac{1}{4}$

5. $\frac{4}{5} \times \frac{1}{2} - \frac{1}{6} \div \frac{2}{5} + \frac{2}{3}$

6. $\frac{3}{5} - \left(\frac{2}{3} - \frac{1}{2}\right) \div \left(\frac{5}{6} \times \frac{3}{2}\right)$

7. $\frac{1}{2} \text{ of } \left(4\frac{2}{5} - 3\frac{7}{10}\right) + \left(3\frac{1}{3} \div \frac{2}{3}\right) - \frac{2}{5}$

8. $\dfrac{6\frac{2}{3} \times 1\frac{2}{5} - \frac{1}{3}}{6\frac{3}{4} \div 1\frac{1}{2}}$

9. $1\frac{1}{3} \times 2\frac{1}{5} \div \frac{2}{5}$

10. $\frac{1}{4} \times \frac{2}{5} - \frac{1}{5} \div \frac{2}{3} + \frac{4}{15}$

11. $\dfrac{\frac{2}{3} + 3\frac{1}{5} \times 2\frac{1}{2} + 1\frac{1}{3}}{8\frac{1}{3} \div 3\frac{1}{3}}$

12. $\frac{1}{13} \text{ of } \left(2\frac{9}{10} - 1\frac{3}{5}\right) + \left(2\frac{1}{3} \div \frac{2}{3}\right) - \frac{3}{4}$

Practice Exercise 9   Multiple-choice questions on fractions (answers on page 443)

Each question has only one correct answer

1. In the fraction $\frac{3}{8}$, the 3 is called the:
   (a) denominator   (b) mixed number
   (c) numerator   (d) LCM

2. Which of the following is a proper fraction?
   (a) $\frac{7}{3}$ (b) $\frac{1}{3}$ (c) $\frac{5}{3}$ (d) $\frac{9}{3}$

3. Expressing the improper fraction $\frac{27}{7}$ as a mixed number is:
   (a) $7\frac{1}{2}$ (b) $2\frac{1}{7}$ (c) $3\frac{6}{7}$ (d) $6\frac{1}{3}$

4. In the fraction $\frac{5}{9}$, the 9 is called the:
   (a) denominator   (b) mixed number
   (c) numerator   (d) HCF

5. The mixed number $4\frac{2}{3}$ expressed as an improper fraction is:
   (a) $\frac{24}{3}$ (b) $\frac{11}{3}$ (c) $\frac{10}{3}$ (d) $\frac{14}{3}$

6. $\frac{3}{4} \div 1\frac{3}{4}$ is equal to:
   (a) $\frac{3}{7}$ (b) $1\frac{9}{16}$ (c) $1\frac{5}{16}$ (d) $2\frac{1}{2}$

7. What fraction of the numbers from 2 to 12 are prime numbers?
    (a) $\dfrac{1}{11}$ (b) $\dfrac{10}{11}$ (c) $\dfrac{5}{11}$ (d) $\dfrac{6}{11}$

8. $\dfrac{72}{3} \div \dfrac{18}{6}$ is equal to:
    (a) 8   (b) 18   (c) 72   (d) 54

9. $\dfrac{11}{3} + \dfrac{8}{3} + \dfrac{17}{3}$ is equal to:
    (a) 11   (b) 12   (c) 14   (d) 15

10. $1\dfrac{1}{3} + 1\dfrac{2}{3} \div 2\dfrac{2}{3} - \dfrac{1}{3}$ is equal to:
    (a) $1\dfrac{5}{8}$ (b) $\dfrac{19}{24}$ (c) $2\dfrac{1}{21}$ (d) $1\dfrac{2}{7}$

11. The value of $\dfrac{2}{5}$ of

    $\left(4\dfrac{1}{2} - 3\dfrac{1}{4}\right) + 5 \div \dfrac{5}{16} - \dfrac{1}{4}$ is:

    (a) $17\dfrac{7}{20}$ (b) $80\dfrac{1}{2}$ (c) $16\dfrac{1}{4}$ (d) 88

12. Which of the fractions $\dfrac{2}{3}, \dfrac{3}{5}, \dfrac{5}{8}, \dfrac{7}{9}$ is the smallest?
    (a) $\dfrac{2}{3}$ (b) $\dfrac{3}{5}$ (c) $\dfrac{5}{8}$ (d) $\dfrac{7}{9}$

13. The reciprocal of $1\dfrac{3}{11}$ is:
    (a) $\dfrac{11}{13}$ (b) $\dfrac{11}{10}$ (c) $\dfrac{11}{9}$ (d) $\dfrac{11}{14}$

14. $\dfrac{1}{2}$ of $\left(\dfrac{3}{4} \div \dfrac{2}{3}\right)$ is equal to:
    (a) $\dfrac{3}{16}$ (b) $\dfrac{9}{16}$ (c) $\dfrac{3}{6}$ (d) $\dfrac{9}{4}$

15. $\dfrac{1}{\pi}\left(1\dfrac{1}{18} + \dfrac{1}{6} - \dfrac{5}{3}\right)$ is equal to:
    (a) $\dfrac{2\pi}{9}$ (b) $\dfrac{2}{9\pi}$ (c) $-\dfrac{16\pi}{9}$ (d) $-\dfrac{4}{9\pi}$

For fully worked solutions to each of the problems in Practice Exercises 6 to 8 in this chapter, go to the website:
www.routledge.com/cw/bird

COMPANION @ WEBSITE

Revision Test 1: Basic arithmetic and fractions

This assignment covers the material contained in Chapters 1 and 2. *The marks available are shown in brackets at the end of each question.*

1. Evaluate
   $1009\,cm - 356\,cm - 742\,cm + 94\,cm.$ (3)

2. Determine £284 × 9 (3)

3. Evaluate
   (a) $-11239 - (-4732) + 9639$
   (b) -164×-12
   (c) 367×-19 (6)

4. Calculate (a) $153 ÷ 9 (b) 1397 g ÷ 11 (4)

5. A small component has a mass of 27 grams. Calculate the mass, in kilograms, of 750 such components. (3)

6. Find (a) the highest common factor and (b) the lowest common multiple of the following numbers: 15 40 75 120 (7)

Evaluate the expressions in questions 7 to 12.

7. $7 + 20 \div (9 - 5)$ (3)

8. $147 - 21(24 \div 3) + 31$ (3)

9. $40 \div (1 + 4) + 7[8 + (3 \times 8) - 27]$ (5)

10. $\dfrac{(7 - 3)(2 - 5)}{3(9 - 5) \div (2 - 6)}$ (3)

11. $\dfrac{(7 + 4 \times 5) \div 3 + 6 \div 2}{2 \times 4 + (5 - 8) - 2^2 + 3}$ (5)

12. $\dfrac{(4^2 \times 5 - 8) \div 3 + 9 \times 8}{4 \times 3^2 - 20 \div 5}$ (5)

13. Simplify
   (a) $\dfrac{3}{4} - \dfrac{7}{15}$
   (b) $1\dfrac{5}{8} - 2\dfrac{1}{3} + 3\dfrac{5}{6}$ (8)

14. A training college has 480 students of which 150 are girls. Express this as a fraction in its simplest form. (2)

15. A tank contains 18 000 litres of oil. Initially, $\dfrac{7}{10}$ of the contents are removed, then $\dfrac{2}{5}$ of the remainder is removed. How much oil is left in the tank? (4)

16. Evaluate
   (a) $1\dfrac{7}{9} \times \dfrac{3}{8} \times 3\dfrac{3}{5}$
   (b) $6\dfrac{2}{3} \div 1\dfrac{1}{3}$
   (c) $1\dfrac{1}{3} \times 2\dfrac{1}{5} \div \dfrac{2}{5}$ (10)

17. Calculate
   (a) $\dfrac{1}{4} \times \dfrac{2}{5} - \dfrac{1}{5} \div \dfrac{2}{3} + \dfrac{4}{15}$
   (b) $\dfrac{\dfrac{2}{3} + 3\dfrac{1}{5} \times 2\dfrac{1}{2} + 1\dfrac{1}{3}}{8\dfrac{1}{3} \div 3\dfrac{1}{3}}$ (8)

18. Simplify $\left\{\dfrac{1}{13} \text{ of } \left(2\dfrac{9}{10} - 1\dfrac{3}{5}\right)\right\} + \left(2\dfrac{1}{3} \div \dfrac{2}{3}\right) - \dfrac{3}{4}$ (8)

Chapter 3

Decimals

Why it is important to understand: Decimals

Engineers and scientists use decimal numbers all the time in calculations. Calculators are able to handle calculations with decimals; however, there will be times when a quick calculation involving addition, subtraction, multiplication and division of decimals is needed. Again, do not spend too much time on this chapter because we deal with the calculator later; however, try to have some idea how to do quick calculations involving decimal numbers in the absence of a calculator. You will feel more confident to deal with decimal numbers in calculations if you can do this.

At the end of this chapter you should be able to:

- convert a decimal number to a fraction and vice versa
- understand and use significant figures and decimal places in calculations
- add and subtract decimal numbers
- multiply and divide decimal numbers

3.1 Introduction

The decimal system of numbers is based on the digits 0 to 9.

There are a number of everyday occurrences in which we use decimal numbers. For example, a radio is, say, tuned to 107.5 MHz FM; 107.5 is an example of a decimal number.

In a shop, a pair of trainers cost, say, £57.95; 57.95 is another example of a decimal number. 57.95 is a decimal fraction, where a decimal point separates the integer, i.e. 57, from the fractional part, i.e. 0.95

57.95 actually means $(5 \times 10) + (7 \times 1)$

$$+ \left(9 \times \frac{1}{10}\right) + \left(5 \times \frac{1}{100}\right)$$

3.2 Converting decimals to fractions and vice versa

Converting decimals to fractions and vice versa is demonstrated below with worked examples.

Problem 1. Convert 0.375 to a proper fraction in its simplest form

(i) 0.375 may be written as $\dfrac{0.375 \times 1000}{1000}$ i.e.

$0.375 = \dfrac{375}{1000}$

(ii) Dividing both numerator and denominator by 5 gives $\dfrac{375}{1000} = \dfrac{75}{200}$

(iii)   Dividing both numerator and denominator by 5 again gives $\dfrac{75}{200} = \dfrac{15}{40}$

(iv)   Dividing both numerator and denominator by 5 again gives $\dfrac{15}{40} = \dfrac{3}{8}$

Since both 3 and 8 are only divisible by 1, we cannot 'cancel' any further, so $\dfrac{3}{8}$ is the 'simplest form' of the fraction.

Hence, **the decimal fraction** $0.375 = \dfrac{3}{8}$ **as a proper fraction**.

Problem 2.   Convert 3.4375 to a mixed number

Initially, the whole number 3 is ignored.

(i)   0.4375 may be written as $\dfrac{0.4375 \times 10000}{10000}$ i.e. $0.4375 = \dfrac{4375}{10000}$

(ii)   Dividing both numerator and denominator by 25 gives $\dfrac{4375}{10000} = \dfrac{175}{400}$

(iii)   Dividing both numerator and denominator by 5 gives $\dfrac{175}{400} = \dfrac{35}{80}$

(iv)   Dividing both numerator and denominator by 5 again gives $\dfrac{35}{80} = \dfrac{7}{16}$

Since both 5 and 16 are only divisible by 1, we cannot 'cancel' any further, so $\dfrac{7}{16}$ is the 'lowest form' of the fraction.

(v)   Hence, $0.4375 = \dfrac{7}{16}$

Thus, **the decimal fraction** $3.4375 = 3\dfrac{7}{16}$ **as a mixed number**.

Problem 3.   Express $\dfrac{7}{8}$ as a decimal fraction

To convert a proper fraction to a decimal fraction, the numerator is divided by the denominator.

$$\begin{array}{r} 0.8\ 7\ 5 \\ 8\overline{)7.0\ 0\ 0} \end{array}$$

(i)   8 into 7 will not go. Place the 0 above the 7

(ii)   Place the decimal point above the decimal point of 7.000

(iii)   8 into 70 goes 8, remainder 6. Place the 8 above the first zero after the decimal point and carry the 6 remainder to the next digit on the right, making it 60

(iv)   8 into 60 goes 7, remainder 4. Place the 7 above the next zero and carry the 4 remainder to the next digit on the right, making it 40

(v)   8 into 40 goes 5, remainder 0. Place 5 above the next zero.

Hence, **the proper fraction** $\dfrac{7}{8} = 0.875$ **as a decimal fraction**.

Problem 4.   Express $5\dfrac{13}{16}$ as a decimal fraction

For mixed numbers it is only necessary to convert the proper fraction part of the mixed number to a decimal fraction.

$$\begin{array}{r} 0.8\ 1\ 2\ 5 \\ 16\overline{)13.0\ 0\ 0\ 0} \end{array}$$

(i)   16 into 13 will not go. Place the 0 above the 3

(ii)   Place the decimal point above the decimal point of 13.0000

(iii)   16 into 130 goes 8, remainder 2. Place the 8 above the first zero after the decimal point and carry the 2 remainder to the next digit on the right, making it 20

(iv)   16 into 20 goes 1, remainder 4. Place the 1 above the next zero and carry the 4 remainder to the next digit on the right, making it 40

(v)   16 into 40 goes 2, remainder 8. Place the 2 above the next zero and carry the 8 remainder to the next digit on the right, making it 80

(vi)   16 into 80 goes 5, remainder 0. Place the 5 above the next zero.

(vii)   Hence, $\dfrac{13}{16} = 0.8125$

Thus, **the mixed number** $5\dfrac{13}{16} = 5.8125$ **as a decimal fraction**.

Now try the following Practice Exercise

Practice Exercise 10   Converting decimals to fractions and vice versa (answers on page 443)

1. Convert 0.65 to a proper fraction.

2. Convert 0.036 to a proper fraction.

3. Convert 0.175 to a proper fraction.

4. Convert 0.048 to a proper fraction.

5. Convert the following to proper fractions.
   (a) 0.66   (b) 0.84   (c) 0.0125
   (d) 0.282   (e) 0.024

6. Convert 4.525 to a mixed number.

7. Convert 23.44 to a mixed number.

8. Convert 10.015 to a mixed number.

9. Convert 6.4375 to a mixed number.

10. Convert the following to mixed numbers.
    (a) 1.82   (b) 4.275   (c) 14.125
    (d) 15.35   (e) 16.2125

11. Express $\frac{5}{8}$ as a decimal fraction.

12. Express $6\frac{11}{16}$ as a decimal fraction.

13. Express $\frac{7}{32}$ as a decimal fraction.

14. Express $11\frac{3}{16}$ as a decimal fraction.

15. Express $\frac{9}{32}$ as a decimal fraction.

3.3 Significant figures and decimal places

A number which can be expressed exactly as a decimal fraction is called a **terminating decimal**. For example,

$$3\frac{3}{16} = 3.1825 \text{ is a terminating decimal}$$

A number which cannot be expressed exactly as a decimal fraction is called a **non-terminating decimal**. For example,

$$1\frac{5}{7} = 1.7142857\ldots \text{ is a non-terminating decimal}$$

The answer to a non-terminating decimal may be expressed in two ways, depending on the accuracy required:

(a) correct to a number of **significant figures**, or

(b) correct to a number of **decimal places** i.e. the number of figures after the decimal point.

The last digit in the answer is unaltered if the next digit on the right is in the group of numbers 0, 1, 2, 3 or 4. For example,

$$1.714285\ldots = \mathbf{1.714} \text{ correct to 4 significant figures}$$
$$= \mathbf{1.714} \text{ correct to 3 decimal places}$$

since the next digit on the right in this example is 2 The last digit in the answer is increased by 1 if the next digit on the right is in the group of numbers 5, 6, 7, 8 or 9. For example,

$$1.7142857\ldots = \mathbf{1.7143} \text{ correct to 5 significant figures}$$
$$= \mathbf{1.7143} \text{ correct to 4 decimal places}$$

since the next digit on the right in this example is 8

Problem 5. Express 15.36815 correct to
(a) 2 decimal places,   (b) 3 significant figures,
(c) 3 decimal places,   (d) 6 significant figures

(a) $15.36815 = \mathbf{15.37}$ correct to 2 decimal places.

(b) $15.36815 = \mathbf{15.4}$ correct to 3 significant figures.

(c) $15.36815 = \mathbf{15.368}$ correct to 3 decimal places.

(d) $15.36815 = \mathbf{15.3682}$ correct to 6 significant figures.

Problem 6. Express 0.004369 correct to
(a) 4 decimal places,   (b) 3 significant figures

(a) $0.004369 = \mathbf{0.0044}$ correct to 4 decimal places.

(b) $0.004369 = \mathbf{0.00437}$ correct to 3 significant figures.

Note that the zeros to the right of the decimal point do not count as significant figures.

Now try the following Practice Exercise

Practice Exercise 11   Significant figures and decimal places (answers on page 443)

1. Express 14.1794 correct to 2 decimal places.

2. Express 2.7846 correct to 4 significant figures.

3. Express 65.3792 correct to 2 decimal places.

4. Express 43.2746 correct to 4 significant figures.

5. Express 1.2973 correct to 3 decimal places.

6. Express 0.0005279 correct to 3 significant figures.

3.4 Adding and subtracting decimal numbers

When adding or subtracting decimal numbers, care needs to be taken to ensure that the decimal points are beneath each other. This is demonstrated in the following worked examples.

Problem 7.   Evaluate $46.8 + 3.06 + 2.4 + 0.09$ and give the answer correct to 3 significant figures

The decimal points are placed under each other as shown. Each column is added, starting from the right.

$$
\begin{array}{r}
46.8 \\
3.06 \\
2.4 \\
+\underline{0.09} \\
\underline{52.35} \\
1\;1\;\;1
\end{array}
$$

(i) $6 + 9 = 15$. Place 5 in the hundredths column. Carry 1 in the tenths column.

(ii) $8 + 0 + 4 + 0 + 1$ (carried) $= 13$. Place the 3 in the tenths column. Carry the 1 into the units column.

(iii) $6 + 3 + 2 + 0 + 1$ (carried) $= 12$. Place the 2 in the units column. Carry the 1 into the tens column.

(iv) $4 + 1$(carried) $= 5$. Place the 5 in the hundreds column.

Hence,

$$46.8 + 3.06 + 2.4 + 0.09 = 52.35$$

$$= 52.4, \text{correct to 3 significant figures}$$

Problem 8.   Evaluate $64.46 - 28.77$ and give the answer correct to 1 decimal place

As with addition, the decimal points are placed under each other as shown.

$$
\begin{array}{r}
64.46 \\
-\underline{28.77} \\
\underline{35.69}
\end{array}
$$

(i) $6 - 7$ is not possible; therefore 'borrow' 1 from the tenths column. This gives $16 - 7 = 9$. Place the 9 in the hundredths column.

(ii) $3 - 7$ is not possible; therefore 'borrow' 1 from the units column. This gives $13 - 7 = 6$. Place the 6 in the tenths column.

(iii) $3 - 8$ is not possible; therefore 'borrow' from the hundreds column. This gives $13 - 8 = 5$. Place the 5 in the units column.

(iv) $5 - 2 = 3$. Place the 3 in the hundreds column.

Hence,

$$64.46 - 28.77 = 35.69$$

$$= 35.7 \text{ correct to 1 decimal place}$$

Problem 9.   Evaluate $312.64 - 59.826 - 79.66 + 38.5$ and give the answer correct to 4 significant figures

The sum of the positive decimal fractions $= 312.64 + 38.5 = 351.14$

The sum of the negative decimal fractions $= 59.826 + 79.66 = 139.486$

Taking the sum of the negative decimal fractions from the sum of the positive decimal fractions gives

$$
\begin{array}{r}
351.140 \\
-\underline{139.486} \\
\underline{211.654}
\end{array}
$$

Hence, $351.140 - 139.486 = \mathbf{211.654} = \mathbf{211.7}$, **correct to 4 significant figures**.

Now try the following Practice Exercise

> **Practice Exercise 12 Adding and subtracting decimal numbers (answers on page 443)**
>
> Determine the following without using a calculator.
>
> 1. Evaluate $37.69 + 42.6$, correct to 3 significant figures.
> 2. Evaluate $378.1 - 48.85$, correct to 1 decimal place.
> 3. Evaluate $68.92 + 34.84 - 31.223$, correct to 4 significant figures.
> 4. Evaluate $67.841 - 249.55 + 56.883$, correct to 2 decimal places.
> 5. Evaluate $483.24 - 120.44 - 67.49$, correct to 4 significant figures.
> 6. Evaluate $738.22 - 349.38 - 427.336 + 56.779$, correct to 1 decimal place.
> 7. Determine the dimension marked x in the length of the shaft shown in Fig. 3.1. The dimensions are in millimetres.

Figure 3.1

3.5 Multiplying and dividing decimal numbers

When multiplying decimal fractions:

(a) the numbers are multiplied as if they were integers, and

(b) the position of the decimal point in the answer is such that there are as many digits to the right of it as the sum of the digits to the right of the decimal points of the two numbers being multiplied together.

This is demonstrated in the following worked examples.

> **Problem 10.** Evaluate 37.6×5.4

$$
\begin{array}{r}
376 \\
\times 54 \\
\hline
1504 \\
18800 \\
\hline
20304 \\
\hline
\end{array}
$$

(i) $376 \times 54 = 20304$

(ii) As there are $1 + 1 = 2$ digits to the right of the decimal points of the two numbers being multiplied together, $37.\underline{6} \times 5.\underline{4}$, then

$$\mathbf{37.6 \times 5.4 = 203.04}$$

> **Problem 11.** Evaluate $44.25 \div 1.2$, correct to (a) 3 significant figures, (b) 2 decimal places

$$44.25 \div 1.2 = \frac{44.25}{1.2}$$

The denominator is multiplied by 10 to change it into an integer. The numerator is also multiplied by 10 to keep the fraction the same. Thus,

$$\frac{44.25}{1.2} = \frac{44.25 \times 10}{1.2 \times 10} = \frac{442.5}{12}$$

The long division is similar to the long division of integers and the steps are as shown.

$$
\begin{array}{r}
36.875 \\
12\overline{\smash{)}442.500} \\
\underline{36} \\
82 \\
\underline{72} \\
105 \\
\underline{96} \\
90 \\
\underline{84} \\
60 \\
\underline{60} \\
0
\end{array}
$$

(i) 12 into 44 goes 3; place the 3 above the second 4 of 442.500

(ii) $3 \times 12 = 36$; place the 36 below the 44 of 442.500

(iii) $44 - 36 = 8$

(iv) Bring down the 2 to give 82

(v) 12 into 82 goes 6; place the 6 above the 2 of 442.500

(vi) 6 × 12 = 72; place the 72 below the 82

(vii) 82 − 72 = 10

(viii) Bring down the 5 to give 105

(ix) 12 into 105 goes 8; place the 8 above the 5 of 442.500

(x) 8 × 12 = 96; place the 96 below the 105

(xi) 105 − 96 = 9

(xii) Bring down the 0 to give 90

(xiii) 12 into 90 goes 7; place the 7 above the first zero of 442.500

(xiv) 7 × 12 = 84; place the 84 below the 90

(xv) 90 − 84 = 6

(xvi) Bring down the 0 to give 60

(xvii) 12 into 60 gives 5 exactly; place the 5 above the second zero of 442.500

(xviii) Hence, $44.25 \div 1.2 = \dfrac{442.5}{12} = \mathbf{36.875}$

So,

(a) $44.25 \div 1.2 = \mathbf{36.9}$, **correct to 3 significant figures**.

(b) $44.25 \div 1.2 = \mathbf{36.88}$, **correct to 2 decimal places**.

Problem 12. Express $7\dfrac{2}{3}$ as a decimal fraction, correct to 4 significant figures

Dividing 2 by 3 gives $\dfrac{2}{3} = 0.666666\ldots$

and $7\dfrac{2}{3} = 7.666666\ldots$

Hence, $7\dfrac{2}{3} = \mathbf{7.667}$ **correct to 4 significant figures**.

Note that 7.6666… is called **7.6 recurring** and is written as $\mathbf{7.\dot{6}}$

Now try the following Practice Exercise

Practice Exercise 13   Multiplying and dividing decimal numbers (answers on page 443)

In Problems 1 to 8, evaluate without using a calculator.

1. Evaluate 3.57 × 1.4

2. Evaluate 67.92 × 0.7

3. Evaluate 167.4 × 2.3

4. Evaluate 342.6 × 1.7

5. Evaluate 548.28 ÷ 1.2

6. Evaluate 478.3 ÷ 1.1, correct to 5 significant figures.

7. Evaluate 563.48 ÷ 0.9, correct to 4 significant figures.

8. Evaluate 2387.4 ÷ 1.5

In Problems 9 to 14, express as decimal fractions to the accuracy stated.

9. $\dfrac{4}{9}$, correct to 3 significant figures.

10. $\dfrac{17}{27}$, correct to 5 decimal places.

11. $1\dfrac{9}{16}$, correct to 4 significant figures.

12. $53\dfrac{5}{11}$, correct to 3 decimal places.

13. $13\dfrac{31}{37}$, correct to 2 decimal places.

14. $8\dfrac{9}{13}$, correct to 3 significant figures.

15. Evaluate 421.8 ÷ 17, (a) correct to 4 significant figures and (b) correct to 3 decimal places.

16. Evaluate $\dfrac{0.0147}{2.3}$, (a) correct to 5 decimal places and (b) correct to 2 significant figures.

17. Evaluate (a) $\dfrac{12.\dot{6}}{1.5}$ (b) $5.\dot{2} \times 12$

18. A tank contains 1800 litres of oil. How many tins containing 0.75 litres can be filled from this tank?

19. Holes are drilled 35.7 mm apart in a metal plate. If a row of 26 holes is drilled, determine the distance, in centimetres, between the centres of the first and last holes.

Practice Exercise 14   Multiple-choice questions on decimals (answers on page 443)

Each question has only one correct answer

1. Correct to 2 significant figures, 3.748 is equal to:
   (a) 3.75   (b) 3.8   (c) 3.74   (d) 3.7

2. $\dfrac{1}{5}$ expressed as a decimal is:
   (a) 0.2   (b) 0.5   (c) 0.002   (d) 0.02

3. $\dfrac{27}{8}$ expressed as a decimal is:
   (a) 3.35   (b) 3.375   (c) 33.75   (d) 337.5

4. $1\dfrac{5}{6}$ expressed as a decimal, correct to 4 significant figures is:
   (a) 1.83   (b) 0.633   (c) 1.833   (d) 1.8333

5. Expressing $\dfrac{1}{11}$ as a decimal, correct to 3 significant figures, is equal to:
   (a) 0.099   (b) 0.909
   (c) 0.0909   (d) 0.009

6. The decimal form of $\dfrac{12}{5}$ is:
   (a) 0.24   (b) 2.4   (c) 1.4   (d) 2.04

7. Correct to 1 decimal place, 14.783 is:
   (a) 14.7   (b) 14.8   (c) 15   (d) 14.78

8. Correct to 2 significant figures, 0.01479 is:
   (a) 0.01   (b) 0.014   (c) 0.0148   (d) 0.015

9. The value of 0.02×0.003 is:
   (a) 0.06   (b) 0.006
   (c) 0.0006   (d) 0.00006

10. $\dfrac{1524 \div 15.24}{0.5}$ is equal to:
    (a) 200   (b) 20   (c) 100   (d) 0.2

For fully worked solutions to each of the problems in Practice Exercises 10 to 13 in this chapter, go to the website:
www.routledge.com/cw/bird

Chapter 4

Using a calculator

Why it is important to understand: Using a calculator

The availability of electronic pocket calculators, at prices which all can afford, has had a considerable impact on engineering education. Engineers and student engineers now use calculators all the time since calculators are able to handle a very wide range of calculations. You will feel more confident to deal with all aspects of engineering studies if you are able to correctly use a calculator accurately.

At the end of this chapter you should be able to:

- use a calculator to add, subtract, multiply and divide decimal numbers
- use a calculator to evaluate square, cube, reciprocal, power, root and $\times 10^x$ functions
- use a calculator to evaluate expressions containing fractions and trigonometric functions
- use a calculator to evaluate expressions containing π and e^x functions
- evaluate formulae, given values

4.1 Introduction

In engineering, calculations often need to be performed. For simple numbers it is useful to be able to use mental arithmetic. However, when numbers are larger an electronic calculator needs to be used.

There are several calculators on the market, many of which will be satisfactory for our needs. It is essential to have a **scientific notation calculator** which will have all the necessary functions needed and more.

This chapter assumes you have a **CASIO fx-991ES PLUS calculator**, or similar, as shown in Fig. 4.1.

Besides straightforward addition, subtraction, multiplication and division, which you will already be able to do, we will check that you can use squares, cubes, powers, reciprocals, roots, fractions and trigonometric functions (the latter in preparation for Chapter 22). There are several other functions on the calculator which we do not need to concern ourselves with at this level.

4.2 Adding, subtracting, multiplying and dividing

Initially, after switching on, press **Mode**.

Of the possibilities, use **Comp**, which is achieved by pressing **1**.

Next, press **Shift** followed by **Setup** and, of the eight possibilities, use **Mth IO**, which is achieved by pressing **1**.

Next, for the **Result Format**, use **MathO**, which is achieved by pressing **1**.

By all means experiment with the other menu options – refer to your 'User's guide'.

All calculators have $+$, $-$, $\times$ **and** $\div$ **functions** and these functions will, no doubt, already have been used in calculations.

Problem 1. Evaluate $364.7 \div 57.5$ correct to 3 decimal places

(i) Type in 364.7

Figure 4.1 A Casio fx-991ES PLUS calculator

(ii)   Press ÷

(iii)   Type in 57.5

(iv)   Press = and the fraction $\dfrac{3647}{575}$ appears.

(v)   Press the $S \Leftrightarrow D$ function and the decimal answer 6.34260869... appears.

Alternatively, after step (iii) press Shift and = and the decimal will appear.

Hence, **364.7 ÷ 57.5 = 6.343 correct to 3 decimal places**.

Problem 2. Evaluate $\dfrac{12.47 \times 31.59}{70.45 \times 0.052}$ correct to 4 significant figures

(i)   Type in 12.47

(ii)   Press ×

(iii)   Type in 31.59

(iv)   Press ÷

(v)   The denominator must have brackets; i.e. press (

(vi)   Type in 70.45 × 0.052 and complete the bracket; i.e.)

(vii)   Press = and the answer 107.530518... appears.

Hence, $\dfrac{\mathbf{12.47 \times 31.59}}{\mathbf{70.45 \times 0.052}} = \mathbf{107.5}$ **correct to 4 significant figures**.

Now try the following Practice Exercise

Practice Exercise 15   Addition, subtraction, multiplication and division using a calculator (answers on page 443)

1. Evaluate $378.37 - 298.651 + 45.64 - 94.562$

2. Evaluate 25.63×465.34 correct to 5 significant figures.

3. Evaluate $562.6 \div 41.3$ correct to 2 decimal places.

4. Evaluate $\dfrac{17.35 \times 34.27}{41.53 \div 3.76}$ correct to 3 decimal places.

5. Evaluate $27.48 + 13.72 \times 4.15$ correct to 4 significant figures.

6. Evaluate $\dfrac{(4.527 + 3.63)}{(452.51 \div 34.75)} + 0.468$ correct to 5 significant figures.

7. Evaluate $52.34 - \dfrac{(912.5 \div 41.46)}{(24.6 - 13.652)}$ correct to 3 decimal places.

8. Evaluate $\dfrac{52.14 \times 0.347 \times 11.23}{19.73 \div 3.54}$ correct to 4 significant figures.

9. Evaluate $\dfrac{451.2}{24.57} - \dfrac{363.8}{46.79}$ correct to 4 significant figures.

10.   Evaluate $\dfrac{45.6 - 7.35 \times 3.61}{4.672 - 3.125}$ correct to 3 decimal places.

4.3   Further calculator functions

Square and cube functions

Locate the x^2 and x^3 functions on your calculator and then check the following worked examples.

Problem 3.   Evaluate 2.4^2

(i)   Type in 2.4

(ii)   Press x^2 and 2.4^2 appears on the screen.

(iii)   Press = and the answer $\dfrac{144}{25}$ appears.

(iv)   Press the $S \Leftrightarrow D$ function and the fraction changes to a decimal 5.76

Alternatively, after step (ii) press Shift and = . Thus, $\mathbf{2.4^2 = 5.76}$

Problem 4.   Evaluate 0.17^2 in engineering form

(i)   Type in 0.17

(ii)   Press x^2 and 0.17^2 appears on the screen.

(iii)   Press Shift and = and the answer 0.0289 appears.

(iv)   Press the ENG function and the answer changes to 28.9×10^{-3}, which is **engineering form**.

Hence, $\mathbf{0.17^2 = 28.9 \times 10^{-3}}$ **in engineering form**. The ENG function is extremely important in engineering calculations.

Problem 5.   Change 348620 into engineering form

(i)   Type in 348620

(ii)   Press = then ENG.

Hence, $\mathbf{348620 = 348.62 \times 10^{3}}$ **in engineering form**.

Problem 6.   Change 0.0000538 into engineering form

(i)   Type in 0.0000538

(ii)   Press = then ENG.

Hence, $\mathbf{0.0000538 = 53.8 \times 10^{-6}}$ **in engineering form**.

Problem 7.   Evaluate 1.4^3

(i)   Type in 1.4

(ii)   Press x^3 and 1.4^3 appears on the screen.

(iii)   Press = and the answer $\dfrac{343}{125}$ appears.

(iv)   Press the $S \Leftrightarrow D$ function and the fraction changes to a decimal: 2.744

Thus, $\mathbf{1.4^3 = 2.744}$

Now try the following Practice Exercise

Practice Exercise 16   Square and cube functions (answers on page 443)

1.   Evaluate 3.5^2

2.   Evaluate 0.19^2

3.   Evaluate 6.85^2 correct to 3 decimal places.

4.   Evaluate $(0.036)^2$ in engineering form.

5.   Evaluate 1.563^2 correct to 5 significant figures.

6.   Evaluate 1.3^3

7.   Evaluate 3.14^3 correct to 4 significant figures.

8.   Evaluate $(0.38)^3$ correct to 4 decimal places.

9.   Evaluate $(6.03)^3$ correct to 2 decimal places.

10.   Evaluate $(0.018)^3$ in engineering form.

Reciprocal and power functions

The reciprocal of 2 is $\dfrac{1}{2}$, the reciprocal of 9 is $\dfrac{1}{9}$ and the reciprocal of x is $\dfrac{1}{x}$, which from indices may be written as x^{-1}. Locate the reciprocal, i.e. x^{-1} on the calculator. Also, locate the power function, i.e. $x^{\square}$, on your calculator and then check the following worked examples.

Problem 8.   Evaluate $\dfrac{1}{3.2}$

(i) Type in 3.2

(ii) Press x^{-1} and 3.2^{-1} appears on the screen.

(iii) Press = and the answer $\dfrac{5}{16}$ appears.

(iv) Press the $S \Leftrightarrow D$ function and the fraction changes to a decimal: 0.3125

Thus, $\dfrac{1}{3.2} = 0.3125$

Problem 9. Evaluate 1.5^5 correct to 4 significant figures

(i) Type in 1.5

(ii) Press $x^{\square}$ and $1.5^{\square}$ appears on the screen.

(iii) Press 5 and 1.5^5 appears on the screen.

(iv) Press Shift and = and the answer 7.59375 appears.

Thus, $1.5^5 = 7.594$ **correct to 4 significant figures**.

Problem 10. Evaluate $2.4^6 - 1.9^4$ correct to 3 decimal places

(i) Type in 2.4

(ii) Press $x^{\square}$ and $2.4^{\square}$ appears on the screen.

(iii) Press 6 and 2.4^6 appears on the screen.

(iv) The cursor now needs to be moved; this is achieved by using the cursor key (the large blue circular function in the top centre of the calculator). Press $\rightarrow$

(v) Press $-$

(vi) Type in 1.9, press $x^{\square}$, then press 4

(vii) Press = and the answer 178.07087... appears.

Thus, $2.4^6 - 1.9^4 = 178.071$ **correct to 3 decimal places**.

Now try the following Practice Exercise

Practice Exercise 17   Reciprocal and power functions (answers on page 443)

1. Evaluate $\dfrac{1}{1.75}$ correct to 3 decimal places.

2. Evaluate $\dfrac{1}{0.0250}$

3. Evaluate $\dfrac{1}{7.43}$ correct to 5 significant figures.

4. Evaluate $\dfrac{1}{0.00725}$ correct to 1 decimal place.

5. Evaluate $\dfrac{1}{0.065} - \dfrac{1}{2.341}$ correct to 4 significant figures.

6. Evaluate 2.1^4

7. Evaluate $(0.22)^5$ correct to 5 significant figures in engineering form.

8. Evaluate $(1.012)^7$ correct to 4 decimal places.

9. Evaluate $(0.05)^6$ in engineering form.

10. Evaluate $1.1^3 + 2.9^4 - 4.4^2$ correct to 4 significant figures.

Root and $\times 10^x$ functions

Locate the square root function $\sqrt{\square}$ and the $\sqrt[\square]{\square}$ function (which is a Shift function located above the $x^{\square}$ function) on your calculator. Also, locate the $\times 10^x$ function and then check the following worked examples.

Problem 11. Evaluate $\sqrt{361}$

(i) Press the $\sqrt{\square}$ function.

(ii) Type in 361 and $\sqrt{361}$ appears on the screen.

(iii) Press = and the answer 19 appears.

Thus, $\sqrt{361} = 19$

Problem 12. Evaluate $\sqrt[4]{81}$

(i) Press the $\sqrt[\square]{\square}$ function.

(ii) Type in 4 and $\sqrt[4]{\square}$ appears on the screen.

(iii) Press $\rightarrow$ to move the cursor and then type in 81 and $\sqrt[4]{81}$ appears on the screen.

(iv) Press = and the answer 3 appears.

Thus, $\sqrt[4]{81} = 3$

Problem 13.   Evaluate $6 \times 10^5 \times 2 \times 10^{-7}$

(i)   Type in 6

(ii)   Press the $\times 10^x$ function (note, you do not have to use $\times$)

(iii)   Type in 5

(iv)   Press $\times$

(v)   Type in 2

(vi)   Press the $\times 10^x$ function.

(vii)   Type in -7

(viii)   Press $=$ and the answer $\dfrac{3}{25}$ appears.

(ix)   Press the $S \Leftrightarrow D$ function and the fraction changes to a decimal: 0.12

Thus, $6 \times 10^5 \times 2 \times 10^{-7} = 0.12$

Now try the following Practice Exercise

Practice Exercise 18   Root and $\times 10^x$ functions (answers on page 443)

1.   Evaluate $\sqrt{4.76}$ correct to 3 decimal places.

2.   Evaluate $\sqrt{123.7}$ correct to 5 significant figures.

3.   Evaluate $\sqrt{34528}$ correct to 2 decimal places.

4.   Evaluate $\sqrt{0.69}$ correct to 4 significant figures.

5.   Evaluate $\sqrt{0.025}$ correct to 4 decimal places.

6.   Evaluate $\sqrt[3]{17}$ correct to 3 decimal places.

7.   Evaluate $\sqrt[4]{773}$ correct to 4 significant figures.

8.   Evaluate $\sqrt[5]{3.12}$ correct to 4 decimal places.

9.   Evaluate $\sqrt[3]{0.028}$ correct to 5 significant figures.

10.   Evaluate $\sqrt[6]{2451} - \sqrt[4]{46}$ correct to 3 decimal places.

Express the answers to questions 11 to 15 in engineering form.

11.   Evaluate $5 \times 10^{-3} \times 7 \times 10^8$

12.   Evaluate $\dfrac{3 \times 10^{-4}}{8 \times 10^{-9}}$

13.   Evaluate $\dfrac{6 \times 10^3 \times 14 \times 10^{-4}}{2 \times 10^6}$

14.   Evaluate $\dfrac{56.43 \times 10^{-3} \times 3 \times 10^4}{8.349 \times 10^3}$ correct to 3 decimal places.

15.   Evaluate $\dfrac{99 \times 10^5 \times 6.7 \times 10^{-3}}{36.2 \times 10^{-4}}$ correct to 4 significant figures.

Fractions

Locate the $\dfrac{\square}{\square}$ and $\square\dfrac{\square}{\square}$ functions on your calculator (the latter function is a Shift function found above the $\dfrac{\square}{\square}$ function) and then check the following worked examples.

Problem 14.   Evaluate $\dfrac{1}{4} + \dfrac{2}{3}$

(i)   Press the $\dfrac{\square}{\square}$ function.

(ii)   Type in 1

(iii)   Press $\downarrow$ on the cursor key and type in 4

(iv) $\dfrac{1}{4}$ appears on the screen.

(v)   Press $\rightarrow$ on the cursor key and type in $+$

(vi)   Press the $\dfrac{\square}{\square}$ function.

(vii)   Type in 2

(viii)   Press $\downarrow$ on the cursor key and type in 3

(ix)   Press $\rightarrow$ on the cursor key.

(x)   Press $=$ and the answer $\dfrac{11}{12}$ appears.

(xi)   Press the $S \Leftrightarrow D$ function and the fraction changes to a decimal 0.9166666...

Thus, $\dfrac{1}{4} + \dfrac{2}{3} = \dfrac{11}{12} = 0.9167$ as a decimal, correct to 4 decimal places.

It is also possible to deal with **mixed numbers** on the calculator. Press Shift then the $\dfrac{\square}{\square}$ function and $\square\dfrac{\square}{\square}$ appears.

Problem 15.   Evaluate $5\frac{1}{5} - 3\frac{3}{4}$

(i)   Press Shift then the $\frac{\square}{\square}$ function and $\square\frac{\square}{\square}$ appears on the screen.

(ii)   Type in 5 then $\rightarrow$ on the cursor key.

(iii)   Type in 1 and $\downarrow$ on the cursor key.

(iv)   Type in 5 and $5\frac{1}{5}$ appears on the screen.

(v)   Press $\rightarrow$ on the cursor key.

(vi)   Type in $-$ and then press Shift then the $\frac{\square}{\square}$ function and $5\frac{1}{5} - \square\frac{\square}{\square}$ appears on the screen.

(vii)   Type in 3 then $\rightarrow$ on the cursor key.

(viii)   Type in 3 and $\downarrow$ on the cursor key.

(ix)   Type in 4 and $5\frac{1}{5} - 3\frac{3}{4}$ appears on the screen.

(x)   Press $\rightarrow$ on the cursor key.

(xi)   Press $=$ and the answer $\frac{29}{20}$ appears.

(xii)   Press $S \Leftrightarrow D$ function and the fraction changes to a decimal 1.45

Thus, $5\frac{1}{5} - 3\frac{3}{4} = \frac{29}{20} = 1\frac{9}{20} = 1.45$ as a decimal.

Now try the following Practice Exercise

Practice Exercise 19   Fractions (answers on page 444)

1.   Evaluate $\frac{4}{5} - \frac{1}{3}$ as a decimal, correct to 4 decimal places.

2.   Evaluate $\frac{2}{3} - \frac{1}{6} + \frac{3}{7}$ as a fraction.

3.   Evaluate $2\frac{5}{6} + 1\frac{5}{8}$ as a decimal, correct to 4 significant figures.

4.   Evaluate $5\frac{6}{7} - 3\frac{1}{8}$ as a decimal, correct to 4 significant figures.

5.   Evaluate $\frac{1}{3} - \frac{3}{4} \times \frac{8}{21}$ as a fraction.

6.   Evaluate $\frac{3}{8} + \frac{5}{6} - \frac{1}{2}$ as a decimal, correct to 4 decimal places.

7.   Evaluate $\frac{3}{4} \times \frac{4}{5} - \frac{2}{3} \div \frac{4}{9}$ as a fraction.

8.   Evaluate $8\frac{8}{9} \div 2\frac{2}{3}$ as a mixed number.

9.   Evaluate $3\frac{1}{5} \times 1\frac{1}{3} - 1\frac{7}{10}$ as a decimal, correct to 3 decimal places.

10.   Evaluate $\dfrac{\left(4\frac{1}{5} - 1\frac{2}{3}\right)}{\left(3\frac{1}{4} \times 2\frac{3}{5}\right)} - \frac{2}{9}$ as a decimal, correct to 3 significant figures.

Trigonometric functions

Trigonometric ratios will be covered in Chapter 22. However, very briefly, there are three functions on your calculator that are involved with trigonometry. They are:

sin which is an abbreviation of **sine**
cos which is an abbreviation of **cosine**, and
tan which is an abbreviation of **tangent**

Exactly what these mean will be explained in Chapter 22.

There are two main ways that angles are measured, i.e. in **degrees** or in **radians**. Pressing Shift, Setup and 3 shows degrees, and Shift, Setup and 4 shows radians. Press 3 and your calculator will be in **degrees mode**, indicated by a small D appearing at the top of the screen. Press 4 and your calculator will be in **radian mode**, indicated by a small R appearing at the top of the screen. Locate the sin, cos and tan functions on your calculator and then check the following worked examples.

Problem 16.   Evaluate $\sin 38°$

(i)   Make sure your calculator is in degrees mode.

(ii)   Press sin function and sin(appears on the screen.

(iii)   Type in 38 and close the bracket with) and sin (38) appears on the screen.

(iv)   Press $=$ and the answer 0.615661475... appears.

Thus, **$\sin 38° = 0.6157$, correct to 4 decimal places**.

Problem 17.   Evaluate $5.3 \tan (2.23 \text{ rad})$

(i)   Make sure your calculator is in radian mode by pressing Shift then Setup then 4 (a small R appears at the top of the screen).

(ii)   Type in 5.3 then press tan function and 5.3 tan(appears on the screen.

(iii)   Type in 2.23 and close the bracket with) and 5.3 tan (2.23) appears on the screen.

(iv)   Press = and the answer −6.84021262... appears.

Thus, $5.3\tan(2.23\,\text{rad}) = -6.8402$, **correct to 4 decimal places.**

Now try the following Practice Exercise

Practice Exercise 20   Trigonometric functions (answers on page 444)

Evaluate the following, each correct to 4 decimal places.

1.   Evaluate $\sin 67°$

2.   Evaluate $\cos 43°$

3.   Evaluate $\tan 71°$

4.   Evaluate $\sin 15.78°$

5.   Evaluate $\cos 63.74°$

6.   Evaluate $\tan 39.55° - \sin 52.53°$

7.   Evaluate $\sin(0.437\,\text{rad})$

8.   Evaluate $\cos(1.42\,\text{rad})$

9.   Evaluate $\tan(5.673\,\text{rad})$

10.   Evaluate $\dfrac{(\sin 42.6°)(\tan 83.2°)}{\cos 13.8°}$

π and e^{x} functions

Press Shift and then press the $\times 10^{x}$ function key and π appears on the screen. Either press Shift and = (or = and $S \Leftrightarrow D$) and the value of π appears in decimal form as 3.14159265...
Press Shift and then press the ln function key and $e^{\square}$ appears on the screen. Enter 1 and then press = and $e^{1} = e = 2.71828182...$
Now check the following worked examples involving π and e^{x} functions.

Problem 18.   Evaluate 3.57π

(i)   Enter 3.57

(ii)   Press Shift and the $\times 10^{x}$ key and 3.57π appears on the screen.

(iii)   Either press Shift and = (or = and $S \Leftrightarrow D$) and the value of 3.57π appears in decimal as 11.2154857...

Hence, $3.57\,\pi = 11.22$ **correct to 4 significant figures.**

Problem 19.   Evaluate $e^{2.37}$

(i)   Press Shift and then press the ln function key and $e^{\square}$ appears on the screen.

(ii)   Enter 2.37 and $e^{2.37}$ appears on the screen.

(iii)   Press Shift and = (or = and $S \Leftrightarrow D$) and the value of $e^{2.37}$ appears in decimal as 10.6973922...

Hence, $e^{2.37} = 10.70$ **correct to 4 significant figures.**

Now try the following Practice Exercise

Practice Exercise 21 π and e^{x} functions (answers on page 444)

Evaluate the following, each correct to 4 significant figures.

1. 1.59π

2. $2.7(\pi - 1)$

3. $\pi^{2}\left(\sqrt{13} - 1\right)$

4. $3e^{\pi}$

5. $8.5e^{-2.5}$

6. $3e^{2.9} - 1.6$

7. $3e^{(2\pi - 1)}$

8. $2\pi e^{\frac{\pi}{3}}$

9. $\sqrt{\left[\dfrac{5.52\pi}{2e^{-2} \times \sqrt{26.73}}\right]}$

10. $\sqrt{\left[\dfrac{e^{\left(2 - \sqrt{3}\right)}}{\pi \times \sqrt{8.57}}\right]}$

4.4   Evaluation of formulae

The statement $y = mx + c$ is called a **formula** for y in terms of m, x and c.
y, m, x and c are called **symbols** or **variables**.
When given values of m, x and c we can evaluate y.
There are a large number of formulae used in engineering and in this section we will insert numbers in place of symbols to evaluate engineering quantities.
Just four examples of important formulae are:

1.   A straight line graph is of the form $y = mx + c$ (see Chapter 17).

2. Ohm's law* states that $V = I \times R$.

3. Velocity is expressed as $v = u + at$.

4. Force is expressed as $F = m \times a$.

Here are some practical examples. Check with your calculator that you agree with the working and answers.

> **Problem 20.** In an electrical circuit the voltage V is given by Ohm's law, i.e. $V = IR$. Find, correct to 4 significant figures, the voltage when $I = 5.36$ A and $R = 14.76\,\Omega$

$$V = IR = (5.36)(14.76)$$

Hence, **voltage $V = 79.11$ V, correct to 4 significant figures**.

> **Problem 21.** The surface area A of a hollow cone is given by $A = \pi rl$. Determine, correct to 1

*Who was Ohm? – **Georg Simon Ohm** (16 March 1789–6 July 1854) was a Bavarian physicist and mathematician who discovered what came to be known as Ohm's law – the direct proportionality between voltage and the resultant electric current. To find out more go to **www.routledge.com/cw/bird**

decimal place, the surface area when $r = 3.0$ cm and $l = 8.5$ cm

$$A = \pi rl = \pi(3.0)(8.5)\,\text{cm}^2$$

Hence, **surface area $A = 80.1\,\text{cm}^2$, correct to 1 decimal place**.

> **Problem 22.** Velocity v is given by $v = u + at$. If $u = 9.54$ m/s, $a = 3.67$ m/s^2 and $t = 7.82$ s, find v, correct to 3 significant figures

$$v = u + at = 9.54 + 3.67 \times 7.82$$
$$= 9.54 + 28.6994 = 38.2394$$

Hence, **velocity $v = 38.2$ m/s, correct to 3 significant figures**.

> **Problem 23.** The area, A, of a circle is given by $A = \pi r^2$. Determine the area correct to 2 decimal places, given radius $r = 5.23$ m

$$A = \pi r^2 = \pi(5.23)^2 = \pi(27.3529)$$

Hence, **area, $A = 85.93\,\text{m}^2$, correct to 2 decimal places**.

> **Problem 24.** Density $= \dfrac{\text{mass}}{\text{volume}}$. Find the density when the mass is 6.45 kg and the volume is $300 \times 10^{-6}\,\text{m}^3$

$$\textbf{Density} = \frac{\text{mass}}{\text{volume}} = \frac{6.45\,\text{kg}}{300 \times 10^{-6}\,\text{m}^3} = \textbf{21500 kg/m}^3$$

> **Problem 25.** The power, P watts, dissipated in an electrical circuit is given by the formula $P = \dfrac{V^2}{R}$. Evaluate the power, correct to 4 significant figures, given that $V = 230$ V and $R = 35.63\,\Omega$

$$P = \frac{V^2}{R} = \frac{(230)^2}{35.63} = \frac{52900}{35.63} = 1484.70390\ldots$$

Press ENG and $1.48470390\ldots \times 10^3$ appears on the screen.

Hence, **power, $P = 1485$ W or 1.485 kW correct to 4 significant figures**.

Now try the following Practice Exercise

Practice Exercise 22 Evaluation of formulae
(answers on page 444)

1. The area A of a rectangular plate is given by the formula $A = lb$. Evaluate the area when $l = 12.4$ cm and $b = 5.37$ cm.

2. The circumference C of a circular disc is given by the formula $C = 2\pi r$. Determine the circumference given $r = 8.40$ mm.

3. A formula used in connection with gases is $R = \dfrac{PV}{T}$. Evaluate R when $P = 1500$, $V = 5$ and $T = 200$

4. The velocity of a body is given by $v = u + at$. The initial velocity u is measured and found to be 12 m/s. If the acceleration a is 9.81 m/s^2 calculate the final velocity, v, when time t is 15 seconds.

5. Calculate the current I in an electrical circuit, where $I = V/R$ amperes when the voltage V is measured and found to be 7.2 V and the resistance R is $17.7\,\Omega$.

6. Find the distance s, given that $s = \dfrac{1}{2}gt^2$ when time $t = 0.032$ seconds and acceleration due to gravity $g = 9.81$ m/s^2. Give the answer in millimetres.

7. The energy stored in a capacitor is given by $E = \dfrac{1}{2}CV^2$ joules. Determine the energy when capacitance $C = 5 \times 10^{-6}$ farads and voltage $V = 240$ V.

8. Find the area A of a triangular plot of land, given $A = \dfrac{1}{2}bh$, when the base length b is 23.42 m and the height h is 53.7 m.

9. Resistance R_2 is given by $R_2 = R_1(1 + \alpha t)$. Find R_2, correct to 4 significant figures, when $R_1 = 220$, $\alpha = 0.00027$ and $t = 75.6$

10. Density $= \dfrac{\text{mass}}{\text{volume}}$. Find the density when the mass is 2.462 kg and the volume is 173 cm^3. Give the answer in units of kg/m^3. (Note that 1 cm$^3 = 10^{-6}$ m^3)

11. Velocity $=$ frequency $\times$ wavelength. Find the velocity when the frequency is 1825 Hz and the wavelength is 0.154 m.

12. Evaluate resistance R_T, given
$\dfrac{1}{R_T} = \dfrac{1}{R_1} + \dfrac{1}{R_2} + \dfrac{1}{R_3}$ when $R_1 = 5.5\,\Omega$, $R_2 = 7.42\,\Omega$ and $R_3 = 12.6\,\Omega$.

Here are some further practical examples. Again, check with your calculator that you agree with the working and answers.

Problem 26. The volume V cm^3 of a right circular cone is given by $V = \dfrac{1}{3}\pi r^2 h$. Given that radius $r = 2.45$ cm and height $h = 18.7$ cm, find the volume, correct to 4 significant figures

$$V = \frac{1}{3}\pi r^2 h = \frac{1}{3}\pi(2.45)^2(18.7)$$
$$= \frac{1}{3} \times \pi \times 2.45^2 \times 18.7$$
$$= 117.544521\ldots$$

Hence, **volume, $V = 117.5$ cm^3, correct to 4 significant figures**.

Problem 27. Force F newtons is given by the formula $F = \dfrac{Gm_1m_2}{d^2}$, where m_1 and m_2 are masses, d their distance apart and G is a constant. Find the value of the force given that $G = 6.67 \times 10^{-11}$, $m_1 = 7.36$, $m_2 = 15.5$ and $d = 22.6$. Express the answer in standard form, correct to 3 significant figures

$$F = \frac{Gm_1m_2}{d^2} = \frac{(6.67 \times 10^{-11})(7.36)(15.5)}{(22.6)^2}$$
$$= \frac{(6.67)(7.36)(15.5)}{(10^{11})(510.76)} = \frac{1.490}{10^{11}}$$

Hence, **force $F = 1.49 \times 10^{-11}$ newtons, correct to 3 significant figures**.

Problem 28. The time of swing, t seconds, of a simple pendulum is given by $t = 2\pi\sqrt{\dfrac{l}{g}}$

Determine the time, correct to 3 decimal places, given that $l = 12.9$ and $g = 9.81$

$$t = 2\pi\sqrt{\frac{l}{g}} = (2\pi)\sqrt{\frac{12.9}{9.81}} = 7.20510343\ldots$$

Hence, **time $t = 7.205$ seconds, correct to 3 decimal places**.

Problem 29. Resistance, $R\,\Omega$, varies with temperature according to the formula $R = R_0(1 + \alpha t)$. Evaluate R, correct to 3 significant figures, given $R_0 = 14.59$, $\alpha = 0.0043$ and $t = 80$

$$R = R_0(1 + \alpha t) = 14.59[1 + (0.0043)(80)]$$
$$= 14.59(1 + 0.344)$$
$$= 14.59(1.344)$$

Hence, **resistance, $R = 19.6\,\Omega$, correct to 3 significant figures**.

Problem 30. The current, I amperes, in an a.c. circuit is given by $I = \dfrac{V}{\sqrt{(R^2 + X^2)}}$. Evaluate the current, correct to 2 decimal places, when $V = 250\,\text{V}$, $R = 25.0\,\Omega$ and $X = 18.0\,\Omega$.

$$I = \frac{V}{\sqrt{(R^2 + X^2)}} = \frac{250}{\sqrt{(25.0^2 + 18.0^2)}} = 8.11534341\ldots$$

Hence, **current, $I = 8.12\,\text{A}$, correct to 2 decimal places**.

Now try the following Practice Exercise

Practice Exercise 23   Evaluation of formulae (answers on page 444)

1. Find the total cost of 37 calculators costing £12.65 each and 19 drawing sets costing £6.38 each.

2. Power $= \dfrac{\text{force} \times \text{distance}}{\text{time}}$. Find the power when a force of $3760\,\text{N}$ raises an object a distance of $4.73\,\text{m}$ in $35\,\text{s}$.

3. The potential difference, V volts, available at battery terminals is given by $V = E - Ir$. Evaluate V when $E = 5.62$, $I = 0.70$ and $r = 4.30$

4. Given force $F = \dfrac{1}{2}m(v^2 - u^2)$, find F when $m = 18.3$, $v = 12.7$ and $u = 8.24$

5. The current I amperes flowing in a number of cells is given by $I = \dfrac{nE}{R + nr}$. Evaluate the current when $n = 36$, $E = 2.20$, $R = 2.80$ and $r = 0.50$

6. The time, t seconds, of oscillation for a simple pendulum is given by $t = 2\pi\sqrt{\dfrac{l}{g}}$. Determine the time when $l = 54.32$ and $g = 9.81$

7. Energy, E joules, is given by the formula $E = \dfrac{1}{2}LI^2$. Evaluate the energy when $L = 5.5$ and $I = 1.2$

8. The current I amperes in an a.c. circuit is given by $I = \dfrac{V}{\sqrt{(R^2 + X^2)}}$. Evaluate the current when $V = 250$, $R = 11.0$ and $X = 16.2$

9. Distance s metres is given by the formula $s = ut + \dfrac{1}{2}at^2$. If $u = 9.50$, $t = 4.60$ and $a = -2.50$, evaluate the distance.

10. The area, A, of any triangle is given by $A = \sqrt{[s(s-a)(s-b)(s-c)]}$ where $s = \dfrac{a+b+c}{2}$. Evaluate the area, given $a = 3.60\,\text{cm}$, $b = 4.00\,\text{cm}$ and $c = 5.20\,\text{cm}$.

11. Given that $a = 0.290$, $b = 14.86$, $c = 0.042$, $d = 31.8$ and $e = 0.650$, evaluate v given that $v = \sqrt{\left(\dfrac{ab}{c} - \dfrac{d}{e}\right)}$

12. Deduce the following information from the train timetable shown in Table 4.1 on page 37.

    (a) At what time should a man catch a train at Fratton to enable him to be in London Waterloo by 14.23 h?

    (b) A girl leaves Cosham at 12.39 h and travels to Woking. How long does the journey take? And, if the distance between Cosham and Woking is 55 miles, calculate the average speed of the train.

    (c) A man living at Havant has a meeting in London at 15.30 h. It takes around 25 minutes on the underground to reach his destination from London Waterloo. What train should he catch from Havant to comfortably make the meeting?

    (d) Nine trains leave Portsmouth harbour between 12.18 h and 13.15 h. Which train should be taken for the shortest journey time?

Practice Exercise 24   Multiple-choice questions on using a calculator (answers on page 444)

Each question has only one correct answer

1. Using the 24-hour clock, the time, in hours and minutes, between 0340 and 2222 is:
   (a) 26 h 2 min   (b) 18 h 42 min
   (c) 18 h 18 min   (d) 6 h 42 min

2. The symbol for pi is:
   (a) Ω (b) ε (c) μ (d) π

3. $106 \times 106 - 94 \times 94$ is equal to:
   (a) 2400   (b) 1904   (c) 1906   (d) 2004

4. The reciprocal of 5 is:
   (a) 0.5   (b) 5   (c) $\dfrac{1}{5}$ (d) -5

5. Which of the following is a square number?
   (a) 2   (b) 4   (c) 6   (d) 8

6. $(0.5)^2 \times (0.1)^3$ is equal to:
   (a) 25   (b) 0.0025   (c) 0.00025   (d) 0.25

7. If $x = \dfrac{57.06 \times 0.0711}{\sqrt{0.0635}}$ cm, which of the following statements is correct?
   (a) $x = 16.09$ cm, correct to 4 significant figures

   (b) $x = 16$ cm, correct to 2 significant figures

   (c) $x = 1.61 \times 10^1$ cm, correct to 3 decimal places

   (d) $x = 16.099$ cm, correct to 3 decimal places

8. $\dfrac{62.91}{0.12} + \sqrt{\left(\dfrac{6.36\pi}{2e^{-3} \times \sqrt{73.81}}\right)}$ is equal to:
   (a) 529.08 correct to 5 significant figures

   (b) 529.082 correct to 3 decimal places

   (c) 5.29×10^2

   (d) 529.0 correct to 1 decimal place

9. The area A of a triangular piece of land of sides a, b and c may be calculated using

   $$A = \sqrt{[s(s-a)(s-b)(s-c)]}$$

   where $s = \dfrac{a+b+c}{2}$

   When $a = 15$ m, $b = 11$ m and $c = 8$ m, the area, correct to the nearest square metre, is:
   (a) 1836 m^2 (b) 648 m^2
   (c) 445 m^2 (d) 43 m^2

10. Evaluating $\dfrac{1}{10}\sqrt{\left(\dfrac{2e^{(1-\sqrt{5})}}{3\pi \times \sqrt{12.94}}\right)}$ correct to 4 significant figures gives:
    (a) 0.0131   (b) 0.131
    (c) 0.1309   (d) 0.01309

For fully worked solutions to each of the problems in Practice Exercises 15 to 23 in this chapter, go to the website:
www.routledge.com/cw/bird

Table 4.1 Train timetable from Portsmouth Harbour to London Waterloo

Saturdays

	OUTWARD	Time	Time	Time	Time	Time	Time	Time	Time	Time
	Train Alterations	S04	S03	S08	S02	S03	S04	S04	S01	S02
Portsmouth Harbour	dep	12:18SW	12:22GW	12:22GW	12:45SW	12:45SW	12:45SW	12:54SW	13:12SN	13:15SW
Portsmouth & Southsea	arr	12:21	12:25	12:25	12:48	12:48	12:48	12:57	13:15	13:18
Portsmouth & Southsea	dep	12:24	12:27	12:27	12:50	12:50	12:50	12:59	13:16	13:20
Fratton	arr	12:27	12:30	12:30	12:53	12:53	12:53	13:02	13:19	13:23
Fratton	dep	12:28	12:31	12:31	12:54	12:54	12:54	13:03	13:20	13:24
Hilsea	arr	12:32						13:07		
Hilsea	dep	12:32						13:07		
Cosham	arr		12:38	12:38				13:12		
Cosham	dep		12:39	12:39				13:12		
Bedhampton	arr	12:37								
Bedhampton	dep	12:37								
Havant	arr	12:39			13:03	13:03	13:02		13:29	13:33
Havant	dep	12:40			13:04	13:04	13:04		13:30	13:34
Rowlands Castle	arr	12:46								
Rowlands Castle	dep	12:46								
Chichester	arr								13:40	
Chichester	dep								13:41	
Barnham	arr								13:48	
Barnham	dep								13:49	
Horsham	arr								14:16	
Horsham	dep								14:20	
Crawley	arr								14:28	
Crawley	dep								14:29	
Three Bridges	arr								14:32	
Three Bridges	dep								14:33	
Gatwick Airport	arr								14:37	
Gatwick Airport	dep								14:38	
Horley	arr								14:41	
Horley	dep								14:41	
Redhill	arr								14:47	
Redhill	dep								14:48	
East Croydon	arr								15:00	
East Croydon	dep								15:00	
Petersfield	arr	12:56			13:17	13:17	13:17			13:47
Petersfield	dep	12:57			13:18	13:18	13:18			13:48
Liss	arr	13:02								
Liss	dep	13:02								
Liphook	arr	13:09								
Liphook	dep	13:09								
Haslemere	arr	13:14C			13:31	13:31	13:30C			14:01
Haslemere	dep	13:20SW R			13:32	13:32	13:36SW R			14:02
Guildford	arr	13:55C			13:45	13:45	14:11C			14:15
Guildford	dep	14:02SW			13:47	13:47	14:17SW R			14:17
Portchester	arr							13:17		
Portchester	dep							13:17		
Fareham	arr		12:46	12:46				13:22		
Fareham	dep		12:47	12:47				13:23		
Southampton Central	arr			13:08C						
Southampton Central	dep			13:30SW						
Botley	arr							13:30		
Botley	dep							13:30		
Hedge End	arr							13:34		
Hedge End	dep							13:35		
Eastleigh	arr		13:00C					13:41		
Eastleigh	dep		13:09SW					13:42		
Southampton Airport Parkway	arr			13:37						
Southampton Airport Parkway	dep			13:38						
Winchester	arr		13:17	13:47				13:53		
Winchester	dep		13:18	13:48				13:54		
Micheldever	arr							14:02		
Micheldever	dep							14:02		
Basingstoke	arr		13:34					14:15		
Basingstoke	dep		13:36					14:17		
Farnborough	arr							14:30		
Farnborough	dep							14:31		
Woking	arr	14:11		14:19	13:57	13:57	14:25	14:40		14:25
Woking	dep	14:12		14:21	13:59	13:59	14:26	14:41		14:26
Clapham Junction	arr	14:31	14:12					15:01	15:11C	
Clapham Junction	dep	14:32	14:13						15:21SW	
Vauxhall	arr								15:26	
Vauxhall	dep								15:26	
London Waterloo	arr	14:40	14:24	14:49	14:23	14:27	14:51	15:13	15:31	14:51

Percentages

At the end of this chapter you should be able to:

- understand the term 'percentage'
- convert decimals to percentages and vice versa
- calculate the percentage of a quantity
- express one quantity as a percentage of another quantity
- calculate percentage error and percentage change

5.1 Introduction

Percentages are used to give a common standard. The use of percentages is very common in many aspects of commercial life, as well as in engineering. Interest rates, sale reductions, pay rises, exams and VAT are all examples of situations in which percentages are used. For this chapter you will need to know about decimals and fractions and be able to use a calculator.

We are familiar with the symbol for percentage, i.e. %. Here are some examples.

- Interest rates indicate the cost at which we can borrow money. If you borrow £8000 at a **1.5% interest rate** for a year, it will cost you 1.5% of the amount borrowed to do so, which will need to be repaid along with the original money you borrowed. If you repay the loan in 1 year, how much interest will you have paid?

- A pair of trainers in a shop cost £60. They are advertised in a sale as **20% off**. How much will you pay?

- If you earn £20000 p.a. and you receive a **2.5% pay rise**, how much extra will you have to spend the following year?

- A book costing £18 can be purchased on the internet for **30% less**. What will be its cost?

When we have completed this chapter on percentages you will be able to understand how to perform the above calculations.

Percentages are fractions having 100 as their denominator. For example, the fraction $\dfrac{40}{100}$ is written as 40% and is read as 'forty per cent'.

The easiest way to understand percentages is to go through some worked examples.

5.2 Percentage calculations

To convert a decimal to a percentage

A decimal number is converted to a percentage by multiplying by 100.

Problem 1. Express 0.015 as a percentage

To express a decimal number as a percentage, merely multiply by 100, i.e.

$$0.015 = 0.015 \times 100\%$$
$$= \mathbf{1.5\%}$$

Multiplying a decimal number by 100 means moving the decimal point 2 places **to the right**.

Problem 2. Express 0.275 as a percentage

$$0.275 = 0.275 \times 100\%$$
$$= \mathbf{27.5\%}$$

To convert a percentage to a decimal

A percentage is converted to a decimal number by dividing by 100.

Problem 3. Express 6.5% as a decimal number

$$6.5\% = \frac{6.5}{100} = \mathbf{0.065}$$

Dividing by 100 means moving the decimal point 2 places **to the left**.

Problem 4. Express 17.5% as a decimal number

$$17.5\% = \frac{17.5}{100}$$
$$= \mathbf{0.175}$$

To convert a fraction to a percentage

A fraction is converted to a percentage by multiplying by 100.

Problem 5. Express $\frac{5}{8}$ as a percentage

$$\frac{5}{8} = \frac{5}{8} \times 100\% = \frac{500}{8}\%$$
$$= \mathbf{62.5\%}$$

Problem 6. Express $\frac{5}{19}$ as a percentage, correct to 2 decimal places

$$\frac{5}{19} = \frac{5}{19} \times 100\%$$
$$= \frac{500}{19}\%$$
$$= 26.3157889\ldots \text{ by calculator}$$
$$= \mathbf{26.32\%} \text{ correct to 2 decimal places}$$

Problem 7. In two successive tests a student gains marks of 57/79 and 49/67. Is the second mark better or worse than the first?

$$57/79 = \frac{57}{79} = \frac{57}{79} \times 100\% = \frac{5700}{79}\%$$
$$= \mathbf{72.15\%} \text{ correct to 2 decimal places}$$
$$49/67 = \frac{49}{67} = \frac{49}{67} \times 100\% = \frac{4900}{67}\%$$
$$= \mathbf{73.13\%} \text{ correct to 2 decimal places}$$

Hence, **the second test is marginally better than the first test**. This question demonstrates how much easier it is to compare two fractions when they are expressed as percentages.

To convert a percentage to a fraction

A percentage is converted to a fraction by dividing by 100 and then, by cancelling, reducing it to its simplest form.

Problem 8. Express 75% as a fraction

$$75\% = \frac{75}{100}$$
$$= \mathbf{\frac{3}{4}}$$

The fraction $\frac{75}{100}$ is reduced to its simplest form by cancelling, i.e. dividing both numerator and denominator by 25.

Problem 9. Express 37.5% as a fraction

$$37.5\% = \frac{37.5}{100}$$

$$= \frac{375}{1000} \qquad \text{by multiplying both numerator and denominator by 10}$$

$$= \frac{15}{40} \qquad \text{by dividing both numerator and denominator by 25}$$

$$= \frac{3}{8} \qquad \text{by dividing both numerator and denominator by 5}$$

Now try the following Practice Exercise

Practice Exercise 25   Percentages (answers on page 444)

In Problems 1 to 5, express the given numbers as percentages.

1. 0.0032

2. 1.734

3. 0.057

4. 0.374

5. 1.285

6. Express 20% as a decimal number.

7. Express 1.25% as a decimal number.

8. Express $\frac{11}{16}$ as a percentage.

9. Express $\frac{5}{13}$ as a percentage, correct to 3 decimal places.

10. Express as percentages, correct to 3 significant figures,

     (a) $\frac{7}{33}$ (b) $\frac{19}{24}$ (c) $1\frac{11}{16}$

11. Place the following in order of size, the smallest first, expressing each as a percentage correct to 1 decimal place.

     (a) $\frac{12}{21}$ (b) $\frac{9}{17}$ (c) $\frac{5}{9}$ (d) $\frac{6}{11}$

12. Express 65% as a fraction in its simplest form.

13. Express 31.25% as a fraction in its simplest form.

14. Express 56.25% as a fraction in its simplest form.

15. Evaluate A to J in the following table.

Decimal number	Fraction	Percentage
0.5	A	B
C	$\frac{1}{4}$	D
E	F	30
G	$\frac{3}{5}$	H
I	J	85

5.3 Further percentage calculations

Finding a percentage of a quantity

To find a percentage of a quantity, convert the percentage to a fraction (by dividing by 100) and remember that 'of' means multiply.

Problem 10. Find 27% of £65

$$27\% \text{ of } £65 = \frac{27}{100} \times 65$$

$$= \textbf{£17.55} \text{ by calculator}$$

Problem 11. In a machine shop, it takes 32 minutes to machine a certain part. Using a new tool, the time can be reduced by 12.5%. Calculate the new time taken

$$12.5\% \text{ of } 32 \text{ minutes} = \frac{12.5}{100} \times 32$$

$$= 4 \text{ minutes}$$

Hence, **new time taken** $= 32 - 4 = $ **28 minutes**.

Alternatively, if the time is reduced by 12.5%, it now takes $100\% - 12.5\% = 87.5\%$ of the original time, i.e.

$$87.5\% \text{ of } 32 \text{ minutes} = \frac{87.5}{100} \times 32$$

$$= \textbf{28 minutes}$$

Problem 12. A resistor has a value of 820 Ω ± 5%. Determine the range of resistance values expected.

$$5\% \text{ of } 820 = \frac{5}{100} \times 820 = 41$$

The lowest value expected is $820 - 5\%$ of 820 i.e. $820 - 41 = 779\,\Omega$

The highest value expected is $820 + 5\%$ of 820 i.e. $820 + 41 = 861\,\Omega$

Hence, **range of values expected is: 779 Ω to 861 Ω**

Problem 13.   A 160 GB iPod is advertised as costing £190 excluding VAT. If VAT is added at 20%, what will be the total cost of the iPod?

$$\text{VAT} = 20\% \text{ of } £190 = \frac{20}{100} \times 190 = £38$$

Total cost of iPod = £190 + £38 = £228

A quicker method to determine the total cost is: $1.20 \times £190 = $ **£228**

Expressing one quantity as a percentage of another quantity

To express one quantity as a percentage of another quantity, divide the first quantity by the second then multiply by 100.

Problem 14.   Express 23 cm as a percentage of 72 cm, correct to the nearest 1%

$$23 \text{ cm as a percentage of } 72 \text{ cm} = \frac{23}{72} \times 100\%$$

$$= 31.94444\ldots\%$$

$$= \textbf{32\%} \text{ correct to the nearest 1\%}$$

Problem 15.   Express 47 minutes as a percentage of 2 hours, correct to 1 decimal place

Note that it is essential that the two quantities are in the **same units**.

Working in minute units, 2 hours $= 2 \times 60$
$$= 120 \text{ minutes}$$

$$47 \text{ minutes as a percentage of } 120 \text{ min} = \frac{47}{120} \times 100\%$$

$$= \textbf{39.2\%} \text{ correct to 1 decimal place}$$

Percentage change

Percentage change is given by
$$\frac{\text{new value} - \text{original value}}{\text{original value}} \times 100\%.$$

Problem 16.   A box of resistors increases in price from £45 to £52. Calculate the percentage change in cost, correct to 3 significant figures

$$\% \text{ change} = \frac{\text{new value} - \text{original value}}{\text{original value}} \times 100\%$$

$$= \frac{52 - 45}{45} \times 100\% = \frac{7}{45} \times 100$$

$$= \textbf{15.6\%} = \textbf{percentage change in cost}$$

Problem 17.   A drilling speed should be set to 400 rev/min. The nearest speed available on the machine is 412 rev/min. Calculate the percentage overspeed

$$\% \text{ overspeed} = \frac{\text{available speed} - \text{correct speed}}{\text{correct speed}} \times 100\%$$

$$= \frac{412 - 400}{400} \times 100\% = \frac{12}{400} \times 100\%$$

$$= \textbf{3\%}$$

Problem 18.   A brick being tested for its water absorption properties is found to have a mass of 2.628 kg when dry and 3.127 kg after soaking in water for a day. Calculate its percentage absorption by mass, correct to the nearest percent.

Percentage absorption

$$= \frac{\text{new value} - \text{original value}}{\text{original value}} \times 100\%$$

$$= \frac{3.127 - 2.628}{2.628} \times 100\%$$

$$= \frac{0.499}{2.628} \times 100\%$$

$$= 18.99\%$$

$$= \textbf{19\%} \text{ correct to the nearest percent}$$

Now try the following Practice Exercise

Practice Exercise 26 Further percentages
(answers on page 444)

1. Calculate 43.6% of 50 kg.

2. Determine 36% of 27 m.

3. Calculate, correct to 4 significant figures,
   (a) 18% of 2758 tonnes
   (b) 47% of 18.42 grams
   (c) 147% of 14.1 seconds.

4. When 1600 bolts are manufactured, 36 are unsatisfactory. Determine the percentage that is unsatisfactory.

5. Express
   (a) 140 kg as a percentage of 1 t.
   (b) 47 s as a percentage of 5 min.
   (c) 13.4 cm as a percentage of 2.5 m.

6. A block of Monel alloy consists of 70% nickel and 30% copper. If it contains 88.2 g of nickel, determine the mass of copper in the block.

7. An athlete runs 5000 m in 15 minutes 20 seconds. With intense training, he is able to reduce this time by 2.5%. Calculate his new time.

8. A copper alloy comprises 89% copper, 1.5% iron and the remainder aluminium. Find the amount of aluminium, in grams, in a 0.8 kg mass of the alloy.

9. A computer is advertised on the internet at £520, exclusive of VAT. If VAT is payable at 17.5%, what is the total cost of the computer?

10. Express 325 mm as a percentage of 867 mm, correct to 2 decimal places.

11. A child sleeps on average 9 hours 25 minutes per day. Express this as a percentage of the whole day, correct to 1 decimal place.

12. Express 408 g as a percentage of 2.40 kg.

13. When signing a new contract, a Premiership footballer's pay increases from £15 500 to £21 500 per week. Calculate the percentage pay increase, correct to 3 significant figures.

14. A metal rod 1.80 m long is heated and its length expands by 48.6 mm. Calculate the percentage increase in length.

15. 12.5% of a length of wood is 70 cm. What is the full length?

16. A metal rod, 1.20 m long, is heated and its length expands by 42 mm. Calculate the percentage increase in length.

17. For each of the following resistors, determine the (i) minimum value, (ii) maximum value:
    (a) 680 $\Omega \pm 20\%$ (b) 47 k$\Omega \pm 5\%$

18. An engine speed is 2400 rev/min. The speed is increased by 8%. Calculate the new speed.

19. A production run produces 4200 components of which 97% are reliable. Calculate the number of unreliable components.

20. The mass, m, of pollutant in a reservoir decreases according to the law

    $$m = m_0 e^{-0.08t}$$

    where t is the time, in days, and m_0 is the initial mass. Determine, as a percentage, correct to 4 decimal places, how much the mass has decreased in 40 days.

5.4 More percentage calculations

Percentage error

$$\text{Percentage error} = \frac{\text{error}}{\text{correct value}} \times 100\%$$

Problem 19. The length of a component is measured incorrectly as 64.5 mm. The actual length is 63 mm. What is the percentage error in the measurement?

$$\% \text{ error} = \frac{\text{error}}{\text{correct value}} \times 100\%$$
$$= \frac{64.5 - 63}{63} \times 100\%$$
$$= \frac{1.5}{63} \times 100\% = \frac{150}{63}\%$$
$$= \mathbf{2.38\%}$$

The percentage measurement error is **2.38% too high**, which is sometimes written as **+ 2.38% error**.

Problem 20. The voltage across a component in an electrical circuit is calculated as 50 V using Ohm's law. When measured, the actual voltage is 50.4 V. Calculate, correct to 2 decimal places, the percentage error in the calculation

$$\% \text{ error} = \frac{\text{error}}{\text{correct value}} \times 100\%$$
$$= \frac{50.4 - 50}{50.4} \times 100\%$$
$$= \frac{0.4}{50.4} \times 100\% = \frac{40}{50.4}\%$$
$$= \mathbf{0.79\%}$$

The percentage error in the calculation is **0.79% too low**, which is sometimes written as **−0.79% error**.

Original value

$$\text{Original value} = \frac{\text{new value}}{100 \pm \% \text{ change}} \times 100\%$$

Problem 21. A man pays £149.50 in a sale for a DVD player which is labelled '35% off'. What was the original price of the DVD player?

In this case, it is a 35% reduction in price, so we use $\dfrac{\text{new value}}{100 - \% \text{ change}} \times 100$, i.e. a minus sign in the denominator.

$$\text{Original price} = \frac{\text{new value}}{100 - \% \text{ change}} \times 100$$
$$= \frac{149.5}{100 - 35} \times 100$$
$$= \frac{149.5}{65} \times 100 = \frac{14\,950}{65}$$
$$= \mathbf{£230}$$

Problem 22. A couple buys a flat and makes an 18% profit by selling it 3 years later for £153 400. Calculate the original cost of the flat

In this case, it is an 18% increase in price, so we use $\dfrac{\text{new value}}{100 + \% \text{ change}} \times 100$, i.e. a plus sign in the denominator.

$$\text{Original cost} = \frac{\text{new value}}{100 + \% \text{ change}} \times 100$$
$$= \frac{153\,400}{100 + 18} \times 100$$
$$= \frac{153\,400}{118} \times 100 = \frac{15\,340\,000}{118}$$
$$= \mathbf{£130\,000}$$

Problem 23. An electrical store makes 40% profit on each widescreen television it sells. If the selling price of a 32 inch HD television is £630, what was the cost to the dealer?

In this case, it is a 40% mark-up in price, so we use $\dfrac{\text{new value}}{100 + \% \text{ change}} \times 100$, i.e. a plus sign in the denominator.

$$\text{Dealer cost} = \frac{\text{new value}}{100 + \% \text{ change}} \times 100$$
$$= \frac{630}{100 + 40} \times 100$$
$$= \frac{630}{140} \times 100 = \frac{63\,000}{140}$$
$$= \mathbf{£450}$$

The dealer buys from the manufacturer for £450 and sells to his customers for £630.

Percentage increase/decrease and interest

$$\text{New value} = \frac{100 + \% \text{ increase}}{100} \times \text{original value}$$

Problem 24. £3600 is placed in an ISA account which pays 1.5% interest per annum. How much is the investment worth after 1 year?

$$\text{Value after 1 year} = \frac{100 + 1.5}{100} \times £3600$$

$$= \frac{101.5}{100} \times £3600$$

$$= 1.015 \times £3600$$

$$= \textbf{£3654}$$

> ⌐ **Problem 25.** The price of a fully installed combination condensing boiler is increased by 6.5%. It originally cost £2400. What is the new price?

$$\text{New price} = \frac{100 + 6.5}{100} \times £2400$$

$$= \frac{106.5}{100} \times £2400 = 1.065 \times £2400$$

$$= \textbf{£2556}$$

Now try the following Practice Exercise

Practice Exercise 27   Further percentages (answers on page 444)

⌐1. A machine part has a length of 36 mm. The length is incorrectly measured as 36.9 mm. Determine the percentage error in the measurement.

⌐2. When a resistor is removed from an electrical circuit the current flowing increases from $450\,\mu A$ to $531\,\mu A$. Determine the percentage increase in the current.

3. In a shoe shop sale, everything is advertised as '40% off'. If a lady pays £186 for a pair of Jimmy Choo shoes, what was their original price?

4. Over a 4 year period a family home increases in value by 22.5% to £214 375. What was the value of the house 4 years ago?

5. An electrical retailer makes a 35% profit on all its products. What price does the retailer pay for a dishwasher which is sold for £351?

6. The cost of a sports car is £24 000 inclusive of VAT at 20%. What is the cost of the car without the VAT added?

7. £8000 is invested in bonds at a building society which is offering a rate of 1.50% per annum. Calculate the value of the investment after 2 years.

8. An electrical contractor earning £36 000 per annum receives a pay rise of 2.5%. He pays 22% of his income as tax and 11% on National Insurance contributions. Calculate the increase he will actually receive per month.

9. Five mates enjoy a meal out. With drinks, the total bill comes to £176. They add a 12.5% tip and divide the amount equally between them. How much does each pay?

10. In December a shop raises the cost of a 40 inch LCD TV costing £920 by 5%. It does not sell and in its January sale it reduces the TV by 5%. What is the sale price of the TV?

11. A man buys a business and makes a 20% profit when he sells it three years later for £222 000. What did he pay originally for the business?

⌐12. A drilling machine should be set to 250 rev/min. The nearest speed available on the machine is 268 rev/min. Calculate the percentage overspeed.

⌐13. Two kilograms of a compound contain 30% of element A, 45% of element B and 25% of element C. Determine the masses of the three elements present.

⌐14. A concrete mixture contains seven parts by volume of ballast, four parts by volume of sand and two parts by volume of cement. Determine the percentage of each of these three constituents correct to the nearest 1% and the mass of cement in a two tonne dry mix, correct to 1 significant figure.

⌐15. In a sample of iron ore, 18% is iron. How much ore is needed to produce 3600 kg of iron?

⌐16. A screw's dimension is $12.5 \pm 8\%$ mm. Calculate the maximum and minimum possible length of the screw.

⌐17. The output power of an engine is 450 kW. If the efficiency of the engine is 75%, determine the power input.

Practice Exercise 28   Multiple-choice questions on percentages (answers on page 444)

Each question has only one correct answer

1. 0.075 as a percentage is:
    (a) 0.075%   (b) 75%   (c) 0.75%   (d) 7.5%

2. 15.8% as a decimal is:
    (a) 0.158   (b) 1.58   (c) 15.8   (d) 158.0

3. 11 mm expressed as a percentage of 41 mm is:
    (a)   2.68, correct to 3 significant figures

    (b)   2.6, correct to 2 significant figures

    (c)   26.83, correct to 2 decimal places

    (d)   0.2682, correct to 4 decimal places

4. 1.24 as a percentage is:
    (a) 0.124%   (b) 1.24%
    (c) 12.4%   (d) 124%

5. The current in a component in an electrical circuit is calculated as 25 mA using Ohm's law. When measured, the actual current is 25.2 mA. Correct to 2 decimal places, the percentage error in the calculation is:
    (a) 0.80%   (b) 1.25%
    (c) 0.79%   (d) 1.26%

6. 0.226 as a percentage is:
    (a) 2.26%   (b) 22.6%
    (c) 0.226%   (d) 226%

7. Given that 12% of P is 48, the value of P is:
    (a) 250   (b) 100   (c) 400   (d) 200

8. 3.5% as a decimal is:
    (a) 0.35   (b) 0.035   (c) 3.5   (d) 350.0

9. Given that $12q = 75\%$ of 336, the value of q is:
    (a) 21   (b) 48   (c) 28   (d) 252

10. 10.7% as a decimal is:
    (a) 0.107   (b) 1.07   (c) 10.7   (d) 107

11. A resistor of 47 kΩ has a tolerance of ±5%. The highest possible value is:
    (a) 52 kΩ   (b) 47.005 kΩ
    (c) 42 kΩ   (d) 49.35 kΩ

12. 65.3% as a decimal is:
    (a) 0.0653   (b) 0.653   (c) 6.53   (d) 65.3

13. Expressing 12 minutes 15 seconds as a percentage of 4 hours gives:
    (a) 5.10%   (b) 3.06%
    (c) 5.06%   (d) 0.051%

14. 9.9% as a decimal is:
    (a) 99.0   (b) 0.099   (c) 9.9   (d) 0.99

15. The length and width of a rectangle are each increased by 10%. The percentage increase in the perimeter is:
    (a) 20   (b) 12.1   (c) 10   (d) 21

For fully worked solutions to each of the problems in Practice Exercises 25 to 27 in this chapter, go to the website:

This assignment covers the material contained in Chapters 3–5. *The marks available are shown in brackets at the end of each question.*

1. Convert 0.048 to a proper fraction. (2)

2. Convert 6.4375 to a mixed number. (3)

3. Express $\dfrac{9}{32}$ as a decimal fraction. (2)

4. Express 0.0784 correct to 2 decimal places. (2)

5. Express 0.0572953 correct to 4 significant figures. (2)

6. Evaluate
   (a) $46.7 + 2.085 + 6.4 + 0.07$
   (b) $68.51 - 136.34$ (4)

7. Determine 2.37×1.2 (3)

8. Evaluate $250.46 \div 1.1$ correct to 1 decimal place. (3)

9. Evaluate $5.\dot{2} \times 12$ (2)

10. Evaluate the following, correct to 4 significant figures: $3.3^2 - 2.7^3 + 1.8^4$ (3)

11. Evaluate $\sqrt{6.72} - \sqrt[3]{2.54}$ correct to 3 decimal places. (3)

12. Evaluate $\dfrac{1}{0.0071} - \dfrac{1}{0.065}$ correct to 4 significant figures. (2)

13. The potential difference, V volts, available at battery terminals is given by $V = E - Ir$. Evaluate V when $E = 7.23, I = 1.37$ and $r = 3.60$ (3)

14. Evaluate $\dfrac{4}{9} + \dfrac{1}{5} - \dfrac{3}{8}$ as a decimal, correct to 3 significant figures. (3)

15. Evaluate $\dfrac{16 \times 10^{-6} \times 5 \times 10^9}{2 \times 10^7}$ in engineering form. (2)

16. Evaluate resistance, R, given
   $\dfrac{1}{R} = \dfrac{1}{R_1} + \dfrac{1}{R_2} + \dfrac{1}{R_3}$ when $R_1 = 3.6\,\text{k}\Omega$, $R_2 = 7.2\,\text{k}\Omega$ and $R_3 = 13.6\,\text{k}\Omega$. (3)

17. Evaluate $6\dfrac{2}{7} - 4\dfrac{5}{9}$ as a mixed number and as a decimal, correct to 3 decimal places. (3)

18. Evaluate, correct to 3 decimal places:
   $\sqrt{\left[\dfrac{2e^{1.7} \times 3.67^3}{4.61 \times \sqrt{3\pi}}\right]}$ (3)

19. If $a = 0.270, b = 15.85, c = 0.038, d = 28.7$ and $e = 0.680$, evaluate v correct to 3 significant figures, given that $v = \sqrt{\left(\dfrac{ab}{c} - \dfrac{d}{e}\right)}$ (4)

20. Evaluate the following, each correct to 2 decimal places.
   (a) $\left(\dfrac{36.2^2 \times 0.561}{27.8 \times 12.83}\right)^3$
   (b) $\sqrt{\left(\dfrac{14.69^2}{\sqrt{17.42} \times 37.98}\right)}$ (4)

21. If $1.6\,\text{km} = 1\,\text{mile}$, determine the speed of 45 miles/hour in kilometres per hour. (2)

22. The area A of a circle is given by $A = \pi r^2$. Find the area of a circle of radius $r = 3.73\,\text{cm}$, correct to 2 decimal places. (3)

23. Evaluate B, correct to 3 significant figures, when $W = 7.20, v = 10.0$ and $g = 9.81$, given that $B = \dfrac{Wv^2}{2g}$ (3)

24. Express 56.25% as a fraction in its simplest form. (3)

25. 12.5% of a length of wood is 90 cm. What is the full length? (3)

26. A metal rod, 1.50 m long, is heated and its length expands by 45 mm. Calculate the percentage increase in length. (2)

27. A man buys a house and makes a 20% profit when he sells it three years later for £312 000. What did he pay for it originally? (3)

For lecturers/instructors/teachers, fully worked solutions to each of the problems in Revision Test 2, together with a full marking scheme, are available at the website:
www.routledge.com/cw/bird

Ratio and proportion

At the end of this chapter you should be able to:

* define ratio
* perform calculations with ratios
* define direct proportion
* perform calculations with direct proportion, including Hooke's law, Charles's law and Ohm's law
* define inverse proportion
* perform calculations with inverse proportion, including Boyle's law

6.1 Introduction

Ratio is a way of comparing amounts of something; it shows how much bigger one thing is than the other. Some practical examples include mixing paint, sand and cement, or screen wash. Gears, map scales, food recipes, scale drawings and metal alloy constituents all use ratios.

Two quantities are in **direct proportion** when they increase or decrease in the **same ratio**. There are several practical engineering laws which rely on direct proportion. Also, calculating currency exchange rates and converting imperial to metric units rely on direct proportion.

Sometimes, as one quantity increases at a particular rate, another quantity decreases at the same rate; this is called **inverse proportion**. For example, the time taken to do a job is inversely proportional to the number of people in a team: double the people, half the time.

When we have completed this chapter on ratio and proportion you will be able to understand, and confidently perform, calculations on the above topics.

For this chapter you will need to know about decimals and fractions and to be able to use a calculator.

6.2 Ratios

Ratios are generally shown as numbers separated by a colon (:) so the ratio of 2 and 7 is written as 2 : 7 and we read it as a ratio of 'two to seven'.

Some practical examples which are familiar include:

- Mixing 1 measure of screen wash to 6 measures of water; i.e. the ratio of screen wash to water is 1 : 6

- Mixing 1 shovel of cement to 4 shovels of sand; i.e. the ratio of cement to sand is 1 : 4

- Mixing 3 parts of red paint to 1 part white, i.e. the ratio of red to white paint is 3 : 1

Ratio is the number of parts to a mix. The paint mix is 4 parts total, with 3 parts red and 1 part white. 3 parts red paint to 1 part white paint means there is

$$\frac{3}{4} \text{ red paint to } \frac{1}{4} \text{ white paint}$$

Here are some worked examples to help us understand more about ratios.

Problem 1. In a class, the ratio of female to male students is 6 : 27. Reduce the ratio to its simplest form

(i) Both 6 and 27 can be divided by 3

(ii) Thus, 6 : 27 is the same as **2 : 9**

6 : 27 and 2 : 9 are called **equivalent ratios**.
It is normal to express ratios in their lowest, or simplest, form. In this example, the simplest form is **2 : 9** which means for every 2 females in the class there are 9 male students.

Problem 2. A gear wheel having 128 teeth is in mesh with a 48-tooth gear. What is the gear ratio?

Gear ratio = 128 : 48

A ratio can be simplified by finding common factors.

(i) 128 and 48 can both be divided by 2, i.e. 128 : 48 is the same as 64 : 24

(ii) 64 and 24 can both be divided by 8, i.e. 64 : 24 is the same as 8 : 3

(iii) There is no number that divides completely into both 8 and 3 so 8 : 3 is the simplest ratio, i.e. **the gear ratio is 8 : 3**

Thus, 128 : 48 is equivalent to 64 : 24 which is equivalent to 8 : 3 and **8 : 3 is the simplest form**.

Problem 3. A wooden pole is 2.08 m long. Divide it in the ratio of 7 to 19

(i) Since the ratio is 7 : 19, the total number of parts is 7 + 19 = 26 parts.

(ii) 26 parts corresponds to 2.08 m = 208 cm, hence, 1 part corresponds to $\frac{208}{26} = 8$

(iii) Thus, 7 parts corresponds to $7 \times 8 = $ **56 cm** and 19 parts corresponds to $19 \times 8 = $ **152 cm**.

Hence, **2.08 m divides in the ratio of 7 : 19 as 56 cm to 152 cm**.
(Check: 56 + 152 must add up to 208, otherwise an error has been made.)

Problem 4. In a competition, prize money of £828 is to be shared among the first three in the ratio 5 : 3 : 1

(i) Since the ratio is 5 : 3 : 1 the total number of parts is 5 + 3 + 1 = 9 parts.

(ii) 9 parts corresponds to £828

(iii) 1 part corresponds to $\frac{828}{9} = $ **£92**, 3 parts corresponds to $3 \times £92 = $ **£276** and 5 parts corresponds to $5 \times £92 = $ **£460**

Hence, **£828 divides in the ratio of 5 : 3 : 1 as £460 to £276 to £92**. (Check: 460 + 276 + 92 must add up to 828, otherwise an error has been made.)

Problem 5. A map scale is 1 : 30 000. On the map the distance between two schools is 6 cm. Determine the actual distance between the schools, giving the answer in kilometres

Actual distance between schools

$$= 6 \times 30\,000\,\text{cm} = 180\,000\,\text{cm}$$

$$= \frac{180,000}{100}\,\text{m} = 1800\,\text{m}$$

$$= \frac{1800}{1000}\,\text{km} = \mathbf{1.80\,km}$$

(1 mile ≈ 1.6 km, hence the schools are just over 1 mile apart.)

Problem 6. A mixture used for making concrete contains cement, sand, and rubble in the proportion 2:5:8. Calculate (a) the mass of sand in 750 kg of mixture (b) the percentage of cement in the mixture.

Since the ratio is 2:3:8, the total number of parts is $2 + 5 + 8 = 15$ parts

(a) Sand is 5 parts out of 15, i.e. $\frac{5}{15}$ of the 750 kg mixture

Hence, **the mass of sand in the mixture is**

$$\frac{5}{15} \times 750 = \mathbf{250\,kg}$$

(b) Cement is 2 parts out of 15, i.e. $\frac{2}{15}$ of the mixture

As a percentage of the mixture, cement is

$$\frac{2}{15} \times 100 = \mathbf{13.33\%}$$

Now try the following Practice Exercise

Practice Exercise 29   Ratios (answers on page 444)

1. In a box of 333 paper clips, 9 are defective. Express the number of non-defective paper clips as a ratio of the number of defective paper clips, in its simplest form.

2. A gear wheel having 84 teeth is in mesh with a 24-tooth gear. Determine the gear ratio in its simplest form.

3. In a box of 2000 nails, 120 are defective. Express the number of non-defective nails as a ratio of the number of defective ones, in its simplest form.

4. A metal pipe 3.36 m long is to be cut into two in the ratio 6 to 15. Calculate the length of each piece.

5. The instructions for cooking a turkey say that it needs to be cooked 45 minutes for every kilogram. How long will it take to cook a 7 kg turkey?

6. In a will, £6440 is to be divided among three beneficiaries in the ratio 4 : 2 : 1. Calculate the amount each receives.

7. A local map has a scale of 1 : 22 500. The distance between two motorways is 2.7 km. How far are they apart on the map?

8. Prize money in a lottery totals £3801 and is shared among three winners in the ratio 4 : 2 : 1. How much does the first prize winner receive?

9. A packet contains 36 screws. The screws are of two sizes: P and Q. The ratio of size P screws to size Q screws is 1:8. Calculate the number of each size of screw.

10. A mass of 56 kg is divided into 3 parts in the ratio 3:5:6. Calculate the mass of each part.

Here are some further worked examples on ratios.

Problem 7. Express 45 p as a ratio of £7.65 in its simplest form

(i) Changing both quantities to the same units, i.e. to pence, gives a ratio of 45 : 765

(ii) Dividing both quantities by 5 gives
$45 : 765 \equiv 9 : 153$

(iii) Dividing both quantities by 3 gives
$9 : 153 \equiv 3 : 51$

(iv)   Dividing both quantities by 3 again gives
$3:51 \equiv 1:17$

Thus, **45 p as a ratio of £7.65 is 1 : 17**
$45:765, 9:153, 3:51$ and $1:17$ are **equivalent ratios** and **1 : 17 is the simplest ratio**.

▶ **Problem 8.**   A glass contains 30 ml of whisky which is 40% alcohol. If 45 ml of water is added and the mixture stirred, what is now the alcohol content?

(i)   The 30 ml of whisky contains 40%
alcohol $= \dfrac{40}{100} \times 30 = 12$ ml.

(ii)   After 45 ml of water is added we have $30 + 45 = 75$ ml of fluid, of which alcohol is 12 ml.

(iii)   Fraction of alcohol present $= \dfrac{12}{75}$

(iv)   Percentage of alcohol present $= \dfrac{12}{75} \times 100\%$

$$= \mathbf{16\%}$$

▶ **Problem 9.**   20 tonnes of a mixture of sand and gravel is 30% sand. How many tonnes of sand must be added to produce a mixture which is 40% gravel?

(i)   Amount of sand in 20 tonnes $= 30\%$ of 20 t
$= \dfrac{30}{100} \times 20 = 6$ t

(ii)   If the mixture has 6 t of sand then amount of gravel $= 20 - 6 = 14$ t

(iii)   We want this 14 t of gravel to be 40% of the new mixture. 1% would be $\dfrac{14}{40}$ t and 100% of the mixture would be $\dfrac{14}{40} \times 100$ t $= 35$ t

(iv)   If there is 14 t of gravel then amount of sand $= 35 - 14 = 21$ t

(v)   We already have 6 t of sand, so **amount of sand to be added to produce a mixture with 40% gravel $= 21 - 6 = 15$ t**

(Note 1 tonne = 1000 kg.)

Now try the following Practice Exercise

Practice Exercise 30   Further ratios (answers on page 444)

▶ 1.   Express 130 g as a ratio of 1.95 kg.

▶ 2.   In a laboratory, acid and water are mixed in the ratio 2 : 5. How much acid is needed to make 266 ml of the mixture?

▶ 3.   A glass contains 30 ml of gin which is 40% alcohol. If 18 ml of water is added and the mixture stirred, determine the new percentage alcoholic content.

▶ 4.   A wooden beam 4 m long weighs 84 kg. Determine the mass of a similar beam that is 60 cm long.

▶ 5.   An alloy is made up of metals P and Q in the ratio $3.25 : 1$ by mass. How much of P has to be added to 4.4 kg of Q to make the alloy?

▶ 6.   15 000 kg of a mixture of sand and gravel is 20% sand. Determine the amount of sand that must be added to produce a mixture with 30% gravel.

6.3   Direct proportion

Two quantities are in **direct proportion** when they increase or decrease in the **same ratio**. For example, if 12 cans of lager have a mass of 4 kg, then 24 cans of lager will have a mass of 8 kg; i.e. if the quantity of cans doubles then so does the mass. This is direct proportion.

In the previous section we had an example of mixing 1 shovel of cement to 4 shovels of sand; i.e. the ratio of cement to sand was 1 : 4. So, if we have a mix of 10 shovels of cement and 40 shovels of sand and we wanted to double the amount of the mix then we would need to double both the cement and sand, i.e. 20 shovels of cement and 80 shovels of sand. This is another example of direct proportion.

Here are three laws in engineering which involve direct proportion:

(a) **Hooke's law*** states that, within the elastic limit of a material, the strain ε produced is directly proportional to the stress σ producing it, i.e. $\varepsilon \propto \sigma$ (note than '$\propto$' means 'is proportional to').

(b) **Charles's law*** states that, for a given mass of gas at constant pressure, the volume V is directly proportional to its thermodynamic temperature T, i.e. $V \propto T$.

(c) **Ohm's law*** states that the current I flowing through a fixed resistance is directly proportional to the applied voltage V, i.e. $I \propto V$.

Here are some worked examples to help us understand more about direct proportion.

Problem 10.   Three energy saving light bulbs cost £7.80. Determine the cost of seven such light bulbs

(i)   3 light bulbs cost £7.80

(ii)   Therefore, 1 light bulb costs $\dfrac{7.80}{3} = £2.60$

Hence, **7 light bulbs cost** $7 \times £2.60 = \textbf{£18.20}$

Problem 11.   If 56 litres of petrol costs £64.40, calculate the cost of 32 litres

(i)   56 litres of petrol costs £64.40

(ii)   Therefore, 1 litre of petrol costs $\dfrac{64.40}{56} = £1.15$

Hence, **32 litres cost** $32 \times 1.15 = \textbf{£36.80}$

Problem 12.   Hooke's law states that stress, σ, is directly proportional to strain, ε, within the elastic limit of a material. When, for mild steel, the stress is 63 MPa, the strain is 0.0003. Determine (a) the value of strain when the stress is 42 MPa, (b) the value of stress when the strain is 0.00072

(a)   Stress is directly proportional to strain.

(i) When the stress is 63 MPa, the strain is 0.0003

(ii) Hence, a stress of 1 MPa corresponds to a strain of $\dfrac{0.0003}{63}$

(iii) Thus, **the value of strain when the stress is 42 MPa** $= \dfrac{0.0003}{63} \times 42 = \textbf{0.0002}$

ROBERT HOOKE
1635-1703

*Who was Hooke? – **Robert Hooke** FRS (28 July 1635–3 March 1703) was an English natural philosopher, architect and polymath who, amongst other things, discovered the law of elasticity. To find out more go to **www.routledge.com/cw/bird**

*Who was Charles? – **Jacques Alexandre César Charles** (12 November 1746–7 April 1823) was a French inventor, scientist, mathematician and balloonist. Charles's law describes how gases expand when heated. To find out more go to **www.routledge.com/cw/bird**

† Who was Ohm? – See page 33. To find out more go to **www. routledge. com/ cw/bird**

(b) Strain is proportional to stress.

(i) When the strain is 0.0003, the stress is 63 MPa.

(ii) Hence, a strain of 0.0001 corresponds to $\frac{63}{3}$ MPa.

(iii) Thus, **the value of stress when the strain is 0.00072** $= \frac{63}{3} \times 7.2 =$ **151.2 MPa**.

Problem 13. Charles's law states that for a given mass of gas at constant pressure, the volume is directly proportional to its thermodynamic temperature. A gas occupies a volume of 2.4 litres at 600 K. Determine (a) the temperature when the volume is 3.2 litres, (b) the volume at 540 K

(a) Volume is directly proportional to temperature.

(i) When the volume is 2.4 litres, the temperature is 600 K.

(ii) Hence, a volume of 1 litre corresponds to a temperature of $\frac{600}{2.4}$ K.

(iii) Thus, **the temperature when the volume is 3.2 litres** $= \frac{600}{2.4} \times 3.2 =$ **800 K**

(b) Temperature is proportional to volume.

(i) When the temperature is 600 K, the volume is 2.4 litres.

(ii) Hence, a temperature of 1 K corresponds to a volume of $\frac{2.4}{600}$ litres.

(iii) Thus, **the volume at a temperature of 540 K** $= \frac{2.4}{600} \times 540 =$ **2.16 litres**.

Now try the following Practice Exercise

Practice Exercise 31   Direct proportion (answers on page 445)

1. Three engine parts cost £208.50. Calculate the cost of eight such parts.

2. If 9 litres of gloss white paint costs £24.75, calculate the cost of 24 litres of the same paint.

3. The total mass of 120 house bricks is 57.6 kg. Determine the mass of 550 such bricks.

4. A simple machine has an effort : load ratio of 3 : 37. Determine the effort, in newtons, to lift a load of 5.55 kN.

5. If 16 cans of lager weighs 8.32 kg, what will 28 cans weigh?

6. Hooke's law states that stress is directly proportional to strain within the elastic limit of a material. When, for copper, the stress is 60 MPa, the strain is 0.000625. Determine (a) the strain when the stress is 24 MPa and (b) the stress when the strain is 0.0005

7. Charles's law states that volume is directly proportional to thermodynamic temperature for a given mass of gas at constant pressure. A gas occupies a volume of 4.8 litres at 330 K. Determine (a) the temperature when the volume is 6.4 litres and (b) the volume when the temperature is 396 K.

8. A machine produces 320 bolts in a day. Calculate the number of bolts produced by 4 machines in 7 days.

Here are some further worked examples on direct proportion.

Problem 14. Some guttering on a house has to decline by 3 mm for every 70 cm to allow rainwater to drain. The gutter spans 8.4 m. How much lower should the low end be?

(i) The guttering has to decline in the ratio 3 : 700 or $\frac{3}{700}$

(ii) If d is the vertical drop in 8.4 m or 8400 mm, then the decline must be in the ratio d : 8400 or $\frac{d}{8400}$

(iii) Now $\frac{d}{8400} = \frac{3}{700}$

(iv) Cross-multiplying gives $700 \times d = 8400 \times 3$ from which, $d = \frac{8400 \times 3}{700}$

i.e. **d = 36 mm, which is how much lower the end should be to allow rainwater to drain**.

Problem 15. Ohm's law states that the current flowing in a fixed resistance is directly proportional to the applied voltage. When 90 mV is applied across a resistor the current flowing is 3 A. Determine (a) the current when the voltage is 60 mV and (b) the voltage when the current is 4.2 A

(a) Current is directly proportional to the voltage.

    (i) When voltage is 90 mV, the current is 3 A.

    (ii) Hence, a voltage of 1 mV corresponds to a current of $\dfrac{3}{90}$ A.

    (iii) Thus, **when the voltage is 60 mV, the current** $= 60 \times \dfrac{3}{90} = \textbf{2A}$.

(b) Voltage is directly proportional to the current.

    (i) When current is 3 A, the voltage is 90 mV.

    (ii) Hence, a current of 1 A corresponds to a voltage of $\dfrac{90}{3}$ mV $= 30$ mV.

    (iii) Thus, **when the current is 4.2 A, the voltage** $= 30 \times 4.2 = \textbf{126 mV}$.

Problem 16. Some approximate imperial to metric conversions are shown in Table 6.1. Use the table to determine

(a) the number of millimetres in 12.5 inches

(b) a speed of 50 miles per hour in kilometres per hour

(c) the number of miles in 300 km

(d) the number of kilograms in 20 pounds weight

(e) the number of pounds and ounces in 56 kilograms (correct to the nearest ounce)

(f) the number of litres in 24 gallons

(g) the number of gallons in 60 litres

Table 6.1

length	1 inch = 2.54 cm	
	1 mile = 1.6 km	
weight	2.2 lb = 1 kg	
	(1 lb = 16 oz)	
capacity	1.76 pints = 1 litre	
	(8 pints = 1 gallon)	

(a) 12.5 inches $= 12.5 \times 2.54$ cm $= 31.75$ cm
31.73 cm $= 31.75 \times 10$ mm $= \textbf{317.5 mm}$

(b) 50 m.p.h. $= 50 \times 1.6$ km/h $= \textbf{80 km/h}$

(c) 300 km $= \dfrac{300}{1.6}$ miles $= \textbf{187.5 miles}$

(d) 20 lb $= \dfrac{20}{2.2}$ kg $= \textbf{9.09 kg}$

(e) 56 kg $= 56 \times 2.2$ lb $= 123.2$ lb

0.2 lb $= 0.2 \times 16$ oz $= 3.2$ oz $= 3$ oz, correct to the nearest ounce.

Thus, 56 kg $= \textbf{123 lb } \textbf{3 oz}$, correct to the nearest ounce.

(f) 24 gallons $= 24 \times 8$ pints $= 192$ pints

192 pints $= \dfrac{192}{1.76}$ litres $= \textbf{109.1 litres}$

(g) 60 litres $= 60 \times 1.76$ pints $= 105.6$ pints

105.6 pints $= \dfrac{105.6}{8}$ gallons $= \textbf{13.2 gallons}$

Problem 17. Currency exchange rates for five countries are shown in Table 6.2. Calculate

(a) how many euros £55 will buy

(b) the number of Japanese yen which can be bought for £23

(c) the number of pounds sterling which can be exchanged for 6405 krone

(d) the number of American dollars which can be purchased for £92.50

(e) the number of pounds sterling which can be exchanged for 2925 Swiss francs

Table 6.2

France	£1 = 1.15 euros
Japan	£1 = 140 yen
Norway	£1 = 10.50 krone
Switzerland	£1 = 1.25 francs
USA	£1 = 1.30 dollars

(a) £1 = 1.25 euros, hence £55 = 55 × 1.15 euros
= **63.25 euros**.

(b) £1 = 140 yen, hence £23 = 23 × 140 yen
= **3220 yen**.

(c) £1 = 10.50 krone, hence 6405 krone = £$\dfrac{6405}{10.50}$
= **£610**

(d) £1 = 1.30 dollars, hence
£92.50 = 92.50 × 1.30 dollars = **\$120.25**

(e) £1 = 1.25 Swiss francs, hence
2925 Swiss francs = £$\dfrac{2925}{1.25}$ = **£2340**

Now try the following Practice Exercise

Practice Exercise 32   Further direct proportion (answers on page 445)

1. Ohm's law states that current is proportional to p.d. in an electrical circuit. When a p.d. of 60 mV is applied across a circuit a current of 24 μA flows. Determine (a) the current flowing when the p.d. is 5 V and (b) the p.d. when the current is 10 mA.

2. The tourist rate for the Swiss franc is quoted in a newspaper as £1 = 1.20 fr. How many francs can be purchased for £310?

3. If 1 inch = 2.54 cm, find the number of millimetres in 27 inches.

4. If 2.2 lb = 1 kg and 1 lb = 16 oz, determine the number of pounds and ounces in 38 kg (correct to the nearest ounce).

5. If 1 litre = 1.76 pints and 8 pints = 1 gallon, determine (a) the number of litres in 35 gallons and (b) the number of gallons in 75 litres.

6. Hooke's law states that stress is directly proportional to strain within the elastic limit of a material. When for brass the stress is 21 MPa, the strain is 0.00025. Determine the stress when the strain is 0.00035

7. If 12 inches = 30.48 cm, find the number of millimetres in 23 inches.

8. The tourist rate for the Canadian dollar is quoted in a newspaper as £1 = \$1.84. How many Canadian dollars can be purchased for £550?

6.4 Inverse proportion

Two variables, x and y, are in inverse proportion to one another if y is proportional to $\dfrac{1}{x}$, i.e. $y \propto \dfrac{1}{x}$ or $y = \dfrac{k}{x}$ or $k = xy$ where k is a constant, called the **coefficient of proportionality**.

Inverse proportion means that, as the value of one variable increases, the value of another decreases, and that their product is always the same.

For example, the time for a journey is inversely proportional to the speed of travel. So, if at 30 m.p.h. a journey is completed in 20 minutes, then at 60 m.p.h. the journey would be completed in 10 minutes. Double the speed, half the journey time. (Note that 30 × 20 = 60 × 10)

In another example, the time needed to dig a hole is inversely proportional to the number of people digging. So, if four men take 3 hours to dig a hole, then two men (working at the same rate) would take 6 hours. Half the men, twice the time. (Note that 4 × 3 = 2 × 6)

Here are some worked examples on inverse proportion.

Problem 18. It is estimated that a team of four designers would take a year to develop an engineering process. How long would three designers take?

If 4 designers take 1 year, then 1 designer would take 4 years to develop the process. Hence, 3 designers would take $\dfrac{4}{3}$ years, i.e. **1 year 4 months**.

Problem 19. A team of five people can deliver leaflets to every house in a particular area in four hours. How long will it take a team of three people?

If 5 people take 4 hours to deliver the leaflets, then 1 person would take $5 \times 4 = 20$ hours. Hence, 3 people would take $\dfrac{20}{3}$ hours, i.e. $6\dfrac{2}{3}$ hours, i.e. **6 hours 40 minutes**.

Problem 20. The electrical resistance R of a piece of wire is inversely proportional to the cross-sectional area A. When $A = 5\,\text{mm}^2$, $R = 7.2$ ohms. Determine (a) the coefficient of proportionality and (b) the cross-sectional area when the resistance is 4 ohms

(a) $R\alpha\dfrac{1}{A}$, i.e. $R = \dfrac{k}{A}$ or $k = RA$. Hence, when $R = 7.2$ and $A = 5$, the

coefficient of proportionality, $k = (7.2)(5) = \mathbf{36}$

(b) Since $k = RA$ then $A = \dfrac{k}{R}$. Hence, when $R = 4$,

the cross sectional area, $A = \dfrac{36}{4} = \mathbf{9\,mm^2}$

Problem 21. Boyle's law* states that, at constant temperature, the volume V of a fixed mass of gas is inversely proportional to its absolute pressure p. If a gas occupies a volume of $0.08\,\text{m}^3$ at a pressure of 1.5×10^6 pascals, determine (a) the coefficient of proportionality and (b) the volume if the pressure is changed to 4×10^6 pascals

(a) $V \propto \dfrac{1}{p}$ i.e. $V = \dfrac{k}{p}$ or $k = pV$. Hence, the

coefficient of proportionality, k

$$= (1.5 \times 10^6)(0.08) = \mathbf{0.12 \times 10^6}$$

(b) **Volume,** $V = \dfrac{k}{p} = \dfrac{0.12 \times 10^6}{4 \times 10^6} = \mathbf{0.03\,m^3}$

Now try the following Practice Exercise

Practice Exercise 33   Inverse proportion (answers on page 445)

1. A 10 kg bag of potatoes lasts for a week with a family of seven people. Assuming all eat the same amount, how long will the potatoes last if there are only two in the family?

2. If eight men take 5 days to build a wall, how long would it take two men?

3. If y is inversely proportional to x and $y = 15.3$ when $x = 0.6$, determine (a) the coefficient of proportionality, (b) the value of y when x is 1.5 and (c) the value of x when y is 27.2

Continued on page 56

*Who was Boyle? – **Robert Boyle** (25 January 1627–31 December 1691) was a natural philosopher, chemist, physicist and inventor. Regarded today as the first modern chemist, he is best known for Boyle's law, which describes the inversely proportional relationship between the absolute pressure and volume of a gas, providing the temperature is kept constant within a closed system. To find out more go to **www.routledge.com/cw/bird**

4. A car travelling at 50 km/h makes a journey in 70 minutes. How long will the journey take at 70 km/h?

5. Boyle's law states that, for a gas at constant temperature, the volume of a fixed mass of gas is inversely proportional to its absolute pressure. If a gas occupies a volume of $1.5\,m^3$ at a pressure of 200×10^3 pascals, determine (a) the constant of proportionality, (b) the volume when the pressure is 800×10^3 pascals and (c) the pressure when the volume is $1.25\,m^3$.

6. The energy received by a surface from a source of heat is inversely proportional to the square of the distance between the heat source and the surface. A surface 1m from the heat source receives 200 J of energy. Calculate (a) the energy received when the distance is changed to 2.5 m, (b) the distance required if the surface is to receive 800 J of energy.

Practice Exercise 34   Multiple-choice questions on ratio and proportion (Answers on page 445)

Each question has only one correct answer

1. Expressing the ratio 12:18 in its simplest form is:
   (a) 3:2   (b) 6:9   (c) 2:3   (d) 4:6

2. Four engineers can complete a task in 5 hours. Assuming the rate of work remains constant, six engineers will complete the task in:
   (a) 126 h   (b) 4 h 48 min
   (c) 7 h 30 min   (d) 3 h 20 min

3. A machine can produce 24 identical face masks per hour. The number of machines needed to produce 144 masks in 20 minutes is:
   (a) 36   (b) 12   (c) 18   (d) 6

4. A map has a scale of 1:200,000. Two engineering sites are 15 cm apart on the map. The actual distance apart of the two sites is:
   (a) 30 km   (b) 300 km
   (c) 15 km   (d) 150 km

5. y varies as the square of x, and $y = 12$ when $x = 2$. When $x = 3$, the value of y is:
   (a) 432   (b) 27   (c) 36   (d) 9

6. The simplest form of 1.5:2.5 is:
   (a) 15:25   (b) 6:10   (c) 3:5   (d) 0.75:1.25

7. A bonus of £7632 is to be divided between three workers A, B and C in the ratio 5:3:1 respectively. The amount that worker B receives is:
   (a) £4240   (b) £848   (c) £3826   (d) £2544

8. How many litres of water needs to be added to 6 litres of a 60% alcohol solution to create a 40% alcohol solution?
   (a) 2 litres   (b) 3 litres
   (c) 6 litres   (d) 9 litres

9. A simple machine has an effort to load ratio of 5:39. The effort, in newtons, to lift a load of 5.46 kN is:
   (a) 700 N   (b) 620.45 N
   (c) 42588 N   (d) 0.7 N

10. In an engineering laboratory, acid and water are mixed in the ratio 2:7. To make 369 ml of the mixture, the amount of acid added is:
    (a) 82 ml   (b) 184.5 ml
    (c) 105.4 ml   (d) 287 ml

For fully worked solutions to each of the problems in Practice Exercises 29 to 33 in this chapter, go to the website:
www.routledge.com/cw/bird

Chapter 7

Powers, roots and laws of indices

Why it is important to understand: **Powers, roots and laws of indices**

Powers and roots are used extensively in mathematics and engineering, so it is important to get a good grasp of what they are and how, and why, they are used. Being able to multiply powers together by adding their indices is particularly useful for disciplines like engineering and electronics, where quantities are often expressed as a value multiplied by some power of ten. In the field of electrical engineering, for example, the relationship between electric current, voltage and resistance in an electrical system is critically important, and yet the typical unit values for these properties can differ by several orders of magnitude. Studying, or working, in an engineering discipline, you very quickly become familiar with powers and roots and laws of indices. This chapter provides an important lead into Chapter 8 which deals with units, prefixes and engineering notation.

At the end of this chapter you should be able to:

- understand the terms base, index and power
- understand square roots
- perform calculations with powers and roots
- state the laws of indices
- perform calculations using the laws of indices

7.1 Introduction

The manipulation of powers and roots is a crucial underlying skill needed in algebra. In this chapter, powers and roots of numbers are explained, together with the laws of indices.

Many worked examples are included to help understanding.

7.2 Powers and roots

7.2.1 Indices

The number 16 is the same as $2 \times 2 \times 2 \times 2$, and $2 \times 2 \times 2 \times 2$ can be abbreviated to 2^4. When written as 2^4, 2 is called the **base** and the 4 is called the **index** or **power**. 2^4 is read as '**two to the power of four**'.

Similarly, 3^5 is read as '**three to the power of 5**'

When the indices are 2 and 3 they are given special names; i.e. 2 is called 'squared' and 3 is called 'cubed'. Thus,

4^2 is called '**four squared**' rather than '4 to the power of 2' and

5^3 is called '**five cubed**' rather than '5 to the power of 3'

When no index is shown, the power is 1. For example, 2 means 2^1

Problem 1. Evaluate (a) 2^6 (b) 3^4

(a) 2^6 means $2 \times 2 \times 2 \times 2 \times 2 \times 2$ (i.e. 2 multiplied by itself 6 times), and $2 \times 2 \times 2 \times 2 \times 2 \times 2 = \mathbf{64}$

i.e. $2^6 = \mathbf{64}$

(b) 3^4 means $3 \times 3 \times 3 \times 3$ (i.e. 3 multiplied by itself 4 times), and $3 \times 3 \times 3 \times 3 = \mathbf{81}$

i.e. $3^4 = \mathbf{81}$

Problem 2. Change the following to index form:
(a) 32 (b) 625

(a) (i) To express 32 in its lowest factors, 32 is initially divided by the lowest prime number, i.e. 2

(ii) $32 \div 2 = 16$, hence $32 = 2 \times 16$

(iii) 16 is also divisible by 2, i.e. $16 = 2 \times 8$. Thus, $32 = 2 \times 2 \times 8$

(iv) 8 is also divisible by 2, i.e. $8 = 2 \times 4$. Thus, $32 = 2 \times 2 \times 2 \times 4$

(v) 4 is also divisible by 2, i.e. $4 = 2 \times 2$. Thus, $32 = 2 \times 2 \times 2 \times 2 \times 2$

(vi) Thus, $\mathbf{32 = 2^5}$

(b) (i) 625 is not divisible by the lowest prime number, i.e. 2. The next prime number is 3 and 625 is not divisible by 3 either. The next prime number is 5

(ii) $625 \div 5 = 125$, i.e. $625 = 5 \times 125$

(iii) 125 is also divisible by 5, i.e. $125 = 5 \times 25$. Thus, $625 = 5 \times 5 \times 25$

(iv) 25 is also divisible by 5, i.e. $25 = 5 \times 5$. Thus, $625 = 5 \times 5 \times 5 \times 5$

(v) Thus, $\mathbf{625 = 5^4}$

Problem 3. Evaluate $3^3 \times 2^2$

$$3^3 \times 2^2 = 3 \times 3 \times 3 \times 2 \times 2$$
$$= 27 \times 4$$
$$= \mathbf{108}$$

7.2.2 Square roots

When a number is multiplied by itself the product is called a square.

For example, the square of 3 is $3 \times 3 = 3^2 = 9$

A square root is the reverse process; i.e. the value of the base which when multiplied by itself gives the number; i.e. the square root of 9 is 3

The symbol $\sqrt{}$ is used to denote a square root. Thus, $\sqrt{9} = 3$. Similarly, $\sqrt{4} = 2$ and $\sqrt{25} = 5$

Because $-3 \times -3 = 9$, $\sqrt{9}$ also equals -3. Thus, $\sqrt{9} = +3$ or -3 which is usually written as $\sqrt{9} = \pm 3$. Similarly, $\sqrt{16} = \pm 4$ and $\sqrt{36} = \pm 6$

The square root of, say, 9 may also be written in index form as $9^{\frac{1}{2}}$.

$$9^{\frac{1}{2}} \equiv \sqrt{9} = \pm 3$$

Problem 4. Evaluate $\dfrac{3^2 \times 2^3 \times \sqrt{36}}{\sqrt{16} \times 4}$ taking only positive square roots

$$\frac{3^2 \times 2^3 \times \sqrt{36}}{\sqrt{16} \times 4} = \frac{3 \times 3 \times 2 \times 2 \times 2 \times 6}{4 \times 4}$$
$$= \frac{9 \times 8 \times 6}{16} = \frac{9 \times 1 \times 6}{2}$$
$$= \frac{9 \times 1 \times 3}{1} \qquad \text{by cancelling}$$
$$= \mathbf{27}$$

Problem 5. Evaluate $\dfrac{10^4 \times \sqrt{100}}{10^3}$ taking the positive square root only

$$\frac{10^4 \times \sqrt{100}}{10^3} = \frac{10 \times 10 \times 10 \times 10 \times 10}{10 \times 10 \times 10}$$
$$= \frac{1 \times 1 \times 1 \times 10 \times 10}{1 \times 1 \times 1} \qquad \text{by cancelling}$$
$$= \frac{100}{1}$$
$$= \mathbf{100}$$

Now try the following Practice Exercise

Practice Exercise 35   Powers and roots (answers on page 445)

Evaluate the following without the aid of a calculator.

1. 3^3

2. 2^7

3. 10^5

4. $2^4 \times 3^2 \times 2 \div 3$

5. Change 16 to index form.

6. $25^{\frac{1}{2}}$

7. $64^{\frac{1}{2}}$

8. $\dfrac{10^5}{10^3}$

9. $\dfrac{10^2 \times 10^3}{10^5}$

10. $\dfrac{2^5 \times 64^{\frac{1}{2}} \times 3^2}{\sqrt{144} \times 3}$ taking positive square roots only.

7.3 Laws of indices

There are six laws of indices.

(1) From earlier, $2^2 \times 2^3 = (2 \times 2) \times (2 \times 2 \times 2)$
$$= 32$$
$$= 2^5$$
Hence, $2^2 \times 2^3 = 2^5$
or $2^2 \times 2^3 = 2^{2+3}$

This is the first law of indices, which demonstrates that **when multiplying two or more numbers having the same base, the indices are added**.

(2) $\dfrac{2^5}{2^3} = \dfrac{2 \times 2 \times 2 \times 2 \times 2}{2 \times 2 \times 2} = \dfrac{1 \times 1 \times 1 \times 2 \times 2}{1 \times 1 \times 1}$
$$= \dfrac{2 \times 2}{1} = 4 = 2^2$$
Hence, $\dfrac{2^5}{2^3} = 2^2$ or $\dfrac{2^5}{2^3} = 2^{5-3}$

This is the second law of indices, which demonstrates that **when dividing two numbers having the same base, the index in the denominator is subtracted from the index in the numerator**.

(3) $(3^5)^2 = 3^{5 \times 2} = 3^{10}$ and $(2^2)^3 = 2^{2 \times 3} = 2^6$

This is the third law of indices, which states that **when a number which is raised to a power is raised to a further power, the indices are multiplied**.

(4) $3^0 = 1$ and $17^0 = 1$

This is the fourth law of indices, which states that **when a number has an index of 0, its value is 1**.

(5) $3^{-4} = \dfrac{1}{3^4}$ and $\dfrac{1}{2^{-3}} = 2^3$

This is the fifth law of indices, which demonstrates that **a number raised to a negative power is the reciprocal of that number raised to a positive power**.

(6) $8^{\frac{2}{3}} = \sqrt[3]{8^2} = (2)^2 = 4$ and
$$25^{\frac{1}{2}} = \sqrt[2]{25^1} = \sqrt{25^1} = \pm 5$$
(Note that $\sqrt{} \equiv \sqrt[2]{}$)

This is the sixth law of indices, which demonstrates that **when a number is raised to a fractional power the denominator of the fraction is the root of the number and the numerator is the power**.

Here are some worked examples using the laws of indices.

Problem 6.   Evaluate in index form $5^3 \times 5 \times 5^2$

$5^3 \times 5 \times 5^2 = 5^3 \times 5^1 \times 5^2$ (Note that 5 means 5^1)
$$= 5^{3+1+2}$$ from law (1)
$$= \mathbf{5^6}$$

Problem 7.   Evaluate $\dfrac{3^5}{3^4}$

$\dfrac{3^5}{3^4} = 3^{5-4}$ from law (2)
$$= 3^1$$
$$= \mathbf{3}$$

Problem 8.   Evaluate $\dfrac{2^4}{2^4}$

$\dfrac{2^4}{2^4} = 2^{4-4}$ from law (2)
$$= 2^0$$
But $\dfrac{2^4}{2^4} = \dfrac{2 \times 2 \times 2 \times 2}{2 \times 2 \times 2 \times 2} = \dfrac{16}{16} = 1$
Hence, $\mathbf{2^0 = 1}$ from law (4)

Any number raised to the power of zero equals 1. For example, $6^0 = 1, 128^0 = 1, 13742^0 = 1$, and so on.

Problem 9. Evaluate $\dfrac{3 \times 3^2}{3^4}$

$$\frac{3 \times 3^2}{3^4} = \frac{3^1 \times 3^2}{3^4} = \frac{3^{1+2}}{3^4} = \frac{3^3}{3^4} = 3^{3-4} = 3^{-1}$$

from laws (1) and (2)

But $\dfrac{3^3}{3^4} = \dfrac{3 \times 3 \times 3}{3 \times 3 \times 3 \times 3} = \dfrac{1 \times 1 \times 1}{1 \times 1 \times 1 \times 3}$

(by cancelling)

$$= \frac{1}{3}$$

Hence, $\dfrac{3 \times 3^2}{3^4} = 3^{-1} = \dfrac{1}{3}$ from law (5)

Similarly, $2^{-1} = \dfrac{1}{2}, 2^{-5} = \dfrac{1}{2^5}, \dfrac{1}{5^4} = 5^{-4}$, and so on.

Problem 10. Evaluate $\dfrac{10^3 \times 10^2}{10^8}$

$$\frac{10^3 \times 10^2}{10^8} = \frac{10^{3+2}}{10^8} = \frac{10^5}{10^8}$$ from law (1)

$$= 10^{5-8} = 10^{-3}$$ from law (2)

$$= \frac{1}{10^{+3}} = \frac{1}{1000}$$ from law (5)

Hence, $\dfrac{10^3 \times 10^2}{10^8} = 10^{-3} = \dfrac{1}{1000} = 0.001$

Understanding powers of ten is important, especially when dealing with prefixes in Chapter 8. For example,

$$10^2 = 100, 10^3 = 1000, 10^4 = 10\,000,$$

$$10^5 = 100\,000, 10^6 = 1\,000\,000$$

$$10^{-1} = \frac{1}{10} = 0.1, 10^{-2} = \frac{1}{10^2} = \frac{1}{100} = 0.01,$$

and so on.

Problem 11. Evaluate (a) $5^2 \times 5^3 \div 5^4$ (b) $(3 \times 3^5) \div (3^2 \times 3^3)$

From laws (1) and (2):

(a) $5^2 \times 5^3 \div 5^4 = \dfrac{5^2 \times 5^3}{5^4} = \dfrac{5^{(2+3)}}{5^4}$

$$= \frac{5^5}{5^4} = 5^{(5-4)} = 5^1 = 5$$

(b) $(3 \times 3^5) \div (3^2 \times 3^3) = \dfrac{3 \times 3^5}{3^2 \times 3^3} = \dfrac{3^{(1+5)}}{3^{(2+3)}}$

$$= \frac{3^6}{3^5} = 3^{6-5} = 3^1 = 3$$

Problem 12. Simplify (a) $(2^3)^4$ (b) $(3^2)^5$, expressing the answers in index form

From law (3):

(a) $(2^3)^4 = 2^{3 \times 4} = 2^{12}$

(b) $(3^2)^5 = 3^{2 \times 5} = 3^{10}$

Problem 13. Evaluate: $\dfrac{(10^2)^3}{10^4 \times 10^2}$

From laws (1) to (4):

$$\frac{(10^2)^3}{10^4 \times 10^2} = \frac{10^{(2 \times 3)}}{10^{(4+2)}} = \frac{10^6}{10^6} = 10^{6-6} = 10^0 = 1$$

Problem 14. Find the value of (a) $\dfrac{2^3 \times 2^4}{2^7 \times 2^5}$ (b) $\dfrac{(3^2)^3}{3 \times 3^9}$

From the laws of indices:

(a) $\dfrac{2^3 \times 2^4}{2^7 \times 2^5} = \dfrac{2^{(3+4)}}{2^{(7+5)}} = \dfrac{2^7}{2^{12}} = 2^{7-12}$

$$= 2^{-5} = \frac{1}{2^5} = \frac{1}{32}$$

(b) $\dfrac{(3^2)^3}{3 \times 3^9} = \dfrac{3^{2 \times 3}}{3^{1+9}} = \dfrac{3^6}{3^{10}} = 3^{6-10}$

$$= 3^{-4} = \frac{1}{3^4} = \frac{1}{81}$$

Problem 15. Evaluate (a) $4^{1/2}$ (b) $16^{3/4}$ (c) $27^{2/3}$ (d) $9^{-1/2}$

(a) $4^{1/2} = \sqrt{4} = \pm 2$

(b) $16^{3/4} = \sqrt[4]{16^3} = (2)^3 = 8$
    (Note that it does not matter whether the 4th root of 16 is found first or whether 16 cubed is found first – the same answer will result.)

(c) $27^{2/3} = \sqrt[3]{27^2} = (3)^2 = 9$

(d) $9^{-1/2} = \dfrac{1}{9^{1/2}} = \dfrac{1}{\sqrt{9}} = \dfrac{1}{\pm 3} = \pm\dfrac{1}{3}$

Now try the following Practice Exercise

Practice Exercise 36   Laws of indices (answers on page 445)

Evaluate the following without the aid of a calculator.

1. $2^2 \times 2 \times 2^4$

2. $3^5 \times 3^3 \times 3$ in index form

3. $\dfrac{2^7}{2^3}$

4. $\dfrac{3^3}{3^5}$

5. 7^0

6. $\dfrac{2^3 \times 2 \times 2^6}{2^7}$

7. $\dfrac{10 \times 10^6}{10^5}$

8. $10^4 \div 10$

9. $\dfrac{10^3 \times 10^4}{10^9}$

10. $5^6 \times 5^2 \div 5^7$

11. $(7^2)^3$ in index form

12. $(3^3)^2$

13. $\dfrac{3^7 \times 3^4}{3^5}$ in index form

14. $\dfrac{(9 \times 3^2)^3}{(3 \times 27)^2}$ in index form

15. $\dfrac{(16 \times 4)^2}{(2 \times 8)^3}$

16. $\dfrac{5^{-2}}{5^{-4}}$

17. $\dfrac{3^2 \times 3^{-4}}{3^3}$

18. $\dfrac{7^2 \times 7^{-3}}{7 \times 7^{-4}}$

19. $\dfrac{2^3 \times 2^{-4} \times 2^5}{2 \times 2^{-2} \times 2^6}$

20. $\dfrac{5^{-7} \times 5^2}{5^{-8} \times 5^3}$

Here are some further worked examples using the laws of indices.

Problem 16. Evaluate $\dfrac{3^3 \times 5^7}{5^3 \times 3^4}$

The laws of indices only apply to terms **having the same base**. Grouping terms having the same base and then applying the laws of indices to each of the groups independently gives

$$\frac{3^3 \times 5^7}{5^3 \times 3^4} = \frac{3^3}{3^4} \times \frac{5^7}{5^3} = 3^{(3-4)} \times 5^{(7-3)}$$

$$= 3^{-1} \times 5^4 = \frac{5^4}{3^1} = \frac{625}{3} = \mathbf{208\frac{1}{3}}$$

Problem 17. Find the value of $\dfrac{2^3 \times 3^5 \times (7^2)^2}{7^4 \times 2^4 \times 3^3}$

$$\frac{2^3 \times 3^5 \times (7^2)^2}{7^4 \times 2^4 \times 3^3} = 2^{3-4} \times 3^{5-3} \times 7^{2\times2-4}$$

$$= 2^{-1} \times 3^2 \times 7^0$$

$$= \frac{1}{2} \times 3^2 \times 1 = \frac{9}{2} = \mathbf{4\frac{1}{2}}$$

Problem 18. Evaluate $\dfrac{4^{1.5} \times 8^{1/3}}{2^2 \times 32^{-2/5}}$

$$4^{1.5} = 4^{3/2} = \sqrt{4^3} = 2^3 = 8, \ 8^{1/3} = \sqrt[3]{8} = 2,$$

$$2^2 = 4, \ 32^{-2/5} = \frac{1}{32^{2/5}} = \frac{1}{\sqrt[5]{32^2}} = \frac{1}{2^2} = \frac{1}{4}$$

Hence, $\dfrac{4^{1.5} \times 8^{1/3}}{2^2 \times 32^{-2/5}} = \dfrac{8 \times 2}{4 \times \dfrac{1}{4}} = \dfrac{16}{1} = \mathbf{16}$

Problem 19. Evaluate $\dfrac{3^2 \times 5^5 + 3^3 \times 5^3}{3^4 \times 5^4}$

Dividing each term by the HCF (highest common factor) of the three terms, i.e. $3^2 \times 5^3$, gives

$$\frac{3^2 \times 5^5 + 3^3 \times 5^3}{3^4 \times 5^4} = \frac{\dfrac{3^2 \times 5^5}{3^2 \times 5^3} + \dfrac{3^3 \times 5^3}{3^2 \times 5^3}}{\dfrac{3^4 \times 5^4}{3^2 \times 5^3}}$$

$$= \frac{3^{(2-2)} \times 5^{(5-3)} + 3^{(3-2)} \times 5^0}{3^{(4-2)} \times 5^{(4-3)}}$$

$$= \frac{3^0 \times 5^2 + 3^1 \times 5^0}{3^2 \times 5^1}$$

$$= \frac{1 \times 25 + 3 \times 1}{9 \times 5} = \mathbf{\frac{28}{45}}$$

Problem 20. Find the value of $\dfrac{3^2 \times 5^5}{3^4 \times 5^4 + 3^3 \times 5^3}$

To simplify the arithmetic, each term is divided by the HCF of all the terms, i.e. $3^2 \times 5^3$. Thus,

$$\frac{3^2 \times 5^5}{3^4 \times 5^4 + 3^3 \times 5^3} = \frac{\dfrac{3^2 \times 5^5}{3^2 \times 5^3}}{\dfrac{3^4 \times 5^4}{3^2 \times 5^3} + \dfrac{3^3 \times 5^3}{3^2 \times 5^3}}$$

$$= \frac{3^{(2-2)} \times 5^{(5-3)}}{3^{(4-2)} \times 5^{(4-3)} + 3^{(3-2)} \times 5^{(3-3)}}$$

$$= \frac{3^0 \times 5^2}{3^2 \times 5^1 + 3^1 \times 5^0}$$

$$= \frac{1 \times 5^2}{3^2 \times 5 + 3 \times 1} = \frac{25}{45 + 3} = \mathbf{\frac{25}{48}}$$

Problem 21. Simplify $\dfrac{7^{-3} \times 3^4}{3^{-2} \times 7^5 \times 5^{-2}}$ expressing the answer in index form with positive indices

Since $7^{-3} = \dfrac{1}{7^3}$, $\dfrac{1}{3^{-2}} = 3^2$ and $\dfrac{1}{5^{-2}} = 5^2$, then

$$\frac{7^{-3} \times 3^4}{3^{-2} \times 7^5 \times 5^{-2}} = \frac{3^4 \times 3^2 \times 5^2}{7^3 \times 7^5}$$

$$= \frac{3^{(4+2)} \times 5^2}{7^{(3+5)}} = \mathbf{\frac{3^6 \times 5^2}{7^8}}$$

Problem 22. Simplify $\dfrac{\left(\dfrac{4}{3}\right)^3 \times \left(\dfrac{3}{5}\right)^{-2}}{\left(\dfrac{2}{5}\right)^{-3}}$ giving the answer with positive indices

Raising a fraction to a power means that both the numerator and the denominator of the fraction are raised to that power, i.e. $\left(\dfrac{4}{3}\right)^3 = \dfrac{4^3}{3^3}$

A fraction raised to a negative power has the same value as the inverse of the fraction raised to a positive power.

Thus, $\left(\dfrac{3}{5}\right)^{-2} = \dfrac{1}{\left(\dfrac{3}{5}\right)^2} = \dfrac{1}{\dfrac{3^2}{5^2}} = 1 \times \dfrac{5^2}{3^2} = \dfrac{5^2}{3^2}$

Similarly, $\left(\dfrac{2}{5}\right)^{-3} = \left(\dfrac{5}{2}\right)^3 = \dfrac{5^3}{2^3}$

Thus, $\dfrac{\left(\dfrac{4}{3}\right)^3 \times \left(\dfrac{3}{5}\right)^{-2}}{\left(\dfrac{2}{5}\right)^{-3}} = \dfrac{\dfrac{4^3}{3^3} \times \dfrac{5^2}{3^2}}{\dfrac{5^3}{2^3}}$

$$= \frac{4^3}{3^3} \times \frac{5^2}{3^2} \times \frac{2^3}{5^3} = \frac{(2^2)^3 \times 2^3}{3^{(3+2)} \times 5^{(3-2)}}$$

$$= \mathbf{\frac{2^9}{3^5 \times 5}}$$

Now try the following Practice Exercise

In Problems 1 to 4, simplify the expressions given, expressing the answers in index form and with positive indices.

1. $\dfrac{3^3 \times 5^2}{5^4 \times 3^4}$

2. $\dfrac{7^{-2} \times 3^{-2}}{3^5 \times 7^4 \times 7^{-3}}$

3. $\dfrac{4^2 \times 9^3}{8^3 \times 3^4}$

4. $\dfrac{8^{-2} \times 5^2 \times 3^{-4}}{25^2 \times 2^4 \times 9^{-2}}$

In Problems 5 to 15, evaluate the expressions given.

5. $\left(\dfrac{1}{3^2}\right)^{-1}$

6. $81^{0.25}$

7. $16^{-\frac{1}{4}}$

8. $\left(\dfrac{4}{9}\right)^{1/2}$

9. $\dfrac{9^2 \times 7^4}{3^4 \times 7^4 + 3^3 \times 7^2}$

10. $\dfrac{3^3 \times 5^2}{2^3 \times 3^2 - 8^2 \times 9}$

11. $\dfrac{3^3 \times 7^2 - 5^2 \times 7^3}{3^2 \times 5 \times 7^2}$

12. $\dfrac{(2^4)^2 - 3^{-2} \times 4^4}{2^3 \times 16^2}$

13. $\dfrac{\left(\dfrac{1}{2}\right)^3 - \left(\dfrac{2}{3}\right)^{-2}}{\left(\dfrac{3}{2}\right)^2}$

14. $\dfrac{\left(\dfrac{4}{3}\right)^4}{\left(\dfrac{2}{9}\right)^2}$

15. $\dfrac{(3^2)^{3/2} \times (8^{1/3})^2}{(3)^2 \times (4^3)^{1/2} \times (9)^{-1/2}}$

Practice Exercise 38   Multiple-choice questions on powers, roots and laws of indices (answers on page 445)

Each question has only one correct answer

1. The value of $\dfrac{2^{-3}}{2^{-4}} - 1$ is equal to:
   (a) $-\dfrac{1}{2}$ (b) 2   (c) $\dfrac{1}{2}$ (d) 1

2. $c^{\frac{1}{3}}$ is equivalent to:
   (a) $\dfrac{1}{c^3}$ (b) $\sqrt[3]{c}$ (c) $\dfrac{1}{\sqrt[3]{c}}$ (d) $\left(\dfrac{1}{c}\right)^3$

3. $7^0 \times 2^0$ is equal to:
   (a) 1   (b) 0   (c) 14   (d) 1400

4. 6^{-2} is equal to:
   (a) -12 (b) $\dfrac{1}{12}$ (c) $\dfrac{1}{36}$ (d) -36

5. In an engineering equation $\dfrac{3^4}{3^r} = \dfrac{1}{9}$. The value of r is:
   (a) -6 (b) 2   (c) 6   (d) -2

6. When $p = 3$, $q = -\dfrac{1}{2}$ and $r = -2$, the engineering expression $2p^2 q^3 r^4$ is equal to:
   (a) -36 (b) 1296   (c) 36   (d) 18

7. $16^{-\frac{3}{4}}$ is equal to:
   (a) 8   (b) $-\dfrac{1}{2^3}$ (c) 4   (d) $\dfrac{1}{8}$

8. $p \times \left(\dfrac{1}{p^{-2}}\right)^{-3}$ is equivalent to:
   (a) $\dfrac{1}{p^{\frac{3}{2}}}$ (b) p^{-5} (c) p^5 (d) $\dfrac{1}{p^{-5}}$

9. $32^{\frac{1}{5}}$ is equal to:
   (a) 16   (b) 18   (c) 2   (d) 160

10. $\dfrac{5^2 \times 5^{-3}}{5^{-4}}$ is equivalent to:
    (a) 5^{-5} (b) 5^{24} (c) 5^3 (d) 5

11. If $2^4 \times 2^b = 64$ the value of b is:
    (a) 1   (b) 2   (c) 3   (d) 4

12. $(16^{-\frac{1}{4}} - 27^{-\frac{2}{3}})$ is equal to:
    (a) -7 (b) $\dfrac{7}{18}$ (c) $1\dfrac{8}{9}$ (d) $-8\dfrac{1}{2}$

13. $a^8 \times a^{12} \div a^3$ simplifies to:
    (a) a^{23} (b) a^{17} (c) $a^{\frac{20}{3}}$ (d) a^{32}

14. $9^{\frac{3}{2}}$ is equal to:
    (a) 27   (b) 18   (c) -18 (d) $\dfrac{1}{27}$

15. The value of $xy^{-1} + yz^{-1}$ when $x = 3, y = 2$ and $c = \dfrac{1}{2}$ is:
    (a) 7   (b) 2.5   (c) 10   (d) 5.5

16. $64^{\frac{1}{2}}$ is equal to:
    (a) 4   (b) 8   (c) 16   (d) 32

17. $\dfrac{a^5}{a^{12}}$ is equivalent to:
    (a) a^{17} (b) $a^{\frac{1}{7}}$ (c) $\dfrac{1}{a^7}$ (d) $\dfrac{5}{a^{12}}$

18. $\left(\dfrac{y^2}{y^3} \times y^4\right)^2$ is equivalent to:
    (a) y^6 (b) y^{-6} (c) y^3 (d) $\dfrac{1}{y}$

19. $(3x)^2$ is equivalent to:
    (a) $6x^2$ (b) $3x^2$ (c) $9x^2$ (d) $3x^3$

20. $64^{-\frac{3}{2}}$ is equal to:
    (a) $\dfrac{1}{96}$ (b) $\dfrac{1}{64}$ (c) 512   (d) $\dfrac{1}{512}$

21. $\left(-\dfrac{2}{5}\right)^3$ is equal to:
    (a) $\dfrac{8}{125}$ (b) $-\dfrac{6}{15}$ (c) $-\dfrac{8}{125}$ (d) $\dfrac{6}{15}$

22. $\left(\dfrac{1}{4}\right)^{-\frac{1}{2}}$ is equal to:
    (a) 2   (b) $\dfrac{1}{16}$ (c) $\sqrt{2}$ (d) $\dfrac{1}{2}$

23. $(-2)^2 \times (-10)^3$ is equal to:
    (a) 400   (b) 4000   (c) -2000 (d) -4000

24. 30 to the power of zero is:
    (a) 1   (b) -30 (c) 0   (d) 30

25. $125^{-\frac{2}{3}}$ is equal to:
    (a) 25   (b) $-\dfrac{1}{25}$ (c) -25 (d) $\dfrac{1}{25}$

For fully worked solutions to each of the problems in Practice Exercises 35 to 37 in this chapter, go to the website:
www.routledge.com/cw/bird

COMPANION @ WEBSITE

Chapter 8

Units, prefixes and engineering notation

Why it is important to understand: Units, prefixes and engineering notation

In engineering there are many different quantities to get used to, and hence many units to become familiar with. For example, force is measured in newtons, electric current is measured in amperes and pressure is measured in pascals. Sometimes the units of these quantities are either very large or very small and hence prefixes are used. For example, 1000 pascals may be written as 10^3 Pa which is written as 1 kPa in prefix form, the k being accepted as a symbol to represent 1000 or 10^3. Studying, or working, in an engineering discipline, you very quickly become familiar with the standard units of measurement, the prefixes used and engineering notation. An electronic calculator is extremely helpful with engineering notation.

At the end of this chapter you should be able to:

- state the seven SI units
- understand derived units
- recognise common engineering units
- understand common prefixes used in engineering
- express decimal numbers in standard form
- use engineering notation and prefix form with engineering units

8.1 Introduction

Of considerable importance in engineering is a knowledge of units of engineering quantities, the prefixes used with units, and engineering notation.
We need to know, for example, that

$$80\,kV = 80 \times 10^3\,V, \text{ which means } 80\,000 \text{ volts}$$

and $25\,mA = 25 \times 10^{-3}\,A,$
which means 0.025 amperes

and $50\,nF = 50 \times 10^{-9}\,F,$
which means 0.000000050 farads

This is explained in this chapter.

8.2 SI units

The system of units used in engineering and science is the Système Internationale d'Unités (**International System of Units**), usually abbreviated to SI units, and is based on the metric system. This was introduced in 1960 and has now been adopted by the majority of countries as the official system of measurement.
The basic seven units used in the SI system are listed in Table 8.1 with their symbols.
There are, of course, many units other than these seven. These other units are called **derived units** and are

Table 8.1 Basic SI units

Quantity	Unit	Symbol
Length	metre	m (1 m = 100 cm = 1000 mm)
Mass	kilogram	kg (1 kg = 1000 g)
Time	second	s
Electric current	ampere	A
Thermodynamic temperature	kelvin	K (K = °C + 273)
Luminous intensity	candela	cd
Amount of substance	mole	mol

defined in terms of the standard units listed in the table. For example, speed is measured in metres per second, therefore using two of the standard units, i.e. length and time.

Some derived units are given **special names**. For example, force = mass × acceleration has units of kilogram metre per second squared, which uses three of the base units, i.e. kilograms, metres and seconds. The unit of kg m/s² is given the special name of a **newton**[*]. Table 8.2 contains a list of some quantities and their units that are common in engineering.

8.3 Common prefixes

SI units may be made larger or smaller by using prefixes which denote multiplication or division by a particular amount.

The most common multiples are listed in Table 8.3. A knowledge of indices is needed since all of the prefixes are powers of 10 with indices that are a multiple of 3.

Here are some examples of prefixes used with engineering units.

A **frequency of 15 GHz** means 15×10^9 Hz, which is 15 000 000 000 hertz[*],

i.e. 15 gigahertz is written as 15 GHz and is equal to 15 thousand million hertz.

*Who was Newton? – **Sir Isaac Newton** PRS MP (25 December 1642–20 March 1727) was an English polymath. Newton showed that the motions of objects are governed by the same set of natural laws, by demonstrating the consistency between Kepler's laws of planetary motion and his theory of gravitation. To find out more go to **www.routledge.com/cw/bird**

*Who was Hertz? – **Heinrich Rudolf Hertz** (22 February 1857–1 January 1894) was the first person to conclusively prove the existence of electromagnetic waves. The scientific unit of frequency – cycles per second – was named the '**hertz**' in his honour. To find out more go to **www.routledge.com/cw/bird**

Table 8.2 Some quantities and their units that are common in engineering

Quantity	Unit	Symbol
Length	metre	m
Area	square metre	m^2
Volume	cubic metre	m^3
Mass	kilogram	kg
Time	second	s
Electric current	ampere	A
Speed, velocity	metre per second	m/s
Acceleration	metre per second squared	m/s^2
Density	kilogram per cubic metre	kg/m^3
Temperature	kelvin or Celsius	K or °C
Angle	radian or degree	rad or °
Angular velocity	radian per second	rad/s
Frequency	hertz	Hz
Force	newton	N
Pressure	pascal	Pa
Energy, work	joule	J
Power	watt	W
Charge, quantity of electricity	coulomb	C
Electric potential	volt	V
Capacitance	farad	F
Electrical resistance	ohm	Ω
Inductance	henry	H
Moment of force	newton metre	N m

Table 8.3 Common SI multiples

Prefix	Name	Meaning	
G	giga	multiply by 10^9	i.e. $\times\ 1\,000\,000\,000$
M	mega	multiply by 10^6	i.e. $\times\ 1\,000\,000$
k	kilo	multiply by 10^3	i.e. $\times\ 1\,000$
m	milli	multiply by 10^{-3}	i.e. $\times\ \dfrac{1}{10^3} = \dfrac{1}{1000} = 0.001$
μ	micro	multiply by 10^{-6}	i.e. $\times\ \dfrac{1}{10^6} = \dfrac{1}{1\,000\,000} = 0.000001$
n	nano	multiply by 10^{-9}	i.e. $\times\ \dfrac{1}{10^9} = \dfrac{1}{1\,000\,000\,000} = 0.000\,000\,001$
p	pico	multiply by 10^{-12}	i.e. $\times\ \dfrac{1}{10^{12}} = \dfrac{1}{1\,000\,000\,000\,000} =$ $0.000\,000\,000\,001$

(Instead of writing $15\,000\,000\,000$ hertz, it is much neater, takes up less space and prevents errors caused by having so many zeros, to write the frequency as 15 GHz.)

A **voltage of 40 MV** means 40×10^6 V, which is $40\,000\,000$ volts,

i.e. 40 megavolts is written as 40 MV and is equal to 40 million volts.

An **inductance of 12 mH** means 12×10^{-3} H or $\dfrac{12}{10^3}$ H or $\dfrac{12}{1000}$ H, which is 0.012 H,

i.e. 12 millihenrys is written as 12 mH and is equal to 12 thousandths of a henry*.

A **time of 150 ns** means 150×10^{-9} s or $\dfrac{150}{10^9}$ s, which is $0.000\,000\,150$ s,

i.e. 150 nanoseconds is written as 150 ns and is equal to 150 thousand millionths of a second.

A **force of 20 kN** means 20×10^3 N, which is $20\,000$ newtons,

i.e. 20 kilonewtons is written as 20 kN and is equal to 20 thousand newtons.

A **charge of 30 μC** means 30×10^{-6} C or $\dfrac{30}{10^6}$ C, which is $0.000\,030$ C,

i.e. 30 microcoulombs is written as 30 μC and is equal to 30 millionths of a coulomb.

A **capacitance of 45 pF** means 45×10^{-12} F or $\dfrac{45}{10^{12}}$ F, which is $0.000\,000\,000\,045$ F,

*Who was **Henry**? – **Joseph Henry** (17 December 1797–13 May 1878) was an American scientist who discovered the electromagnetic phenomenon of self-inductance. He also discovered mutual inductance independently of Michael Faraday, though Faraday was the first to publish his results. Henry was the inventor of a precursor to the electric doorbell and electric relay. The SI unit of inductance, the **henry**, is named in his honour. To find out more go to **www.routledge.com/cw/bird**

i.e. 45 picofarads is written as 45 pF and is equal to 45 million millionths of farad (named after Michael Faraday*).

In engineering it is important to understand what such quantities as 15 GHz, 40 MV, 12 mH, 150 ns, 20 kN, 30 µC and 45 pF mean.

Now try the following Practice Exercise

Practice Exercise 39 SI units and common prefixes (answers on page 445)

1. State the SI unit of volume.

2. State the SI unit of capacitance.

3. State the SI unit of area.

4. State the SI unit of velocity.

5. State the SI unit of density.

6. State the SI unit of energy.

7. State the SI unit of charge.

8. State the SI unit of power.

9. State the SI unit of angle.

10. State the SI unit of electric potential.

11. State which quantity has the unit kg.

12. State which quantity has the unit symbol Ω.

13. State which quantity has the unit Hz.

14. State which quantity has the unit m/s^2.

15. State which quantity has the unit symbol A.

16. State which quantity has the unit symbol H.

17. State which quantity has the unit symbol m.

18. State which quantity has the unit symbol K.

19. State which quantity has the unit Pa.

20. State which quantity has the unit rad/s.

21. What does the prefix G mean?

22. What is the symbol and meaning of the prefix milli?

23. What does the prefix p mean?

24. What is the symbol and meaning of the prefix mega?

*Who was Faraday? – **Michael Faraday**, FRS (22 September 1791–25 August 1867) was an English scientist who contributed to the fields of electromagnetism and electrochemistry. His main discoveries include those of electromagnetic induction, diamagnetism and electrolysis. To find out more go to www.routledge.com/cw/bird

8.4 Standard form

A number written with one digit to the left of the decimal point and multiplied by 10 raised to some power is said to be written in **standard form**.

For example, $43\,645 = 4.3645 \times 10^4$

in standard form

and $0.0534 = 5.34 \times 10^{-2}$

in standard form

Problem 1. Express in standard form (a) 38.71 (b) 3746 (c) 0.0124

For a number to be in standard form, it is expressed with only one digit to the left of the decimal point. Thus,

(a) 38.71 must be divided by 10 to achieve one digit to the left of the decimal point and it must also be multiplied by 10 to maintain the equality, i.e.

$$38.71 = \frac{38.71}{10} \times 10 = \textbf{3.871} \times \textbf{10} \text{ in standard form}$$

(b) $3746 = \frac{3746}{1000} \times 1000 = \textbf{3.746} \times \textbf{10}^{\textbf{3}}$ in standard form.

(c) $0.0124 = 0.0124 \times \frac{100}{100} = \frac{1.24}{100} = \textbf{1.24} \times \textbf{10}^{-\textbf{2}}$ in standard form.

Problem 2. Express the following numbers, which are in standard form, as decimal numbers:

(a) 1.725×10^{-2} (b) 5.491×10^{4} (c) 9.84×10^{0}

(a) $1.725 \times 10^{-2} = \frac{1.725}{100} = \textbf{0.01725}$ (i.e. move the decimal point 2 places to the left).

(b) $5.491 \times 10^{4} = 5.491 \times 10000 = \textbf{54910}$ (i.e. move the decimal point 4 places to the right).

(c) $9.84 \times 10^{0} = 9.84 \times 1 = \textbf{9.84}$ (since $10^{0} = 1$).

Problem 3. Express in standard form, correct to 3 significant figures, (a) $\frac{3}{8}$ (b) $19\frac{2}{3}$ (c) $741\frac{9}{16}$

(a) $\frac{3}{8} = 0.375$, and expressing it in standard form gives

$$0.375 = \textbf{3.75} \times \textbf{10}^{-\textbf{1}}$$

(b) $19\frac{2}{3} = 19.\dot{6} = \textbf{1.97} \times \textbf{10}$ in standard form, correct to 3 significant figures.

(c) $741\frac{9}{16} = 741.5625 = \textbf{7.42} \times \textbf{10}^{\textbf{2}}$ in standard form, correct to 3 significant figures.

Problem 4. Express the following numbers, given in standard form, as fractions or mixed numbers, (a) 2.5×10^{-1} (b) 6.25×10^{-2} (c) 1.354×10^{2}

(a) $2.5 \times 10^{-1} = \frac{2.5}{10} = \frac{25}{100} = \frac{\textbf{1}}{\textbf{4}}$

(b) $6.25 \times 10^{-2} = \frac{6.25}{100} = \frac{625}{10000} = \frac{\textbf{1}}{\textbf{16}}$

(c) $1.354 \times 10^{2} = 135.4 = 135\frac{4}{10} = \textbf{135}\frac{\textbf{2}}{\textbf{5}}$

Problem 5. Evaluate (a) $(3.75 \times 10^{3})(6 \times 10^{4})$
(b) $\frac{3.5 \times 10^{5}}{7 \times 10^{2}}$, expressing the answers in standard form

(a) $(3.75 \times 10^{3})(6 \times 10^{4}) = (3.75 \times 6)(10^{3+4})$
$$= 22.50 \times 10^{7}$$
$$= \textbf{2.25} \times \textbf{10}^{\textbf{8}}$$

(b) $\frac{3.5 \times 10^{5}}{7 \times 10^{2}} = \frac{3.5}{7} \times 10^{5-2} = 0.5 \times 10^{3} = \textbf{5} \times \textbf{10}^{\textbf{2}}$

Now try the following Practice Exercise

Practice Exercise 40   Standard form (answers on page 445)

In Problems 1 to 5, express in standard form.

1.   (a)   73.9   (b)   28.4   (c)   197.62

2.   (a)   2748   (b)   33170   (c)   274218

3.   (a)   0.2401   (b)   0.0174   (c)   0.00923

4.   (a)   1702.3   (b)   10.04   (c)   0.0109

5.   (a) $\frac{1}{2}$ (b) $11\frac{7}{8}$
    (c) $\frac{1}{32}$ (d) $130\frac{3}{5}$

In Problems 6 and 7, express the numbers given as integers or decimal fractions.

6.   (a) 1.01×10^{3} (b) 9.327×10^{2}
    (c) 5.41×10^{4} (d) 7×10^{0}

7.   (a) 3.89×10^{-2} (b) 6.741×10^{-1}
    (c) 8×10^{-3}

In Problems 8 and 9, evaluate the given expressions, stating the answers in standard form.

8.   (a) $(4.5 \times 10^{-2})(3 \times 10^{3})$
    (b) $2 \times (5.5 \times 10^{4})$

9.   (a) $\frac{6 \times 10^{-3}}{3 \times 10^{-5}}$
    (b) $\frac{(2.4 \times 10^{3})(3 \times 10^{-2})}{(4.8 \times 10^{4})}$

10.   Write the following statements in standard form.
    (a)   The density of aluminium is 2710 kg m^{-3}.

(b)   Poisson's ratio for gold is 0.44

(c)   The impedance of free space is 376.73 Ω.

(d)   The electron rest energy is 0.511 MeV.

(e)   Proton charge–mass ratio is 95789700 $C \, kg^{-1}$.

(f)   The normal volume of a perfect gas is $0.02241 \, m^3 \, mol^{-1}$.

8.5   Engineering notation

In engineering, standard form is not as important as engineering notation. **Engineering notation** is similar to standard form except that the power of 10 **is always a multiple of 3**.

For example, $43\,645 = 43.645 \times 10^3$

in engineering notation

and $0.0534 = 53.4 \times 10^{-3}$

in engineering notation

From the list of engineering prefixes on page 62 it is apparent that all prefixes involve powers of 10 that are multiples of 3.

For example, a force of 43 645 N can be rewritten as 43.645×10^3 N and from the list of prefixes can then be expressed as 43.645 kN.

Thus, $\mathbf{43\,645 \, N \equiv 43.645 \, kN}$

To help further, on your calculator is an 'ENG' button. Enter the number 43 645 into your calculator and then press =. Now press the ENG button and the answer is 43.645×10^3. We then have to appreciate that 10^3 is the prefix 'kilo', giving **43 645 N ≡ 43.645 kN**.

In another example, let a current be 0.0745 A. Enter 0.0745 into your calculator. Press =. Now press ENG and the answer is 74.5×10^{-3}. We then have to appreciate that 10^{-3} is the prefix 'milli', giving **0.0745 A ≡ 74.5 mA**.

> **Problem 6.**   Express the following in engineering notation and in prefix form:
>
>   (a) 300 000 W   (b) 0.000068 H

(a)   Enter 300 000 into the calculator. Press =

Now press ENG and the answer is 300×10^3

From the table of prefixes on page 62, 10^3 corresponds to kilo.

Hence, 300 000 W = $\mathbf{300 \times 10^3 \, W}$ in engineering notation

= **300 kW** in prefix form.

(b)   Enter 0.000068 into the calculator. Press =

Now press ENG and the answer is 68×10^{-6}

From the table of prefixes on page 62, 10^{-6} corresponds to micro.

Hence,   0.000068 H = $\mathbf{68 \times 10^{-6} \, H}$ in engineering notation

= **68 μH** in prefix form.

> **Problem 7.**   Express the following in engineering notation and in prefix form:
>
>   (a) $42 \times 10^5 \, \Omega$ (b) $47 \times 10^{-10} F$

(a)   Enter 42×10^5 into the calculator. Press =

Now press ENG and the answer is 4.2×10^6

From the table of prefixes on page 62, 10^6 corresponds to mega.

Hence, $42 \times 10^5 \, \Omega = \mathbf{4.2 \times 10^6 \, \Omega}$ in engineering notation

= **4.2 MΩ** in prefix form.

(b)   Enter $47 \div 10^{10} = \dfrac{47}{10\,000\,000\,000}$ into the calculator. Press =

Now press ENG and the answer is 4.7×10^{-9}

From the table of prefixes on page 62, 10^{-9} corresponds to nano.

Hence, $47 \div 10^{10} \, F = \mathbf{4.7 \times 10^{-9} \, F}$ in engineering notation

= **4.7 nF** in prefix form.

> **Problem 8.**   Rewrite (a) 0.056 mA in μA (b) 16 700 kHz as MHz

(a)   Enter $0.056 \div 1000$ into the calculator (since milli means $\div 1000$). Press =

Now press ENG and the answer is 56×10^{-6}

From the table of prefixes on page 62, 10^{-6} corresponds to micro.

Hence, $0.056 \, \text{mA} = \dfrac{0.056}{1000} \, \text{A} = 56 \times 10^{-6} \, \text{A}$

$\qquad\qquad\qquad\qquad = \mathbf{56 \, \mu A}$.

(b)   Enter $16\,700 \times 1000$ into the calculator (since kilo means $\times 1000$). Press $=$

Now press ENG and the answer is 16.7×10^6

From the table of prefixes on page 62, 10^6 corresponds to mega.

Hence, $16\,700 \, \text{kHz} = 16\,700 \times 1000 \, \text{Hz}$

$\qquad\qquad\qquad = 16.7 \times 10^6 \, \text{Hz}$

$\qquad\qquad\qquad = \mathbf{16.7 \, MHz}$

Problem 9.   Rewrite (a) 63×10^4 V in kV (b) $3100 \, \text{pF}$ in nF

(a)   Enter 63×10^4 into the calculator. Press $=$

Now press ENG and the answer is 630×10^3

From the table of prefixes on page 62, 10^3 corresponds to kilo.

Hence, $63 \times 10^4 \, \text{V} = 630 \times 10^3 \, \text{V} = \mathbf{630 \, kV}$.

(b)   Enter 3100×10^{-12} into the calculator. Press $=$

Now press ENG and the answer is 3.1×10^{-9}

From the table of prefixes on page 62, 10^{-9} corresponds to nano.

Hence, $3100 \, \text{pF} = 3100 \times 10^{-12} \, \text{F} = 3.1 \times 10^{-9} \, \text{F}$

$\qquad\qquad\qquad\qquad\qquad = \mathbf{3.1 \, nF}$

Problem 10.   Rewrite (a) $14\,700 \, \text{mm}$ in metres (b) $276 \, \text{cm}$ in metres (c) $3.375 \, \text{kg}$ in grams

(a) $1 \, \text{m} = 1000 \, \text{mm}$, hence

$1 \, \text{mm} = \dfrac{1}{1000} = \dfrac{1}{10^3} = 10^{-3} \, \text{m}$.

Hence, $14\,700 \, \text{mm} = 14\,700 \times 10^{-3} \, \text{m} = \mathbf{14.7 \, m}$.

(b) $1 \, \text{m} = 100 \, \text{cm}$, hence $1 \, \text{cm} = \dfrac{1}{100} = \dfrac{1}{10^2} = 10^{-2} \, \text{m}$.

Hence, $276 \, \text{cm} = 276 \times 10^{-2} \, \text{m} = \mathbf{2.76 \, m}$.

(c) $1 \, \text{kg} = 1000 \, \text{g} = 10^3 \, \text{g}$

Hence, $3.375 \, \text{kg} = 3.375 \times 10^3 \, \text{g} = \mathbf{3375 \, g}$.

Problem 11.   The heat h generated in an electrical circuit is given by: $h = I^2 Rt$ joules, where

I is the current in amperes, R the resistance in ohms and t the time in seconds. Calculate the heat generated when a current of 300 mA is maintained through a $2 \, \text{k}\Omega$ resistor for 30 minutes.

Current, $I = 300 \, \text{mA} = 300 \times 10^{-3} A$, resistance $R = 2 \, \text{k}\Omega = 2 \times 10^3 \Omega$, and time $t = 30 \times 60 \, \text{s}$

Hence, heat generated,

$h = I^2 Rt = (300 \times 10^{-3})^2 \times (2 \times 10^3) \times (30 \times 60)$

i.e. **heat generated**, $h = \mathbf{324000 \, J = 324 \times 10^3 = 324 \, kJ}$

Problem 12.   A solid disc of uniform thickness has a diameter of 0.3 m and thickness, $t = 0.08$ m. The density, ρ, of its material of construction is $7860 \, \text{kg/m}^3$. The mass moment of inertia about its centroid, I, is given by: $I = \rho \pi R^2 t \times \dfrac{R^2}{2}$ where R is its radius. Calculate the mass moment of inertia for the disc.

If the diameter $= 0.3$ m then the radius,

$\qquad\qquad\qquad\qquad R = 0.3/2 = 0.15 \, \text{m}$

Mass moment of inertia, $I = \rho \pi R^2 t \times \dfrac{R^2}{2} =$

$7860 \, \text{kg/m}^3 \times \pi \times 0.15^2 \, \text{m}^2 \times 0.08 \, \text{m} \times \dfrac{0.15^2}{2} \, \text{m}^2$

$\qquad\qquad\qquad\qquad = \mathbf{0.50 \, kg \, m^2}$

Problem 13.   Buoyancy $= \rho \times V \times g$ newtons, where $g = 9.81 \, \text{m/s}^2$ and the density of water, $\rho = 1000 \, \text{kg/m}^3$. A barge of length 30 m and width 8 m floats on an even keel at a depth of 3 m. Calculate the value of its buoyancy.

Volume below the water line $= 30 \times 8 \times 3 = 720 \, \text{m}^3$

Buoyancy $= \rho \times V \times g$

$\qquad = 1020 \, \text{kg/m}^3 \times 720 \, \text{m}^3 \times 9.81 \, \text{m/s}^2$

$\qquad = 7204464 \text{N} = \mathbf{7.20 \, MN}$ (Note that $1 \, \text{kg/s}^2 = 1\text{N}$)

Now try the following Practice Exercise

Practice Exercise 41   Engineering notation (answers on page 446)

In Problems 1 to 12, express in engineering notation in prefix form.

1. $60\,000 \, \text{Pa}$

2. $0.00015 \, \text{W}$

3. 5×10^7 V

4. 5.5×10^{-8} F

5. 100 000 W

6. 0.00054 A

7. 15×10^5 Ω

8. 225×10^{-4} V

9. 35 000 000 000 Hz

10. 1.5×10^{-11} F

11. 0.000017 A

12. 46 200 Ω

13. Rewrite 0.003 mA in μA

14. Rewrite 2025 kHz as MHz

15. Rewrite 5×10^4 N in kN

16. Rewrite 300 pF in nF

17. Rewrite 6250 cm in metres

18. Rewrite 34.6 g in kg

In Problems 19 and 20, use a calculator to evaluate in engineering notation.

19. $4.5 \times 10^{-7} \times 3 \times 10^4$

20. $\dfrac{(1.6 \times 10^{-5})(25 \times 10^3)}{(100 \times 10^{-6})}$

21. The distance from Earth to the moon is around 3.8×10^8 m. State the distance in kilometres.

22. The radius of a hydrogen atom is 0.53×10^{-10} m. State the radius in nanometres.

23. The tensile stress acting on a rod is 5600000 Pa. Write this value in engineering notation.

24. The expansion of a rod is 0.0043 m. Write this value in engineering notation.

25. The strength value, V, of a rivet in double shear is given by the formula: $V = 2f_s\left(\dfrac{1}{4}\pi d^2\right)$ Determine the strength of a 20 mm diameter, d, rivet in double shear, given that $f_s = 100$ N/mm^2. State the answer in kN, correct to 4 significant figures.

26. The mass of a steel bolt is 4.50×10^{-4} kg. Calculate, in grams, the mass of 1600 bolts.

27. A concrete pillar, which is reinforced with steel rods, supports a compressive axial load, P, of 1 MN. The compressive stress in the steel, σ_s, is given by: $\sigma_s = \dfrac{-P \times E_s}{(A_s E_s + A_c E_c)}$ Calculate the stress in the steel given that the cross-sectional area of the steel, $A_s = 3 \times 10^{-3}$ m^2, the cross-sectional area of the concrete, $A_c = 0.1$ m^2, and Young's modulus for steel and concrete respectively is $E_s = 2 \times 10^{11}$ N/m^2 and $E_c = 2 \times 10^{10}$ N/m^2. Give the answer in MN correct to 4 significant figures.

8.6 Metric conversions

Length in metric units

$$1\text{ m} = 100\text{ cm} = 1000\text{ mm}$$

$$1\text{ cm} = \frac{1}{100}\text{ m} = \frac{1}{10^2}\text{ m} = 10^{-2}\text{ m}$$

$$1\text{ mm} = \frac{1}{1000}\text{ m} = \frac{1}{10^3}\text{ m} = 10^{-3}\text{ m}$$

Problem 14. Rewrite 12400 mm in metres

1 m = 1000 mm hence, 1 mm $= \dfrac{1}{1000} = \dfrac{1}{10^3} = 10^{-3}$ m

Hence, 12400 mm $= 12400 \times 10^{-3}$ m $= \mathbf{12.4\ m}$

Problem 15. Rewrite 358 cm in metres

1 m = 100 cm hence, 1 cm $= \dfrac{1}{100} = \dfrac{1}{10^2} = 10^{-2}$ m

Hence, 358 cm $= 358 \times 10^{-2}$ m $= \mathbf{3.58\ m}$

Now try the following Practice Exercise

Practice Exercise 42 Length in metric units (Answers on page 446)

1. State 2.45 m in millimetres

2. State 1.675 m in centimetres

3. State the number of millimetres in 65.8 cm

4. Rewrite 25,400 mm in metres

5. Rewrite 5632 cm in metres

6. State the number of millimetres in 4.356 m

7. How many centimetres are there in 0.875 m?

8. State a length of 465 cm in (a) mm (b) m

9. State a length of 5040 mm in (a) cm (b) m

10. A machine part is measured as 15.0 cm ± 1%. Between what two values would the measurement be? Give the answer in millimetres.

Areas in metric units

Area is a measure of the size or extent of a plane surface.
Area is measured in **square units** such as mm^2, cm^2 and m^2.

The area of the above square is 1 m^2

$$1\,m^2 = 100\,cm \times 100\,cm = 10000\,cm^2 = 10^4 cm^2$$

i.e. to change from square metres to square centimetres, multiply by 10^4

Hence, $2.5\,m^2 = 2.5 \times 10^4\,cm^2$
and $0.75\,m^2 = 0.75 \times 10^4\,cm^2$

Since $1\,m^2 = 10^4\,cm^2$ then $1cm^2 = \dfrac{1}{10^4}m^2 = 10^{-4}m^2$

i.e. to change from square centimetres to square metres, multiply by 10^{-4}

Hence, $52\,cm^2 = 52 \times 10^{-4}\,m^2$
and $643\,cm^2 = 643 \times 10^{-4}\,m^2$

The area of the above square is $1m^2$

$$1\,m^2 = 1000\,mm \times 1000\,mm = 1000000\,mm^2 = 10^6 mm^2$$

i.e. to change from square metres to square millimetres, multiply by 10^6

Hence, $7.5\,m^2 = 7.5 \times 10^6\,mm^2$
and $0.63\,m^2 = 0.63 \times 10^6\,mm^2$

Since $1\,m^2 = 10^6\,mm^2$
then $1\,mm^2 = \dfrac{1}{10^6}m^2 = 10^{-6}\,m^2$

i.e. to change from square millimetres to square metres, multiply by 10^{-6}

Hence, $235\,mm^2 = 235 \times 10^{-6}m^2$
and $47\,mm^2 = 47 \times 10^{-6}m^2$

The area of the above square is 1 cm^2

$$1\,cm^2 = 10\,mm \times 10\,mm = 100\,mm^2 = 10^2 mm^2$$

i.e. to change from square centimetres to square millimetres, multiply by 100 or 10^2

Hence, $3.5\,cm^2 = 3.5 \times 10^2 mm^2 = 350\,mm^2$
and $0.75\,cm^2 = 0.75 \times 10^2 mm^2 = 75\,mm^2$

Since $1\,cm^2 = 10^2 mm^2$
then $1\,mm^2 = \dfrac{1}{10^2}cm^2 = 10^{-2}\,cm^2$

i.e. to change from square millimetres to square centimetres, multiply by 10^{-2}

Hence, $250\,mm^2 = 250 \times 10^{-2}cm^2 = 2.5\,cm^2$
and $85\,mm^2 = 85 \times 10^{-2}cm^2 = 0.85\,cm^2$

Problem 16. Rewrite $12\,m^2$ in square centimetres

$1\,m^2 = 10^4 cm^2$ hence, $12\,m^2 = 12 \times 10^4 cm^2$

Problem 17. Rewrite $50\,cm^2$ in square metres

$1\,cm^2 = 10^{-4}m^2$ hence, $50\,cm^2 = 50 \times 10^{-4}m^2$

Problem 18. Rewrite $2.4\,m^2$ in square millimetres

$1 \, m^2 = 10^6 mm^2$ hence, $\mathbf{2.4 \, m^2 = 2.4 \times 10^6 mm^2}$

> **Problem 19.** Rewrite 147 mm² in square metres

$1 \, mm^2 = 10^{-6} m^2$ hence, $\mathbf{147 \, mm^2 = 147 \times 10^{-6} m^2}$

> **Problem 20.** Rewrite 34.5 cm² in square millimetres

$1 \, cm^2 = 10^2 mm^2$
hence, $\mathbf{34.5 \, cm^2 = 34.5 \times 10^2 mm^2 = 3450 \, mm^2}$

> **Problem 21.** Rewrite 400 mm² in square centimetres

$1 \, mm^2 = 10^{-2} cm^2$
hence, $\mathbf{400 \, mm^2 = 400 \times 10^{-2} cm^2 = 4 \, cm^2}$

> **Problem 22.** The top of a small rectangular table is 800 mm long and 500 mm wide. Determine its area in (a) mm² (b) cm² (c) m²

(a) **Area of rectangular table top** $= l \times b = 800 \times 500$
$$= \mathbf{400,000 \, mm^2}$$

(b) Since 1 cm = 10 mm then $1 \, cm^2 = 1 \, cm \times 1 \, cm = 10 \, mm \times 10 \, mm = 100 \, mm^2$

  or $\qquad 1 mm^2 = \dfrac{1}{100} = 0.01 \, cm^2$

  Hence, $\mathbf{400,000 \, mm^2} = 400,000 \times 0.01 \, cm^2$
  $$= \mathbf{4000 \, cm^2}$$

(c) $1 \, cm^2 = 10^{-4} m^2$
  hence, $\mathbf{4000 \, cm^2} = 4000 \times 10^{-4} m^2 = \mathbf{0.4 \, m^2}$

Now try the following Practice Exercise

Practice Exercise 43   Areas in metric units (Answers on page 446)

1. Rewrite 8 m² in square centimetres

2. Rewrite 240 cm² in square metres

3. Rewrite 3.6 m² in square millimetres

4. Rewrite 350 mm² in square metres

5. Rewrite 50 cm² in square millimetres

6. Rewrite 250 mm² in square centimetres

7. A rectangular piece of metal is 720 mm long and 400 mm wide. Determine its area in

  (a) mm²   (b) cm²   (c) m²

Volumes in metric units

The **volume** of any solid is a measure of the space occupied by the solid.
Volume is measured in **cubic units** such as mm³, cm³ and m³.

The volume of the cube shown is 1 m³

$$\mathbf{1 \, m^3 = 100 \, cm \times 100 \, cm \times 100 \, cm}$$
$$\mathbf{= 1000000 \, cm^2 = 10^6 cm^3}$$
$$\mathbf{1 \, litre = 1000 \, cm^3}$$

i.e. to change from cubic metres to cubic centimetres, multiply by 10^6
Hence, $\mathbf{3.2 \, m^3 = 3.2 \times 10^6 cm^3}$
and $\quad \mathbf{0.43 \, m^3 = 0.43 \times 10^6 cm^3}$
Since $\mathbf{1 \, m^3 = 10^6 cm^3}$ then $\mathbf{1 \, cm^3 = \dfrac{1}{10^6} m^3 = 10^{-6} \, m^3}$
i.e. to change from cubic centimetres to cubic metres, multiply by 10^{-6}
Hence, $\mathbf{140 \, cm^3 = 140 \times 10^{-6} m^3}$
and $\quad \mathbf{2500 \, cm^3 = 2500 \times 10^{-6} m^3}$

The volume of the cube shown is 1 m³

$$\mathbf{1 \, m^3 = 1000 \, mm \times 1000 \, mm \times 1000 \, mm}$$
$$\mathbf{= 1000000000 \, mm^3 = 10^9 mm^3}$$

i.e. to change from cubic metres to cubic millimetres, multiply by 10^9
Hence, $\mathbf{4.5 \, m^3 = 4.5 \times 10^9 mm^3}$

and $\quad$ **$0.25\,m^3 = 0.25 \times 10^9\,mm^3$**

Since $\quad$ $1\,m^3 = 10^9\,mm^2$

then $\quad$ $1\,mm^3 = \dfrac{1}{10^9}\,m^3 = 10^{-9}\,m^3$

i.e. to change from cubic millimetres to cubic metres, multiply by 10^{-9}

Hence, **$500\,mm^3 = 500 \times 10^{-9}\,m^3$**

and **$4675\,mm^3 = 4675 \times 10^{-9}\,m^3$ or $4.675 \times 10^{-6}\,m^3$**

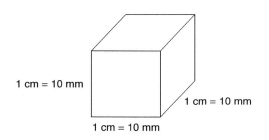

The volume of the cube shown is $1\,cm^3$

$$1\,cm^3 = 10\,mm \times 10\,mm \times 10\,mm$$
$$= 1000\,mm^3 = 10^3\,mm^3$$

i.e. to change from cubic centimetres to cubic millimetres, multiply by 1000 or 10^3

Hence, **$5\,cm^3 = 5 \times 10^3\,mm^3 = 5000\,mm^3$**

and **$0.35\,cm^3 = 0.35 \times 10^3\,mm^3 = 350\,mm^3$**

Since $\quad$ $1\,cm^3 = 10^3\,mm^3$

then $\quad$ $1\,mm^3 = \dfrac{1}{10^3}\,cm^3 = 10^{-3}\,cm^3$

i.e. to change from cubic millimetres to cubic centimetres, multiply by 10^{-3}

Hence, **$650\,mm^3 = 650 \times 10^{-3}\,cm^3 = 0.65\,cm^3$**

and $\quad$ **$75\,mm^3 = 75 \times 10^{-3}\,cm^3 = 0.075\,cm^3$**

◤ Problem 23. $\quad$ Rewrite $1.5\,m^3$ in cubic centimetres

$1\,m^3 = 10^6\,cm^3$ hence, **$1.5\,m^3 = 1.5 \times 10^6\,cm^3$**

◤ Problem 24. $\quad$ Rewrite $300\,cm^3$ in cubic metres

$1\,cm^3 = 10^{-6}\,m^3$ hence, **$300\,cm^3 = 300 \times 10^{-6}\,m^3$**

◤ Problem 25. $\quad$ Rewrite $0.56\,m^3$ in cubic millimetres

$1\,m^3 = 10^9\,mm^3$

hence, **$0.56\,m^3 = 0.56 \times 10^9\,mm^3$ or $560 \times 10^6\,mm^3$**

◤ Problem 26. $\quad$ Rewrite $1250\,mm^3$ in cubic metres

$1\,mm^3 = 10^{-9}\,m^3$

hence, $\quad$ **$1250\,mm^3 = 1250 \times 10^{-9}\,m^3$** $\quad$ or **$1.25 \times 10^{-6}\,m^3$**

◤ Problem 27. $\quad$ Rewrite $8\,cm^3$ in cubic millimetres

$1\,cm^3 = 10^3\,mm^3$

hence, **$8\,cm^3 = 8 \times 10^3\,mm^3 = 8000\,mm^3$**

◤ Problem 28. $\quad$ Rewrite $600\,mm^3$ in cubic centimetres

$1\,mm^3 = 10^{-3}\,cm^3$

hence, **$600\,mm^3 = 600 \times 10^{-3}\,cm^3 = 0.6\,cm^3$**

◤ Problem 29. $\quad$ A water tank is in the shape of a rectangular prism having length 1.2 m, breadth 50 cm and height 250 mm. Determine the capacity of the tank (a) m^3 (b) cm^3 (c) litres

Capacity means volume. When dealing with liquids, the word capacity is usually used.

(a) $\quad$ Capacity of water tank $= l \times b \times h$
where l = 1.2 m, b = 50 cm and h = 250 mm.
To use this formula, all dimensions **must** be in the same units.
Thus, l = 1.2 m, b = 0.50 m and h = 0.25 m (since 1 m = 100 cm = 1000 mm)
Hence, **capacity of tank** $= 1.2 \times 0.50 \times 0.25$
$$= 0.15\ m^3$$

(b) $\quad$ $1\,m^3 = 10^6\,cm^3$
Hence, **capacity** $= 0.15\ m^3 = 0.15 \times 10^6\,cm^3$
$$= 150{,}000\ cm^3$$

(c) $\quad$ 1 litre = 1000 cm^3
Hence, **$150{,}000\ cm^3 = \dfrac{150{,}000}{1000} = 150$ litres**

Now try the following Practice Exercise

Practice Exercise 44   Volumes in metric units (Answers on page 446)

1. Rewrite $2.5\,m^3$ in cubic centimetres

2. Rewrite $400\,cm^3$ in cubic metres

3. Rewrite $0.87\,m^3$ in cubic millimetres

4. Change a volume of $2,400,000\ cm^3$ to cubic metres.

5. Rewrite $1500\,mm^3$ in cubic metres

6. Rewrite $400\,mm^3$ in cubic centimetres

7. Rewrite $6.4\,cm^3$ in cubic millimetres

8. Change a volume of $7500\ mm^3$ to cubic centimetres.

9. An oil tank is in the shape of a rectangular prism having length 1.5 m, breadth 60 cm and height 200 mm. Determine the capacity of the tank in (a) m^3 (b) cm^3 (c) litres

8.7   Metric – US/Imperial conversions

The Imperial System (which uses yards, feet, inches, etc. to measure length) was developed over hundreds of years in the UK, then the French developed the Metric System (metres) in 1670, which soon spread through Europe, even to the UK itself in 1960. But the USA and a few other countries still prefer feet and inches.

When converting from metric to imperial units, or vice versa, one of the following tables (8.4 to 8.11) should help.

Table 8.4   Metric to imperial length

Metric	US or Imperial
1 millimetre, mm	0.03937 inch
1 centimetre, cm = 10 mm	0.3937 inch
1 metre, m = 100 cm	1.0936 yard
1 kilometre, km = 1000 m	0.6214 mile

Problem 30. Calculate the number of inches in 350 mm, correct to 2 decimal places

350 mm = 350 × 0.03937 inches = **13.78 inches** from Table 8.4

Problem 31. Calculate the number of inches in 52 cm, correct to 4 significant figures

52 cm = 52 × 0.3937 inches = **20.47 inches** from Table 8.4

Problem 32. Calculate the number of yards in 74 m, correct to 2 decimal places

74 m = 74 × 1.0936 yards = **80.93 yds** from Table 8.4

Problem 33. Calculate the number of miles in 12.5 km, correct to 3 significant figures

12.5 km = 12.5 × 0.6214 miles = **7.77 miles** from Table 8.4

Table 8.5   Imperial to metric length

US or Imperial	Metric
1 inch, in	2.54 cm
1 foot, ft = 12 in	0.3048 m
1 yard, yd = 3 ft	0.9144 m
1 mile = 1760 yd	1.6093 km
1 nautical mile = 2025.4 yd	1.853 km

Problem 34. Calculate the number of centimetres in 35 inches, correct to 1 decimal places

35 inches = 35 × 2.54 cm = **88.9 cm** from Table 8.5

Problem 35. Calculate the number of metres in 66 inches, correct to 2 decimal places

66 inches = $\dfrac{66}{12}$ feet = $\dfrac{66}{12}$ × 0.3048 m = **1.68 m** from Table 8.5

Problem 36. Calculate the number of metres in 50 yards, correct to 2 decimal places

50 yards = 50 × 0.9144 m = **45.72 m** from Table 8.5

> **Problem 37.** Calculate the number of kilometres in 7.2 miles, correct to 2 decimal places

7.2 miles = 7.2 × 1.6093 km = **11.59 km** from Table 8.5

> **Problem 38.** Calculate the number of (a) yards (b) kilometres in 5.2 nautical miles

(a) 5.2 nautical miles = 5.2 × 2025.4 yards
= **10532 yards** from Table 8.5

(b) 5.2 nautical miles = 5.2 × 1.853 km
= **9.636 km** from Table 8.5

Table 8.6 Metric to imperial area

Metric	US or Imperial
1 cm^2 = 100 mm^2	0.1550 in^2
1 m^2 = 10,000 cm^2	1.1960 yd^2
1 hectare, ha = 10,000 m^2	2.4711 acres
1 km^2 = 100 ha	0.3861 mile^2

> **Problem 39.** Calculate the number of square inches in 47 cm^2, correct to 4 significant figures

47 cm^2 = 47 × 0.1550 in^2 = **7.285 in^2** from Table 8.6

> **Problem 40.** Calculate the number of square yards in 20 m^2, correct to 2 decimal places

20 m^2 = 20 × 1.1960 yd^2 = **23.92 yd^2** from Table 8.6

> **Problem 41.** Calculate the number of acres in 23 hectares of land, correct to 2 decimal places

23 hectares = 23 × 2.4711 acres = **56.84 acres** from Table 8.6

> **Problem 42.** Calculate the number of square miles in a field of 15 km^2 area, correct to 2 decimal places

15 km^2 = 15 × 0.3861 mile^2 = **5.79 mile^2** from Table 8.6

Table 8.7 Imperial to metric area

US or Imperial	Metric
1 in^2	6.4516 cm^2
1 ft^2 = 144 in^2	0.0929 m^2
1 yd^2 = 9 ft^2	0.8361 m^2
1 acre = 4840 yd^2	4046.9 m^2
1 mile^2 = 640 acres	2.59 km^2

> **Problem 43.** Calculate the number of square centimetres in 17.5 in^2, correct to the nearest square centimetre

17.5 in^2 = 17.5 × 6.4516 cm^2 = **113 cm^2** from Table 8.7

> **Problem 44.** Calculate the number of square metres in 205 ft^2, correct to 2 decimal places

205 ft^2 = 205 × 0.0929 m^2 = **19.04 m^2** from Table 8.7

> **Problem 45.** Calculate the number of square metres in 11.2 acres, correct to the nearest square metre

11.2 acres = 11.2 × 4046.9 m^2 = **45325 m^2** from Table 8.7

> **Problem 46.** Calculate the number of square kilometres in 12.6 mile^2, correct to 2 decimal places

12.6 mile^2 = 12.6 × 2.59 km^2 = **32.63 km^2** from Table 8.7

Table 8.8 Metric to imperial volume/capacity

Metric	US or Imperial
1 cm^3	0.0610 in^3
1 dm^3 = 1000 cm^3	0.0353 ft^3
1 m^3 = 1000 dm^3	1.3080 yd^3
1 litre = 1 dm^3 = 1000 cm^3	2.113 fluid pt = 1.7598 pt

Problem 47. Calculate the number of cubic inches in 123.5 cm^3, correct to 2 decimal places

123.5 cm^3 = 123.5 × 0.0610 cm^3 = **7.53 cm^3** from Table 8.8

Problem 48. Calculate the number of cubic feet in 144 dm^3, correct to 3 decimal places

144 dm^3 = 144 × 0.0353 ft^3 = **5.083 ft^3** from Table 8.8

Problem 49. Calculate the number of cubic yards in 5.75 m^3, correct to 4 significant figures

5.75 m^3 = 5.75 × 1.3080 yd^3 = **7.521 yd^3** from Table 8.8

Problem 50. Calculate the number of US fluid pints in 6.34 litres of oil, correct to 1 decimal place

6.34 litre = 6.34 × 2.113 US fluid pints = **13.4 US fluid pints** from Table 8.8

Table 8.9 Imperial to metric volume/capacity

US or Imperial	Metric
1 in^3	16.387 cm^3
1 ft^3	0.02832 m^3
1 US fl oz = 1.0408 UK fl oz	0.0296 litres
1 US pint (16 fl oz) = 0.8327 UK pt	0.4732 litre
1 US gal (231 in^3) = 0.8327 UK gal	3.7854 litre

Problem 51. Calculate the number of cubic centimetres in 3.75 in^3, correct to 2 decimal places

3.75 in^3 = 3.75 × 16.387 cm^3 = **61.45 cm^3** from Table 8.9

Problem 52. Calculate the number of cubic metres in 210 ft^3, correct to 3 significant figures

210 ft^3 = 210 × 0.02832 m^3 = **5.95 m^3** from Table 8.9

Problem 53. Calculate the number of litres in 4.32 US pints, correct to 3 decimal places

4.32 US pints = 4.32 × 0.4732 litres = **2.044 litres** from Table 8.9

Problem 54. Calculate the number of litres in 8.62 US gallons, correct to 2 decimal places

8.62 US gallons = 8.62 × 3.7854 litre = **32.63 litre** from Table 8.9

Table 8.10 Metric to imperial mass

Metric	US or Imperial
1 g = 1000 mg	0.0353 oz
1 kg = 1000 g	2.2046 lb
1 tonne, t, = 1000 kg	1.1023 short ton
1 tonne, t, = 1000 kg	0.9842 long ton

The British ton is the long ton, which is 2240 pounds, and the US ton is the short ton which is 2000 pounds.

Problem 55. Calculate the number of ounces in a mass of 1346 g, correct to 2 decimal places

1346 g = 1346 × 0.0353 oz = **47.51 oz** from Table 8.10

Problem 56. Calculate the mass, in pounds, in a 210.4 kg mass, correct to 4 significant figures

210.4 kg = 210.4 × 2.2046 lb = **463.8 lb** from Table 8.10

Problem 57. Calculate the number of short tons in 5000 kg, correct to 2 decimal places

5000 kg = 5 t = 5 × 1.1023 short tons = **5.51 short tons** from Table 8.10

Problem 58. Calculate the number of grams in 5.63 oz, correct to 4 significant figures

5.63 oz = 5.63 × 28.35 g = **159.6 g** from Table 8.11

Table 8.11 Imperial to metric mass

US or Imperial	Metric
1 oz = 437.5 grain	28.35 g
1 lb = 16 oz	0.4536 kg
1 stone = 14 lb	6.3503 kg
1 hundredweight, cwt = 112 lb	50.802 kg
1 short ton	0.9072 tonne
1 long ton	1.0160 tonne

Problem 59. Calculate the number of kilograms in 75 oz, correct to 3 decimal places

$75 \text{ oz} = \dfrac{75}{16}\text{lb} = \dfrac{75}{16} \times 0.4536 \text{ kg} = \textbf{2.126 kg}$ from Table 8.11

Problem 60. Convert 3.25 cwt into (a) pounds (b) kilograms

(a) $3.25 \text{ cwt} = 3.25 \times 112 \text{ lb} = \textbf{364 lb}$ from Table 8.11

(b) $3.25 \text{ cwt} = 3.25 \times 50.802 \text{ kg} = \textbf{165.1 kg}$ from Table 8.11

Temperature

To convert from Celsius to Fahrenheit, first multiply by 9/5, then add 32.
To convert from Fahrenheit to Celsius, first subtract 32, then multiply by 5/9

Problem 61. Convert 35° C to degrees Fahrenheit

$F = \dfrac{9}{5}C + 32$ hence $35° \text{ C} = \dfrac{9}{5}(35) + 32 = 63 + 32$
$\qquad\qquad\qquad\qquad\qquad\qquad = \textbf{95° F}$

Problem 62. Convert 113° F to degrees Celsius

$C = \dfrac{5}{9}(F - 32)$ hence $113° \text{ F} = \dfrac{5}{9}(113 - 32) = \dfrac{5}{9}(81)$
$\qquad\qquad\qquad\qquad\qquad\qquad = \textbf{45° C}$

Now try the following Practice Exercise

Practice Exercise 45   Metric/Imperial conversions (answers on page 446)

In the following Problems, use the metric/imperial conversions in Tables 8.4 to 8.11

1.   Calculate the number of inches in 476 mm, correct to 2 decimal places.

2.   Calculate the number of inches in 209 cm, correct to 4 significant figures.

3.   Calculate the number of yards in 34.7 m, correct to 2 decimal places.

4.   Calculate the number of miles in 29.55 km, correct to 2 decimal places.

5.   Calculate the number of centimetres in 16.4 inches, correct to 2 decimal places.

6.   Calculate the number of metres in 78 inches, correct to 2 decimal places.

7.   Calculate the number of metres in 15.7 yards, correct to 2 decimal places.

8.   Calculate the number of kilometres in 3.67 miles, correct to 2 decimal places.

9.   Calculate the number of (a) yards (b) kilometres in 11.23 nautical miles.

10.   Calculate the number of square inches in 62.5 cm^2, correct to 4 significant figures.

11.   Calculate the number of square yards in 15.2 m^2, correct to 2 decimal places.

12.   Calculate the number of acres in 12.5 hectares, correct to 2 decimal places.

13.   Calculate the number of square miles in 56.7 km^2, correct to 2 decimal places.

14.   Calculate the number of square centimetres in 6.37 in^2, correct to the nearest square centimetre.

15.   Calculate the number of square metres in 308.6 ft^2, correct to 2 decimal places.

16.   Calculate the number of square metres in 2.5 acres, correct to the nearest square metre.

17. Calculate the number of square kilometres in 21.3 mile^2, correct to 2 decimal places.

18. Calculate the number of cubic inches in 200.7 cm^3, correct to 2 decimal places.

19. Calculate the number of cubic feet in 214.5 dm^3, correct to 3 decimal places.

20. Calculate the number of cubic yards in 13.45 m^3, correct to 4 significant figures.

21. Calculate the number of US fluid pints in 15 litres, correct to 1 decimal place.

22. Calculate the number of cubic centimetres in 2.15 in^3, correct to 2 decimal places.

23. Calculate the number of cubic metres in 175 ft^3, correct to 4 significant figures.

24. Calculate the number of litres in 7.75 US pints, correct to 3 decimal places.

25. Calculate the number of litres in 12.5 US gallons, correct to 2 decimal places.

26. Calculate the number of ounces in 980 g, correct to 2 decimal places.

27. Calculate the mass, in pounds, in 55 kg, correct to 4 significant figures.

28. Calculate the number of short tons in 4000 kg, correct to 3 decimal places.

29. Calculate the number of grams in 7.78 oz, correct to 4 significant figures.

30. Calculate the number of kilograms in 57.5 oz, correct to 3 decimal places.

31. Convert 2.5 cwt into (a) pounds (b) kilograms.

32. Convert 55 °C to degrees Fahrenheit.

33. Convert 167 °F to degrees Celsius.

Practice Exercise 46   Multiple-choice questions on units, prefixes and engineering notation (Answers on page 446)

Each question has only one correct answer

1. The unit of the quantity of electricity is the:
(a) volt   (b) joule   (c) ohm   (d) coulomb

2. 0.088 m in centimetres is:
(a) 88 cm   (b) 8.8 cm
(c) 880 cm   (d) 0.88 cm

3. The joule is a unit of:
(a) power   (b) voltage
(c) energy   (d) quantity of electricity

4. 5 g written in kg is:
(a) 5000 kg   (b) 0.05 kg
(c) 0.005 kg   (d) 0.0005 kg

5. The ohm is the unit of:
(a) charge   (b) current
(c) resistance   (d) power

6. The number of cm^2 in 2 m^2 is:
(a) 20000   (b) 200   (c) 0.002   (d) 2000

7. The unit of current is the:
(a) volt   (b) coulomb
(c) joule   (d) ampere

8. 650 mg written in g is:
(a) 0.65 g   (b) 65 g
(c) 6500 g   (d) 0.065 g

9. Converting 500 litres into cubic metres gives:
(a) 0.5 m^3 (b) 5 m^3
(c) 50 m^3 (d) 0.005 m^3

10. Converting 1.2 m^2 into cm^2 gives:
(a) 120 cm^2 (b) 12000 cm^2
(c) 1200 cm^2 (d) 0.012 cm^2

11. 17 cm is the same as:
(a) 1.7 m   (b) 0.017 m
(c) 0.17 m   (d) 170 m

12. 0.000095 H is the same as:
(a) 9.5×10^{-5} H in engineering notation
(b) 95×10^{-6} in prefix form
(c) 95×10^{-4} H in engineering notation
(d) 95 μH in prefix form

13. Four masses are 1850 mg, 1.65 g, 0.002 kg and 2.5×10^3 mg. The lightest of these is:
(a) 1850 mg   (b) 0.002 kg
(c) 1.65 g   (d) 2.5×10^3 mg

14. The number of millimetres in 1752 km is:
(a) 1752000000 mm   (b) 17520 mm
(c) 1.752 mm   (d) 17.52 mm

15. 3.485 kg is the same mass as:
    (a) 0.003485 g   (b) 3.485×10^{-3} g
    (c) 348.5 g   (d) 3485 g

16. $(5.5 \times 10^2)(2 \times 10^3)$ cm in standard form is equal to:
    (a) 11×10^6 cm   (b) 1.1×10^6 cm
    (c) 11×10^5 cm   (d) 1.1×10^5 cm

17. Resistance, R, is given by the formula $R = \dfrac{\rho \ell}{A}$. When resistivity $\rho = 0.017\,\mu\Omega\text{m}$, length $\ell = 5$ km and cross-sectional area $A = 25$ mm^2, the resistance is:
    (a) 3.4 mΩ (b) 0.034 Ω
    (c) 3.4 kΩ (d) 3.4 Ω

18. What percentage of 4 kg is 64 g?
    (a) 1.6%   (b) 6.25%   (c) 64%   (d) 16%

19. The unit of inductance is the:
    (a) farad   (b) henry
    (c) coulomb   (d) pascal

20. Modulus of elasticity E is given by the equation $E = \dfrac{F\ell}{\pi r^2 x}$, where F is measured in newtons, and ℓ, r and x are measured in metres. If $F = 3.2$ kN, $\ell = 0.56$ m, $r = 5$ mm, and $x = 2$ mm, the value of E is:
    (a) 1141 MN/m^2 (b) 11.41 GN/m^2
    (c) 114.1 MN/m^2 (d) 1.141×10^8 N/m^2

For fully worked solutions to each of the problems in Practice Exercises 39 to 45 in this chapter, go to the website:
www.routledge.com/cw/bird

This assignment covers the material contained in Chapters 6–8. *The marks available are shown in brackets at the end of each question.*

1. In a box of 1500 nails, 125 are defective. Express the non-defective nails as a ratio of the defective ones, in its simplest form. (3)

2. Prize money in a lottery totals £4500 and is shared among three winners in the ratio 5 : 3 : 1. How much does the first prize winner receive? (3)

3. A simple machine has an effort : load ratio of 3 : 41. Determine the effort, in newtons, to lift a load of 6.15 kN. (3)

4. If 15 cans of lager weigh 7.8 kg, what will 24 cans weigh? (3)

5. Hooke's law states that stress is directly proportional to strain within the elastic limit of a material. When for brass the stress is 21 MPa, the strain is 250×10^{-6}. Determine the stress when the strain is 350×10^{-6}. (3)

6. If 12 inches $= 30.48$ cm, find the number of millimetres in 17 inches. (3)

7. If x is inversely proportional to y and $x = 12$ when $y = 0.4$, determine
   (a) the value of x when y is 3.
   (b) the value of y when $x = 2$. (3)

8. Evaluate
   (a) $3 \times 2^3 \times 2^2$
   (b) $49^{\frac{1}{2}}$ (4)

9. Evaluate $\dfrac{3^2 \times \sqrt{36} \times 2^2}{3 \times 81^{\frac{1}{2}}}$ taking positive square roots only. (3)

10. Evaluate $6^4 \times 6 \times 6^2$ in index form. (3)

11. Evaluate
    (a) $\dfrac{2^7}{2^2}$ (b) $\dfrac{10^4 \times 10 \times 10^5}{10^6 \times 10^2}$ (4)

12. Evaluate
    (a) $\dfrac{2^3 \times 2 \times 2^2}{2^4}$ (b) $\dfrac{\left(2^3 \times 16\right)^2}{\left(8 \times 2\right)^3}$
    (c) $\left(\dfrac{1}{4^2}\right)^{-1}$ (7)

13. Evaluate
    (a) $(27)^{-\frac{1}{3}}$ (b) $\dfrac{\left(\dfrac{3}{2}\right)^{-2} - \dfrac{2}{9}}{\left(\dfrac{2}{3}\right)^2}$ (5)

14. State the SI unit of (a) capacitance (b) electrical potential (c) work. (3)

15. State the quantity that has an SI unit of (a) kilograms (b) henrys (c) hertz (d) m^3. (4)

16. Express the following in engineering notation in prefix form.
    (a) 250 000 J (b) 0.05 H
    (c) 2×10^8 W (d) 750×10^{-8} F (4)

17. Rewrite (a) 0.0067 mA in μA (b) 40×10^4 kV as MV. (2)

18. Rewrite 32 cm^2 in square millimetres. (1)

19. A rectangular tabletop is 1500 mm long and 800 mm wide. Determine its area in
    (a) mm^2 (b) cm^2 (c) m^2 (3)

20. Rewrite 0.065 m^3 in cubic millimetres. (1)

21. Rewrite 20000 mm^3 in cubic metres. (1)

22. Rewrite 8.3 cm^3 in cubic millimetres. (1)

23. A petrol tank is in the shape of a rectangular prism having length 1.0 m, breadth 75 cm and height 120 mm. Determine the capacity of the tank in
    (a) m^3 (b) cm^3 (c) litres (3)

For lecturers/instructors/teachers, fully worked solutions to each of the problems in Revision Test 3, together with a full marking scheme, are available at the website:
www.routledge.com/cw/bird

Chapter 9

Basic algebra

Why it is important to understand: Basic algebra

Algebra is one of the most fundamental tools for engineers because it allows them to determine the value of something (length, material constant, temperature, mass and so on) given values that they do know (possibly other length, material properties, mass). Although the types of problems that mechanical, chemical, civil, environmental or electrical engineers deal with vary, all engineers use algebra to solve problems. An example where algebra is frequently used is in simple electrical circuits, where the resistance is proportional to voltage. Using Ohm's law, or $V = IR$, an engineer simply multiplies the current in a circuit by the resistance to determine the voltage across the circuit. Engineers and scientists use algebra in many ways, and so frequently that they don't even stop the think about it. Depending on what type of engineer you choose to be, you will use varying degrees of algebra, but in all instances algebra lays the foundation for the mathematics you will need to become an engineer.

At the end of this chapter you should be able to:

- understand basic operations in algebra
- add, subtract, multiply and divide using letters instead of numbers
- state the laws of indices in letters instead of numbers
- simplify algebraic expressions using the laws of indices

9.1 Introduction

We are already familiar with evaluating formulae using a calculator from Chapter 4.

For example, if the length of a football pitch is L and its width is b, then the formula for the area A is given by

$$A = L \times b$$

This is an **algebraic equation**.
If $L = 120$ m and $b = 60$ m, then the area
$A = 120 \times 60 = 7200$ m^2.

The total resistance, R_T, of resistors R_1, R_2 and R_3 connected in series is given by

$$R_T = R_1 + R_2 + R_3$$

This is an **algebraic equation**.
If $R_1 = 6.3$ kΩ, $R_2 = 2.4$ kΩ and $R_3 = 8.5$ kΩ, then

$$R_T = 6.3 + 2.4 + 8.5 = 17.2 \text{ k}\Omega$$

The temperature in Fahrenheit, F, is given by

$$F = \frac{9}{5}C + 32$$

where C is the temperature in Celsius*. This is an **algebraic equation**.

If $C = 100°C$, then $F = \dfrac{9}{5} \times 100 + 32$

$$= 180 + 32 = 212°F.$$

If you can cope with evaluating formulae then you will be able to cope with algebra.

9.2 Basic operations

Algebra merely uses letters to represent numbers.
If, say, a, b, c and d represent any four numbers then in algebra:

(a) $a + a + a + a = 4a$. For example, if $a = 2$, then $2 + 2 + 2 + 2 = 4 \times 2 = 8$

(b) $5b$ means $5 \times b$. For example, if $b = 4$, then $5b = 5 \times 4 = 20$

(c) $2a + 3b + a - 2b = 2a + a + 3b - 2b = 3a + b$

Only similar terms can be combined in algebra. The $2a$ and the $+a$ can be combined to give $3a$

*Who was **Celsius**? – **Anders Celsius** (27 November 1701–25 April 1744) was the Swedish astronomer that proposed the Celsius temperature scale in 1742 which takes his name. To find out more go to **www.routledge.com/cw/bird**

and the $3b$ and $-2b$ can be combined to give $1b$, which is written as b.

In addition, with terms separated by $+$ and $-$ signs, the order in which they are written does not matter. In this example, $2a + 3b + a - 2b$ is the same as $2a + a + 3b - 2b$, which is the same as $3b + a + 2a - 2b$, and so on. (Note that the first term, i.e. $2a$, means $+2a$)

(d) $4abcd = 4 \times a \times b \times c \times d$

For example, if $a = 3, b = -2, c = 1$ and $d = -5$, then $4abcd = 4 \times 3 \times -2 \times 1 \times -5 = 120$. (Note that $- \times - = +$)

(e) $(a)(c)(d)$ means $a \times c \times d$

Brackets are often used instead of multiplication signs.
For example, $(2)(5)(3)$ means $2 \times 5 \times 3 = 30$

(f) $ab = ba$
If $a = 2$ and $b = 3$ then 2×3 is exactly the same as 3×2, i.e. 6

(g) $b^2 = b \times b$. For example, if $b = 3$, then $3^2 = 3 \times 3 = 9$

(h) $a^3 = a \times a \times a$ For example, if $a = 2$, then $2^3 = 2 \times 2 \times 2 = 8$

Here are some worked examples to help get a feel for basic operations in this introduction to algebra.

Addition and subtraction

Problem 1. Find the sum of $4x, 3x, -2x$ and $-x$

$$4x + 3x + -2x + -x = 4x + 3x - 2x - x$$

(Note that $+ \times - = -$)

$$= 4x$$

Problem 2. Find the sum of $5x, 3y, z, -3x, -4y$ and $6z$

$$5x + 3y + z + -3x + -4y + 6z$$

$$= 5x + 3y + z - 3x - 4y + 6z$$

$$= 5x - 3x + 3y - 4y + z + 6z$$

$$= 2x - y + 7z$$

Note that the order can be changed when terms are separated by $+$ and $-$ signs. Only similar terms can be combined.

Problem 3. Simplify $4x^2 - x - 2y + 5x + 3y$

$$4x^2 - x - 2y + 5x + 3y = 4x^2 + 5x - x + 3y - 2y$$
$$= \mathbf{4x^2 + 4x + y}$$

Problem 4. Simplify $3xy - 7x + 4xy + 2x$

$$3xy - 7x + 4xy + 2x = 3xy + 4xy + 2x - 7x$$
$$= \mathbf{7xy - 5x}$$

Now try the following Practice Exercise

Practice Exercise 47 Addition and subtraction in algebra (answers on page 446)

1. Find the sum of $4a, -2a, 3a$ and $-8a$
2. Find the sum of $2a, 5b, -3c, -a, -3b$ and $7c$
3. Simplify $2x - 3x^2 - 7y + x + 4y - 2y^2$
4. Simplify $5ab - 4a + ab + a$
5. Simplify $2x - 3y + 5z - x - 2y + 3z + 5x$
6. Simplify $3 + x + 5x - 2 - 4x$
7. Add $x - 2y + 3$ to $3x + 4y - 1$
8. Subtract $a - 2b$ from $4a + 3b$
9. From $a + b - 2c$ take $3a + 2b - 4c$
10. From $x^2 + xy - y^2$ take $xy - 2x^2$

Multiplication and division

Problem 5. Simplify $bc \times abc$

$$bc \times abc = a \times b \times b \times c \times c$$
$$= a \times b^2 \times c^2$$
$$= \mathbf{ab^2c^2}$$

Problem 6. Simplify $-2p \times -3p$

$$- \times - = + \text{ hence, } -2p \times -3p = \mathbf{6p^2}$$

Problem 7. Simplify $ab \times b^2c \times a$

$$ab \times b^2c \times a = a \times a \times b \times b \times b \times c$$
$$= a^2 \times b^3 \times c$$
$$= \mathbf{a^2b^3c}$$

Problem 8. Evaluate $3ab + 4bc - abc$ when $a = 3, b = 2$ and $c = 5$

$$3ab + 4bc - abc = 3 \times a \times b + 4 \times b \times c - a \times b \times c$$
$$= 3 \times 3 \times 2 + 4 \times 2 \times 5 - 3 \times 2 \times 5$$
$$= 18 + 40 - 30$$
$$= \mathbf{28}$$

Problem 9. Determine the value of $5pq^2r^3$, given that $p = 2, q = \dfrac{2}{5}$ and $r = 2\dfrac{1}{2}$

$$5pq^2r^3 = 5 \times p \times q^2 \times r^3$$
$$= 5 \times 2 \times \left(\frac{2}{5}\right)^2 \times \left(2\frac{1}{2}\right)^3$$
$$= 5 \times 2 \times \left(\frac{2}{5}\right)^2 \times \left(\frac{5}{2}\right)^3 \quad \text{since} \quad 2\frac{1}{2} = \frac{5}{2}$$
$$= \frac{5}{1} \times \frac{2}{1} \times \frac{2}{5} \times \frac{2}{5} \times \frac{5}{2} \times \frac{5}{2} \times \frac{5}{2}$$
$$= \frac{1}{1} \times \frac{1}{1} \times \frac{1}{1} \times \frac{1}{1} \times \frac{1}{1} \times \frac{5}{1} \times \frac{5}{1} \quad \text{by cancelling}$$
$$= 5 \times 5$$
$$= \mathbf{25}$$

Problem 10. Multiply $2a + 3b$ by $a + b$

Each term in the first expression is multiplied by a, then each term in the first expression is multiplied by b and the two results are added. The usual layout is shown below.

$$
\begin{array}{r}
2a + 3b \\
a + b \\
\hline
\end{array}
$$

Multiplying by a gives $\quad 2a^2 + 3ab$

Multiplying by b gives $\qquad\quad 2ab + 3b^2$

Adding gives $\qquad\quad \mathbf{2a^2 + 5ab + 3b^2}$

Thus, $\mathbf{(2a + 3b)(a + b) = 2a^2 + 5ab + 3b^2}$

Problem 11. Multiply $3x - 2y^2 + 4xy$ by $2x - 5y$

$$
\begin{array}{l}
3x \quad - 2y^2 \quad + 4xy \\
2x \quad - 5y \\
\hline
\end{array}
$$

Multiplying by $2x \rightarrow \quad 6x^2 - 4xy^2 \quad + 8x^2y$

Multiplying by $-5y \rightarrow \qquad -20xy^2 \qquad -15xy + 10y^3$

Adding gives $\quad \mathbf{6x^2 - 24xy^2 + 8x^2y - 15xy + 10y^3}$

Thus, $(3x - 2y^2 + 4xy)(2x - 5y)$
$$= \mathbf{6x^2 - 24xy^2 + 8x^2y - 15xy + 10y^3}$$

Problem 12. Simplify $2x \div 8xy$

$2x \div 8xy$ means $\dfrac{2x}{8xy}$

$$
\begin{aligned}
\frac{2x}{8xy} &= \frac{2 \times x}{8 \times x \times y} \\
&= \frac{1 \times 1}{4 \times 1 \times y} \qquad \text{by cancelling} \\
&= \frac{1}{4y}
\end{aligned}
$$

Problem 13. Simplify $\dfrac{9a^2bc}{3ac}$

$$
\begin{aligned}
\frac{9a^2bc}{3ac} &= \frac{9 \times a \times a \times b \times c}{3 \times a \times c} \\
&= 3 \times a \times b \\
&= \mathbf{3ab}
\end{aligned}
$$

Problem 14. Divide $2x^2 + x - 3$ by $x - 1$

(i) $2x^2 + x - 3$ is called the **dividend** and $x - 1$ the **divisor**. The usual layout is shown below with the dividend and divisor both arranged in descending powers of the symbols.

$$
\begin{array}{r}
2x + 3 \\
x - 1\overline{)2x^2 + x - 3} \\
\underline{2x^2 - 2x} \\
3x - 3 \\
\underline{3x - 3} \\
\cdot \quad \cdot
\end{array}
$$

(ii) Dividing the first term of the dividend by the first term of the divisor, i.e. $\dfrac{2x^2}{x}$ gives $2x$, which is put above the first term of the dividend as shown.

(iii) The divisor is then multiplied by $2x$, i.e. $2x(x - 1) = 2x^2 - 2x$, which is placed under the dividend as shown. Subtracting gives $3x - 3$

(iv) The process is then repeated, i.e. the first term of the divisor, x, is divided into $3x$, giving $+3$, which is placed above the dividend as shown.

(v) Then $3(x - 1) = 3x - 3$, which is placed under the $3x - 3$. The remainder, on subtraction, is zero, which completes the process.

Thus, $\mathbf{(2x^2 + x - 3) \div (x - 1) = (2x + 3)}$

(A check can be made on this answer by multiplying $(2x + 3)$ by $(x - 1)$, which equals $2x^2 + x - 3$)

Problem 15. Simplify $\dfrac{x^3 + y^3}{x + y}$

$$
\begin{array}{r}
\text{(i) (iv) (vii)} \\
x^2 - xy + y^2 \\
x + y\overline{)x^3 + 0 \quad + 0 \quad + y^3} \\
\underline{x^3 + x^2y} \\
-x^2y \qquad + y^3 \\
\underline{-x^2y - xy^2} \\
xy^2 + y^3 \\
\underline{xy^2 + y^3} \\
\cdot \quad \cdot
\end{array}
$$

(i) x into x^3 goes x^2. Put x^2 above x^3.

(ii) $x^2(x + y) = x^3 + x^2y$

(iii) Subtract.

(iv) x into $-x^2y$ goes $-xy$. Put $-xy$ above the dividend.

(v) $-xy(x+y) = -x^2y - xy^2$

(vi) Subtract.

(vii) x into xy^2 goes y^2. Put y^2 above the dividend.

(viii) $y^2(x+y) = xy^2 + y^3$

(ix) Subtract.

Thus, $\dfrac{x^3+y^3}{x+y} = x^2 - xy + y^2$

The zeros shown in the dividend are not normally shown, but are included to clarify the subtraction process and to keep similar terms in their respective columns.

Problem 16. Divide $4a^3 - 6a^2b + 5b^3$ by $2a - b$

$$
\begin{array}{r}
2a^2 - 2ab - b^2 \\
2a-b{\overline{\smash{\big)}\,4a^3 - 6a^2b \qquad + 5b^3}} \\
\underline{4a^3 - 2a^2b} \\
-4a^2b \qquad + 5b^3 \\
\underline{-4a^2b + 2ab^2} \\
-2ab^2 + 5b^3 \\
\underline{-2ab^2 + b^3} \\
4b^3
\end{array}
$$

Thus, $\dfrac{4a^3 - 6a^2b + 5b^3}{2a-b} = 2a^2 - 2ab - b^2$, remainder $4b^3$.

Alternatively, the answer may be expressed as

$$\frac{4a^3 - 6a^2b + 5b^3}{2a-b} = 2a^2 - 2ab - b^2 + \frac{4b^3}{2a-b}$$

Now try the following Practice Exercise

Practice Exercise 48   Basic operations in algebra (answers on page 446)

1. Simplify $pq \times pq^2r$

2. Simplify $-4a \times -2a$

3. Simplify $3 \times -2q \times -q$

4. Evaluate $3pq - 5qr - pqr$ when $p=3$, $q=-2$ and $r=4$

5. Determine the value of $3x^2yz^3$, given that $x=2, y=1\frac{1}{2}$ and $z=\frac{2}{3}$

6. If $x=5$ and $y=6$, evaluate $\dfrac{23(x-y)}{y+xy+2x}$

7. If $a=4, b=3, c=5$ and $d=6$, evaluate $\dfrac{3a+2b}{3c-2d}$

8. Simplify $2x \div 14xy$

9. Simplify $\dfrac{25x^2yz^3}{5xyz}$

10. Multiply $3a - b$ by $a + b$

11. Multiply $2a - 5b + c$ by $3a + b$

12. Simplify $3a \div 9ab$

13. Simplify $4a^2b \div 2a$

14. Divide $6x^2y$ by $2xy$

15. Divide $2x^2 + xy - y^2$ by $x + y$

16. Divide $3p^2 - pq - 2q^2$ by $p - q$

17. Simplify $(a+b)^2 + (a-b)^2$

9.3 Laws of indices

The laws of indices with numbers were covered in Chapter 7; the laws of indices in algebraic terms are as follows:

(1) $a^m \times a^n = a^{m+n}$
   For example, $a^3 \times a^4 = a^{3+4} = a^7$

(2) $\dfrac{a^m}{a^n} = a^{m-n}$ For example, $\dfrac{c^5}{c^2} = c^{5-2} = c^3$

(3) $(a^m)^n = a^{mn}$ For example, $(d^2)^3 = d^{2\times3} = d^6$

(4) $a^{\frac{m}{n}} = \sqrt[n]{a^m}$ For example, $x^{\frac{4}{3}} = \sqrt[3]{x^4}$

(5) $a^{-n} = \dfrac{1}{a^n}$ For example, $3^{-2} = \dfrac{1}{3^2} = \dfrac{1}{9}$

(6) $a^0 = 1$ For example, $17^0 = 1$

Here are some worked examples to demonstrate these laws of indices.

Problem 17. Simplify $a^2b^3c \times ab^2c^5$

$$a^2b^3c \times ab^2c^5 = a^2 \times b^3 \times c \times a \times b^2 \times c^5$$
$$= a^2 \times b^3 \times c^1 \times a^1 \times b^2 \times c^5$$

Grouping together like terms gives

$$a^2 \times a^1 \times b^3 \times b^2 \times c^1 \times c^5$$

Using law (1) of indices gives

$$a^{2+1} \times b^{3+2} \times c^{1+5} = a^3 \times b^5 \times c^6$$

i.e. $$a^2b^3c \times ab^2c^5 = a^3 b^5 c^6$$

Problem 18. Simplify $a^{\frac{1}{3}} b^{\frac{3}{2}} c^{-2} \times a^{\frac{1}{6}} b^{\frac{1}{2}} c$

Using law (1) of indices,

$$a^{\frac{1}{3}} b^{\frac{3}{2}} c^{-2} \times a^{\frac{1}{6}} b^{\frac{1}{2}} c = a^{\frac{1}{3}+\frac{1}{6}} \times b^{\frac{3}{2}+\frac{1}{2}} \times c^{-2+1}$$
$$= a^{\frac{1}{2}} b^2 c^{-1}$$

Problem 19. Simplify $\dfrac{x^5 y^2 z}{x^2 y z^3}$

$$\frac{x^5 y^2 z}{x^2 y z^3} = \frac{x^5 \times y^2 \times z}{x^2 \times y \times z^3}$$
$$= \frac{x^5}{x^2} \times \frac{y^2}{y^1} \times \frac{z}{z^3}$$
$$= x^{5-2} \times y^{2-1} \times z^{1-3} \quad \text{by law (2) of indices}$$
$$= x^3 \times y^1 \times z^{-2}$$
$$= x^3 y z^{-2} \text{ or } \frac{x^3 y}{z^2}$$

Problem 20. Simplify $\dfrac{a^3 b^2 c^4}{abc^{-2}}$ and evaluate when $a = 3, b = \dfrac{1}{4}$ and $c = 2$

Using law (2) of indices,

$$\frac{a^3}{a} = a^{3-1} = a^2, \frac{b^2}{b} = b^{2-1} = b \text{ and}$$
$$\frac{c^4}{c^{-2}} = c^{4--2} = c^6$$

Thus, $\dfrac{a^3 b^2 c^4}{abc^{-2}} = a^2bc^6$

When $a = 3, b = \dfrac{1}{4}$ and $c = 2$,

$$a^2bc^6 = (3)^2 \left(\frac{1}{4}\right)(2)^6 = (9)\left(\frac{1}{4}\right)(64) = \mathbf{144}$$

Problem 21. Simplify $(p^3)^2(q^2)^4$

Using law (3) of indices gives

$$(p^3)^2(q^2)^4 = p^{3\times2} \times q^{2\times4}$$
$$= p^6 q^8$$

Problem 22. Simplify $\dfrac{(mn^2)^3}{(m^{1/2}n^{1/4})^4}$

The brackets indicate that each letter in the bracket must be raised to the power outside. Using law (3) of indices gives

$$\frac{(mn^2)^3}{(m^{1/2}n^{1/4})^4} = \frac{m^{1\times3}n^{2\times3}}{m^{(1/2)\times4}n^{(1/4)\times4}} = \frac{m^3 n^6}{m^2 n^1}$$

Using law (2) of indices gives

$$\frac{m^3 n^6}{m^2 n^1} = m^{3-2}n^{6-1} = mn^5$$

Problem 23. Simplify $(a^3b)(a^{-4}b^{-2})$, expressing the answer with positive indices only

Using law (1) of indices gives $a^{3+-4}b^{1+-2} = a^{-1}b^{-1}$

Using law (5) of indices gives $a^{-1}b^{-1} = \dfrac{1}{a^{+1}b^{+1}} = \dfrac{1}{ab}$

Now try the following Practice Exercise

Practice Exercise 49   Laws of indices (answers on page 447)

In Problems 1 to 18, simplify the following, giving each answer as a power.

1. $z^2 \times z^6$ 2. $a \times a^2 \times a^5$

3. $n^8 \times n^{-5}$ 4. $b^4 \times b^7$

5. $b^2 \div b^5$ 6. $c^5 \times c^3 \div c^4$

7. $\dfrac{m^5 \times m^6}{m^4 \times m^3}$ 8. $\dfrac{(x^2)(x)}{x^6}$

9. $\left(x^3\right)^4$

10. $\left(y^2\right)^{-3}$

11. $\left(t \times t^3\right)^2$

12. $\left(c^{-7}\right)^{-2}$

13. $\left(\dfrac{a^2}{a^5}\right)^3$

14. $\left(\dfrac{1}{b^3}\right)^4$

15. $\left(\dfrac{b^2}{b^7}\right)^{-2}$

16. $\dfrac{1}{\left(s^3\right)^3}$

17. $p^3qr^2 \times p^2q^5r \times pqr^2$

18. $\dfrac{x^3y^2z}{x^5yz^3}$

19. Simplify $(x^2y^3z)(x^3yz^2)$ and evaluate when $x=\dfrac{1}{2}, y=2$ and $z=3$

20. Simplify $\dfrac{a^5bc^3}{a^2b^3c^2}$ and evaluate when $a=\dfrac{3}{2}, b=\dfrac{1}{2}$ and $c=\dfrac{2}{3}$

Here are some further worked examples on the laws of indices.

Problem 24. Simplify $\dfrac{x^2y^3 + xy^2}{xy}$

Algebraic expressions of the form $\dfrac{a+b}{c}$ can be split into $\dfrac{a}{c} + \dfrac{b}{c}$. Thus,

$$\dfrac{x^2y^3 + xy^2}{xy} = \dfrac{x^2y^3}{xy} + \dfrac{xy^2}{xy} = x^{2-1}y^{3-1} + x^{1-1}y^{2-1}$$

$$= \boldsymbol{xy^2 + y}$$

$$\text{(since } x^0 = 1, \text{ from law (6) of indices).}$$

Problem 25. Simplify $\dfrac{x^2y}{xy^2 - xy}$

The highest common factor (HCF) of each of the three terms comprising the numerator and denominator is xy. Dividing each term by xy gives

$$\dfrac{x^2y}{xy^2 - xy} = \dfrac{\dfrac{x^2y}{xy}}{\dfrac{xy^2}{xy} - \dfrac{xy}{xy}} = \dfrac{\boldsymbol{x}}{\boldsymbol{y-1}}$$

Problem 26. Simplify $\dfrac{a^2b}{ab^2 - a^{1/2}b^3}$

The HCF of each of the three terms is $a^{1/2}b$. Dividing each term by $a^{1/2}b$ gives

$$\dfrac{a^2b}{ab^2 - a^{1/2}b^3} = \dfrac{\dfrac{a^2b}{a^{1/2}b}}{\dfrac{ab^2}{a^{1/2}b} - \dfrac{a^{1/2}b^3}{a^{1/2}b}} = \dfrac{\boldsymbol{a^{3/2}}}{\boldsymbol{a^{1/2}b - b^2}}$$

Now try the following Practice Exercise

Practice Exercise 50   Laws of indices (answers on page 447)

1. Simplify $\left(a^{3/2}bc^{-3}\right)\left(a^{1/2}b^{-1/2}c\right)$ and evaluate when $a=3, b=4$ and $c=2$.

In Problems 2 to 8, simplify the given expressions.

2. $\dfrac{a^2b + a^3b}{a^2b^2}$

3. $(a^2)^{1/2}(b^2)^3\left(c^{1/2}\right)^3$

4. $\dfrac{(abc)^2}{(a^2b^{-1}c^{-3})^3}$

5. $\dfrac{p^3q^2}{pq^2 - p^2q}$

6. $(\sqrt{x}\sqrt{y^3}\sqrt[3]{z^2})(\sqrt{x}\sqrt{y^3}\sqrt{z^3})$

7. $(e^2f^3)(e^{-3}f^{-5})$, expressing the answer with positive indices only.

8. $\dfrac{(a^3b^{1/2}c^{-1/2})(ab)^{1/3}}{(\sqrt{a^3}\sqrt{b}\,c)}$

Practice Exercise 51   Multiple-choice questions on basic algebra (answers on page 447)

Each question has only one correct answer

1. When $a=2, b=1$ and $c=-3$ the value of $2ab - bc + abc$ is:
   (a) 13   (b) -5 (c) 1   (d) 7

2. $8a \times 2b \times c^2$ is equivalent to:
   (a) $10abc$ (b) $16abc^2$
   (c) $10(abc)^2$ (d) $16(abc)^2$

3. $(3x - y - 2x - y)$ is equivalent to:
   (a) $(x - y)$ (b) $(x - 2y)$
   (c) $(5x - 2y)$ (d) $(x - 2y)$

4. $\left(\dfrac{p + 2p}{p}\right) \times 3p - 2p$ simplifies to:
   (a) $3p$ (b) $6p$ (c) $9p^2 - 2p$ (d) $7p$

5. $(\sqrt{x})(y^{\frac{3}{2}})(x^2 y)$ is equivalent to:
   (a) $\sqrt{(xy)^5}$ (b) $x^{\sqrt{2}} y^{\frac{5}{2}}$ (c) xy^5 (d) $x\sqrt{y^3}$

6. $p \times \left(\dfrac{1}{p^{-2}}\right)^{-3}$ is equivalent to:
   (a) $\dfrac{1}{p^{\frac{3}{2}}}$ (b) p^{-5} (c) p^5 (d) $\dfrac{1}{p^{-5}}$

7. Given that $3(5t - 2y) = 0$ then:
   (a) $15t = 6y$ (b) $15t - 2y = 0$
   (c) $5t - 2y = -3$ (d) $8t - 5y = 0$

8. Given that $7x + 8 - 2y = 12y - 4 - x$ then:
   (a) $y = \dfrac{4x + 6}{7}$ (b) $x = \dfrac{7y + 2}{4}$
   (c) $y = \dfrac{3x - 6}{5}$ (d) $x = \dfrac{5y - 6}{4}$

9. If $x = -6$ and $y = -1$ the value of $\dfrac{2x}{3} - \dfrac{6}{y}$ is:
   (a) -10 (b) -2 (c) -9 (d) 2

10. $\dfrac{2}{5xy} \div \dfrac{4}{15x^2}$ is equivalent to:
    (a) $\dfrac{2}{3x^2 y}$ (b) $\dfrac{2y}{3x}$ (c) $\dfrac{8}{75x^3 y}$ (d) $\dfrac{3x}{2y}$

11. $\dfrac{(3x^2 y)^2}{6xy^2}$ simplifies to:
    (a) $\dfrac{x}{y}$ (b) $\dfrac{3x^2}{2}$ (c) $\dfrac{3x^3}{2}$ (d) $\dfrac{x}{2y}$

12. $\dfrac{3}{k} + \dfrac{1}{2k}$ simplifies to:
    (a) $\dfrac{4k}{3}$ (b) $\dfrac{7}{2k}$ (c) $\dfrac{4}{3k}$ (d) $\dfrac{3}{2k^2}$

13. The value of $x^3 + 4x + 12$ when $x = -1$ is:
    (a) 13 (b) 7 (c) 9 (d) 11

14. $\dfrac{p + q}{q}$ is equivalent to:
    (a) p (b) $\dfrac{p}{q} + q$ (c) $\dfrac{p}{q} + 1$ (d) $\dfrac{p}{q} + p$

15. $\dfrac{(pq^2 r)^3}{(p^{-2} qr^2)^2}$ simplifies to:
    (a) $\dfrac{p^3 q^4}{r}$ (b) $\dfrac{p^3}{pr}$ (c) $\dfrac{p^7 q^4}{r}$ (d) $\dfrac{1}{pqr}$

Chapter 10

Further algebra

Why it is important to understand: **Further algebra**

Algebra is a form of mathematics that allows you to work with unknowns. If you do not know what a number is, arithmetic does not allow you to use it in calculations. Algebra has variables. Variables are labels for numbers and measurements you do not yet know. Algebra lets you use these variables in equations and formulae. A basic form of mathematics, algebra is nevertheless among the most commonly used forms of mathematics in the workforce. Although relatively simple, algebra possesses a powerful problem-solving tool used in many fields of engineering. For example, in designing a rocket to go to the moon, an engineer must use algebra to solve for flight trajectory, how long to burn each thruster and at what intensity, and at what angle to lift off. An engineer uses mathematics all the time – and in particular, algebra. Becoming familiar with algebra will make all engineering mathematics studies so much easier.

At the end of this chapter you should be able to:

- use brackets with basic operations in algebra
- understand factorisation
- factorise simple algebraic expressions
- use the laws of precedence to simplify algebraic expressions

10.1 Introduction

In this chapter, the use of brackets and factorisation with algebra is explained, together with further practice with the laws of precedence. Understanding of these topics is often necessary when solving and transposing equations.

10.2 Brackets

With algebra

(a) $2(a + b) = 2a + 2b$

(b) $(a + b)(c + d) = a(c + d) + b(c + d)$
$$= ac + ad + bc + bd$$

Here are some worked examples to help understanding of brackets with algebra.

Problem 1. Determine $2b(a - 5b)$

$2b(a - 5b) = 2b \times a + 2b \times -5b$
$$= 2ba - 10b^2$$
$$= \mathbf{2ab - 10b^2} \qquad \text{(Note that } 2ba \text{ is the same as } 2ab\text{)}$$

Problem 2.   Determine $(3x + 4y)(x - y)$

$$
\begin{aligned}
(3x + 4y)(x - y) &= 3x(x - y) + 4y(x - y) \\
&= 3x^2 - 3xy + 4yx - 4y^2 \\
&= 3x^2 - 3xy + 4xy - 4y^2 \\
&\quad \text{(Note that } 4yx \text{ is the same as } 4xy) \\
&= \mathbf{3x^2 + xy - 4y^2}
\end{aligned}
$$

Problem 3.   Simplify $3(2x - 3y) - (3x - y)$

$$3(2x - 3y) - (3x - y) = 3 \times 2x - 3 \times 3y - 3x - -y$$

(Note that $-(3x - y) = -1(3x - y)$ and the -1 multiplies **both** terms in the bracket)

$$
\begin{aligned}
&= 6x - 9y - 3x + y \\
&\qquad \text{(Note: } - \times - = +) \\
&= 6x - 3x + y - 9y \\
&= \mathbf{3x - 8y}
\end{aligned}
$$

Problem 4.   Remove the brackets and simplify the expression $(a - 2b) + 5(b - c) - 3(c + 2d)$

$$
\begin{aligned}
&(a - 2b) + 5(b - c) - 3(c + 2d) \\
&= a - 2b + 5 \times b + 5 \times -c - 3 \times c - 3 \times 2d \\
&= a - 2b + 5b - 5c - 3c - 6d \\
&= \mathbf{a + 3b - 8c - 6d}
\end{aligned}
$$

Problem 5.   Simplify $(p + q)(p - q)$

$$
\begin{aligned}
(p + q)(p - q) &= p(p - q) + q(p - q) \\
&= p^2 - pq + qp - q^2 \\
&= \mathbf{p^2 - q^2}
\end{aligned}
$$

Problem 6.   Simplify $(2x - 3y)^2$

$$
\begin{aligned}
(2x - 3y)^2 &= (2x - 3y)(2x - 3y) \\
&= 2x(2x - 3y) - 3y(2x - 3y) \\
&= 2x \times 2x + 2x \times -3y - 3y \times 2x \\
&\qquad\qquad\qquad\qquad\qquad -3y \times -3y \\
&= 4x^2 - 6xy - 6xy + 9y^2 \\
&\quad \text{(Note: } + \times - = - \text{ and } - \times - = +) \\
&= \mathbf{4x^2 - 12xy + 9y^2}
\end{aligned}
$$

Problem 7.   Remove the brackets from the expression and simplify $2[x^2 - 3x(y + x) + 4xy]$

$$2[x^2 - 3x(y + x) + 4xy] = 2[x^2 - 3xy - 3x^2 + 4xy]$$

(Whenever more than one type of brackets is involved, always **start with the inner brackets**)

$$
\begin{aligned}
&= 2[-2x^2 + xy] \\
&= -4x^2 + 2xy \\
&= \mathbf{2xy - 4x^2}
\end{aligned}
$$

Problem 8.   Remove the brackets and simplify the expression $2a - [3\{2(4a - b) - 5(a + 2b)\} + 4a]$

(i)   Removing the innermost brackets gives
$$2a - [3\{8a - 2b - 5a - 10b\} + 4a]$$

(ii)   Collecting together similar terms gives
$$2a - [3\{3a - 12b\} + 4a]$$

(iii)   Removing the 'curly' brackets gives
$$2a - [9a - 36b + 4a]$$

(iv)   Collecting together similar terms gives
$$2a - [13a - 36b]$$

(v)   Removing the outer brackets gives
$$2a - 13a + 36b$$

(vi)   i.e. $\mathbf{-11a + 36b}$ or $\mathbf{36b - 11a}$

Now try the following Practice Exercise

Practice Exercise 52   Brackets (answers on page 447)

Expand the brackets in Problems 1 to 28.

1. $(x + 2)(x + 3)$ 2. $(x + 4)(2x + 1)$

3. $(2x + 3)^2$ 4. $(2j - 4)(j + 3)$

5. $(2x + 6)(2x + 5)$ 6. $(pq + r)(r + pq)$

7. $(a + b)(a + b)$ 8. $(x + 6)^2$

9. $(a - c)^2$ 10. $(5x + 3)^2$

11. $(2x - 6)^2$ 12. $(2x - 3)(2x + 3)$

13. $(8x + 4)^2$ 14. $(rs + t)^2$

15. $3a(b - 2a)$ 16. $2x(x - y)$

17. $(2a - 5b)(a + b)$

18. $3(3p - 2q) - (q - 4p)$

19. $(3x - 4y) + 3(y - z) - (z - 4x)$

20. $(2a + 5b)(2a - 5b)$

21. $(x - 2y)^2$ 22. $(3a - b)^2$

23. $2x + [y - (2x + y)]$

24. $3a + 2[a - (3a - 2)]$

25. $4[a^2 - 3a(2b + a) + 7ab]$

26. $3[x^2 - 2x(y + 3x) + 3xy(1 + x)]$

27. $2 - 5[a(a - 2b) - (a - b)^2]$

28. $24p - [2\{3(5p - q) - 2(p + 2q)\} + 3q]$

10.3 Factorisation

The **factors** of 8 are 1, 2, 4 and 8 because 8 divides by 1, 2, 4 and 8

The factors of 24 are 1, 2, 3, 4, 6, 8, 12 and 24 because 24 divides by 1, 2, 3, 4, 6, 8, 12 and 24

The **common factors** of 8 and 24 are 1, 2, 4 and 8 since 1, 2, 4 and 8 are factors of both 8 and 24

The **highest common factor (HCF)** is the largest number that divides into two or more terms.

Hence, the HCF of 8 and 24 is 8, as explained in Chapter 1.

When two or more terms in an algebraic expression contain a common factor, then this factor can be shown outside of a bracket. For example,

$$df + dg = d(f + g)$$

which is just the reverse of

$$d(f + g) = df + dg$$

This process is called **factorisation**.

Here are some worked examples to help understanding of factorising in algebra.

Problem 9. Factorise $ab - 5ac$

a is common to both terms ab and $-5ac$. a is therefore taken outside of the bracket. What goes inside the bracket?

(i) What multiplies a to make ab? Answer: b

(ii) What multiplies a to make $-5ac$? Answer: $-5c$

Hence, $b - 5c$ appears in the bracket. Thus,

$$ab - 5ac = a(b - 5c)$$

Problem 10. Factorise $2x^2 + 14xy^3$

For the numbers 2 and 14, the highest common factor (HCF) is 2 (i.e. 2 is the largest number that divides into both 2 and 14).

For the x terms, x^2 and x, the HCF is x

Thus, the HCF of $2x^2$ and $14xy^3$ is $2x$

$2x$ is therefore taken outside of the bracket. What goes inside the bracket?

(i) What multiplies $2x$ to make $2x^2$? Answer: x

(ii) What multiplies $2x$ to make $14xy^3$? Answer: $7y^3$

Hence $x + 7y^3$ appears inside the bracket. Thus,

$$2x^2 + 14xy^3 = 2x(x + 7y^3)$$

Problem 11. Factorise $3x^3y - 12xy^2 + 15xy$

For the numbers 3, 12 and 15, the highest common factor is 3 (i.e. 3 is the largest number that divides into 3, 12 and 15).

For the x terms, x^3, x and x, the HCF is x

For the y terms, y, y^2 and y, the HCF is y

Thus, the HCF of $3x^3y$ and $12xy^2$ and $15xy$ is $3xy$

$3xy$ is therefore taken outside of the bracket. What goes inside the bracket?

(i) What multiplies $3xy$ to make $3x^3y$? Answer: x^2

(ii) What multiplies $3xy$ to make $-12xy^2$? Answer: $-4y$

(iii) What multiplies $3xy$ to make $15xy$? Answer: 5

Hence, $x^2 - 4y + 5$ appears inside the bracket. Thus,

$$3x^3y - 12xy^2 + 15xy = 3xy(x^2 - 4y + 5)$$

Problem 12. Factorise $25a^2b^5 - 5a^3b^2$

For the numbers 25 and 5, the highest common factor is 5 (i.e. 5 is the largest number that divides into 25 and 5).

For the a terms, a^2 and a^3, the HCF is a^2

For the b terms, b^5 and b^2, the HCF is b^2

Thus, the HCF of $25a^2b^5$ and $5a^3b^2$ is $5a^2b^2$

$5a^2b^2$ is therefore taken outside of the bracket. What goes inside the bracket?

(i) What multiplies $5a^2b^2$ to make $25a^2b^5$? Answer: $5b^3$

(ii) What multiplies $5a^2b^2$ to make $-5a^3b^2$? Answer: $-a$

Hence, $5b^3 - a$ appears in the bracket. Thus,

$$25a^2b^5 - 5a^3b^2 = 5a^2b^2(5b^3 - a)$$

Problem 13. Factorise $ax - ay + bx - by$

The first two terms have a common factor of a and the last two terms a common factor of b. Thus,

$$ax - ay + bx - by = a(x - y) + b(x - y)$$

The two newly formed terms have a common factor of $(x - y)$. Thus,

$$a(x - y) + b(x - y) = (x - y)(a + b)$$

Problem 14. Factorise $2ax - 3ay + 2bx - 3by$

a is a common factor of the first two terms and b a common factor of the last two terms. Thus,

$$2ax - 3ay + 2bx - 3by = a(2x - 3y) + b(2x - 3y)$$

$(2x - 3y)$ is now a common factor. Thus,

$$a(2x - 3y) + b(2x - 3y) = (2x - 3y)(a + b)$$

Alternatively, $2x$ is a common factor of the original first and third terms and $-3y$ is a common factor of the second and fourth terms. Thus,

$$2ax - 3ay + 2bx - 3by = 2x(a + b) - 3y(a + b)$$

$(a + b)$ is now a common factor. Thus,

$$2x(a + b) - 3y(a + b) = (a + b)(2x - 3y)$$

as before

Problem 15. Factorise $x^3 + 3x^2 - x - 3$

x^2 is a common factor of the first two terms. Thus,

$$x^3 + 3x^2 - x - 3 = x^2(x + 3) - x - 3$$

-1 is a common factor of the last two terms. Thus,

$$x^2(x + 3) - x - 3 = x^2(x + 3) - 1(x + 3)$$

$(x + 3)$ is now a common factor. Thus,

$$x^2(x + 3) - 1(x - 3) = (x + 3)(x^2 - 1)$$

Now try the following Practice Exercise

Practice Exercise 53   Factorisation (answers on page 447)

Factorise and simplify the following.

1. $2x + 4$ 2. $2xy - 8xz$

3. $pb + 2pc$ 4. $2x + 4xy$

5. $4d^2 - 12df^5$ 6. $4x + 8x^2$

7. $2q^2 + 8qn$ 8. $rs + rp + rt$

9. $x + 3x^2 + 5x^3$ 10. $abc + b^3c$

11. $3x^2y^4 - 15xy^2 + 18xy$

12. $4p^3q^2 - 10pq^3$ 13. $21a^2b^2 - 28ab$

14. $2xy^2 + 6x^2y + 8x^3y$

15. $2x^2y - 4xy^3 + 8x^3y^4$

16. $28y + 7y^2 + 14xy$

17. $\dfrac{3x^2 + 6x - 3xy}{xy + 2y - y^2}$ 18. $\dfrac{abc + 2ab}{2c + 4} - \dfrac{abc}{2c}$

19. $\dfrac{5rs + 15r^3t + 20r}{6r^2t^2 + 8t + 2ts} - \dfrac{r}{2t}$

20. $ay + by + a + b$ 21. $px + qx + py + qy$

22. $ax - ay + bx - by$

23. $2ax + 3ay - 4bx - 6by$

24. $\dfrac{A^3}{p^2g^3} - \dfrac{A^2}{pg^2} + \dfrac{A^5}{pg}$

10.4 Laws of precedence

Sometimes addition, subtraction, multiplication, division, powers and brackets can all be involved in an algebraic expression. With mathematics there is a definite order of precedence (first met in Chapter 1) which we need to adhere to.

With the **laws of precedence** the order is

Brackets

Order (or p**O**wer)

Division

Multiplication

Addition

Subtraction

The first letter of each word spells **BODMAS**. Here are some examples to help understanding of BODMAS with algebra.

Problem 16. Simplify $2x + 3x \times 4x - x$

$$
\begin{aligned}
2x + 3x \times 4x - x &= 2x + 12x^2 - x \quad &\text{(M)} \\
&= 2x - x + 12x^2 \\
&= \mathbf{x + 12x^2} \quad &\text{(S)} \\
&\text{or } \mathbf{x(1 + 12x)} \quad &\text{by factorising}
\end{aligned}
$$

Problem 17. Simplify $(y + 4y) \times 3y - 5y$

$$
\begin{aligned}
(y + 4y) \times 3y - 5y &= 5y \times 3y - 5y \quad &\text{(B)} \\
&= \mathbf{15y^2 - 5y} \quad &\text{(M)} \\
&\text{or } \mathbf{5y(3y - 1)} \quad &\text{by factorising}
\end{aligned}
$$

Problem 18. Simplify $p + 2p \times (4p - 7p)$

$$
\begin{aligned}
p + 2p \times (4p - 7p) &= p + 2p \times -3p \quad &\text{(B)} \\
&= \mathbf{p - 6p^2} \quad &\text{(M)} \\
&\text{or } \mathbf{p(1 - 6p)} \quad &\text{by factorising}
\end{aligned}
$$

Problem 19. Simplify $t \div 2t + 3t - 5t$

$$
\begin{aligned}
t \div 2t + 3t - 5t &= \frac{t}{2t} + 3t - 5t \quad &\text{(D)} \\
&= \frac{1}{2} + 3t - 5t \quad &\text{by cancelling} \\
&= \mathbf{\frac{1}{2} - 2t} \quad &\text{(S)}
\end{aligned}
$$

Problem 20. Simplify $x \div (4x + x) - 3x$

$$
\begin{aligned}
x \div (4x + x) - 3x &= x \div 5x - 3x \quad &\text{(B)} \\
&= \frac{x}{5x} - 3x \quad &\text{(D)} \\
&= \mathbf{\frac{1}{5} - 3x} \quad &\text{by cancelling}
\end{aligned}
$$

Problem 21. Simplify $2y \div (6y + 3y - 5y)$

$$
\begin{aligned}
2y \div (6y + 3y - 5y) &= 2y \div 4y \quad &\text{(B)} \\
&= \frac{2y}{4y} \quad &\text{(D)} \\
&= \mathbf{\frac{1}{2}} \quad &\text{by cancelling}
\end{aligned}
$$

Problem 22. Simplify
$$5a + 3a \times 2a + a \div 2a - 7a$$

$$
\begin{aligned}
5a + 3a &\times 2a + a \div 2a - 7a \\
&= 5a + 3a \times 2a + \frac{a}{2a} - 7a \quad &\text{(D)} \\
&= 5a + 3a \times 2a + \frac{1}{2} - 7a \quad &\text{by cancelling} \\
&= 5a + 6a^2 + \frac{1}{2} - 7a \quad &\text{(M)} \\
&= -2a + 6a^2 + \frac{1}{2} \quad &\text{(S)} \\
&= \mathbf{6a^2 - 2a + \frac{1}{2}}
\end{aligned}
$$

Problem 23. Simplify
$$(4y + 3y)2y + y \div 4y - 6y$$

$$
\begin{aligned}
(4y + 3y)&2y + y \div 4y - 6y \\
&= 7y \times 2y + y \div 4y - 6y \quad &\text{(B)} \\
&= 7y \times 2y + \frac{y}{4y} - 6y \quad &\text{(D)} \\
&= 7y \times 2y + \frac{1}{4} - 6y \quad &\text{by cancelling} \\
&= \mathbf{14y^2 + \frac{1}{4} - 6y} \quad &\text{(M)}
\end{aligned}
$$

Problem 24. Simplify
$$5b + 2b \times 3b + b \div (4b - 7b)$$

$$
\begin{aligned}
5b + 2b &\times 3b + b \div (4b - 7b) \\
&= 5b + 2b \times 3b + b \div -3b \quad &\text{(B)} \\
&= 5b + 2b \times 3b + \frac{b}{-3b} \quad &\text{(D)} \\
&= 5b + 2b \times 3b + \frac{1}{-3} \quad &\text{by cancelling} \\
&= 5b + 2b \times 3b - \frac{1}{3} \\
&= \mathbf{5b + 6b^2 - \frac{1}{3}} \quad &\text{(M)}
\end{aligned}
$$

Problem 25. Simplify

$$(5p + p)(2p + 3p) \div (4p - 5p)$$

$(5p + p)(2p + 3p) \div (4p - 5p)$

$$= (6p)(5p) \div (-p) \qquad \text{(B)}$$

$$= 6p \times 5p \div -p$$

$$= 6p \times \frac{5p}{-p} \qquad \text{(D)}$$

$$= 6p \times \frac{5}{-1} \qquad \text{by cancelling}$$

$$= 6p \times -5$$

$$= -30p$$

Now try the following Practice Exercise

Practice Exercise 54   Laws of precedence (answers on page 447)

Simplify the following.

1. $3x + 2x \times 4x - x$

2. $(2y + y) \times 4y - 3y$

3. $4b + 3b \times (b - 6b)$

4. $8a \div 2a + 6a - 3a$

5. $6x \div (3x + x) - 4x$

6. $4t \div (5t - 3t + 2t)$

7. $3y + 2y \times 5y + 2y \div 8y - 6y$

8. $(x + 2x)3x + 2x \div 6x - 4x$

9. $5a + 2a \times 3a + a \div (2a - 9a)$

10. $(3t + 2t)(5t + t) \div (t - 3t)$

11. $x \div 5x - x + (2x - 3x)x$

12. $3a + 2a \times 5a + 4a \div 2a - 6a$

Practice Exercise 55   Multiple-choice questions on further algebra (answers on page 447)

Each question has only one correct answer

1. $2x^2 - (x - xy) - x(2y - x)$ simplifies to:
    (a) $x(3x - 1 - y)$ (b) $x^2 - 3xy - xy$
    (c) $x(xy - y - 1)$ (d) $3x^2 - x + xy$

2. $\left(\dfrac{a^2}{a^{-3} \times a}\right)^{-2}$ is equivalent to:
    (a) a^{-8} (b) $\dfrac{1}{a^{-10}}$ (c) $\dfrac{1}{a^{-8}}$ (d) a^{-10}

3. $3d + 2d \times 4d + d \div (6d - 2d)$ simplifies to:
    (a) $\dfrac{25d}{4}$ (b) $\dfrac{1}{4} + d(3 + 8d)$
    (c) $1 + 2d$ (d) $20d^2 + \dfrac{1}{4}$

4. $5a \div a + 2a - 3a \times a$ simplifies to:
    (a) $4a^2$ (b) $5 + 2a - 3a^2$
    (c) $5 - a^2$ (d) $\dfrac{5}{3} - 3a^2$

5. $(2 + \sqrt{2})(2 - \sqrt{2})$ is equal to:
    (a) 0   (b) 2   (c) 4   (d) 8

6. $(2e - 3f)(e + f)$ is equal to:
    (a) $2e^2 - 3f^2$ (b) $2e^2 - 5ef - 3f^2$
    (c) $2e^2 + 3f^2$ (d) $2e^2 - ef - 3f^2$

7. $(2x - y)^2$ is equal to:
    (a) $4x^2 + y^2$ (b) $2x^2 - 2xy + y^2$
    (c) $4x^2 - y^2$ (d) $4x^2 - 4xy + y^2$

8. $\left(\sqrt{5} + \sqrt{2}\right)\left(\sqrt{5} - \sqrt{2}\right)$ is equal to:
    (a) $\sqrt{3}$ (b) 7   (c) 3   (d) 10

9. Factorising $2xy^2 + 6x^3y - 8x^3y^2$ gives:
    (a) $2x(y^2 + 3x^2 - 4x^2y)$
    (b) $2xy(y + 3x^2y - 4x^2y^2)$
    (c) $96x^7y^5$
    (d) $2xy(y + 3x^2 - 4x^2y)$

10. Expanding $(m - n)^2$ gives:
    (a) $m^2 + n^2$ (b) $m^2 - 2mn - n^2$
    (c) $m^2 - 2mn + n^2$ (d) $m^2 - n^2$

For fully worked solutions to each of the problems in Practice Exercises 52 to 54 in this chapter, go to the website:
www.routledge.com/cw/bird

Chapter 11

Solving simple equations

Why it is important to understand: *Solving simple equations*

In mathematics, engineering and science, formulae are used to relate physical quantities to each other. They provide rules so that if we know the values of certain quantities, we can calculate the values of others. Equations occur in all branches of engineering. Simple equations always involve one unknown quantity which we try to find when we solve the equation. In reality, we all solve simple equations in our heads all the time without even noticing it. If, for example, you have bought two CDs, for the same price, and a DVD, and know that you spent £25 in total and that the DVD was £11, then you actually solve the linear equation $2x + 11 = 25$ to find out that the price of each CD was £7. It is probably true to say that there is no branch of engineering, physics, economics, chemistry and computer science which does not require the solution of simple equations. The ability to solve simple equations is another stepping stone on the way to having confidence to handle engineering mathematics.

At the end of this chapter you should be able to:

- distinguish between an algebraic expression and an algebraic equation
- maintain the equality of a given equation whilst applying arithmetic operations
- solve linear equations in one unknown including those involving brackets and fractions
- form and solve linear equations involved with practical situations
- evaluate an unknown quantity in a formula by substitution of data

11.1 Introduction

$3x - 4$ is an example of an **algebraic expression**.
$3x - 4 = 2$ is an example of an **algebraic equation** (i.e. it contains an '=' sign).

An equation is simply a statement that two expressions are equal.

Hence, $A = \pi r^2$ (where A is the area of a circle of radius r)

$$F = \frac{9}{5}C + 32 \text{ (which relates Fahrenheit and}$$

Celsius temperatures)

and $y = 3x + 2$ (which is the equation of a straight line graph)

are all examples of equations.

11.2 Solving equations

To '**solve an equation**' means '**to find the value of the unknown**'. For example, solving $3x - 4 = 2$ means that the value of x is required.

In this example, $x = 2$. How did we arrive at $x = 2$? This is the purpose of this chapter – to show how to solve such equations.

Many equations occur in engineering and it is essential that we can solve them when needed.

Here are some examples to demonstrate how simple equations are solved.

Problem 1.   Solve the equation $4x = 20$

Dividing each side of the equation by 4 gives

$$\frac{4x}{4} = \frac{20}{4}$$

i.e. $x = 5$ by cancelling, which is the solution to the equation $4x = 20$

The same operation **must** be applied to both sides of an equation so that the equality is maintained.

We can do anything we like to an equation, **as long as we do the same to both sides.** This is, in fact, the only rule to remember when solving simple equations (and also when transposing formulae, which we do in Chapter 12).

Problem 2.   Solve the equation $\dfrac{2x}{5} = 6$

Multiplying both sides by 5 gives $5\left(\dfrac{2x}{5}\right) = 5(6)$

Cancelling and removing brackets gives $2x = 30$

Dividing both sides of the equation by 2 gives

$$\frac{2x}{2} = \frac{30}{2}$$

Cancelling gives $x = 15$

which is the solution of the equation $\dfrac{2x}{5} = 6$

Problem 3.   Solve the equation $a - 5 = 8$

Adding 5 to both sides of the equation gives

$$a - 5 + 5 = 8 + 5$$

i.e. $a = 8 + 5$

i.e. $a = 13$

which is the solution of the equation $a - 5 = 8$

Note that adding 5 to both sides of the above equation results in the -5 moving from the LHS to the RHS, but the sign is changed to $+$

Problem 4.   Solve the equation $x + 3 = 7$

Subtracting 3 from both sides gives $x + 3 - 3 = 7 - 3$

i.e. $x = 7 - 3$

i.e. $x = 4$

which is the solution of the equation $x + 3 = 7$

Note that subtracting 3 from both sides of the above equation results in the $+3$ moving from the LHS to the RHS, but the sign is changed to $-$. So, we can move straight from $x + 3 = 7$ to $x = 7 - 3$

Thus, a term can be moved from one side of an equation to the other **as long as a change in sign is made**.

Problem 5.   Solve the equation $6x + 1 = 2x + 9$

In such equations the terms containing x are grouped on one side of the equation and the remaining terms grouped on the other side of the equation. As in Problems 3 and 4, changing from one side of an equation to the other must be accompanied by a change of sign.

Since $6x + 1 = 2x + 9$

then $6x - 2x = 9 - 1$

i.e. $4x = 8$

Dividing both sides by 4 gives $\dfrac{4x}{4} = \dfrac{8}{4}$

Cancelling gives $x = 2$

which is the solution of the equation $6x + 1 = 2x + 9$.

In the above examples, the solutions can be checked. Thus, in Problem 5, where $6x + 1 = 2x + 9$, if $x = 2$, then

$$\text{LHS of equation} = 6(2) + 1 = 13$$

$$\text{RHS of equation} = 2(2) + 9 = 13$$

Since the left hand side (LHS) equals the right hand side (RHS) then $x = 2$ must be the correct solution of the equation.

When solving simple equations, always check your answers by substituting your solution back into the original equation.

Problem 6.   Solve the equation $4 - 3p = 2p - 11$

In order to keep the p term positive the terms in p are moved to the RHS and the constant terms to the LHS. Similar to Problem 5, if $4 - 3p = 2p - 11$

then $4 + 11 = 2p + 3p$

i.e. $15 = 5p$

Dividing both sides by 5 gives $\dfrac{15}{5} = \dfrac{5p}{5}$

Cancelling gives $3 = p$ or $p = 3$

which is the solution of the equation $4 - 3p = 2p - 11$.

By substituting $p = 3$ into the original equation, the solution may be checked.

$$LHS = 4 - 3(3) = 4 - 9 = -5$$

$$RHS = 2(3) - 11 = 6 - 11 = -5$$

Since LHS = RHS, the solution $p = 3$ must be correct. If, in this example, the unknown quantities had been grouped initially on the LHS instead of the RHS, then $-3p - 2p = -11 - 4$

i.e. $-5p = -15$

from which, $\dfrac{-5p}{-5} = \dfrac{-15}{-5}$

and $p = 3$

as before.

It is often easier, however, to work with positive values where possible.

Problem 7. Solve the equation $3(x - 2) = 9$

Removing the bracket gives $3x - 6 = 9$

Rearranging gives $3x = 9 + 6$

i.e. $3x = 15$

Dividing both sides by 3 gives $x = 5$

which is the solution of the equation $3(x - 2) = 9$. The equation may be checked by substituting $x = 5$ back into the original equation.

Problem 8. Solve the equation
$$4(2r - 3) - 2(r - 4) = 3(r - 3) - 1$$

Removing brackets gives

$$8r - 12 - 2r + 8 = 3r - 9 - 1$$

Rearranging gives $8r - 2r - 3r = -9 - 1 + 12 - 8$

i.e. $3r = -6$

Dividing both sides by 3 gives $r = \dfrac{-6}{3} = -2$

which is the solution of the equation

$$4(2r - 3) - 2(r - 4) = 3(r - 3) - 1$$

The solution may be checked by substituting $r = -2$ back into the original equation.

$$LHS = 4(-4 - 3) - 2(-2 - 4) = -28 + 12 = -16$$

$$RHS = 3(-2 - 3) - 1 = -15 - 1 = -16$$

Since LHS = RHS then $r = -2$ is the correct solution.

Now try the following Practice Exercise

Practice Exercise 56   Solving simple equations (answers on page 447)

Solve the following equations.

1. $2x + 5 = 7$

2. $8 - 3t = 2$

3. $\dfrac{2}{3}c - 1 = 3$

4. $2x - 1 = 5x + 11$

5. $7 - 4p = 2p - 5$

6. $2.6x - 1.3 = 0.9x + 0.4$

7. $2a + 6 - 5a = 0$

8. $3x - 2 - 5x = 2x - 4$

9. $20d - 3 + 3d = 11d + 5 - 8$

10. $2(x - 1) = 4$

11. $16 = 4(t + 2)$

12. $5(f - 2) - 3(2f + 5) + 15 = 0$

13. $2x = 4(x - 3)$

14. $6(2 - 3y) - 42 = -2(y - 1)$

15. $2(3g - 5) - 5 = 0$

16. $4(3x + 1) = 7(x + 4) - 2(x + 5)$

17. $11 + 3(r - 7) = 16 - (r + 2)$

18. $8 + 4(x - 1) - 5(x - 3) = 2(5 - 2x)$

Here are some further worked examples on solving simple equations.

Problem 9. Solve the equation $\dfrac{4}{x} = \dfrac{2}{5}$

The lowest common multiple (LCM) of the denominators, i.e. the lowest algebraic expression that both x and 5 will divide into, is $5x$

Multiplying both sides by $5x$ gives

$$5x\left(\dfrac{4}{x}\right) = 5x\left(\dfrac{2}{5}\right)$$

Cancelling gives $\quad\quad 5(4) = x(2)$

i.e. $\quad\quad\quad\quad\quad\quad 20 = 2x \quad\quad\quad\quad (1)$

Dividing both sides by 2 gives $\quad \dfrac{20}{2} = \dfrac{2x}{2}$

Cancelling gives $\quad\quad\quad\quad \mathbf{10 = x}$ or $\mathbf{x = 10}$

which is the solution of the equation $\dfrac{4}{x} = \dfrac{2}{5}$

When there is just one fraction on each side of the equation as in this example, there is a quick way to arrive at equation (1) without needing to find the LCM of the denominators.

We can move from $\dfrac{4}{x} = \dfrac{2}{5}$ to $4 \times 5 = 2 \times x$ by what is called '**cross-multiplication**'.

In general, if $\dfrac{a}{b} = \dfrac{c}{d}$ then $ad = bc$

We can use cross-multiplication when there is one fraction only on each side of the equation.

Problem 10. Solve the equation $\dfrac{3}{t-2} = \dfrac{4}{3t+4}$

Cross-multiplication gives $\quad 3(3t+4) = 4(t-2)$

Removing brackets gives $\quad\quad 9t + 12 = 4t - 8$

Rearranging gives $\quad\quad\quad 9t - 4t = -8 - 12$

i.e. $\quad\quad\quad\quad\quad\quad\quad 5t = -20$

Dividing both sides by 5 gives $\quad t = \dfrac{-20}{5} = \mathbf{-4}$

which is the solution of the equation $\dfrac{3}{t-2} = \dfrac{4}{3t+4}$

Problem 11. Solve the equation

$$\frac{2y}{5} + \frac{3}{4} + 5 = \frac{1}{20} - \frac{3y}{2}$$

The lowest common multiple (LCM) of the denominators is 20; i.e. the lowest number that 4, 5, 20 and 2 will divide into.

Multiplying each term by 20 gives

$$20\left(\frac{2y}{5}\right) + 20\left(\frac{3}{4}\right) + 20(5) = 20\left(\frac{1}{20}\right) - 20\left(\frac{3y}{2}\right)$$

Cancelling gives $\quad 4(2y) + 5(3) + 100 = 1 - 10(3y)$

i.e. $\quad\quad\quad\quad 8y + 15 + 100 = 1 - 30y$

Rearranging gives $\quad\quad\quad 8y + 30y = 1 - 15 - 100$

i.e. $\quad\quad\quad\quad\quad\quad\quad 38y = -114$

Dividing both sides by 38 gives $\quad \dfrac{38y}{38} = \dfrac{-114}{38}$

Cancelling gives $\quad\quad\quad\quad\quad \mathbf{y = -3}$

which is the solution of the equation

$$\frac{2y}{5} + \frac{3}{4} + 5 = \frac{1}{20} - \frac{3y}{2}$$

Problem 12. Solve the equation $\sqrt{x} = 2$

Whenever square root signs are involved in an equation, both sides of the equation must be squared.

Squaring both sides gives $\quad \left(\sqrt{x}\right)^2 = (2)^2$

i.e. $\quad\quad\quad\quad\quad\quad\quad \mathbf{x = 4}$

which is the solution of the equation $\sqrt{x} = 2$

Problem 13. Solve the equation $2\sqrt{d} = 8$

Whenever square roots are involved in an equation, the square root term needs to be isolated on its own before squaring both sides.

Cross-multiplying gives $\quad\quad \sqrt{d} = \dfrac{8}{2}$

Cancelling gives $\quad\quad\quad\quad \sqrt{d} = 4$

Squaring both sides gives $\quad \left(\sqrt{d}\right)^2 = (4)^2$

i.e. $\quad\quad\quad\quad\quad\quad\quad \mathbf{d = 16}$

which is the solution of the equation $2\sqrt{d} = 8$

Problem 14. Solve the equation $\left(\dfrac{\sqrt{b}+3}{\sqrt{b}}\right) = 2$

Cross-multiplying gives $\quad \sqrt{b} + 3 = 2\sqrt{b}$

Rearranging gives $\quad\quad\quad\quad 3 = 2\sqrt{b} - \sqrt{b}$

i.e. $\quad\quad\quad\quad\quad\quad\quad 3 = \sqrt{b}$

Squaring both sides gives $\quad\quad \mathbf{9 = b}$

which is the solution of the equation $\left(\dfrac{\sqrt{b}+3}{\sqrt{b}}\right) = 2$

Problem 15. Solve the equation $x^2 = 25$

Whenever a square term is involved, the square root of both sides of the equation must be taken.

Taking the square root of both sides gives $\sqrt{x^2} = \sqrt{25}$

i.e. $x = \pm 5$

which is the solution of the equation $x^2 = 25$

> **Problem 16.** Solve the equation $\dfrac{15}{4t^2} = \dfrac{2}{3}$

We need to rearrange the equation to get the t^2 term on its own.

Cross-multiplying gives $15(3) = 2(4t^2)$

i.e. $45 = 8t^2$

Dividing both sides by 8 gives $\dfrac{45}{8} = \dfrac{8t^2}{8}$

By cancelling $5.625 = t^2$

or $t^2 = 5.625$

Taking the square root of both sides gives

$$\sqrt{t^2} = \sqrt{5.625}$$

i.e. $t = \pm 2.372$

correct to 4 significant figures, which is the solution of the equation $\dfrac{15}{4t^2} = \dfrac{2}{3}$

Now try the following Practice Exercise

> **Practice Exercise 57   Solving simple equations (answers on page 448)**
>
> Solve the following equations.
>
> 1. $\dfrac{1}{5}d + 3 = 4$
> 2. $2 + \dfrac{3}{4}y = 1 + \dfrac{2}{3}y + \dfrac{5}{6}$
> 3. $\dfrac{1}{4}(2x - 1) + 3 = \dfrac{1}{2}$
> 4. $\dfrac{1}{5}(2f - 3) + \dfrac{1}{6}(f - 4) + \dfrac{2}{15} = 0$
> 5. $\dfrac{1}{3}(3m - 6) - \dfrac{1}{4}(5m + 4) + \dfrac{1}{5}(2m - 9) = -3$
> 6. $\dfrac{x}{3} - \dfrac{x}{5} = 2$
> 7. $1 - \dfrac{y}{3} = 3 + \dfrac{y}{3} - \dfrac{y}{6}$
> 8. $\dfrac{2}{a} = \dfrac{3}{8}$
> 9. $\dfrac{1}{3n} + \dfrac{1}{4n} = \dfrac{7}{24}$
> 10. $\dfrac{x+3}{4} = \dfrac{x-3}{5} + 2$
> 11. $\dfrac{3t}{20} = \dfrac{6-t}{12} + \dfrac{2t}{15} - \dfrac{3}{2}$
> 12. $\dfrac{y}{5} + \dfrac{7}{20} = \dfrac{5-y}{4}$
> 13. $\dfrac{v-2}{2v-3} = \dfrac{1}{3}$
> 14. $\dfrac{2}{a-3} = \dfrac{3}{2a+1}$
> 15. $\dfrac{x}{4} - \dfrac{x+6}{5} = \dfrac{x+3}{2}$
> 16. $3\sqrt{t} = 9$
> 17. $2\sqrt{y} = 5$
> 18. $4 = \sqrt{\left(\dfrac{3}{a}\right)} + 3$
> 19. $\dfrac{3\sqrt{x}}{1-\sqrt{x}} = -6$
> 20. $10 = 5\sqrt{\left(\dfrac{x}{2} - 1\right)}$
> 21. $16 = \dfrac{t^2}{9}$
> 22. $\sqrt{\left(\dfrac{y+2}{y-2}\right)} = \dfrac{1}{2}$
> 23. $\dfrac{6}{a} = \dfrac{2a}{3}$
> 24. $\dfrac{11}{2} = 5 + \dfrac{8}{x^2}$

11.3 Practical problems involving simple equations

There are many practical situations in engineering where solving equations is needed. Here are some worked examples to demonstrate typical practical situations.

Problem 17. Applying the principle of moments to a beam results in the equation

$$F \times 3 = (7.5 - F) \times 2$$

where F is the force in newtons. Determine the value of F

Removing brackets gives $\qquad 3F = 15 - 2F$

Rearranging gives $\qquad 3F + 2F = 15$

i.e. $\qquad\qquad\qquad 5F = 15$

Dividing both sides by 5 gives $\quad \dfrac{5F}{5} = \dfrac{15}{5}$

from which, force, $F = \mathbf{3N}$

Problem 18. A copper wire has a length L of 1.5 km, a resistance R of 5Ω and a resistivity ρ of $17.2 \times 10^{-6}\,\Omega$ mm. Find the cross-sectional area, a, of the wire, given that $R = \dfrac{\rho L}{a}$

Since $R = \dfrac{\rho L}{a}$ then

$$5\Omega = \frac{(17.2 \times 10^{-6}\,\Omega\,\mathrm{mm})(1500 \times 10^3\,\mathrm{mm})}{a}.$$

From the units given, a is measured in mm^2.

Thus, $\quad 5a = 17.2 \times 10^{-6} \times 1500 \times 10^3$

and $\qquad a = \dfrac{17.2 \times 10^{-6} \times 1500 \times 10^3}{5}$

$$= \frac{17.2 \times 1500 \times 10^3}{10^6 \times 5} = \frac{17.2 \times 15}{10 \times 5} = 5.16$$

Hence, the cross-sectional area of the wire is **5.16 mm^2**.

Problem 19. $PV = mRT$ is the characteristic gas equation. Find the value of gas constant R when pressure $P = 3 \times 10^6$ Pa, volume $V = 0.90\,\mathrm{m}^3$, mass $m = 2.81$ kg and temperature $T = 231$ K

Dividing both sides of $PV = mRT$ by mT gives

$$\frac{PV}{mT} = \frac{mRT}{mT}$$

Cancelling gives $\qquad \dfrac{PV}{mT} = R$

Substituting values gives $\qquad R = \dfrac{\left(3 \times 10^6\right)\left(0.90\right)}{\left(2.81\right)\left(231\right)}$

Using a calculator, **gas constant**, $R = \mathbf{4160\,J/(kg\,K)}$, correct to 4 significant figures.

Problem 20. A rectangular box with square ends has its length 15 cm greater than its breadth and the total length of its edges is 2.04 m. Find the width of the box and its volume

Let x cm = width = height of box. Then the length of the box is $(x + 15)$ cm, as shown in Fig. 11.1.

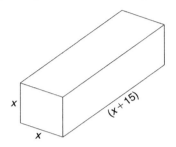

Figure 11.1

The length of the edges of the box is $2(4x) + 4(x + 15)$ cm, which equals 2.04 m or 204 cm.

Hence, $\qquad\qquad 204 = 2(4x) + 4(x + 15)$

$$204 = 8x + 4x + 60$$

$$204 - 60 = 12x$$

i.e. $\qquad\qquad 144 = 12x$

and $\qquad\qquad x = 12\,\mathrm{cm}$

Hence, **the width of the box is 12 cm**.

Volume of box $= \text{length} \times \text{width} \times \text{height}$

$$= (x + 15)(x)(x) = (12 + 15)(12)(12)$$

$$= (27)(12)(12)$$

$$= \mathbf{3888\,cm^3}$$

Problem 21. The temperature coefficient of resistance α may be calculated from the formula $R_t = R_0(1 + \alpha t)$. Find α, given $R_t = 0.928$, $R_0 = 0.80$ and $t = 40$

Since $R_t = R_0(1 + \alpha t)$, then

$$0.928 = 0.80[1 + \alpha(40)]$$

$$0.928 = 0.80 + (0.8)(\alpha)(40)$$

$$0.928 - 0.80 = 32\alpha$$

$$0.128 = 32\alpha$$

Hence, $$\alpha = \frac{0.128}{32} = \mathbf{0.004}$$

Problem 22. The distance s metres travelled in time t seconds is given by the formula $s = ut + \frac{1}{2}at^2$, where u is the initial velocity in m/s and a is the acceleration in m/s^2. Find the acceleration of the body if it travels 168 m in 6 s, with an initial velocity of 10 m/s

$$s = ut + \frac{1}{2}at^2, \text{ and } s = 168, u = 10 \text{ and } t = 6$$

Hence, $$168 = (10)(6) + \frac{1}{2}a(6)^2$$

$$168 = 60 + 18a$$

$$168 - 60 = 18a$$

$$108 = 18a$$

$$a = \frac{108}{18} = 6$$

Hence, **the acceleration of the body is 6 m/s^2**

Problem 23. When three resistors in an electrical circuit are connected in parallel the total resistance R_T is given by $\frac{1}{R_T} = \frac{1}{R_1} + \frac{1}{R_2} + \frac{1}{R_3}$. Find the total resistance when $R_1 = 5\,\Omega, R_2 = 10\,\Omega$ and $R_3 = 30\,\Omega$

$$\frac{1}{R_T} = \frac{1}{5} + \frac{1}{10} + \frac{1}{30} = \frac{6+3+1}{30} = \frac{10}{30} = \frac{1}{3}$$

Taking the reciprocal of both sides gives $\mathbf{R_T = 3\,\Omega}$

Alternatively, if $\frac{1}{R_T} = \frac{1}{5} + \frac{1}{10} + \frac{1}{30}$, the LCM of the denominators is $30R_T$

Hence, $$30R_T\left(\frac{1}{R_T}\right) = 30R_T\left(\frac{1}{5}\right)$$
$$+ 30R_T\left(\frac{1}{10}\right) + 30R_T\left(\frac{1}{30}\right)$$

Cancelling gives $30 = 6R_T + 3R_T + R_T$

i.e. $30 = 10R_T$

and $$R_T = \frac{30}{10} = \mathbf{3\,\Omega}, \text{ as above.}$$

Now try the following Practice Exercise

Practice Exercise 58   Practical problems involving simple equations (answers on page 448)

1. A formula used for calculating resistance of a cable is $R = \frac{\rho L}{a}$. Given $R = 1.25, L = 2500$ and $a = 2 \times 10^{-4}$, find the value of ρ.

2. Force F newtons is given by $F = ma$, where m is the mass in kilograms and a is the acceleration in metres per second squared. Find the acceleration when a force of 4 kN is applied to a mass of 500 kg.

3. $PV = mRT$ is the characteristic gas equation. Find the value of m when $P = 100 \times 10^3, V = 3.00, R = 288$ and $T = 300$

4. When three resistors R_1, R_2 and R_3 are connected in parallel, the total resistance R_T is determined from $\frac{1}{R_T} = \frac{1}{R_1} + \frac{1}{R_2} + \frac{1}{R_3}$
    (a) Find the total resistance when $R_1 = 3\,\Omega, R_2 = 6\,\Omega$ and $R_3 = 18\,\Omega$.
    (b) Find the value of R_3 given that $R_T = 3\,\Omega, R_1 = 5\,\Omega$ and $R_2 = 10\,\Omega$.

5. Six digital camera batteries and three camcorder batteries cost £96. If a camcorder battery costs £5 more than a digital camera battery, find the cost of each.

6. Ohm's law may be represented by $I = V/R$, where I is the current in amperes, V is the voltage in volts and R is the resistance in ohms. A soldering iron takes a current of 0.30 A from a 240 V supply. Find the resistance of the element.

7. The distance, s, travelled in time t seconds is given by the formula $s = ut + \frac{1}{2}at^2$ where u is the initial velocity in m/s and a is the acceleration in m/s^2. Calculate the acceleration of the body if it travels 165 m in 3 s, with an initial velocity of 10 m/s.

8. The stress, σ pascals, acting on the reinforcing rod in a concrete column is given

in the following equation:
$$500 \times 10^{-6}\sigma + 2.67 \times 10^5 = 3.55 \times 10^5$$
Find the value of the stress in MPa.

Here are some further worked examples on solving simple equations in practical situations.

Problem 24. The extension x m of an aluminium tie bar of length l m and cross-sectional area A m^2 when carrying a load of F newtons is given by the modulus of elasticity $E = Fl/Ax$. Find the extension of the tie bar (in mm) if $E = 70 \times 10^9$ N/m^2, $F = 20 \times 10^6$ N, $A = 0.1$m^2 and $l = 1.4$ m

$E = Fl/Ax$, hence

$$70 \times 10^9 \frac{N}{m^2} = \frac{(20 \times 10^6 \text{N})(1.4\text{m})}{(0.1\text{m}^2)(x)}$$

(the unit of x is thus metres)

$$70 \times 10^9 \times 0.1 \times x = 20 \times 10^6 \times 1.4$$

$$x = \frac{20 \times 10^6 \times 1.4}{70 \times 10^9 \times 0.1}$$

Cancelling gives $\quad x = \frac{2 \times 1.4}{7 \times 100}$ m

$$= \frac{2 \times 1.4}{7 \times 100} \times 1000\text{mm}$$

$$= 4\text{mm}$$

Hence, **the extension of the tie bar, $x = 4$ mm**.

Problem 25. Power in a d.c. circuit is given by $P = \frac{V^2}{R}$ where V is the supply voltage and R is the circuit resistance. Find the supply voltage if the circuit resistance is 1.25 Ω and the power measured is 320 W

Since $P = \frac{V^2}{R}$, then $320 = \frac{V^2}{1.25}$

$$(320)(1.25) = V^2$$

i.e. $\quad V^2 = 400$

Supply voltage, $\quad V = \sqrt{400} = \pm 20$ V

Problem 26. A painter is paid £6.30 per hour for a basic 36 hour week and overtime is paid at one and a third times this rate. Determine how many hours the painter has to work in a week to earn £319.20

Basic rate per hour $= £6.30$ and overtime rate per hour $= 1\frac{1}{3} \times £6.30 = £8.40$

Let the number of overtime hours worked $= x$

Then, $\quad (36)(6.30) + (x)(8.40) = 319.20$

$$226.80 + 8.40x = 319.20$$

$$8.40x = 319.20 - 226.80 = 92.40$$

$$x = \frac{92.40}{8.40} = 11$$

Thus, 11 hours overtime would have to be worked to earn £319.20 per week. Hence, **the total number of hours worked** is $36 + 11$, i.e. **47 hours**.

Problem 27. A formula relating initial and final states of pressures, P_1 and P_2, volumes, V_1 and V_2, and absolute temperatures, T_1 and T_2, of an ideal gas is $\frac{P_1 V_1}{T_1} = \frac{P_2 V_2}{T_2}$. Find the value of P_2 given $P_1 = 100 \times 10^3, V_1 = 1.0, V_2 = 0.266, T_1 = 423$ and $T_2 = 293$

Since $\quad \frac{P_1 V_1}{T_1} = \frac{P_2 V_2}{T_2}$

then $\quad \frac{(100 \times 10^3)(1.0)}{423} = \frac{P_2(0.266)}{293}$

Cross-multiplying gives

$$(100 \times 10^3)(1.0)(293) = P_2(0.266)(423)$$

$$P_2 = \frac{(100 \times 10^3)(1.0)(293)}{(0.266)(423)}$$

Hence, $\quad$ **$P_2 = 260 \times 10^3$ or 2.6×10^5**

Problem 28. The stress, f, in a material of a thick cylinder can be obtained from $\frac{D}{d} = \sqrt{\left(\frac{f+p}{f-p}\right)}$. Calculate the stress, given that $D = 21.5, d = 10.75$ and $p = 1800$

Since $\frac{D}{d} = \sqrt{\left(\frac{f+p}{f-p}\right)}$ then $\frac{21.5}{10.75} = \sqrt{\left(\frac{f+1800}{f-1800}\right)}$

i.e. $$2 = \sqrt{\left(\frac{f+1800}{f-1800}\right)}$$

Squaring both sides gives
$$4 = \frac{f + 1800}{f - 1800}$$

Cross-multiplying gives

$$4(f - 1800) = f + 1800$$

$$4f - 7200 = f + 1800$$

$$4f - f = 1800 + 7200$$

$$3f = 9000$$

$$f = \frac{9000}{3} = 3000$$

Hence,   **stress, $f = 3000$**

▶ **Problem 29.**   The reaction moment M of a cantilever carrying three point loads is given by:

$$M = 3.5x + 2.0(x - 1.8) + 4.2(x - 2.6)$$

where x is the length of the cantilever in metres. If $M = 55.32$ kN m, calculate the value of x.

If $M = 3.5x + 2.0(x - 1.8) + 4.2(x - 2.6)$
and $M = 55.32$
then $55.32 = 3.5x + 2.0(x - 1.8) + 4.2(x - 2.6)$
i.e. $55.32 = 3.5x + 2.0x - 3.6 + 4.2x - 10.92$
and $55.32 + 3.6 + 10.92 = 9.7x$
i.e. $69.84 = 9.7x$
from which, **the length of the cantilever,**

$$x = \frac{69.84}{9.7} = \textbf{7.2 m}$$

Now try the following Practice Exercise

Practice Exercise 59   Practical problems involving simple equations (answers on page 448)

▶ 1.   A rectangle has a length of 20 cm and a width b cm. When its width is reduced by 4 cm its area becomes 160 cm². Find the original width and area of the rectangle.

▶ 2.   Given $R_2 = R_1(1 + \alpha t)$, find α given $R_1 = 5.0, R_2 = 6.03$ and $t = 51.5$

▶ 3.   If $v^2 = u^2 + 2as$, find u given $v = 24$, $a = -40$ and $s = 4.05$

▶ 4.   The relationship between the temperature on a Fahrenheit scale and that on a Celsius scale is given by $F = \frac{9}{5}C + 32$. Express $113°$F in degrees Celsius.

▶ 5.   If $t = 2\pi\sqrt{\dfrac{w}{Sg}}$, find the value of S given $w = 1.219, g = 9.81$ and $t = 0.3132$

▶ 6.   Two joiners and five mates earn £1824 between them for a particular job. If a joiner earns £72 more than a mate, calculate the earnings for a joiner and for a mate.

▶ 7.   An alloy contains 60% by weight of copper, the remainder being zinc. How much copper must be mixed with 50 kg of this alloy to give an alloy containing 75% copper?

▶ 8.   A rectangular laboratory has a length equal to one and a half times its width and a perimeter of 40 m. Find its length and width.

▶ 9.   Applying the principle of moments to a beam results in the following equation:

$$F \times 3 = (5 - F) \times 7$$

where F is the force in newtons. Determine the value of F.

Practice Exercise 60   Multiple-choice questions on solving simple equations (answers on page 448)

Each question has only one correct answer

1.   Solving the equation $-\dfrac{x}{3} = 3$ for x gives:
   (a) 18   (b) 12   (c) -9 (d) 8

2.   Solving the equation $-8t = 4.4$ for t gives:
   (a) 0.5   (b) -0.55 (c) 5   (d) 0.55

3.   Solving the equation $y - \dfrac{2}{3} = 7$ for y gives:
   (a) $7\dfrac{2}{3}$ (b) $10\dfrac{1}{2}$ (c) $6\dfrac{1}{3}$ (d) $-4\dfrac{2}{3}$

4. Solving the equation $\dfrac{x}{9} = \dfrac{7}{72}$ for x gives:

   (a) $1\dfrac{1}{7}$ (b) $9\dfrac{7}{72}$ (c) $\dfrac{7}{8}$ (d) $8\dfrac{65}{72}$

5. The solution of the equation $2x + 1 = x + 3$ is:

   (a) 1 (b) 2 (c) 3 (d) 4

6. Solving the equation $3 - 2x = 3x + 8$ for x gives:

   (a) -1 (b) 5 (c) 1 (d) -5

7. Given $4 + \dfrac{x}{2} = \dfrac{7}{2}$, the value of x is:

   (a) 2 (b) 1 (c) -2 (d) -1

8. The value of x in the equation $5x - 27 + 3x = 4 + 9 - 2x$ is:

   (a) -4 (b) 5 (c) 4 (d) 6

9. Solving the equation $1 + 3x = 2(x - 1)$ gives:

   (a) $x = -1$ (b) $x = -2$
   (c) $x = 1$ (d) $x = -3$

10. Solving the equation $17 + 19(x + y) = 19(y - x) - 21$ for x gives:

    (a) -1 (b) -2 (c) -3 (d) -4

For fully worked solutions to each of the problems in Practice Exercises 56 to 59 in this chapter, go to the website:
www.routledge.com/cw/bird

This assignment covers the material contained in Chapters 9–11. *The marks available are shown in brackets at the end of each question.*

1. Evaluate $3pqr^3 - 2p^2qr + pqr$ when $p = \dfrac{1}{2}$, $q = -2$ and $r = 1$ (3)

   In Problems 2 to 7, simplify the expressions.

2. $\dfrac{9p^2qr^3}{3pq^2r}$ (3)

3. $2(3x - 2y) - (4y - 3x)$ (3)

4. $(x - 2y)(2x + y)$ (3)

5. $p^2q^{-3}r^4 \times pq^2r^{-3}$ (3)

6. $(3a - 2b)^2$ (3)

7. $\dfrac{a^4b^2c}{ab^3c^2}$ (3)

8. Factorise

   (a) $2x^2y^3 - 10xy^2$

   (b) $21ab^2c^3 - 7a^2bc^2 + 28a^3bc^4$ (5)

9. Factorise and simplify

   $\dfrac{2x^2y + 6xy^2}{x + 3y} - \dfrac{x^3y^2}{x^2y}$ (5)

10. Remove the brackets and simplify
    $10a - [3(2a - b) - 4(b - a) + 5b]$ (4)

11. Simplify $x \div 5x - x + (2x - 3x)x$ (4)

12. Simplify $3a + 2a \times 5a + 4a \div 2a - 6a$ (4)

13. Solve the equations

    (a) $3a = 39$

    (b) $2x - 4 = 9$ (3)

14. Solve the equations

    (a) $\dfrac{4}{9}y = 8$

    (b) $6x - 1 = 4x + 5$ (4)

15. Solve the equation
    $5(t - 2) - 3(4 - t) = 2(t + 3) - 40$ (4)

16. Solve the equations

    (a) $\dfrac{3}{2x + 1} = \dfrac{1}{4x - 3}$

    (b) $2x^2 = 162$ (7)

17. Kinetic energy is given by the formula, $E_k = \dfrac{1}{2}mv^2$ joules, where m is the mass in kilograms and v is the velocity in metres per second. Evaluate the velocity when $E_k = 576 \times 10^{-3}$ J and the mass is 5 kg. (4)

18. An approximate relationship between the number of teeth T on a milling cutter, the diameter of the cutter D and the depth of cut d is given by $T = \dfrac{12.5D}{D + 4d}$. Evaluate d when $T = 10$ and $D = 32$ (5)

19. The modulus of elasticity E is given by the formula $E = \dfrac{FL}{xA}$ where F is force in newtons, L is the length in metres, x is the extension in metres and A the cross-sectional area in square metres. Evaluate A, in square centimetres, when $E = 80 \times 10^9$ N/m², $x = 2$ mm, $F = 100 \times 10^3$ N and $L = 2.0$ m. (5)

Chapter 12

Transposing formulae

Why it is important to understand: Transposing formulae

As was mentioned in the last chapter, formulae are used frequently in almost all aspects of engineering in order to relate a physical quantity to one or more others. Many well-known physical laws are described using formulae – for example, Ohm's law, $V = I \times R$, or Newton's second law of motion, $F = m \times a$. In an everyday situation, imagine you buy five identical items for £20. How much did each item cost? If you divide £20 by 5 to get an answer of £4, you are actually applying transposition of a formula. Transposing formulae is a basic skill required in all aspects of engineering. The ability to transpose formulae is yet another stepping stone on the way to having confidence to handle engineering mathematics.

At the end of this chapter you should be able to:

- define 'subject of the formula'
- transpose equations whose terms are connected by plus and/or minus signs
- transpose equations that involve fractions
- transpose equations that contain a root or power
- transpose equations in which the subject appears in more than one term

12.1 Introduction

In the formula $I = \dfrac{V}{R}$, I is called the **subject of the formula**.

Similarly, in the formula $y = mx + c$, y is the subject of the formula.

When a symbol other than the subject is required to be the subject, the formula needs to be rearranged to make a new subject. This rearranging process is called **transposing the formula** or **transposition**.

For example, in the above formulae,

$$\text{if } I = \frac{V}{R} \text{ then } V = IR$$

$$\text{and if } y = mx + c \text{ then } x = \frac{y - c}{m}$$

How did we arrive at these transpositions? This is the purpose of this chapter – to show how to transpose formulae. A great many equations occur in engineering and it is essential that we can transpose them when needed.

12.2 Transposing formulae

There are no new rules for transposing formulae. The same rules as were used for simple equations in Chapter 11 are used; i.e. **the balance of an equation must be maintained**: whatever is done to one side of an equation must be done to the other.

It is best that you cover simple equations before trying this chapter.

Here are some worked examples to help understanding of transposing formulae.

Problem 1. Transpose $p = q + r + s$ to make r the subject

The object is to obtain r on its own on the LHS of the equation. Changing the equation around so that r is on the LHS gives

$$q + r + s = p \qquad (1)$$

From Chapter 11 on simple equations, a term can be moved from one side of an equation to the other side as long as the sign is changed.
Rearranging gives $r = p - q - s$
Mathematically, we have subtracted $q + s$ from both sides of equation (1).

Problem 2. If $a + b = w - x + y$, express x as the subject

As stated in Problem 1, a term can be moved from one side of an equation to the other side but with a change of sign.
Hence, rearranging gives $x = w + y - a - b$

Problem 3. Transpose $v = f\lambda$ to make λ the subject

$v = f\lambda$ relates velocity v, frequency f and wavelength λ

Rearranging gives $\qquad f\lambda = v$

Dividing both sides by f gives $\dfrac{f\lambda}{f} = \dfrac{v}{f}$

Cancelling gives $\qquad \lambda = \dfrac{v}{f}$

Problem 4. When a body falls freely through a height h, the velocity v is given by $v^2 = 2gh$. Express this formula with h as the subject

Rearranging gives $\qquad 2gh = v^2$

Dividing both sides by $2g$ gives $\dfrac{2gh}{2g} = \dfrac{v^2}{2g}$

Cancelling gives $\qquad h = \dfrac{v^2}{2g}$

Problem 5. If $I = \dfrac{V}{R}$, rearrange to make V the subject

$I = \dfrac{V}{R}$ is Ohm's law, where I is the current, V is the voltage and R is the resistance.

Rearranging gives $\qquad \dfrac{V}{R} = I$

Multiplying both sides by R gives $\quad R\left(\dfrac{V}{R}\right) = R(I)$

Cancelling gives $\qquad V = IR$

Problem 6. Transpose $a = \dfrac{F}{m}$ for m

$a = \dfrac{F}{m}$ relates acceleration a, force F and mass m.

Rearranging gives $\qquad \dfrac{F}{m} = a$

Multiplying both sides by m gives $\quad m\left(\dfrac{F}{m}\right) = m(a)$

Cancelling gives $\qquad F = ma$

Rearranging gives $\qquad ma = F$

Dividing both sides by a gives $\qquad \dfrac{ma}{a} = \dfrac{F}{a}$

i.e. $\qquad m = \dfrac{F}{a}$

Problem 7. Rearrange the formula $R = \dfrac{\rho L}{A}$ to make (a) A the subject and (b) L the subject

$R = \dfrac{\rho L}{A}$ relates resistance R of a conductor, resistivity ρ, conductor length L and conductor cross-sectional area A.

(a) Rearranging gives $\qquad \dfrac{\rho L}{A} = R$

Multiplying both sides by A gives

$$A\left(\dfrac{\rho L}{A}\right) = A(R)$$

Cancelling gives $\qquad \rho L = AR$

Rearranging gives $\qquad AR = \rho l$

Dividing both sides by R gives $\qquad \dfrac{AR}{R} = \dfrac{\rho L}{R}$

Cancelling gives $\qquad A = \dfrac{\rho L}{R}$

(b) Multiplying both sides of $\dfrac{\rho L}{A} = R$ by A gives

$$\rho L = AR$$

Dividing both sides by ρ gives $\quad \dfrac{\rho L}{\rho} = \dfrac{AR}{\rho}$

Cancelling gives $\quad L = \dfrac{AR}{\rho}$

Problem 8. Transpose $y = mx + c$ to make m the subject

$y = mx + c$ is the equation of a straight line graph, where y is the vertical axis variable, x is the horizontal axis variable, m is the gradient of the graph and c is the y-axis intercept.

Subtracting c from both sides gives $\quad y - c = mx$

or $\quad mx = y - c$

Dividing both sides by x gives $\quad m = \dfrac{y - c}{x}$

Now try the following Practice Exercise

Practice Exercise 61   Transposing formulae (answers on page 448)

Make the symbol indicated the subject of each of the formulae shown and express each in its simplest form.

1. $a + b = c - d - e$ $\qquad$ (d)
2. $y = 7x$ $\qquad$ (x)
3. $pv = c$ $\qquad$ (v)
4. $v = u + at$ $\qquad$ (a)
5. $V = IR$ $\qquad$ (R)
6. $x + 3y = t$ $\qquad$ (y)
7. $c = 2\pi r$ $\qquad$ (r)
8. $y = mx + c$ $\qquad$ (x)
9. $I = PRT$ $\qquad$ (T)
10. $Q = mc\Delta T$ $\qquad$ (c)
11. $X_L = 2\pi f L$ $\qquad$ (L)
12. $I = \dfrac{E}{R}$ $\qquad$ (R)
13. $y = \dfrac{x}{a} + 3$ $\qquad$ (x)
14. $F = \dfrac{9}{5}C + 32$ $\qquad$ (C)
15. $X_C = \dfrac{1}{2\pi f C}$ $\qquad$ (f)
16. $pV = mRT$ $\qquad$ (R)

12.3   Further transposing of formulae

Here are some more transposition examples to help us further understand how more difficult formulae are transposed.

Problem 9. Transpose the formula $v = u + \dfrac{Ft}{m}$ to make F the subject

$v = u + \dfrac{Ft}{m}$ relates final velocity v, initial velocity u, force F, mass m and time t. $\left(\dfrac{F}{m} \text{ is acceleration } a.\right)$

Rearranging gives $\quad u + \dfrac{Ft}{m} = v$

and $\quad \dfrac{Ft}{m} = v - u$

Multiplying each side by m gives

$$m\left(\dfrac{Ft}{m}\right) = m(v - u)$$

Cancelling gives $\quad Ft = m(v - u)$

Dividing both sides by t gives $\quad \dfrac{Ft}{t} = \dfrac{m(v - u)}{t}$

Cancelling gives $\quad F = \dfrac{m(v - u)}{t}$ or $F = \dfrac{m}{t}(v - u)$

This shows two ways of expressing the answer. There is often more than one way of expressing a transposed answer. In this case, these equations for F are equivalent; neither one is more correct than the other.

Problem 10. The final length L_2 of a piece of wire heated through $\theta°C$ is given by the formula $L_2 = L_1(1 + \alpha\theta)$ where L_1 is the original length. Make the coefficient of expansion α the subject

Rearranging gives $\quad L_1(1 + \alpha\theta) = L_2$

Removing the bracket gives $\quad L_1 + L_1\alpha\theta = L_2$

Rearranging gives $\quad L_1\alpha\theta = L_2 - L_1$

Dividing both sides by $L_1\theta$ gives $\dfrac{L_1\alpha\theta}{L_1\theta} = \dfrac{L_2 - L_1}{L_1\theta}$

Cancelling gives $\alpha = \dfrac{L_2 - L_1}{L_1\theta}$

An alternative method of transposing $L_2 = L_1\left(1 + \alpha\theta\right)$ for α is:

Dividing both sides by L_1 gives $\dfrac{L_2}{L_1} = 1 + \alpha\theta$

Subtracting 1 from both sides gives $\dfrac{L_2}{L_1} - 1 = \alpha\theta$

or $\alpha\theta = \dfrac{L_2}{L_1} - 1$

Dividing both sides by θ gives $\alpha = \dfrac{\dfrac{L_2}{L_1} - 1}{\theta}$

The two answers $\alpha = \dfrac{L_2 - L_1}{L_1\theta}$ and $\alpha = \dfrac{\dfrac{L_2}{L_1} - 1}{\theta}$ look quite different. They are, however, equivalent. The first answer looks tidier but is no more correct than the second answer.

> **Problem 11.** A formula for the distance s moved by a body is given by $s = \dfrac{1}{2}(v + u)t$. Rearrange the formula to make u the subject

Rearranging gives $\dfrac{1}{2}(v + u)t = s$

Multiplying both sides by 2 gives $(v + u)t = 2s$

Dividing both sides by t gives $\dfrac{(v + u)t}{t} = \dfrac{2s}{t}$

Cancelling gives $v + u = \dfrac{2s}{t}$

Rearranging gives $u = \dfrac{2s}{t} - v$ or $u = \dfrac{2s - vt}{t}$

> **Problem 12.** A formula for kinetic energy is $k = \dfrac{1}{2}mv^2$. Transpose the formula to make v the subject

Rearranging gives $\dfrac{1}{2}mv^2 = k$

Whenever the prospective new subject is a squared term, that term is isolated on the LHS and then the square root of both sides of the equation is taken.

Multiplying both sides by 2 gives $mv^2 = 2k$

Dividing both sides by m gives $\dfrac{mv^2}{m} = \dfrac{2k}{m}$

Cancelling gives $v^2 = \dfrac{2k}{m}$

Taking the square root of both sides gives

$$\sqrt{v^2} = \sqrt{\left(\dfrac{2k}{m}\right)}$$

i.e. $$v = \sqrt{\left(\dfrac{2k}{m}\right)}$$

> **Problem 13.** In a right-angled triangle having sides x, y and hypotenuse z, Pythagoras' theorem* states $z^2 = x^2 + y^2$. Transpose the formula to find x

*Who was **Pythagoras**? – **Pythagoras of Samos** (born about 570 BC and died about 495 BC) was an Ionian Greek philosopher and mathematician. He is best known for the Pythagorean theorem, which states that in a right-angled triangle $a^2 + b^2 = c^2$. To find out more go to **www.routledge.com/cw/bird**

Rearranging gives $\qquad x^2 + y^2 = z^2$

and $\qquad\qquad\qquad x^2 = z^2 - y^2$

Taking the square root of both sides gives

$$x = \sqrt{z^2 - y^2}$$

Problem 14. Transpose $y = \dfrac{ML^2}{8EI}$ to make L the subject

Multiplying both sides by $8EI$ gives $\quad 8EIy = ML^2$

Dividing both sides by M gives $\qquad \dfrac{8EIy}{M} = L^2$

or $\qquad\qquad\qquad\qquad L^2 = \dfrac{8EIy}{M}$

Taking the square root of both sides gives

$$\sqrt{L^2} = \sqrt{\dfrac{8EIy}{M}}$$

i.e. $\qquad\qquad L = \sqrt{\dfrac{8EIy}{M}}$

Problem 15. Given $t = 2\pi\sqrt{\dfrac{l}{g}}$, find g in terms of t, l and π

Whenever the prospective new subject is within a square root sign, it is best to isolate that term on the LHS and then to square both sides of the equation.

Rearranging gives $\qquad\qquad 2\pi\sqrt{\dfrac{l}{g}} = t$

Dividing both sides by 2π gives $\qquad \sqrt{\dfrac{l}{g}} = \dfrac{t}{2\pi}$

Squaring both sides gives $\qquad \dfrac{l}{g} = \left(\dfrac{t}{2\pi}\right)^2 = \dfrac{t^2}{4\pi^2}$

Cross-multiplying, (i.e. multiplying

each term by $4\pi^2 g$), gives $\qquad 4\pi^2 l = gt^2$

or $\qquad\qquad\qquad\qquad gt^2 = 4\pi^2 l$

Dividing both sides by t^2 gives $\qquad \dfrac{gt^2}{t^2} = \dfrac{4\pi^2 l}{t^2}$

Cancelling gives $\qquad\qquad g = \dfrac{4\pi^2 l}{t^2}$

Problem 16. The impedance Z of an a.c. circuit is given by $Z = \sqrt{R^2 + X^2}$ where R is the resistance. Make the reactance, X, the subject

Rearranging gives $\qquad \sqrt{R^2 + X^2} = Z$

Squaring both sides gives $\qquad R^2 + X^2 = Z^2$

Rearranging gives $\qquad\qquad X^2 = Z^2 - R^2$

Taking the square root of both sides gives

$$X = \sqrt{Z^2 - R^2}$$

Problem 17. The volume V of a hemisphere of radius r is given by $V = \dfrac{2}{3}\pi r^3$. (a) Find r in terms of V. (b) Evaluate the radius when $V = 32\,\text{cm}^3$

(a)   Rearranging gives $\qquad\qquad \dfrac{2}{3}\pi r^3 = V$

Multiplying both sides by 3 gives $\quad 2\pi r^3 = 3V$

Dividing both sides by 2π gives $\quad \dfrac{2\pi r^3}{2\pi} = \dfrac{3V}{2\pi}$

Cancelling gives $\qquad\qquad\qquad r^3 = \dfrac{3V}{2\pi}$

Taking the cube root of both sides gives

$$\sqrt[3]{r^3} = \sqrt[3]{\left(\dfrac{3V}{2\pi}\right)}$$

i.e. $\qquad\qquad r = \sqrt[3]{\left(\dfrac{3V}{2\pi}\right)}$

(b)   When $V = 32\,\text{cm}^3$,

$$\textbf{radius } r = \sqrt[3]{\left(\dfrac{3V}{2\pi}\right)} = \sqrt[3]{\left(\dfrac{3 \times 32}{2\pi}\right)} = \textbf{2.48 cm}.$$

Now try the following Practice Exercise

Practice Exercise 62   Further transposing formulae (answers on page 448)

Make the symbol indicated the subject of each of the formulae shown in Problems 1 to 15 and express each in its simplest form.

1. $S = \dfrac{a}{1-r}$ $\qquad (r)$

2. $y = \dfrac{\lambda(x-d)}{d}$ $\qquad (x)$

3. $A = \dfrac{3(F-f)}{L}$ (f)

4. $y = \dfrac{AB^2}{5CD}$ (D)

5. $R = R_0(1 + \alpha t)$ (t)

6. $\dfrac{1}{R} = \dfrac{1}{R_1} + \dfrac{1}{R_2}$ (R_2)

7. $I = \dfrac{E - e}{R + r}$ (R)

8. $\dfrac{P_1 V_1}{T_1} = \dfrac{P_2 V_2}{T_2}$ (V_2)

9. $y = 4ab^2 c^2$ (b)

10. $\dfrac{a^2}{x^2} + \dfrac{b^2}{y^2} = 1$ (x)

11. $t = 2\pi\sqrt{\dfrac{L}{g}}$ (L)

12. $v^2 = u^2 + 2as$ (u)

13. $A = \dfrac{\pi R^2 \theta}{360}$ (R)

14. $\dfrac{P_1 V_1}{T_1} = \dfrac{P_2 V_2}{T_2}$ (T_2)

15. $N = \sqrt{\left(\dfrac{a+x}{y}\right)}$ (a)

16.   Transpose $Z = \sqrt{R^2 + (2\pi fL)^2}$ for L and evaluate L when $Z = 27.82, R = 11.76$ and $f = 50$.

17.   The lift force, L, on an aircraft is given by: $L = \dfrac{1}{2}\rho v^2 a c$ where ρ is the density, v is the velocity, a is the area and c is the lift coefficient. Transpose the equation to make the velocity the subject.

18.   The angular deflection θ of a beam of electrons due to a magnetic field is given by: $\theta = k\left(\dfrac{HL}{V^{\frac{1}{2}}}\right)$ Transpose the equation for V.

19.   A radar has a wavelength, λ, of 40 mm. The radar emits and receives electromagnetic waves which have a speed, v, of 300×10^8 m/s. Given that $v = f\lambda$, where f is the frequency in hertz, calculate the frequency.

20.   The velocity of a particle moving with simple harmonic motion is: $V = W\sqrt{(a^2 - x^2)}$ Rearrange the formula to make x the subject.

12.4   More difficult transposing of formulae

Here are some more transposition examples to help us further understand how more difficult formulae are transposed.

Problem 18.   (a) Transpose $S = \sqrt{\dfrac{3d(L-d)}{8}}$ to make L the subject. (b) Evaluate L when $d = 1.65$ and $S = 0.82$

The formula $S = \sqrt{\dfrac{3d(L-d)}{8}}$ represents the sag S at the centre of a wire.

(a)   Squaring both sides gives $S^2 = \dfrac{3d(L-d)}{8}$

Multiplying both sides by 8 gives

$$8S^2 = 3d(L-d)$$

Dividing both sides by $3d$ gives $\dfrac{8S^2}{3d} = L - d$

Rearranging gives $L = d + \dfrac{8S^2}{3d}$

(b)   When $d = 1.65$ and $S = 0.82$,

$$L = d + \dfrac{8S^2}{3d} = 1.65 + \dfrac{8 \times 0.82^2}{3 \times 1.65} = \mathbf{2.737}$$

Problem 19.   Transpose the formula $p = \dfrac{a^2 x + a^2 y}{r}$ to make a the subject

Rearranging gives $\dfrac{a^2 x + a^2 y}{r} = p$

Multiplying both sides by r gives

$$a^2 x + a^2 y = rp$$

Factorising the LHS gives $a^2(x + y) = rp$

Dividing both sides by $(x + y)$ gives

$$\frac{a^2(x + y)}{(x + y)} = \frac{rp}{(x + y)}$$

Cancelling gives $a^2 = \dfrac{rp}{(x + y)}$

Taking the square root of both sides gives

$$a = \sqrt{\left(\frac{rp}{x + y}\right)}$$

Whenever the letter required as the new subject occurs more than once in the original formula, after rearranging, **factorising** will always be needed.

Problem 20. Make b the subject of the formula $a = \dfrac{x-y}{\sqrt{bd+be}}$

Rearranging gives $\dfrac{x-y}{\sqrt{bd+be}} = a$

Multiplying both sides by $\sqrt{bd+be}$ gives

$$x - y = a\sqrt{bd+be}$$

or $$a\sqrt{bd+be} = x-y$$

Dividing both sides by a gives $\sqrt{bd+be} = \dfrac{x-y}{a}$

Squaring both sides gives $bd+be = \left(\dfrac{x-y}{a}\right)^2$

Factorising the LHS gives $b(d+e) = \left(\dfrac{x-y}{a}\right)^2$

Dividing both sides by $(d+e)$ gives

$$b = \dfrac{\left(\dfrac{x-y}{a}\right)^2}{(d+e)} \quad \text{or} \quad b = \dfrac{(x-y)^2}{a^2(d+e)}$$

Problem 21. If $a = \dfrac{b}{1+b}$, make b the subject of the formula

Rearranging gives $\dfrac{b}{1+b} = a$

Multiplying both sides by $(1+b)$ gives

$$b = a(1+b)$$

Removing the bracket gives $b = a + ab$

Rearranging to obtain terms in b on the LHS gives

$$b - ab = a$$

Factorising the LHS gives $b(1-a) = a$

Dividing both sides by $(1-a)$ gives $b = \dfrac{a}{1-a}$

Problem 22. Transpose the formula $V = \dfrac{Er}{R+r}$ to make r the subject

Rearranging gives $\dfrac{Er}{R+r} = V$

Multiplying both sides by $(R+r)$ gives

$$Er = V(R+r)$$

Removing the bracket gives $Er = VR + Vr$

Rearranging to obtain terms in r on the LHS gives

$$Er - Vr = VR$$

Factorising gives $r(E-V) = VR$

Dividing both sides by $(E-V)$ gives $r = \dfrac{VR}{E-V}$

Problem 23. Transpose the formula $y = \dfrac{pq^2}{r+q^2} - t$ to make q the subject

Rearranging gives $\dfrac{pq^2}{r+q^2} - t = y$

and $\dfrac{pq^2}{r+q^2} = y+t$

Multiplying both sides by $(r+q^2)$ gives

$$pq^2 = (r+q^2)(y+t)$$

Removing brackets gives $pq^2 = ry + rt + q^2y + q^2t$

Rearranging to obtain terms in q on the LHS gives

$$pq^2 - q^2y - q^2t = ry + rt$$

Factorising gives $q^2(p-y-t) = r(y+t)$

Dividing both sides by $(p-y-t)$ gives

$$q^2 = \dfrac{r(y+t)}{(p-y-t)}$$

Taking the square root of both sides gives

$$q = \sqrt{\left(\dfrac{r(y+t)}{p-y-t}\right)}$$

Problem 24. Given that $\dfrac{D}{d} = \sqrt{\left(\dfrac{f+p}{f-p}\right)}$ express p in terms of D, d and f

Rearranging gives $\sqrt{\left(\dfrac{f+p}{f-p}\right)} = \dfrac{D}{d}$

Squaring both sides gives $\left(\dfrac{f+p}{f-p}\right) = \dfrac{D^2}{d^2}$

Cross-multiplying, i.e. multiplying each term by $d^2(f-p)$, gives

$$d^2(f+p) = D^2(f-p)$$

Removing brackets gives $\quad d^2f + d^2p = D^2f - D^2p$

Rearranging, to obtain terms in p on the LHS gives

$$d^2p + D^2p = D^2f - d^2f$$

Factorising gives $\quad p(d^2 + D^2) = f(D^2 - d^2)$

Dividing both sides by $(d^2 + D^2)$ gives

$$p = \frac{f(D^2 - d^2)}{(d^2 + D^2)}$$

Now try the following Practice Exercise

Practice Exercise 63   Further transposing formulae (answers on page 448)

Make the symbol indicated the subject of each of the formulae shown in Problems 1 to 7 and express each in its simplest form.

1. $y = \dfrac{a^2m - a^2n}{x}$ $\quad$ (a)

2. $M = \pi(R^4 - r^4)$ $\quad$ (R)

3. $x + y = \dfrac{r}{3 + r}$ $\quad$ (r)

4. $m = \dfrac{\mu L}{L + rCR}$ $\quad$ (L)

5. $a^2 = \dfrac{b^2 - c^2}{b^2}$ $\quad$ (b)

6. $\dfrac{x}{y} = \dfrac{1 + r^2}{1 - r^2}$ $\quad$ (r)

7. $\dfrac{p}{q} = \sqrt{\left(\dfrac{a + 2b}{a - 2b}\right)}$ $\quad$ (b)

8. A formula for the focal length, f, of a convex lens is $\dfrac{1}{f} = \dfrac{1}{u} + \dfrac{1}{v}$. Transpose the formula to make v the subject and evaluate v when $f = 5$ and $u = 6$

9. The quantity of heat, Q, is given by the formula $Q = mc(t_2 - t_1)$. Make t_2 the subject of the formula and evaluate t_2 when $m = 10$, $t_1 = 15$, $c = 4$ and $Q = 1600$

10. The velocity, v, of water in a pipe appears in the formula $h = \dfrac{0.03Lv^2}{2dg}$. Express v as the subject of the formula and evaluate v when $h = 0.712, L = 150, d = 0.30$ and $g = 9.81$

11. The sag, S, at the centre of a wire is given by the formula $S = \sqrt{\left(\dfrac{3d(L - d)}{8}\right)}$. Make L the subject of the formula and evaluate L when $d = 1.75$ and $S = 0.80$

12. In an electrical alternating current circuit the impedance Z is given by
$$Z = \sqrt{\left\{R^2 + \left(\omega L - \dfrac{1}{\omega C}\right)^2\right\}}.$$
Transpose the formula to make C the subject and hence evaluate C when $Z = 130, R = 120, \omega = 314$ and $L = 0.32$

13. An approximate relationship between the number of teeth, T, on a milling cutter, the diameter of cutter, D, and the depth of cut, d, is given by $T = \dfrac{12.5D}{D + 4d}$. Determine the value of D when $T = 10$ and $d = 4$ mm.

14. Make λ, the wavelength of X-rays, the subject of the following formula: $\dfrac{\mu}{\rho} = \dfrac{CZ^4\sqrt{\lambda^5}\,n}{a}$

15. A simply supported beam of length L has a centrally applied load F and a uniformly distributed load of w per metre length of beam. The reaction at the beam support is given by:

$$R = \frac{1}{2}(F + wL)$$

Rearrange the equation to make w the subject. Hence determine the value of w when $L = 4$ m, $F = 8$ kN and $R = 10$ kN

16. The rate of heat conduction through a slab of material, Q, is given by the formula $Q = \dfrac{kA(t_1 - t_2)}{d}$ where t_1 and t_2 are the temperatures of each side of the material, A is the area of the slab, d is the thickness of

the slab, and k is the thermal conductivity of the material. Rearrange the formula to obtain an expression for t_2

17. The slip, s, of a vehicle is given by: $s = \left(1 - \dfrac{r\omega}{v}\right) \times 100\%$ where r is the tyre radius, ω is the angular velocity and v the velocity. Transpose to make r the subject of the formula.

18. The critical load, F newtons, of a steel column may be determined from the formula $L\sqrt{\dfrac{F}{EI}} = n\pi$ where L is the length, EI is the flexural rigidity, and n is a positive integer. Transpose for F and hence determine the value of F when $n = 1$, $E = 0.25 \times 10^{12}\,\text{N/m}^2$, $I = 6.92 \times 10^{-6}\,\text{m}^4$ and $L = 1.12\,\text{m}$

19. The flow of slurry along a pipe in a coal processing plant is given by: $V = \dfrac{\pi p r^4}{8\eta\ell}$ Transpose the equation for r

20. The deflection head H of a metal structure is given by: $H = \sqrt{\dfrac{I\rho^4 D^2 \ell^{\frac{3}{2}}}{20g}}$ Transpose the formula for length ℓ

21. Young's modulus, $E = \dfrac{\text{stress, } \sigma}{\text{strain, } \varepsilon}$ and strain, $\varepsilon = \dfrac{\text{length increase, } \delta}{\text{length, } \ell}$ For a 50 mm length of a steel bolt under tensile stress, its length increases by 1.25×10^{-4} m. If $E = 2 \times 10^{11}\,\text{N/m}^2$, calculate the stress.

22. A thermodynamic equation of state, the van der Waals' equation, states that for one mole of gas:

$$\left(P + \dfrac{a}{V^2}\right)(V - b) = RT$$

Given that constant, $a = 5.083$, volume, $V = 1.300$, constant, $b = 0.0380$, $R = 0.1700$ and $T = 297.0$, determine the value of pressure, P, correct to 3 significant figures.

Practice Exercise 64   Multiple-choice questions on transposing formulae (answers on page 449)

Each question has only one correct answer

1. Rearranging the formula $A = \dfrac{B}{C}$ to make B the subject gives:

    (a) $\dfrac{A}{C}$ (b) $A - C$ (c) AC (d) $\dfrac{C}{A}$

2. Transposing $I = \dfrac{V}{R}$ for resistance R gives:

    (a) $I - V$ (b) $\dfrac{V}{I}$ (c) $\dfrac{I}{V}$ (d) VI

3. Rearranging the formula $A = BCD$ to make D the subject gives:

    (a) $\dfrac{A}{B} - C$ (b) $A - BC$ (c) ABC (d) $\dfrac{A}{BC}$

4. Transposing $v = f\lambda$ to make wavelength λ the subject gives:

    (a) $\dfrac{v}{f}$ (b) $v + f$ (c) $f - v$ (d) $\dfrac{f}{v}$

5. The relationship between the temperature in degrees Fahrenheit (F) and the temperature in degrees Celsius (C) is given by:

    $\text{F} = \dfrac{9}{5}\text{C} + 32$. $135°\text{F}$ is equivalent to:

    (a) $43°\text{C}$ (b) $57.2°\text{C}$
    (c) $185.4°\text{C}$ (d) $184°\text{C}$

6. A formula for the focal length f of a convex lens is $\dfrac{1}{f} = \dfrac{1}{u} + \dfrac{1}{v}$
    When $f = 4$ and $u = 6$, v is:

    (a) -2 (b) $\dfrac{1}{12}$ (c) 12 (d) $-\dfrac{1}{2}$

7. Electrical resistance $R = \dfrac{\rho\ell}{a}$; transposing this equation for ℓ gives:

    (a) $\dfrac{\rho a}{R}$ (b) $\dfrac{R}{a\rho}$ (c) $\dfrac{a}{R\rho}$ (d) $\dfrac{Ra}{\rho}$

8. $PV = mRT$ is the characteristic gas equation. When $P = 100 \times 10^3$, $V = 4.0$, $R = 288$ and $T = 300$, the value of m is:

    (a) 4.630 (b) 313600
    (c) 0.216 (d) 100592

9. When two resistors R_1 and R_2 are connected in parallel the formula $\dfrac{1}{R_T} = \dfrac{1}{R_1} + \dfrac{1}{R_2}$ is used to determine the total resistance R_T. If $R_1 = 470\Omega$ and $R_2 = 2.7\ k\Omega$, R_T (correct to 3 significant figures) is equal to:
    (a) 2.68 Ω   (b) 400 Ω
    (c) 473 Ω   (d) 3170 Ω

10. The height s of a mass projected vertically upwards at time t is given by: $s = ut - \dfrac{1}{2}gt^2$. When $g = 10, t = 1.5$ and $s = 3.75$, the value of u is:
    (a) 10   (b) −5   (c) +5   (d) −10

11. The quantity of heat Q is given by the formula $Q = mc(t_2 - t_1)$. When $m = 5, t_1 = 20$, $c = 8$ and $Q = 1200$, the value of t_2 is:
    (a) 10   (b) 1.5   (c) 21.5   (d) 50

12. Resistance R ohms varies with temperature t according to the formula $R = R_0(1 + \alpha t)$. Given $R = 21\Omega$, $\alpha = 0.004$ and $t = 100$, R_0 has a value of:
    (a) 21.4 Ω   (b) 29.4 Ω
    (c) 15 Ω   (d) 0.067 Ω

13. The final length l_2 of a piece of wire heated through $\theta°C$ is given by the formula $\ell_2 = \ell_1(1 + \alpha\theta)$. Transposing, the coefficient of expansion, α, is given by:
    (a) $\dfrac{\ell_2}{\ell_1} - \dfrac{1}{\theta}$ (b) $\dfrac{\ell_2 - \ell_1}{\ell_1\theta}$
    (c) $\ell_2 - \ell_1 - \ell_1\theta$ (d) $\dfrac{\ell_1 - \ell_2}{\ell_1\theta}$

14. Current I in an electrical circuit is given by $I = \dfrac{E - e}{R + r}$. Transposing for R gives:
    (a) $\dfrac{E - e - Ir}{I}$ (b) $\dfrac{E - e}{I + r}$
    (c) $(E - e)(I + r)$ (d) $\dfrac{E - e}{Ir}$

15. Transposing $t = 2\pi\sqrt{\dfrac{\ell}{g}}$ for g gives:
    (a) $\dfrac{(t - 2\pi)^2}{\ell}$ (b) $\left(\dfrac{2\pi}{t}\right)\ell^2$
    (c) $\dfrac{\sqrt{\dfrac{t}{2\pi}}}{\ell}$ (d) $\dfrac{4\pi^2\ell}{t^2}$

For fully worked solutions to each of the problems in Practice Exercises 61 to 63 in this chapter, go to the website:
www.routledge.com/cw/bird

Chapter 13

Solving simultaneous equations

Why it is important to understand: Solving simultaneous equations

Simultaneous equations arise a great deal in engineering and science, some applications including theory of structures, data analysis, electrical circuit analysis and air traffic control. Systems that consist of a small number of equations can be solved analytically using standard methods from algebra (as explained in this chapter). Systems of large numbers of equations require the use of numerical methods and computers. Solving simultaneous equations is an important skill required in all aspects of engineering.

At the end of this chapter you should be able to:

- solve simultaneous equations in two unknowns by substitution
- solve simultaneous equations in two unknowns by elimination
- solve simultaneous equations involving practical situations
- solve simultaneous equations in three unknowns

13.1 Introduction

Only one equation is necessary when finding the value of a **single unknown quantity** (as with simple equations in Chapter 11). However, when an equation contains **two unknown quantities** it has an infinite number of solutions. When two equations are available connecting the same two unknown values then a unique solution is possible. Similarly, for three unknown quantities it is necessary to have three equations in order to solve for a particular value of each of the unknown quantities, and so on.

Equations which have to be solved together to find the unique values of the unknown quantities, which are true for each of the equations, are called **simultaneous equations**.

Two methods of solving simultaneous equations analytically are:

(a) by **substitution**, and

(b) by **elimination**.

(A graphical solution of simultaneous equations is shown in Chapter 19. In later studies, matrices and determinants will be used to solve simultaneous equations.)

13.2 Solving simultaneous equations in two unknowns

The method of solving simultaneous equations is demonstrated in the following worked problems.

Problem 1. Solve the following equations for x and y, (a) by substitution and (b) by elimination

$$x + 2y = -1 \qquad (1)$$
$$4x - 3y = 18 \qquad (2)$$

(a) By substitution

From equation (1): $x = -1 - 2y$
Substituting this expression for x into equation (2) gives

$$4(-1 - 2y) - 3y = 18$$

This is now a simple equation in y.
Removing the bracket gives

$$-4 - 8y - 3y = 18$$
$$-11y = 18 + 4 = 22$$
$$y = \frac{22}{-11} = -2$$

Substituting $y = -2$ into equation (1) gives

$$x + 2(-2) = -1$$
$$x - 4 = -1$$
$$x = -1 + 4 = 3$$

Thus, $x = 3$ and $y = -2$ is the solution to the simultaneous equations.
Check: in equation (2), since $x = 3$ and $y = -2$,

$$\text{LHS} = 4(3) - 3(-2) = 12 + 6 = 18 = \text{RHS}$$

(b) By elimination

$$x + 2y = -1 \qquad (1)$$
$$4x - 3y = 18 \qquad (2)$$

If equation (1) is multiplied throughout by 4, the coefficient of x will be the same as in equation (2), giving

$$4x + 8y = -4 \qquad (3)$$

Subtracting equation (3) from equation (2) gives

$$4x - 3y = 18 \qquad (2)$$
$$\underline{4x + 8y = -4 \qquad (3)}$$
$$0 - 11y = 22$$

Hence, $y = \dfrac{22}{-11} = \mathbf{-2}$

(Note: in the above subtraction,
$$18 - -4 = 18 + 4 = 22)$$

Substituting $y = -2$ into either equation (1) or equation (2) will give $x = 3$ as in method (a). The solution $x = 3, y = -2$ is the only pair of values that satisfies both of the original equations.

Problem 2. Solve, by a substitution method, the simultaneous equations

$$3x - 2y = 12 \qquad (1)$$
$$x + 3y = -7 \qquad (2)$$

From equation (2), $x = -7 - 3y$
Substituting for x in equation (1) gives

$$3(-7 - 3y) - 2y = 12$$

i.e. $\qquad -21 - 9y - 2y = 12$
$$-11y = 12 + 21 = 33$$

Hence, $y = \dfrac{33}{-11} = -3$

Substituting $y = -3$ in equation (2) gives

$$x + 3(-3) = -7$$

i.e. $\qquad x - 9 = -7$

Hence $\qquad x = -7 + 9 = 2$

Thus, $x = 2, y = -3$ is the solution of the simultaneous equations. (Such solutions should always be checked by substituting values into each of the original two equations.)

Problem 3. Use an elimination method to solve the following simultaneous equations

$$3x + 4y = 5 \qquad (1)$$
$$2x - 5y = -12 \qquad (2)$$

If equation (1) is multiplied throughout by 2 and equation (2) by 3, the coefficient of x will be the same in the newly formed equations. Thus,

$2 \times$ equation (1) gives $\qquad 6x + 8y = 10 \qquad (3)$

$3 \times$ equation (2) gives $\qquad 6x - 15y = -36 \qquad (4)$

Equation (3) – equation (4) gives

$$0 + 23y = 46$$

i.e. $\qquad y = \dfrac{46}{23} = \mathbf{2}$

(Note $+8y - -15y = 8y + 15y = 23y$ and $10 - -36 = 10 + 36 = 46)$

Substituting $y = 2$ in equation (1) gives

$$3x + 4(2) = 5$$

from which $\quad 3x = 5 - 8 = -3$

and $\qquad\qquad x = -1$

Checking, by substituting $x = -1$ and $y = 2$ in equation (2), gives

$$\text{LHS} = 2(-1) - 5(2) = -2 - 10 = -12 = \text{RHS}$$

Hence, $x = -1$ and $y = 2$ is the solution of the simultaneous equations.

The elimination method is the most common method of solving simultaneous equations.

Problem 4.   Solve

$$7x - 2y = 26 \qquad (1)$$
$$6x + 5y = 29 \qquad (2)$$

When equation (1) is multiplied by 5 and equation (2) by 2, the coefficients of y in each equation are numerically the same, i.e. 10, but are of opposite sign.

$5 \times$ equation (1) gives $\quad 35x - 10y = 130 \qquad (3)$

$2 \times$ equation (2) gives $\quad 12x + 10y = 58 \qquad (4)$

Adding equations (3)
and (4) gives $\qquad\qquad 47x + 0 = 188$

Hence, $\qquad\qquad x = \dfrac{188}{47} = 4$

Note that when the signs of common coefficients are **different** the two equations are **added** and when the signs of common coefficients are the **same** the two equations are **subtracted** (as in Problems 1 and 3).

Substituting $x = 4$ in equation (1) gives

$$7(4) - 2y = 26$$
$$28 - 2y = 26$$
$$28 - 26 = 2y$$
$$2 = 2y$$

Hence, $\qquad\qquad y = 1$

Checking, by substituting $x = 4$ and $y = 1$ in equation (2), gives

$$\text{LHS} = 6(4) + 5(1) = 24 + 5 = 29 = \text{RHS}$$

Thus, the solution is $x = 4, y = 1$

Now try the following Practice Exercise

Solve the following simultaneous equations and verify the results.

1. $2x - y = 6$
   $x + y = 6$
2. $2x - y = 2$
   $x - 3y = -9$
3. $x - 4y = -4$
   $5x - 2y = 7$
4. $3x - 2y = 10$
   $5x + y = 21$
5. $5p + 4q = 6$
   $2p - 3q = 7$
6. $7x + 2y = 11$
   $3x - 5y = -7$
7. $2x - 7y = -8$
   $3x + 4y = 17$
8. $a + 2b = 8$
   $b - 3a = -3$
9. $a + b = 7$
   $a - b = 3$
10. $2x + 5y = 7$
    $x + 3y = 4$
11. $3s + 2t = 12$
    $4s - t = 5$
12. $3x - 2y = 13$
    $2x + 5y = -4$
13. $5m - 3n = 11$
    $3m + n = 8$
14. $8a - 3b = 51$
    $3a + 4b = 14$
15. $5x = 2y$
    $3x + 7y = 41$
16. $5c = 1 - 3d$
    $2d + c + 4 = 0$

13.3 Further solving of simultaneous equations

Here are some further worked problems on solving simultaneous equations.

Problem 5.   Solve

$$3p = 2q \qquad (1)$$
$$4p + q + 11 = 0 \qquad (2)$$

Rearranging gives

$$3p - 2q = 0 \qquad (3)$$
$$4p + q = -11 \qquad (4)$$

Multiplying equation (4) by 2 gives

$$8p + 2q = -22 \qquad (5)$$

Adding equations (3) and (5) gives

$$11p + 0 = -22$$

$$p = \frac{-22}{11} = -2$$

Substituting $p = -2$ into equation (1) gives

$$3(-2) = 2q$$

$$-6 = 2q$$

$$q = \frac{-6}{2} = -3$$

Checking, by substituting $p = -2$ and $q = -3$ into equation (2), gives

$$\text{LHS} = 4(-2) + (-3) + 11 = -8 - 3 + 11 = 0 = \text{RHS}$$

Hence, the solution is $p = -2, q = -3$

> **Problem 6.** Solve
>
> $$\frac{x}{8} + \frac{5}{2} = y \qquad (1)$$
>
> $$13 - \frac{y}{3} = 3x \qquad (2)$$

Whenever fractions are involved in simultaneous equations it is often easier to firstly remove them. Thus, multiplying equation (1) by 8 gives

$$8\left(\frac{x}{8}\right) + 8\left(\frac{5}{2}\right) = 8y$$

i.e. $\qquad\qquad x + 20 = 8y \qquad (3)$

Multiplying equation (2) by 3 gives

$$39 - y = 9x \qquad (4)$$

Rearranging equations (3) and (4) gives

$$x - 8y = -20 \qquad (5)$$

$$9x + y = 39 \qquad (6)$$

Multiplying equation (6) by 8 gives

$$72x + 8y = 312 \qquad (7)$$

Adding equations (5) and (7) gives

$$73x + 0 = 292$$

$$x = \frac{292}{73} = 4$$

Substituting $x = 4$ into equation (5) gives

$$4 - 8y = -20$$

$$4 + 20 = 8y$$

$$24 = 8y$$

$$y = \frac{24}{8} = 3$$

Checking, substituting $x = 4$ and $y = 3$ in the original equations, gives

(1): $\quad \text{LHS} = \frac{4}{8} + \frac{5}{2} = \frac{1}{2} + 2\frac{1}{2} = 3 = y = \text{RHS}$

(2): $\quad \text{LHS} = 13 - \frac{3}{3} = 13 - 1 = 12$

$$\text{RHS} = 3x = 3(4) = 12$$

Hence, the solution is $x = 4, y = 3$

> **Problem 7.** Solve
>
> $$2.5x + 0.75 - 3y = 0$$
>
> $$1.6x = 1.08 - 1.2y$$

It is often easier to remove decimal fractions. Thus, multiplying equations (1) and (2) by 100 gives

$$250x + 75 - 300y = 0 \qquad (1)$$

$$160x = 108 - 120y \qquad (2)$$

Rearranging gives

$$250x - 300y = -75 \qquad (3)$$

$$160x + 120y = 108 \qquad (4)$$

Multiplying equation (3) by 2 gives

$$500x - 600y = -150 \qquad (5)$$

Multiplying equation (4) by 5 gives

$$800x + 600y = 540 \qquad (6)$$

Adding equations (5) and (6) gives

$$1300x + 0 = 390$$

$$x = \frac{390}{1300} = \frac{39}{130} = \frac{3}{10} = 0.3$$

Substituting $x = 0.3$ into equation (1) gives

$$250(0.3) + 75 - 300y = 0$$

$$75 + 75 = 300y$$

$$150 = 300y$$

$$y = \frac{150}{300} = \mathbf{0.5}$$

Checking, by substituting $x = 0.3$ and $y = 0.5$ in equation (2), gives

$$\text{LHS} = 160(0.3) = 48$$

$$\text{RHS} = 108 - 120(0.5) = 108 - 60 = 48$$

Hence, the solution is $\boldsymbol{x = 0.3, y = 0.5}$

Now try the following Practice Exercise

Practice Exercise 66   Solving simultaneous equations (answers on page 449)

Solve the following simultaneous equations and verify the results.

1. $7p + 11 + 2q = 0$
   $-1 = 3q - 5p$

2. $\dfrac{x}{2} + \dfrac{y}{3} = 4$
   $\dfrac{x}{6} - \dfrac{y}{9} = 0$

3. $\dfrac{a}{2} - 7 = -2b$
   $12 = 5a + \dfrac{2}{3}b$

4. $\dfrac{3}{2}s - 2t = 8$
   $\dfrac{s}{4} + 3t = -2$

5. $\dfrac{x}{5} + \dfrac{2y}{3} = \dfrac{49}{15}$
   $\dfrac{3x}{7} - \dfrac{y}{2} + \dfrac{5}{7} = 0$

6. $v - 1 = \dfrac{u}{12}$
   $u + \dfrac{v}{4} - \dfrac{25}{2} = 0$

7. $1.5x - 2.2y = -18$
   $2.4x + 0.6y = 33$

8. $3b - 2.5a = 0.45$
   $1.6a + 0.8b = 0.8$

13.4 Solving more difficult simultaneous equations

Here are some further worked problems on solving more difficult simultaneous equations.

Problem 8. Solve

$$\frac{2}{x} + \frac{3}{y} = 7 \tag{1}$$

$$\frac{1}{x} - \frac{4}{y} = -2 \tag{2}$$

In this type of equation the solution is easier if a substitution is initially made. Let $\dfrac{1}{x} = a$ and $\dfrac{1}{y} = b$

Thus equation (1) becomes $2a + 3b = 7$ $\qquad$ (3)

and equation (2) becomes $a - 4b = -2$ $\qquad$ (4)

Multiplying equation (4) by 2 gives

$$2a - 8b = -4 \tag{5}$$

Subtracting equation (5) from equation (3) gives

$$0 + 11b = 11$$

i.e. $\qquad\qquad\qquad\qquad \boldsymbol{b = 1}$

Substituting $b = 1$ in equation (3) gives

$$2a + 3 = 7$$

$$2a = 7 - 3 = 4$$

i.e. $\qquad\qquad\qquad\qquad \boldsymbol{a = 2}$

Checking, substituting $a = 2$ and $b = 1$ in equation (4), gives

$$\text{LHS} = 2 - 4(1) = 2 - 4 = -2 = \text{RHS}$$

Hence, $\boldsymbol{a = 2}$ and $\boldsymbol{b = 1}$

However, since $\qquad \dfrac{1}{x} = a, \qquad x = \dfrac{1}{a} = \dfrac{1}{2}$ or 0.5

and since $\qquad \dfrac{1}{y} = b, \qquad y = \dfrac{1}{b} = \dfrac{1}{1} = 1$

Hence, the solution is $\boldsymbol{x = 0.5, y = 1}$

Problem 9. Solve

$$\frac{1}{2a} + \frac{3}{5b} = 4 \tag{1}$$

$$\frac{4}{a} + \frac{1}{2b} = 10.5 \tag{2}$$

Let $\dfrac{1}{a} = x$ $\quad$ and $\quad \dfrac{1}{b} = y$

then $\qquad\qquad \dfrac{x}{2} + \dfrac{3}{5}y = 4 \qquad\qquad$ (3)

$$4x + \frac{1}{2}y = 10.5 \tag{4}$$

To remove fractions, equation (3) is multiplied by 10, giving

$$10\left(\frac{x}{2}\right) + 10\left(\frac{3}{5}y\right) = 10(4)$$

i.e. $\qquad 5x + 6y = 40 \qquad$ (5)

Multiplying equation (4) by 2 gives

$$8x + y = 21 \qquad (6)$$

Multiplying equation (6) by 6 gives

$$48x + 6y = 126 \qquad (7)$$

Subtracting equation (5) from equation (7) gives

$$43x + 0 = 86$$

$$x = \frac{86}{43} = 2$$

Substituting $x = 2$ into equation (3) gives

$$\frac{2}{2} + \frac{3}{5}y = 4$$

$$\frac{3}{5}y = 4 - 1 = 3$$

$$y = \frac{5}{3}(3) = 5$$

Since $\qquad \dfrac{1}{a} = x, \quad a = \dfrac{1}{x} = \dfrac{1}{2}$ or 0.5

and since $\quad \dfrac{1}{b} = y, \quad b = \dfrac{1}{y} = \dfrac{1}{5}$ or 0.2

Hence, the solution is $a = 0.5$, $b = 0.2$, which may be checked in the original equations.

Problem 10. Solve

$$\frac{1}{x+y} = \frac{4}{27} \qquad (1)$$

$$\frac{1}{2x-y} = \frac{4}{33} \qquad (2)$$

To eliminate fractions, both sides of equation (1) are multiplied by $27(x + y)$, giving

$$27(x+y)\left(\frac{1}{x+y}\right) = 27(x+y)\left(\frac{4}{27}\right)$$

i.e. $\qquad 27(1) = 4(x+y)$

$$27 = 4x + 4y \qquad (3)$$

Similarly, in equation (2) $33 = 4(2x - y)$

i.e. $\qquad 33 = 8x - 4y \qquad (4)$

Equation (3) + equation (4) gives

$$60 = 12x \text{ and } x = \frac{60}{12} = 5$$

Substituting $x = 5$ in equation (3) gives

$$27 = 4(5) + 4y$$

from which $\qquad 4y = 27 - 20 = 7$

and $\qquad y = \dfrac{7}{4} = 1\dfrac{3}{4}$ or **1.75**

Hence, $x = 5$, $y = 1.75$ is the required solution, which may be checked in the original equations.

Problem 11. Solve

$$\frac{x-1}{3} + \frac{y+2}{5} = \frac{2}{15} \qquad (1)$$

$$\frac{1-x}{6} + \frac{5+y}{2} = \frac{5}{6} \qquad (2)$$

Before equations (1) and (2) can be simultaneously solved, the fractions need to be removed and the equations rearranged.

Multiplying equation (1) by 15 gives

$$15\left(\frac{x-1}{3}\right) + 15\left(\frac{y+2}{5}\right) = 15\left(\frac{2}{15}\right)$$

i.e. $\qquad 5(x-1) + 3(y+2) = 2$

$$5x - 5 + 3y + 6 = 2$$

$$5x + 3y = 2 + 5 - 6$$

Hence, $\qquad 5x + 3y = 1 \qquad (3)$

Multiplying equation (2) by 6 gives

$$6\left(\frac{1-x}{6}\right) + 6\left(\frac{5+y}{2}\right) = 6\left(\frac{5}{6}\right)$$

i.e. $\qquad (1-x) + 3(5+y) = 5$

$$1 - x + 15 + 3y = 5$$

$$-x + 3y = 5 - 1 - 15$$

Hence, $\qquad -x + 3y = -11 \qquad (4)$

Thus the initial problem containing fractions can be expressed as

$$5x + 3y = 1 \qquad (3)$$
$$-x + 3y = -11 \qquad (4)$$

Subtracting equation (4) from equation (3) gives

$$6x + 0 = 12$$
$$x = \frac{12}{6} = 2$$

Substituting $x = 2$ into equation (3) gives

$$5(2) + 3y = 1$$
$$10 + 3y = 1$$
$$3y = 1 - 10 = -9$$
$$y = \frac{-9}{3} = -3$$

Checking, substituting $x = 2$, $y = -3$ in equation (4) gives

$$\text{LHS} = -2 + 3(-3) = -2 - 9 = -11 = \text{RHS}$$

Hence, the solution is $x = 2, y = -3$

Now try the following Practice Exercise

Practice Exercise 67   Solving more difficult simultaneous equations (answers on page 449)

In Problems 1 to 7, solve the simultaneous equations and verify the results

1. $\frac{3}{x} + \frac{2}{y} = 14$
    $\frac{5}{x} - \frac{3}{y} = -2$

2. $\frac{4}{a} - \frac{3}{b} = 18$
    $\frac{2}{a} + \frac{5}{b} = -4$

3. $\frac{1}{2p} + \frac{3}{5q} = 5$
    $\frac{5}{p} - \frac{1}{2q} = \frac{35}{2}$

4. $\frac{5}{x} + \frac{3}{y} = 1.1$
    $\frac{3}{x} - \frac{7}{y} = -1.1$

5. $\frac{c+1}{4} - \frac{d+2}{3} + 1 = 0$
    $\frac{1-c}{5} + \frac{3-d}{4} + \frac{13}{20} = 0$

6. $\frac{3r+2}{5} - \frac{2s-1}{4} = \frac{11}{5}$
    $\frac{3+2r}{4} + \frac{5-s}{3} = \frac{15}{4}$

7. $\frac{5}{x+y} = \frac{20}{27}$
    $\frac{4}{2x-y} = \frac{16}{33}$

8. If $5x - \frac{3}{y} = 1$ and $x + \frac{4}{y} = \frac{5}{2}$, find the value of $\frac{xy+1}{y}$

13.5   Practical problems involving simultaneous equations

There are a number of situations in engineering and science in which the solution of simultaneous equations is required. Some are demonstrated in the following worked problems.

Problem 12. The law connecting friction F and load L for an experiment is of the form $F = aL + b$ where a and b are constants. When $F = 5.6\,\text{N}$, $L = 8.0\,\text{N}$ and when $F = 4.4\,\text{N}$, $L = 2.0\,\text{N}$. Find the values of a and b and the value of F when $L = 6.5\,\text{N}$

Substituting $F = 5.6$ and $L = 8.0$ into $F = aL + b$ gives

$$5.6 = 8.0a + b \qquad (1)$$

Substituting $F = 4.4$ and $L = 2.0$ into $F = aL + b$ gives

$$4.4 = 2.0a + b \qquad (2)$$

Subtracting equation (2) from equation (1) gives

$$1.2 = 6.0a$$
$$a = \frac{1.2}{6.0} = \frac{1}{5} \text{ or } 0.2$$

Substituting $a = \dfrac{1}{5}$ into equation (1) gives

$$5.6 = 8.0 \left(\dfrac{1}{5} \right) + b$$

$$5.6 = 1.6 + b$$

$$5.6 - 1.6 = b$$

i.e. $\qquad\qquad\qquad b = 4$

Checking, substituting $a = \dfrac{1}{5}$ and $b = 4$ in equation (2), gives

$$\text{RHS} = 2.0 \left(\dfrac{1}{5} \right) + 4 = 0.4 + 4 = 4.4 = \text{LHS}$$

Hence, $a = \dfrac{1}{5}$ and $b = 4$

When $L = 6.5$, $F = aL + b = \dfrac{1}{5}(6.5) + 4 = 1.3 + 4$, i.e. $F = \mathbf{5.30\,N}$.

Problem 13. The equation of a straight line, of gradient m and intercept on the y-axis c, is $y = mx + c$. If a straight line passes through the point where $x = 1$ and $y = -2$, and also through the point where $x = 3.5$ and $y = 10.5$, find the values of the gradient and the y-axis intercept

Substituting $x = 1$ and $y = -2$ into $y = mx + c$ gives

$$-2 = m + c \qquad\qquad (1)$$

Substituting $x = 3.5$ and $y = 10.5$ into $y = mx + c$ gives

$$10.5 = 3.5m + c \qquad\qquad (2)$$

Subtracting equation (1) from equation (2) gives

$$12.5 = 2.5m, \quad \text{from which,} \quad m = \dfrac{12.5}{2.5} = 5$$

Substituting $m = 5$ into equation (1) gives

$$-2 = 5 + c$$

$$c = -2 - 5 = -7$$

Checking, substituting $m = 5$ and $c = -7$ in equation (2), gives

$$\text{RHS} = (3.5)(5) + (-7) = 17.5 - 7 = 10.5 = \text{LHS}$$

Hence, the **gradient $m = 5$** and the **y-axis intercept $c = -7$**

Problem 14. When Kirchhoff's laws* are applied to the electrical circuit shown in Fig. 13.1, the currents I_1 and I_2 are connected by the equations

$$27 = 1.5I_1 + 8(I_1 - I_2) \qquad\qquad (1)$$

$$-26 = 2I_2 - 8(I_1 - I_2) \qquad\qquad (2)$$

Figure 13.1

Solve the equations to find the values of currents I_1 and I_2

*Who was **Kirchhoff**? – **Gustav Robert Kirchhoff** (12 March 1824–17 October 1887) was a German physicist. Concepts in circuit theory and thermal emission are named 'Kirchhoff's laws' after him, as well as a law of thermochemistry. To find out more go to **www.routledge.com/cw/bird**

Removing the brackets from equation (1) gives

$$27 = 1.5I_1 + 8I_1 - 8I_2$$

Rearranging gives

$$9.5I_1 - 8I_2 = 27 \qquad (3)$$

Removing the brackets from equation (2) gives

$$-26 = 2I_2 - 8I_1 + 8I_2$$

Rearranging gives

$$-8I_1 + 10I_2 = -26 \qquad (4)$$

Multiplying equation (3) by 5 gives

$$47.5I_1 - 40I_2 = 135 \qquad (5)$$

Multiplying equation (4) by 4 gives

$$-32I_1 + 40I_2 = -104 \qquad (6)$$

Adding equations (5) and (6) gives

$$15.5I_1 + 0 = 31$$

$$I_1 = \frac{31}{15.5} = 2$$

Substituting $I_1 = 2$ into equation (3) gives

$$9.5(2) - 8I_1 = 27$$
$$19 - 8I_2 = 27$$
$$19 - 27 = 8I_2$$
$$-8 = 8I_2$$

and

$$I_2 = -1$$

Hence, the solution is $I_1 = 2$ and $I_2 = -1$ (which may be checked in the original equations).

Problem 15. The distance s metres from a fixed point of a vehicle travelling in a straight line with constant acceleration, a m/s^2, is given by $s = ut + \frac{1}{2}at^2$, where u is the initial velocity in m/s and t the time in seconds. Determine the initial velocity and the acceleration given that $s = 42$ m when $t = 2$ s, and $s = 144$ m when $t = 4$ s. Also find the distance travelled after 3 s

Substituting $s = 42$ and $t = 2$ into $s = ut + \frac{1}{2}at^2$ gives

$$42 = 2u + \frac{1}{2}a(2)^2$$

i.e. $\qquad 42 = 2u + 2a \qquad (1)$

Substituting $s = 144$ and $t = 4$ into $s = ut + \frac{1}{2}at^2$ gives

$$144 = 4u + \frac{1}{2}a(4)^2$$

i.e. $\qquad 144 = 4u + 8a \qquad (2)$

Multiplying equation (1) by 2 gives

$$84 = 4u + 4a \qquad (3)$$

Subtracting equation (3) from equation (2) gives

$$60 = 0 + 4a$$

and $\qquad a = \frac{60}{4} = 15$

Substituting $a = 15$ into equation (1) gives

$$42 = 2u + 2(15)$$
$$42 - 30 = 2u$$
$$u = \frac{12}{2} = 6$$

Substituting $a = 15$ and $u = 6$ in equation (2) gives

$$\text{RHS} = 4(6) + 8(15) = 24 + 120 = 144 = \text{LHS}$$

Hence, **the initial velocity $u = 6$ m/s and the acceleration $a = 15$ m/s^2**.

Distance travelled after 3 s is given by $s = ut + \frac{1}{2}at^2$ where $t = 3, u = 6$ and $a = 15$

Hence, $s = (6)(3) + \frac{1}{2}(15)(3)^2 = 18 + 67.5$

i.e. **distance travelled after 3 s = 85.5 m**.

Problem 16. The resistance $R\,\Omega$ of a length of wire at $t°$C is given by $R = R_0(1 + \alpha t)$, where R_0 is the resistance at $0°$C and α is the temperature coefficient of resistance in $/°$C. Find the values of α and R_0 if $R = 30\,\Omega$ at $50°$C and $R = 35\,\Omega$ at $100°$C

Substituting $R = 30$ and $t = 50$ into $R = R_0(1 + \alpha t)$ gives

$$30 = R_0(1 + 50\alpha) \qquad (1)$$

Substituting $R = 35$ and $t = 100$ into $R = R_0(1 + \alpha t)$ gives

$$35 = R_0(1 + 100\alpha) \qquad (2)$$

Although these equations may be solved by the conventional substitution method, an easier way is to eliminate

R_0 by division. Thus, dividing equation (1) by equation (2) gives

$$\frac{30}{35} = \frac{R_0(1+50\alpha)}{R_0(1+100\alpha)} = \frac{1+50\alpha}{1+100\alpha}$$

Cross-multiplying gives

$$30(1+100\alpha) = 35(1+50\alpha)$$
$$30 + 3000\alpha = 35 + 1750\alpha$$
$$3000\alpha - 1750\alpha = 35 - 30$$
$$1250\alpha = 5$$

i.e.

$$\alpha = \frac{5}{1250} = \frac{1}{250} \quad \text{or} \quad \textbf{0.004}$$

Substituting $\alpha = \frac{1}{250}$ into equation (1) gives

$$30 = R_0\left\{1 + (50)\left(\frac{1}{250}\right)\right\}$$
$$30 = R_0(1.2)$$
$$R_0 = \frac{30}{1.2} = \textbf{25}$$

Checking, substituting $\alpha = \frac{1}{250}$ and $R_0 = 25$ in equation (2), gives

$$\text{RHS} = 25\left\{1 + (100)\left(\frac{1}{250}\right)\right\}$$
$$= 25(1.4) = 35 = \text{LHS}$$

Thus, the solution is $\alpha = \textbf{0.004}/°\textbf{C}$ and $R_0 = \textbf{25}\,\Omega$.

Problem 17. The molar heat capacity of a solid compound is given by the equation $c = a + bT$, where a and b are constants. When $c = 52, T = 100$ and when $c = 172, T = 400$. Determine the values of a and b

When $c = 52$, $T = 100$, hence

$$52 = a + 100b \tag{1}$$

When $c = 172$, $T = 400$, hence

$$172 = a + 400b \tag{2}$$

Equation (2) – equation (1) gives

$$120 = 300b$$

from which, $$b = \frac{120}{300} = \textbf{0.4}$$

Substituting $b = 0.4$ in equation (1) gives

$$52 = a + 100(0.4)$$
$$a = 52 - 40 = \textbf{12}$$

Hence, $a = \textbf{12}$ and $b = \textbf{0.4}$

Now try the following Practice Exercise

Practice Exercise 68   Practical problems involving simultaneous equations (answers on page 449)

1. In a system of pulleys, the effort P required to raise a load W is given by $P = aW + b$, where a and b are constants. If $W = 40$ when $P = 12$ and $W = 90$ when $P = 22$, find the values of a and b.

2. Applying Kirchhoff's laws to an electrical circuit produces the following equations:

    $$5 = 0.2I_1 + 2(I_1 - I_2)$$
    $$12 = 3I_2 + 0.4I_2 - 2(I_1 - I_2)$$

    Determine the values of currents I_1 and I_2

3. Velocity v is given by the formula $v = u + at$. If $v = 20$ when $t = 2$ and $v = 40$ when $t = 7$, find the values of u and a. Then, find the velocity when $t = 3.5$

4. Three new cars and four new vans supplied to a dealer together cost £97 700 and five new cars and two new vans of the same models cost £103 100. Find the respective costs of a car and a van.

5. $y = mx + c$ is the equation of a straight line of slope m and y-axis intercept c. If the line passes through the point where $x = 2$ and $y = 2$, and also through the point where $x = 5$ and $y = 0.5$, find the slope and y-axis intercept of the straight line.

6. The resistance R ohms of copper wire at $t°C$ is given by $R = R_0(1 + \alpha t)$, where R_0 is the resistance at $0°C$ and α is the temperature coefficient of resistance. If $R = 25.44\,\Omega$ at $30°C$ and $R = 32.17\,\Omega$ at $100°C$, find α and R_0

7. The molar heat capacity of a solid compound is given by the equation $c = a + bT$. When $c = 70, T = 110$ and when $c = 160$, $T = 290$. Find the values of a and b.

8. In an engineering process, two variables p and q are related by $q = ap + b/p$, where a and b are constants. Evaluate a and b if $q = 13$ when $p = 2$ and $q = 22$ when $p = 5$

9. In a system of forces, the relationship between two forces F_1 and F_2 is given by

$$5F_1 + 3F_2 + 6 = 0$$
$$3F_1 + 5F_2 + 18 = 0$$

Solve for F_1 and F_2

10. For a balanced beam, the equilibrium of forces is given by: $R_1 + R_2 = 12.0 \text{kN}$
As a result of taking moments:
$0.2R_1 + 7 \times 0.3 + 3 \times 0.6 = 0.8R_2$
Determine the values of the reaction forces R_1 and R_1

11. In an engineering scenario involving reciprocal motion, the following simultaneous equations resulted: $\omega\sqrt{(r^2 - 0.07^2)} = 8$ and $\omega\sqrt{(r^2 - 0.25^2)} = 2$ Calculate the value of radius r and angular velocity ω, each correct to 3 significant figures.

13.6 Solving simultaneous equations in three unknowns

Equations containing three unknowns may be solved using exactly the same procedures as those used with two equations and two unknowns, providing that there are three equations to work with. The method is demonstrated in the following worked problem.

Problem 18. Solve the simultaneous equations.

$$x + y + z = 4 \qquad (1)$$
$$2x - 3y + 4z = 33 \qquad (2)$$
$$3x - 2y - 2z = 2 \qquad (3)$$

There are a number of ways of solving these equations. One method is shown below.

The initial object is to produce two equations with two unknowns. For example, multiplying equation (1) by 4 and then subtracting this new equation from equation (2) will produce an equation with only x and y involved. Multiplying equation (1) by 4 gives

$$4x + 4y + 4z = 16 \qquad (4)$$

Equation (2) – equation (4) gives

$$-2x - 7y = 17 \qquad (5)$$

Similarly, multiplying equation (3) by 2 and then adding this new equation to equation (2) will produce another equation with only x and y involved.

Multiplying equation (3) by 2 gives

$$6x - 4y - 4z = 4 \qquad (6)$$

Equation (2) + equation (6) gives

$$8x - 7y = 37 \qquad (7)$$

Rewriting equation (5) gives

$$-2x - 7y = 17 \qquad (5)$$

Now we can use the previous method for solving simultaneous equations in two unknowns.

Equation (7) – equation (5) gives $10x = 20$

from which, $x = 2$

(Note that $8x - -2x = 8x + 2x = 10x$)

Substituting $x = 2$ into equation (5) gives

$$-4 - 7y = 17$$

from which, $-7y = 17 + 4 = 21$

and $y = -3$

Substituting $x = 2$ and $y = -3$ into equation (1) gives

$$2 - 3 + z = 4$$

from which, $z = 5$

Hence, the solution of the simultaneous equations is $x = 2, y = -3$ and $z = 5$

Now try the following Practice Exercise

Practice Exercise 69   Simultaneous equations in three unknowns (answers on page 449)

In Problems 1 to 9, solve the simultaneous equations in 3 unknowns.

1. $x + 2y + 4z = 16$
     $2x - y + 5z = 18$
     $3x + 2y + 2z = 14$

2. $2x + y - z = 0$
     $3x + 2y + z = 4$
     $5x + 3y + 2z = 8$

3. $3x + 5y + 2z = 6$
     $x - y + 3z = 0$
     $2x + 7y + 3z = -3$

4. $2x + 4y + 5z = 23$
     $3x - y - 2z = 6$
     $4x + 2y + 5z = 31$

5. $2x + 3y + 4z = 36$
     $3x + 2y + 3z = 29$
     $x + y + z = 11$

6. $4x + y + 3z = 31$
     $2x - y + 2z = 10$
     $3x + 3y - 2z = 7$

7. $5x + 5y - 4z = 37$
     $2x - 2y + 9z = 20$
     $-4x + y + z = -14$

8. $6x + 7y + 8z = 13$
     $3x + y - z = -11$
     $2x - 2y - 2z = -18$

9. $3x + 2y + z = 14$
     $7x + 3y + z = 22.5$
     $4x - 4y - z = -8.5$

10.   Kirchhoff's laws are used to determine the current equations in an electrical network and result in the following:

$$i_1 + 8i_2 + 3i_3 = -31$$

$$3i_1 - 2i_2 + i_3 = -5$$

$$2i_1 - 3i_2 + 2i_3 = 6$$

Determine the values of i_1, i_2 and i_3

11.   The forces in three members of a framework are F_1, F_2 and F_3. They are related by the following simultaneous equations.

$$1.4F_1 + 2.8F_2 + 2.8F_3 = 5.6$$

$$4.2F_1 - 1.4F_2 + 5.6F_3 = 35.0$$

$$4.2F_1 + 2.8F_2 - 1.4F_3 = -5.6$$

Find the values of F_1, F_2 and F_3

It is possible, with a modern scientific calculator, to solve simultaneous equations – choose 'Equations' from the 'Mode' menu and follow the rules for your calculator. Naturally, answers may be obtained very much more quickly than by the analytical means shown in this chapter.

Practice Exercise 70   Multiple-choice questions on solving simultaneous equations (answers on page 449)

Each question has only one correct answer

1.   The solution of the simultaneous equations $3x - 2y = 13$ and $2x + 5y = -4$ is:
(a) $x = -2, y = 3$ (b) $x = 1, y = -5$
(c) $x = 3, y = -2$ (d) $x = -7, y = 2$

2.   In a system of pulleys, the effort P required to raise a load W is given by $P = aW + b$, where a and b are constants. If $W = 40$ when $P = 12$ and $W = 90$ when $P = 22$, the values of a and b are:
(a) $a = 5, b = \dfrac{1}{4}$ (b) $a = 1, b = -28$
(c) $a = \dfrac{1}{3}, b = -8$ (d) $a = \dfrac{1}{5}, b = 4$

3.   The cost of 8 pencils and 16 pens is £3.52, and the cost of 4 pens and 4 pencils is 96 p. The cost of a pen is:
(a) 28 p   (b) 36 p   (c) 20 p   (d) 32 p

4.   A pair of simultaneous equations are:
$x + 2y - 8 = 0$ and $2x + 4y = 16$.
The number of possible solutions is:
(a) 0   (b) 1   (c) 2   (d) infinite

5.   Given two simultaneous equations
$x - 2y = 5$ and $2y + 5x = 7$,
then the value of 9x is:
(a) 39   (b) 18   (c) -13.5 (d) -3

6.   If $x + 2y = 1$ and $3x - y = -11$ then the value of x is:
(a) -3 (b) 4.2   (c) 2   (d) -1.6

7.   If $4x - 3y = 7$ and $2x + 5y = -1$ then the value of y is:
(a) -1 (b) $-\dfrac{9}{13}$ (c) 2   (d) $\dfrac{59}{26}$

8. If $5x - 3y = 17.5$ and $2x + y = 4.25$ then the value of $4x + 3y$ is:
   (a) 7.25   (b) 8.5   (c) 12   (d) 14.75

9. Given that $3x = 4y$ and $5x + y - 23 = 0$, then the value of $2x - y$ is:
   (a) 2   (b) 5   (c) 1.5   (d) 1

10. Velocity is given by the formula $v = u + at$. If $v = 18$ when $t = 2$ and $v = 38$ when $t = 6$, the value of the velocity when $t = 7.5$ is:
    (a) 45.5   (b) 65   (c) 29.5   (d) 55

For fully worked solutions to each of the problems in Practice Exercises 65 to 69 in this chapter, go to the website:
www.routledge.com/cw/bird

This assignment covers the material contained in Chapters 12 and 13. *The marks available are shown in brackets at the end of each question.*

1. Transpose $p - q + r = a - b$ for b. (2)

2. Make π the subject of the formula $r = \dfrac{c}{2\pi}$ (2)

3. Transpose $V = \dfrac{1}{3}\pi r^2 h$ for h. (2)

4. Transpose $I = \dfrac{E - e}{R + r}$ for E. (3)

5. Transpose $k = \dfrac{b}{ad - 1}$ for d. (4)

6. Make g the subject of the formula $t = 2\pi\sqrt{\dfrac{L}{g}}$ (3)

7. Transpose $A = \dfrac{\pi R^2 \theta}{360}$ for R. (2)

8. Make r the subject of the formula $x + y = \dfrac{r}{3 + r}$ (5)

9. Make L the subject of the formula $m = \dfrac{\mu L}{L + rCR}$ (5)

10. The surface area A of a rectangular prism is given by the formula $A = 2(bh + hl + lb)$. Evaluate b when $A = 11\,750\,\text{mm}^2, h = 25\,\text{mm}$ and $l = 75\,\text{mm}$. (4)

11. The velocity v of water in a pipe appears in the formula $h = \dfrac{0.03\,L v^2}{2\,dg}$. Evaluate v when $h = 0.384, d = 0.20, L = 80$ and $g = 10$ (5)

12. A formula for the focal length f of a convex lens is $\dfrac{1}{f} = \dfrac{1}{u} + \dfrac{1}{v}$. Evaluate v when $f = 4$ and $u = 20$ (4)

13. Impedance in an a.c. circuit is given by:
$$Z = \sqrt{\left[R^2 + \left(2\pi nL - \dfrac{1}{2\pi nC}\right)^2\right]}$$ where R is the resistance in ohms, L is the inductance in henrys, C is the capacitance in farads and n is the frequency of oscillations per second. Given $n = 50$, $R = 20$, $L = 0.40$ and $Z = 25$, determine the value of capacitance. (9)

In Problems 14 and 15, solve the simultaneous equations.

14. (a) $2x + y = 6$
       $5x - y = 22$

    (b) $4x - 3y = 11$
        $3x + 5y = 30$ (9)

15. (a) $3a - 8 + \dfrac{b}{8} = 0$

        $b + \dfrac{a}{2} = \dfrac{21}{4}$

    (b) $\dfrac{2p + 1}{5} - \dfrac{1 - 4q}{2} = \dfrac{5}{2}$

        $\dfrac{1 - 3p}{7} + \dfrac{2q - 3}{5} + \dfrac{32}{35} = 0$ (18)

16. In an engineering process two variables x and y are related by the equation $y = ax + \dfrac{b}{x}$, where a and b are constants. Evaluate a and b if $y = 15$ when $x = 1$ and $y = 13$ when $x = 3$ (5)

17. Kirchhoff's laws are used to determine the current equations in an electrical network and result in the following:

    $$i_1 + 8i_2 + 3i_3 = -31$$
    $$3i_1 - 2i_2 + i_3 = -5$$
    $$2i_1 - 3i_2 + 2i_3 = 6$$

    Determine the values of i_1, i_2 and i_3 (10)

18. The forces acting on a beam are given by:
    $$R_1 + R_2 = 3.3\text{kN}$$
    and $22 \times 2.7 + 61 \times 0.4 - 12R_1 = 46R_2$
    Calculate the reaction forces R_1 and R_2 (8)

For lecturers/instructors/teachers, fully worked solutions to each of the problems in Revision Test 5, together with a full marking scheme, are available at the website:
www.routledge.com/cw/bird

Chapter 14

Solving quadratic equations

Why it is important to understand: Solving quadratic equations

Quadratic equations have many applications in engineering and science; they are used in describing the trajectory of a ball, determining the height of a throw, and in the concept of acceleration, velocity, ballistics and stopping power. In addition, the quadratic equation has been found to be widely evident in a number of natural processes; some of these include the processes by which light is reflected off a lens, water flows down a rocky stream or even the manner in which fur, spots or stripes develop on wild animals. When traffic police arrive at the scene of a road accident, they measure the length of the skid marks and assess the road conditions. They can then use a quadratic equation to calculate the speed of the vehicles and hence reconstruct exactly what happened. The U-shape of a parabola can describe the trajectories of water jets in a fountain and a bouncing ball, or be incorporated into structures like the parabolic reflectors that form the base of satellite dishes and car headlights. Quadratic functions can help plot the course of moving objects and assist in determining minimum and maximum values. Most of the objects we use every day, from cars to clocks, would not exist if someone somewhere hadn't applied quadratic functions to their design. Solving quadratic equations is an important skill required in all aspects of engineering.

At the end of this chapter you should be able to:

- define a quadratic equation
- solve quadratic equations by factorisation
- solve quadratic equations by 'completing the square'
- solve quadratic equations involving practical situations
- solve linear and quadratic equations simultaneously

14.1 Introduction

As stated in Chapter 11, an **equation** is a statement that two quantities are equal and to **'solve an equation'** means 'to find the value of the unknown'. The value of the unknown is called the **root** of the equation.

A **quadratic equation** is one in which the highest power of the unknown quantity is 2. For example, $x^2 - 3x + 1 = 0$ is a quadratic equation.

There are four methods of **solving quadratic equations**. These are:

(a) by factorisation (where possible),

(b) by 'completing the square',

(c) by using the 'quadratic formula', or

(d) graphically (see Chapter 19).

14.2 Solution of quadratic equations by factorisation

Multiplying out $(x + 1)(x - 3)$ gives $x^2 - 3x + x - 3$ i.e. $x^2 - 2x - 3$. The reverse process of moving from $x^2 - 2x - 3$ to $(x + 1)(x - 3)$ is called **factorising**.

If the quadratic expression can be factorised, this provides the simplest method of solving a quadratic equation.

For example, if $x^2 - 2x - 3 = 0$, then, by factorising

$$(x + 1)(x - 3) = 0$$

Hence, either $\quad (x + 1) = 0, \quad$ i.e. $\quad x = -1$

or $\qquad\qquad (x - 3) = 0, \quad$ i.e. $\quad x = 3$

Hence, $x = -1$ and $x = 3$ are the roots of the quadratic equation $x^2 - 2x - 3 = 0$

The technique of factorising is often one of trial and error.

Problem 1. Solve the equation $x^2 + x - 6 = 0$ by factorisation

The factors of x^2 are x and x. These are placed in brackets: $\quad (x \quad)(x \quad)$

The factors of -6 are $+6$ and -1, or -6 and $+1$, or $+3$ and -2, or -3 and $+2$

The only combination to give a middle term of $+x$ is $+3$ and -2,

i.e. $\qquad x^2 + x - 6 = (x + 3)(x - 2)$

The quadratic equation $x^2 + x - 6 = 0$ thus becomes

$$(x + 3)(x - 2) = 0$$

Since the only way that this can be true is for either the first or the second or both factors to be zero, then

either $\qquad (x + 3) = 0, \quad$ i.e. $\quad x = -3$

or $\qquad\qquad (x - 2) = 0, \quad$ i.e. $\quad x = 2$

Hence, **the roots of $x^2 + x - 6 = 0$ are $x = -3$** and **$x = 2$**

Problem 2. Solve the equation $x^2 + 2x - 8 = 0$ by factorisation

The factors of x^2 are x and x. These are placed in brackets: $\quad (x \quad)(x \quad)$

The factors of -8 are $+8$ and -1, or -8 and $+1$, or $+4$ and -2, or -4 and $+2$

The only combination to give a middle term of $+2x$ is $+4$ and -2,

i.e. $\qquad x^2 + 2x - 8 = (x + 4)(x - 2)$

(Note that the product of the two inner terms $(4x)$ added to the product of the two outer terms $(-2x)$ must equal the middle term, $+2x$ in this case.)

The quadratic equation $x^2 + 2x - 8 = 0$ thus becomes

$$(x + 4)(x - 2) = 0$$

Since the only way that this can be true is for either the first or the second or both factors to be zero,

either $\qquad (x + 4) = 0, \quad$ i.e. $\quad x = -4$

or $\qquad\qquad (x - 2) = 0, \quad$ i.e. $\quad x = 2$

Hence, **the roots of $x^2 + 2x - 8 = 0$ are $x = -4$** and **$x = 2$**

Problem 3. Determine the roots of $x^2 - 6x + 9 = 0$ by factorisation

$x^2 - 6x + 9 = (x - 3)(x - 3), \qquad$ i.e. $(x - 3)^2 = 0$

The LHS is known as a **perfect square**.

Hence, **$x = 3$** is the only root of the equation $x^2 - 6x + 9 = 0$

Problem 4. Solve the equation $x^2 - 4x = 0$

Factorising gives $\quad x(x - 4) = 0$

If $\qquad\qquad\qquad x(x - 4) = 0,$

either $\qquad\qquad\qquad x = 0 \quad$ or $\quad x - 4 = 0$

i.e. $\qquad\qquad\qquad$ **$x = 0$** or **$x = 4$**

These are the two roots of the given equation. Answers can always be checked by substitution into the original equation.

Problem 5. Solve the equation $x^2 + 3x - 4 = 0$

Factorising gives $\qquad (x - 1)(x + 4) = 0$

Hence, either $\qquad x - 1 = 0 \quad$ or $\quad x + 4 = 0$

i.e. $\qquad\qquad$ **$x = 1$** or **$x = -4$**

Problem 6. Determine the roots of $4x^2 - 25 = 0$ by factorisation

The LHS of $4x^2 - 25 = 0$ is **the difference of two squares**, $(2x)^2$ and $(5)^2$
By factorising, $\mathbf{4x^2 - 25 = (2x + 5)(2x - 5)}$,
i.e. $(2x + 5)(2x - 5) = 0$

Hence, either $(2x + 5) = 0$, i.e. $x = -\dfrac{5}{2} = -2.5$

or $(2x - 5) = 0$, i.e. $x = \dfrac{5}{2} = 2.5$

Problem 7. Solve the equation $x^2 - 5x + 6 = 0$

Factorising gives $(x - 3)(x - 2) = 0$

Hence, either $x - 3 = 0$ or $x - 2 = 0$

i.e. $x = 3$ or $x = 2$

Problem 8. Solve the equation $x^2 = 15 - 2x$

Rearranging gives $x^2 + 2x - 15 = 0$

Factorising gives $(x + 5)(x - 3) = 0$

Hence, either $x + 5 = 0$ or $x - 3 = 0$

i.e. $x = -5$ or $x = 3$

Problem 9. Solve the equation $3x^2 - 11x - 4 = 0$ by factorisation

The factors of $3x^2$ are $3x$ and x. These are placed in brackets: $(3x \quad)(x \quad)$
The factors of -4 are -4 and $+1$, or $+4$ and -1, or -2 and 2.
Remembering that the product of the two inner terms added to the product of the two outer terms must equal $-11x$, the only combination to give this is $+1$ and -4,

i.e. $\mathbf{3x^2 - 11x - 4 = (3x + 1)(x - 4)}$

The quadratic equation $3x^2 - 11x - 4 = 0$ thus becomes $(3x + 1)(x - 4) = 0$

Hence, either $(3x + 1) = 0$, i.e. $x = -\dfrac{1}{3}$

or $(x - 4) = 0$, i.e. $x = 4$

and both solutions may be checked in the original equation.

Problem 10. Solve the quadratic equation $4x^2 + 8x + 3 = 0$ by factorising

The factors of $4x^2$ are $4x$ and x, or $2x$ and $2x$

The factors of 3 are 3 and 1, or -3 and -1
Remembering that the product of the inner terms added to the product of the two outer terms must equal $+8x$, the only combination that is true (by trial and error) is

$$(4x^2 + 8x + 3) = (2x + 3)(2x + 1)$$

Hence, $(2x + 3)(2x + 1) = 0$, from which either $(2x + 3) = 0$ or $(2x + 1) = 0$

Thus, $2x = -3$, from which $x = -\dfrac{3}{2}$ or -1.5

or $2x = -1$, from which $x = -\dfrac{1}{2}$ or -0.5

which may be checked in the original equation.

Problem 11. Solve the quadratic equation $15x^2 + 2x - 8 = 0$ by factorising

The factors of $15x^2$ are $15x$ and x or $5x$ and $3x$
The factors of -8 are -4 are $+2$, or 4 and -2, or -8 and $+1$, or 8 and -1
By trial and error the only combination that works is

$$15x^2 + 2x - 8 = (5x + 4)(3x - 2)$$

Hence, $(5x + 4)(3x - 2) = 0$, from which either $5x + 4 = 0$ or $3x - 2 = 0$

Hence, $x = -\dfrac{4}{5}$ or $x = \dfrac{2}{3}$

which may be checked in the original equation.

Problem 12. The roots of a quadratic equation are $\dfrac{1}{3}$ and -2. Determine the equation in x

If the roots of a quadratic equation are, say, α and β, then $(x - \alpha)(x - \beta) = 0$

Hence, if $\alpha = \dfrac{1}{3}$ and $\beta = -2$,

$$\left(x - \frac{1}{3}\right)(x - (-2)) = 0$$

$$\left(x - \frac{1}{3}\right)(x + 2) = 0$$

$$x^2 - \frac{1}{3}x + 2x - \frac{2}{3} = 0$$

$$x^2 + \frac{5}{3}x - \frac{2}{3} = 0$$

or $$3x^2 + 5x - 2 = 0$$

Problem 13.   Find the equation in x whose roots are 5 and -5

If 5 and -5 are the roots of a quadratic equation then

$$(x-5)(x+5) = 0$$

i.e. $$x^2 - 5x + 5x - 25 = 0$$

i.e. $$\mathbf{x^2 - 25 = 0}$$

Problem 14.   Find the equation in x whose roots are 1.2 and -0.4

If 1.2 and -0.4 are the roots of a quadratic equation then

$$(x-1.2)(x+0.4) = 0$$

i.e. $$x^2 - 1.2x + 0.4x - 0.48 = 0$$

i.e. $$\mathbf{x^2 - 0.8x - 0.48 = 0}$$

Now try the following Practice Exercise

Practice Exercise 71   Solving quadratic equations by factorisation (answers on page 449)

In Problems 1 to 30, solve the given equations by factorisation.

1. $x^2 - 16 = 0$
2. $x^2 + 4x - 32 = 0$
3. $(x+2)^2 = 16$
4. $4x^2 - 9 = 0$
5. $3x^2 + 4x = 0$
6. $8x^2 - 32 = 0$
7. $x^2 - 8x + 16 = 0$
8. $x^2 + 10x + 25 = 0$
9. $x^2 - 2x + 1 = 0$
10. $x^2 + 5x + 6 = 0$
11. $x^2 + 10x + 21 = 0$
12. $x^2 - x - 2 = 0$
13. $y^2 - y - 12 = 0$
14. $y^2 - 9y + 14 = 0$
15. $x^2 + 8x + 16 = 0$
16. $x^2 - 4x + 4 = 0$
17. $x^2 + 6x + 9 = 0$
18. $x^2 - 9 = 0$
19. $3x^2 + 8x + 4 = 0$
20. $4x^2 + 12x + 9 = 0$
21. $4z^2 - \dfrac{1}{16} = 0$
22. $x^2 + 3x - 28 = 0$
23. $2x^2 - x - 3 = 0$
24. $6x^2 - 5x + 1 = 0$
25. $10x^2 + 3x - 4 = 0$
26. $21x^2 - 25x = 4$
27. $8x^2 + 13x - 6 = 0$
28. $5x^2 + 13x - 6 = 0$
29. $6x^2 - 5x - 4 = 0$
30. $8x^2 + 2x - 15 = 0$

In Problems 31 to 36, determine the quadratic equations in x whose roots are

31. 3 and 1
32. 2 and -5
33. -1 and -4
34. 2.5 and -0.5
35. 6 and -6
36. 2.4 and -0.7

14.3   Solution of quadratic equations by 'completing the square'

An expression such as x^2 or $(x+2)^2$ or $(x-3)^2$ is called a **perfect square**.

If $x^2 = 3$ then $x = \pm\sqrt{3}$

If $(x+2)^2 = 5$ then $x+2 = \pm\sqrt{5}$ and $x = -2 \pm \sqrt{5}$

If $(x-3)^2 = 8$ then $x-3 = \pm\sqrt{8}$ and $x = 3 \pm \sqrt{8}$

Hence, if a quadratic equation can be rearranged so that one side of the equation is a perfect square and the other side of the equation is a number, then the solution of the equation is readily obtained by taking the square roots of each side as in the above examples. The process of rearranging one side of a quadratic equation into a perfect square before solving is called 'completing the square'.

$$(x+a)^2 = x^2 + 2ax + a^2$$

Thus, in order to make the quadratic expression $x^2 + 2ax$ into a perfect square, it is necessary to add (half the coefficient of x)^2, i.e. $\left(\dfrac{2a}{2}\right)^2$ or a^2

For example, $x^2 + 3x$ becomes a perfect square by adding $\left(\dfrac{3}{2}\right)^2$, i.e.

$$x^2 + 3x + \left(\dfrac{3}{2}\right)^2 = \left(x + \dfrac{3}{2}\right)^2$$

The method of completing the square is demonstrated in the following worked problems.

Problem 15.   Solve $2x^2 + 5x = 3$ by completing the square

The procedure is as follows.

(i)   Rearrange the equation so that all terms are on the same side of the equals sign (and the coefficient of the x^2 term is positive). Hence,

$$2x^2 + 5x - 3 = 0$$

(ii) Make the coefficient of the x^2 term unity. In this case this is achieved by dividing throughout by 2. Hence,

$$\frac{2x^2}{2} + \frac{5x}{2} - \frac{3}{2} = 0$$

i.e. $$x^2 + \frac{5}{2}x - \frac{3}{2} = 0$$

(iii) Rearrange the equations so that the x^2 and x terms are on one side of the equals sign and the constant is on the other side. Hence,

$$x^2 + \frac{5}{2}x = \frac{3}{2}$$

(iv) Add to both sides of the equation (half the coefficient of x)^2. In this case the coefficient of x is $\frac{5}{2}$

Half the coefficient squared is therefore $\left(\frac{5}{4}\right)^2$

Thus,

$$x^2 + \frac{5}{2}x + \left(\frac{5}{4}\right)^2 = \frac{3}{2} + \left(\frac{5}{4}\right)^2$$

The LHS is now a perfect square, i.e.

$$\left(x + \frac{5}{4}\right)^2 = \frac{3}{2} + \left(\frac{5}{4}\right)^2$$

(v) Evaluate the RHS. Thus,

$$\left(x + \frac{5}{4}\right)^2 = \frac{3}{2} + \frac{25}{16} = \frac{24 + 25}{16} = \frac{49}{16}$$

(vi) Take the square root of both sides of the equation (remembering that the square root of a number gives a $\pm$ answer). Thus,

$$\sqrt{\left(x + \frac{5}{4}\right)^2} = \sqrt{\left(\frac{49}{16}\right)}$$

i.e. $$x + \frac{5}{4} = \pm\frac{7}{4}$$

(vii) Solve the simple equation. Thus,

$$x = -\frac{5}{4} \pm \frac{7}{4}$$

i.e. $$x = -\frac{5}{4} + \frac{7}{4} = \frac{2}{4} = \frac{1}{2} \text{ or } 0.5$$

or $$x = -\frac{5}{4} - \frac{7}{4} = -\frac{12}{4} = -3$$

Hence, $x = 0.5$ or $x = -3$; i.e. **the roots of the equation $2x^2 + 5x = 3$ are 0.5 and −3**

Problem 16. Solve $2x^2 + 9x + 8 = 0$, correct to 3 significant figures, by completing the square

Making the coefficient of x^2 unity gives

$$x^2 + \frac{9}{2}x + 4 = 0$$

Rearranging gives $$x^2 + \frac{9}{2}x = -4$$

Adding to both sides (half the coefficient of x)^2 gives

$$x^2 + \frac{9}{2}x + \left(\frac{9}{4}\right)^2 = \left(\frac{9}{4}\right)^2 - 4$$

The LHS is now a perfect square. Thus,

$$\left(x + \frac{9}{4}\right)^2 = \frac{81}{16} - 4 = \frac{81}{16} - \frac{64}{16} = \frac{17}{16}$$

Taking the square root of both sides gives

$$x + \frac{9}{4} = \sqrt{\left(\frac{17}{16}\right)} = \pm 1.031$$

Hence, $$x = -\frac{9}{4} \pm 1.031$$

i.e. $x = -1.22$ or -3.28, correct to 3 significant figures.

Problem 17. By completing the square, solve the quadratic equation $4.6y^2 + 3.5y - 1.75 = 0$, correct to 3 decimal places

$$4.6y^2 + 3.5y - 1.75 = 0$$

Making the coefficient of y^2 unity gives

$$y^2 + \frac{3.5}{4.6}y - \frac{1.75}{4.6} = 0$$

and rearranging gives $$y^2 + \frac{3.5}{4.6}y = \frac{1.75}{4.6}$$

Adding to both sides (half the coefficient of y)^2 gives

$$y^2 + \frac{3.5}{4.6}y + \left(\frac{3.5}{9.2}\right)^2 = \frac{1.75}{4.6} + \left(\frac{3.5}{9.2}\right)^2$$

The LHS is now a perfect square. Thus,

$$\left(y + \frac{3.5}{9.2}\right)^2 = 0.5251654$$

Taking the square root of both sides gives

$$y + \frac{3.5}{9.2} = \sqrt{0.5251654} = \pm 0.7246830$$

Hence,

$$y = -\frac{3.5}{9.2} \pm 0.7246830$$

i.e. $y = 0.344$ or -1.105

Now try the following Practice Exercise

Practice Exercise 72   Solving quadratic equations by completing the square (answers on page 450)

Solve the following equations correct to 3 decimal places by completing the square.

1. $x^2 + 4x + 1 = 0$ 2. $2x^2 + 5x - 4 = 0$
3. $3x^2 - x - 5 = 0$ 4. $5x^2 - 8x + 2 = 0$
5. $4x^2 - 11x + 3 = 0$ 6. $2x^2 + 5x = 2$

14.4 Solution of quadratic equations by formula

Let the general form of a quadratic equation be given by $ax^2 + bx + c = 0$, where a, b and c are constants. Dividing $ax^2 + bx + c = 0$ by a gives

$$x^2 + \frac{b}{a}x + \frac{c}{a} = 0$$

Rearranging gives $x^2 + \frac{b}{a}x = -\frac{c}{a}$

Adding to each side of the equation the square of half the coefficient of the term in x to make the LHS a perfect square gives

$$x^2 + \frac{b}{a}x + \left(\frac{b}{2a}\right)^2 = \left(\frac{b}{2a}\right)^2 - \frac{c}{a}$$

Rearranging gives $\left(x + \frac{b}{a}\right)^2 = \frac{b^2}{4a^2} - \frac{c}{a} = \frac{b^2 - 4ac}{4a^2}$

Taking the square root of both sides gives

$$x + \frac{b}{2a} = \sqrt{\left(\frac{b^2 - 4ac}{4a^2}\right)} = \frac{\pm\sqrt{b^2 - 4ac}}{2a}$$

Hence, $x = -\frac{b}{2a} \pm \frac{\sqrt{b^2 - 4ac}}{2a}$

i.e. the quadratic formula is

$$x = \frac{-b \pm \sqrt{b^2 - 4ac}}{2a}$$

(This method of obtaining the formula is completing the square – as shown in the previous section.)
In summary,

if $ax^2 + bx + c = 0$ then $x = \dfrac{-b \pm \sqrt{b^2 - 4ac}}{2a}$

This is known as the **quadratic formula**.

Problem 18.   Solve $x^2 + 2x - 8 = 0$ by using the quadratic formula

Comparing $x^2 + 2x - 8 = 0$ with $ax^2 + bx + c = 0$ gives $a = 1, b = 2$ and $c = -8$
Substituting these values into the quadratic formula

$$x = \frac{-b \pm \sqrt{b^2 - 4ac}}{2a}$$

gives

$$x = \frac{-2 \pm \sqrt{2^2 - 4(1)(-8)}}{2(1)} = \frac{-2 \pm \sqrt{4 + 32}}{2}$$

$$= \frac{-2 \pm \sqrt{36}}{2} = \frac{-2 \pm 6}{2}$$

$$= \frac{-2 + 6}{2} \text{ or } \frac{-2 - 6}{2}$$

Hence, $x = \dfrac{4}{2}$ or $\dfrac{-8}{2}$, i.e. $x = 2$ or $x = -4$

Problem 19.   Solve $3x^2 - 11x - 4 = 0$ by using the quadratic formula

Comparing $3x^2 - 11x - 4 = 0$ with $ax^2 + bx + c = 0$ gives $a = 3, b = -11$ and $c = -4$. Hence,

$$x = \frac{-(-11) \pm \sqrt{(-11)^2 - 4(3)(-4)}}{2(3)}$$

$$= \frac{+11 \pm \sqrt{121 + 48}}{6} = \frac{11 \pm \sqrt{169}}{6}$$

$$= \frac{11 \pm 13}{6} = \frac{11 + 13}{6} \text{ or } \frac{11 - 13}{6}$$

Hence, $x = \dfrac{24}{6}$ or $\dfrac{-2}{6}$, i.e. $x = 4$ or $x = -\dfrac{1}{3}$

Problem 20.   Solve $4x^2 + 7x + 2 = 0$ giving the roots correct to 2 decimal places

Comparing $4x^2 + 7x + 2 = 0$ with $ax^2 + bx + c$ gives $a = 4, b = 7$ and $c = 2$. Hence,

$$x = \frac{-7 \pm \sqrt{7^2 - 4(4)(2)}}{2(4)}$$

$$= \frac{-7 \pm \sqrt{17}}{8} = \frac{-7 \pm 4.123}{8}$$

$$= \frac{-7 + 4.123}{8} \text{ or } \frac{-7 - 4.123}{8}$$

Hence, $x = -0.36$ or -1.39, correct to 2 decimal places.

Problem 21. Use the quadratic formula to solve $\dfrac{x+2}{4} + \dfrac{3}{x-1} = 7$ correct to 4 significant figures

---

Multiplying throughout by $4(x-1)$ gives

$$4(x-1)\frac{(x+2)}{4} + 4(x-1)\frac{3}{(x-1)} = 4(x-1)(7)$$

Cancelling gives $\quad (x-1)(x+2) + (4)(3) = 28(x-1)$

$$x^2 + x - 2 + 12 = 28x - 28$$

Hence, $\quad\quad\quad\quad\quad x^2 - 27x + 38 = 0$

Using the quadratic formula,

$$x = \frac{-(-27) \pm \sqrt{(-27)^2 - 4(1)(38)}}{2}$$

$$= \frac{27 \pm \sqrt{577}}{2} = \frac{27 \pm 24.0208}{2}$$

Hence, $\quad\quad x = \dfrac{27 + 24.0208}{2} = 25.5104$

or $\quad\quad\quad x = \dfrac{27 - 24.0208}{2} = 1.4896$

Hence, $x = \mathbf{25.51}$ or $\mathbf{1.490}$, correct to 4 significant figures.

Now try the following Practice Exercise

Practice Exercise 73   Solving quadratic equations by formula (answers on page 450)

Solve the following equations by using the quadratic formula, correct to 3 decimal places.

1. $2x^2 + 5x - 4 = 0$

2. $5.76x^2 + 2.86x - 1.35 = 0$

3. $2x^2 - 7x + 4 = 0$

4. $4x + 5 = \dfrac{3}{x}$

5. $(2x + 1) = \dfrac{5}{x - 3}$

6. $3x^2 - 5x + 1 = 0$

7. $4x^2 + 6x - 8 = 0$

8. $5.6x^2 - 11.2x - 1 = 0$

9. $3x(x + 2) + 2x(x - 4) = 8$

10. $4x^2 - x(2x + 5) = 14$

11. $\dfrac{5}{x - 3} + \dfrac{2}{x - 2} = 6$

12. $\dfrac{3}{x - 7} + 2x = 7 + 4x$

13. $\dfrac{x + 1}{x - 1} = x - 3$

14.5 Practical problems involving quadratic equations

There are many **practical problems** in which a quadratic equation has first to be obtained, from given information, before it is solved.

Problem 22. The area of a rectangular plate is $23.6\,\text{cm}^2$ and its width is $3.10\,\text{cm}$ shorter than its length. Determine the dimensions of the rectangle, correct to 3 significant figures

---

Let the length of the rectangle be $x\,\text{cm}$. Then the width is $(x - 3.10)\,\text{cm}$.

$$\text{Area} = \text{length} \times \text{width} = x(x - 3.10) = 23.6$$

i.e. $\quad\quad\quad\quad x^2 - 3.10x - 23.6 = 0$

Using the quadratic formula,

$$x = \frac{-(-3.10) \pm \sqrt{(-3.10)^2 - 4(1)(-23.6)}}{2(1)}$$

$$= \frac{3.10 \pm \sqrt{9.61 + 94.4}}{2} = \frac{3.10 \pm 10.20}{2}$$

$$= \frac{13.30}{2} \text{ or } \frac{-7.10}{2}$$

Hence, $x = 6.65\,\text{cm}$ or $-3.55\,\text{cm}$. The latter solution is neglected since length cannot be negative.
Thus, length $x = 6.65\,\text{cm}$ and width $= x - 3.10 = 6.65 - 3.10 = 3.55\,\text{cm}$, i.e. **the dimensions of the rectangle are 6.65 cm by 3.55 cm.**
(Check: Area $= 6.65 \times 3.55 = 23.6\,\text{cm}^2$, correct to 3 significant figures.)

Problem 23. Calculate the diameter of a solid cylinder which has a height of 82.0 cm and a total surface area of 2.0 m^2

Total surface area of a cylinder
$$= \text{curved surface area} + 2 \text{ circular ends}$$
$$= 2\pi rh + 2\pi r^2 \text{ (where } r = \text{radius and } h = \text{height)}$$

Since the total surface area $= 2.0$ m^2 and the height $h = 82$ cm or 0.82 m,

$$2.0 = 2\pi r(0.82) + 2\pi r^2$$

i.e.
$$2\pi r^2 + 2\pi r(0.82) - 2.0 = 0$$

Dividing throughout by 2π gives $r^2 + 0.82r - \dfrac{1}{\pi} = 0$
Using the quadratic formula,

$$r = \frac{-0.82 \pm \sqrt{(0.82)^2 - 4(1)\left(-\frac{1}{\pi}\right)}}{2(1)}$$

$$= \frac{-0.82 \pm \sqrt{1.94564}}{2} = \frac{-0.82 \pm 1.39486}{2}$$

$$= 0.2874 \text{ or } -1.1074$$

Thus, the radius r of the cylinder is 0.2874 m (the negative solution being neglected).

Hence, the diameter of the cylinder

$$= 2 \times 0.2874$$

$$= \mathbf{0.5748\,m} \text{ or } \mathbf{57.5\,cm}$$

correct to 3 significant figures.

Problem 24. The height s metres of a mass projected vertically upwards at time t seconds is $s = ut - \dfrac{1}{2}gt^2$. Determine how long the mass will take after being projected to reach a height of 16 m (a) on the ascent and (b) on the descent, when $u = 30$ m/s and $g = 9.81$ m/s^2

When height $s = 16$ m, $16 = 30t - \dfrac{1}{2}(9.81)t^2$

i.e.
$$4.905t^2 - 30t + 16 = 0$$

Using the quadratic formula,

$$t = \frac{-(-30) \pm \sqrt{(-30)^2 - 4(4.905)(16)}}{2(4.905)}$$

$$= \frac{30 \pm \sqrt{586.1}}{9.81} = \frac{30 \pm 24.21}{9.81} = 5.53 \text{ or } 0.59$$

Hence, the mass will reach a height of 16 m after 0.59 s on the ascent and after 5.53 s on the descent.

Problem 25. A shed is 4.0 m long and 2.0 m wide. A concrete path of constant width is laid all the way around the shed. If the area of the path is 9.50 m^2, calculate its width to the nearest centimetre

Fig. 14.1 shows a plan view of the shed with its surrounding path of width t metres.

Figure 14.1

$$\text{Area of path} = 2(2.0 \times t) + 2t(4.0 + 2t)$$

i.e.
$$9.50 = 4.0t + 8.0t + 4t^2$$

or
$$4t^2 + 12.0t - 9.50 = 0$$

Hence,

$$t = \frac{-(12.0) \pm \sqrt{(12.0)^2 - 4(4)(-9.50)}}{2(4)}$$

$$= \frac{-12.0 \pm \sqrt{296.0}}{8} = \frac{-12.0 \pm 17.20465}{8}$$

i.e. $t = 0.6506$ m or -3.65058 m.

Neglecting the negative result, which is meaningless, the width of the path, $t = \mathbf{0.651\,m}$ or $\mathbf{65\,cm}$ correct to the nearest centimetre.

Problem 26. If the total surface area of a solid metal cone is 486.2 cm^2 and its slant height is 15.3 cm, determine its base diameter.

From Chapter 28, page 299, the total surface area A of a solid cone is given by $A = \pi r l + \pi r^2$, where l is the slant height and r the base radius.
If $A = 482.2$ and $l = 15.3$, then
$$482.2 = \pi r(15.3) + \pi r^2$$

i.e. $\pi r^2 + 15.3\pi r - 482.2 = 0$

or $r^2 + 15.3r - \dfrac{482.2}{\pi} = 0$

Using the quadratic formula,

$$r = \dfrac{-15.3 \pm \sqrt{\left[(15.3)^2 - 4\left(\dfrac{-482.2}{\pi}\right)\right]}}{2}$$

$$= \dfrac{-15.3 \pm \sqrt{848.0461}}{2} = \dfrac{-15.3 \pm 29.12123}{2}$$

Hence, radius $r = 6.9106$ cm (or -22.21 cm, which is meaningless and is thus ignored).

Thus, the **diameter of the base** $= 2r = 2(6.9106)$

$$= \mathbf{13.82\,cm}.$$

Now try the following Practice Exercise

Practice Exercise 74   Practical problems involving quadratic equations (answers on page 450)

1. The angle a rotating shaft turns through in t seconds is given by $\theta = \omega t + \dfrac{1}{2}\alpha t^2$. Determine the time taken to complete 4 radians if ω is 3.0 rad/s and α is 0.60 rad/s^2.

2. The power P developed in an electrical circuit is given by $P = 10I - 8I^2$, where I is the current in amperes. Determine the current necessary to produce a power of 2.5 watts in the circuit.

3. The area of a triangle is 47.6 cm^2 and its perpendicular height is 4.3 cm more than its base length. Determine the length of the base correct to 3 significant figures.

4. The sag, l, in metres in a cable stretched between two supports, distance x m apart, is given by $l = \dfrac{12}{x} + x$. Determine the distance between the supports when the sag is 20 m.

5. The acid dissociation constant K_a of ethanoic acid is 1.8×10^{-5} mol dm^{-3} for a particular solution. Using the Ostwald dilution law, $K_a = \dfrac{x^2}{v(1-x)}$, determine x, the degree of ionisation, given that $v = 10$ dm^3.

6. A rectangular building is 15 m long by 11 m wide. A concrete path of constant width is laid all the way around the building. If the area of the path is 60.0 m^2, calculate its width correct to the nearest millimetre.

7. The total surface area of a closed cylindrical container is 20.0 m^2. Calculate the radius of the cylinder if its height is 2.80 m.

8. The bending moment M at a point in a beam is given by $M = \dfrac{3x(20-x)}{2}$, where x metres is the distance from the point of support. Determine the value of x when the bending moment is 50 N m.

9. A tennis court measures 24 m by 11 m. In the layout of a number of courts an area of ground must be allowed for at the ends and at the sides of each court. If a border of constant width is allowed around each court and the total area of the court and its border is 950 m^2, find the width of the borders.

10. Two resistors, when connected in series, have a total resistance of 40 ohms. When connected in parallel their total resistance is 8.4 ohms. If one of the resistors has a resistance of R_x, ohms,

    (a) show that $R_x^2 - 40R_x + 336 = 0$ and

    (b) calculate the resistance of each.

11. When a ball is thrown vertically upwards its height h varies with time t according to the equation $h = 25t - 4t^2$. Determine the times, correct to 3 significant figures, when the height is 12 m.

12. In an RLC electrical circuit, reactance X is given by: $X = \omega L - \dfrac{1}{\omega C}$

    $X = 220\ \Omega$, inductance $L = 800$ mH and capacitance $C = 25\ \mu$F. The angular velocity ω is measured in radians per second. Calculate the value of ω.

13. A point of contraflexure from the left-hand end of a 6 m beam is given by the value of x in the following equation:

$$-x^2 + 11.25x - 22.5 = 0$$

Determine the point of contraflexure.

14. The vertical height, h, and the horizontal distance travelled, x, of a projectile fired at an angle of $45°$ at an initial velocity, v_0, are related by the equation:

$$h = x - \frac{gx^2}{v_0{}^2}$$

If the projectile has an initial velocity of 120 m/s, calculate the values of x when the projectile is at a height of 200 m, assuming that $g = 9.81$ m/s^2.

15. The stress, ρ, set up in a bar of length, ℓ, cross-sectional area, A, by a mass W, falling a distance h is given by the formula:

$$\rho^2 - \frac{2W}{A}\rho - \frac{2WEh}{A\ell} = 0$$

Given that $E = 12500$, $\ell = 110$, $h = 0.5$, $W = 2.25$ and $A = 3.20$, calculate the positive value of ρ correct to 3 significant figures.

14.6 Solution of linear and quadratic equations simultaneously

Sometimes a linear equation and a quadratic equation need to be solved simultaneously. An algebraic method of solution is shown in Problem 27; a graphical solution of a similar problem is shown in Chapter 19, page 194.

> **Problem 27.** Determine the values of x and y which simultaneously satisfy the equations $y = 5x - 4 - 2x^2$ and $y = 6x - 7$

For a simultaneous solution the values of y must be equal, hence the RHS of each equation is equated.

Thus, $\qquad 5x - 4 - 2x^2 = 6x - 7$

Rearranging gives $\quad 5x - 4 - 2x^2 - 6x + 7 = 0$

i.e. $\qquad\qquad\qquad -x + 3 - 2x^2 = 0$

or $\qquad\qquad\qquad 2x^2 + x - 3 = 0$

Factorising gives $(2x + 3)(x - 1) = 0$

i.e. $\qquad\qquad x = -\frac{3}{2}$ or $x = 1$

In the equation $y = 6x - 7$,

when $x = -\frac{3}{2}, \qquad y = 6\left(-\frac{3}{2}\right) - 7 = -16$

and when $x = 1, \quad y = 6 - 7 = -1$

(Checking the result in $y = 5x - 4 - 2x^2$:

when $x = -\frac{3}{2}, \qquad y = 5\left(-\frac{3}{2}\right) - 4 - 2\left(-\frac{3}{2}\right)^2$

$$= -\frac{15}{2} - 4 - \frac{9}{2} = -16, \text{as above,}$$

and when $x = 1, \quad y = 5 - 4 - 2 = -1$, as above.)

Hence, the simultaneous solutions occur when

$$x = -\frac{3}{2}, y = -16 \text{ and when } x = 1, y = -1$$

Now try the following Practice Exercise

> **Practice Exercise 75 Solving linear and quadratic equations simultaneously (answers on page 450)**
>
> Determine the solutions of the following simultaneous equations.
>
> 1. $y = x^2 + x + 1$
> $y = 4 - x$
> 2. $y = 15x^2 + 21x - 11$
> $y = 2x - 1$
> 3. $2x^2 + y = 4 + 5x$
> $x + y = 4$

It is possible, with a modern scientific calculator, to solve quadratic equations – choose 'Equations' from the 'Mode' menu and follow the rules for your calculator. Naturally, answers may be obtained very much more quickly than by the analytical means shown in this chapter.

Practice Exercise 76   Multiple-choice questions on solving quadratic equations (answers on page 450)

Each question has only one correct answer

1. A quadratic equation has degree:
(a) 0   (b) 1   (c) 3   (d) 2

2. Which of the following equations is not a quadratic equation?
(a) $x^2 + 4x - 5 = 0$ (b) $5 + 2x - x^2 = 0$
(c) $x^2 - 16 = 0$ (d) $x^2 + x^3 - 7 = 0$

3. The quadratic equation in x whose roots are -2 and $+5$ is:
(a) $x^2 - 3x - 10 = 0$ (b) $x^2 + 7x + 10 = 0$
(c) $x^2 + 3x - 10 = 0$ (d) $x^2 - 7x - 10 = 0$

4. If the equation $ax^2 + bx + c = 0$ has equal roots, then c is equal to:
(a) $\dfrac{b}{2a}$ (b) $-\dfrac{b^2}{2a}$ (c) $\dfrac{b^2}{4a}$ (d) $-\dfrac{b}{2a}$

5. $8x^2 + 13x - 6 = (x+p)(qx-3)$. The values of p and q are:
(a) $p = 2, q = 8$ (b) $p = 3, q = 2$
(c) $p = -2, q = 4$ (d) $p = 1, q = 8$

6. The roots of the quadratic equation $8x^2 + 10x - 3 = 0$ are:
(a) $-\dfrac{1}{4}$ and $\dfrac{3}{2}$ (b) 4 and $\dfrac{2}{3}$
(c) $-\dfrac{3}{2}$ and $\dfrac{1}{4}$ (d) $\dfrac{2}{3}$ and -4

7. For the equation $(x^2 + 10x + p)$ to be a perfect square, the value of p will be:
(a) 5   (b) 25   (c) 100   (d) 10

8. The roots of the quadratic equation $2x^2 - 5x + 1 = 0$, correct to 2 decimal places, are:
(a) -0.22 and -2.28 (b) 2.69 and -0.19
(c) 0.19 and -2.69 (d) 2.28 and 0.22

9. The quadratic equation whose roots are 4 and -2 is given by:
(a) $x^2 - 2x + 8 = 0$ (b) $x^2 - 2x - 8 = 0$
(c) $x^2 + 2x + 8 = 0$ (d) $x^2 + 2x - 8 = 0$

10. The roots of the quadratic equation $3x^2 - 7x - 6 = 0$ are:
(a) 2 and -5 (b) 2 and -9
(c) 3 and $-\dfrac{2}{3}$ (d) 3 and -6

11. The quadratic equation whose roots are 1 and 0 is given by:
(a) $x^2 - 2x + 1 = 0$ (b) $x^2 - x = 0$
(c) $x^2 + 2x + 1 = 0$ (d) $x^2 + x = 0$

12. For the equation $x^2 + kx + 9 = 0$ to be a perfect square, the value of k will be:
(a) 6   (b) -3 (c) -6 (d) 3

13. The roots of the quadratic equation $x^2 + 7x + 10 = 0$ are:
(a) -2 and -5 (b) 2 and 5
(c) 2 and -5 (d) -2 and 5

14. The equation $x^2 - x - 5 = 0$ has:
(a) no roots   (b) imaginary roots
(c) real and equal roots
(d) real and different roots

15. The height S metres of a mass thrown vertically upwards at time t seconds is given by $S = 80t - 16t^2$. To reach a height of 50 metres on the descent will take the mass:
(a) 0.73 s   (b) 5.56 s
(c) 4.27 s   (d) 81.77 s

For fully worked solutions to each of the problems in Practice Exercises 71 to 75 in this chapter, go to the website:
www.routledge.com/cw/bird

Chapter 15

Logarithms

Why it is important to understand: Logarithms

All types of engineers use natural and common logarithms. Chemical engineers use them to measure radioactive decay and pH solutions, both of which are measured on a logarithmic scale. The Richter scale which measures earthquake intensity is a logarithmic scale. Biomedical engineers use logarithms to measure cell decay and growth, and also to measure light intensity for bone mineral density measurements. In electrical engineering, a dB (decibel) scale is very useful for expressing attenuations in radio propagation and circuit gains, and logarithms are used for implementing arithmetic operations in digital circuits. Logarithms are especially useful when dealing with the graphical analysis of non-linear relationships and logarithmic scales are used to linearise data to make data analysis simpler. Understanding and using logarithms is clearly important in all branches of engineering.

At the end of this chapter you should be able to:

- define base, power, exponent and index
- define a logarithm
- distinguish between common and Napierian (i.e. hyperbolic or natural) logarithms
- evaluate logarithms to any base
- state the laws of logarithms
- simplify logarithmic expressions
- solve equations involving logarithms
- solve indicial equations
- sketch graphs of $\log_{10} x$ and $\log_e x$

15.1 Introduction to logarithms

With the use of calculators firmly established, logarithmic tables are now rarely used for calculation. However, the theory of logarithms is important, for there are several scientific and engineering laws that involve the rules of logarithms.

From Chapter 7, we know that $\qquad 16 = 2^4$

The number 4 is called the **power** or the **exponent** or

the **index**. In the expression 2^4, the number 2 is called the **base**.

In another example, we know that $\qquad 64 = 8^2$

In this example, 2 is the power, or exponent, or index. The number 8 is the base.

What is a logarithm?

Consider the expression $16 = 2^4$

An alternative, yet equivalent, way of writing this expression is $\log_2 16 = 4$

This is stated as 'log to the base 2 of 16 equals 4'
We see that the logarithm is the same as the power or index in the original expression. It is the base in the original expression that becomes the base of the logarithm.

$$\text{The two statements } 16 = 2^4 \text{ and}$$

$$\log_2 16 = 4 \text{ are equivalent}$$

If we write either of them, we are automatically implying the other.
In general, if a number y can be written in the form a^x, then the index x is called the 'logarithm of y to the base of a', i.e.

$$\text{if } y = a^x \text{ then } x = \log_a y$$

In another example, if we write down that $64 = 8^2$ then the equivalent statement using logarithms is $\log_8 64 = 2$.
In another example, if we write down that $\log_3 27 = 3$ then the equivalent statement using powers is $3^3 = 27$. So the two sets of statements, one involving powers and one involving logarithms, are equivalent.

Common logarithms

From the above, if we write down that $1000 = 10^3$, then $3 = \log_{10} 1000$. This may be checked using the 'log' button on your calculator.
Logarithms having a base of 10 are called **common logarithms** and $\log_{10}$ is usually abbreviated to lg. The following values may be checked using a calculator.

$$\lg 27.5 = 1.4393\ldots$$

$$\lg 378.1 = 2.5776\ldots$$

$$\lg 0.0204 = -1.6903\ldots$$

Napierian logarithms

Logarithms having a base of e (where e is a mathematical constant approximately equal to 2.7183) are called **hyperbolic, Napierian** or **natural logarithms**, and $\log_e$ is usually abbreviated to ln. The following values may be checked using a calculator.

$$\ln 3.65 = 1.2947\ldots$$

$$\ln 417.3 = 6.0338\ldots$$

$$\ln 0.182 = -1.7037\ldots$$

Napierian logarithms are explained further in Chapter 16, following.

Here are some worked problems to help understanding of logarithms.

Problem 1. Evaluate $\log_3 9$

Let $x = \log_3 9$ then $3^x = 9$ from the definition of a logarithm,

i.e. $3^x = 3^2$, from which $x = 2$

Hence, $\log_3 9 = 2$

Problem 2. Evaluate $\log_{10} 10$

Let $x = \log_{10} 10$ then $10^x = 10$ from the definition of a logarithm,

i.e. $10^x = 10^1$, from which $x = 1$

Hence, $\log_{10} 10 = 1$ (which may be checked using a calculator).

Problem 3. Evaluate $\log_{16} 8$

Let $x = \log_{16} 8$ then $16^x = 8$ from the definition of a logarithm,

i.e. $(2^4)^x = 2^3$ i.e. $2^{4x} = 2^3$ from the laws of indices,

from which, $4x = 3$ and $x = \dfrac{3}{4}$

Hence, $\log_{16} 8 = \dfrac{3}{4}$

Problem 4. Evaluate $\lg 0.001$

Let $x = \lg 0.001 = \log_{10} 0.001$ then $10^x = 0.001$

i.e. $10^x = 10^{-3}$

from which, $x = -3$

Hence, $\lg 0.001 = -3$ (which may be checked using a calculator)

Problem 5. Evaluate $\ln e$

Let $x = \ln e = \log_e e$ then $e^x = e$

i.e. $e^x = e^1$, from which $x = 1$

Hence, $\ln e = 1$ (which may be checked by a calculator)

Problem 6. Evaluate $\log_3 \frac{1}{81}$

Let $x = \log_3 \frac{1}{81}$ then $3^x = \frac{1}{81} = \frac{1}{3^4} = 3^{-4}$

from which $x = -4$

Hence, $\log_3 \frac{1}{81} = -4$

Problem 7. Solve the equation $\lg x = 3$

If $\lg x = 3$ then $\log_{10} x = 3$
and $x = 10^3$ i.e. $x = 1000$

Problem 8. Solve the equation $\log_2 x = 5$

If $\log_2 x = 5$ then $x = 2^5 = 32$

Problem 9. Solve the equation $\log_5 x = -2$

If $\log_5 x = -2$ then $x = 5^{-2} = \frac{1}{5^2} = \frac{1}{25}$

Now try the following Practice Exercise

Practice Exercise 77 Laws of logarithms (answers on page 450)

In Problems 1 to 11, evaluate the given expressions.

1. $\log_{10} 10000$ 2. $\log_2 16$ 3. $\log_5 125$

4. $\log_2 \frac{1}{8}$ 5. $\log_8 2$ 6. $\log_7 343$

7. $\lg 100$ 8. $\lg 0.01$ 9. $\log_4 8$

10. $\log_{27} 3$ 11. $\ln e^2$

In Problems 12 to 18, solve the equations.

12. $\log_{10} x = 4$ 13. $\lg x = 5$

14. $\log_3 x = 2$ 15. $f_4 x = -2\frac{1}{2}$

16. $\lg x = -2$ 17. $\log_8 x = -\frac{4}{3}$

18. $\ln x = 3$

15.2 Laws of logarithms

There are three laws of logarithms, which apply to any base:

(1) To multiply two numbers:

$$\log(A \times B) = \log A + \log B$$

The following may be checked by using a calculator.

$\lg 10 = 1$

Also, $\lg 5 + \lg 2 = 0.69897\ldots + 0.301029\ldots = 1$

Hence, $\lg(5 \times 2) = \lg 10 = \lg 5 + \lg 2$

(2) To divide two numbers:

$$\log\left(\frac{A}{B}\right) = \log A - \log B$$

The following may be checked using a calculator.

$\ln\left(\frac{5}{2}\right) = \ln 2.5 = 0.91629\ldots$

Also, $\ln 5 - \ln 2 = 1.60943\ldots - 0.69314\ldots$

$= 0.91629\ldots$

Hence, $\ln\left(\frac{5}{2}\right) = \ln 5 - \ln 2$

(3) To raise a number to a power:

$$\log A^n = n \log A$$

The following may be checked using a calculator.

$\lg 5^2 = \lg 25 = 1.39794\ldots$

Also, $2\lg 5 = 2 \times 0.69897\ldots = 1.39794\ldots$

Hence, $\lg 5^2 = 2\lg 5$

Here are some worked problems to help understanding of the laws of logarithms.

Problem 10. Write $\log 4 + \log 7$ as the logarithm of a single number

$\log 4 + \log 7 = \log(7 \times 4)$ by the first law of logarithms

$= \log 28$

Problem 11. Write $\log 16 - \log 2$ as the logarithm of a single number

$\log 16 - \log 2 = \log\left(\dfrac{16}{2}\right)$ by the second law of logarithms

$= \mathbf{log8}$

Problem 12. Write $2\log 3$ as the logarithm of a single number

$2\log 3 = \log 3^2$ by the third law of logarithms

$= \mathbf{log9}$

Problem 13. Write $\dfrac{1}{2}\log 25$ as the logarithm of a single number

$\dfrac{1}{2}\log 25 = \log 25^{\frac{1}{2}}$ by the third law of logarithms

$= \log\sqrt{25} = \mathbf{log5}$

Problem 14. Simplify $\log 64 - \log 128 + \log 32$

$64 = 2^6, 128 = 2^7$ and $32 = 2^5$

Hence, $\log 64 - \log 128 + \log 32$

$= \log 2^6 - \log 2^7 + \log 2^5$

$= 6\log 2 - 7\log 2 + 5\log 2$

   by the third law of logarithms

$= \mathbf{4log2}$

Problem 15. Write $\dfrac{1}{2}\log 16 + \dfrac{1}{3}\log 27 - 2\log 5$ as the logarithm of a single number

$\dfrac{1}{2}\log 16 + \dfrac{1}{3}\log 27 - 2\log 5$

$= \log 16^{\frac{1}{2}} + \log 27^{\frac{1}{3}} - \log 5^2$

   by the third law of logarithms

$= \log\sqrt{16} + \log\sqrt[3]{27} - \log 25$

   by the laws of indices

$= \log 4 + \log 3 - \log 25$

$= \log\left(\dfrac{4\times 3}{25}\right)$ by the first and second laws of logarithms

$= \log\left(\dfrac{12}{25}\right) = \mathbf{log0.48}$

Problem 16. Evaluate
$$\dfrac{\log 25 - \log 125 + \dfrac{1}{2}\log 625}{3\log 5}$$

$\dfrac{\log 25 - \log 125 + \dfrac{1}{2}\log 625}{3\log 5}$

$= \dfrac{\log 5^2 - \log 5^3 + \dfrac{1}{2}\log 5^4}{3\log 5}$

$= \dfrac{2\log 5 - 3\log 5 + \dfrac{4}{2}\log 5}{3\log 5}$

$= \dfrac{1\log 5}{3\log 5} = \dfrac{\mathbf{1}}{\mathbf{3}}$

Problem 17. Solve the equation
$\log(x-1) + \log(x+8) = 2\log(x+2)$

$\text{LHS} = \log(x-1) + \log(x+8) = \log(x-1)(x+8)$

   from the first law of logarithms

$= \log(x^2 + 7x - 8)$

$\text{RHS} = 2\log(x+2) = \log(x+2)^2$

   from the first law of logarithms

$= \log(x^2 + 4x + 4)$

Hence, $\log(x^2 + 7x - 8) = \log(x^2 + 4x + 4)$

from which, $x^2 + 7x - 8 = x^2 + 4x + 4$

i.e. $7x - 8 = 4x + 4$

i.e. $3x = 12$

and $\mathbf{x = 4}$

Problem 18. Solve the equation $\dfrac{1}{2}\log 4 = \log x$

$\dfrac{1}{2}\log 4 = \log 4^{\frac{1}{2}}$ from the third law of logarithms

$= \log\sqrt{4}$ from the laws of indices

Hence, $\dfrac{1}{2}\log 4 = \log x$

becomes $\log \sqrt{4} = \log x$

i.e. $\log 2 = \log x$

from which, $2 = x$

i.e. the solution of the equation is **$x = 2$**

Problem 19.   Solve the equation
$$\log(x^2 - 3) - \log x = \log 2$$

$\log(x^2 - 3) - \log x = \log\left(\dfrac{x^2 - 3}{x}\right)$ from the second

law of logarithms

Hence, $\log\left(\dfrac{x^2 - 3}{x}\right) = \log 2$

from which, $\dfrac{x^2 - 3}{x} = 2$

Rearranging gives $x^2 - 3 = 2x$

and $x^2 - 2x - 3 = 0$

Factorising gives $(x - 3)(x + 1) = 0$

from which, $x = 3$ or $x = -1$

$x = -1$ is not a valid solution since the logarithm of a negative number has no real root.

Hence, the solution of the equation is **$x = 3$**

Now try the following Practice Exercise

Practice Exercise 78   Laws of logarithms (answers on page 450)

In Problems 1 to 11, write as the logarithm of a single number.

1. $\log 2 + \log 3$ 2. $\log 3 + \log 5$
3. $\log 3 + \log 4 - \log 6$
4. $\log 7 + \log 21 - \log 49$
5. $2 \log 2 + \log 3$ 6. $2 \log 2 + 3 \log 5$
7. $2 \log 5 - \dfrac{1}{2}\log 81 + \log 36$
8. $\dfrac{1}{3}\log 8 - \dfrac{1}{2}\log 81 + \log 27$
9. $\dfrac{1}{2}\log 4 - 2 \log 3 + \log 45$
10. $\dfrac{1}{4}\log 16 + 2 \log 3 - \log 18$
11. $2 \log 2 + \log 5 - \log 10$

Simplify the expressions given in Problems 12 to 14.

12. $\log 27 - \log 9 + \log 81$
13. $\log 64 + \log 32 - \log 128$
14. $\log 8 - \log 4 + \log 32$

Evaluate the expressions given in Problems 15 and 16.

15. $\dfrac{\dfrac{1}{2}\log 16 - \dfrac{1}{3}\log 8}{\log 4}$

16. $\dfrac{\log 9 - \log 3 + \dfrac{1}{2}\log 81}{2 \log 3}$

Solve the equations given in Problems 17 to 22.

17. $\log x^4 - \log x^3 = \log 5x - \log 2x$
18. $\log 2t^3 - \log t = \log 16 + \log t$
19. $2 \log b^2 - 3 \log b = \log 8b - \log 4b$
20. $\log(x + 1) + \log(x - 1) = \log 3$
21. $\dfrac{1}{3}\log 27 = \log(0.5a)$
22. $\log(x^2 - 5) - \log x = \log 4$

15.3   Indicial equations

The laws of logarithms may be used to solve certain equations involving powers, called **indicial equations**.

For example, to solve, say, $3^x = 27$, logarithms to a base of 10 are taken of both sides,

i.e. $\log_{10} 3^x = \log_{10} 27$

and $x \log_{10} 3 = \log_{10} 27$

by the third law of logarithms

Rearranging gives $x = \dfrac{\log_{10} 27}{\log_{10} 3} = \dfrac{1.43136\ldots}{0.47712\ldots}$

$= \mathbf{3}$ which may be readily checked.

$\left(\text{Note, } \dfrac{\log 27}{\log 3} \text{ is **not** equal to } \log\dfrac{27}{3}\right)$

Problem 20.   Solve the equation $2^x = 5$, correct to 4 significant figures

Taking logarithms to base 10 of both sides of $2^x = 5$ gives

$$\log_{10} 2^x = \log_{10} 5$$

i.e. $\qquad x \log_{10} 2 = \log_{10} 5$

by the third law of logarithms

Rearranging gives $x = \dfrac{\log_{10} 5}{\log_{10} 2} = \dfrac{0.6989700\dots}{0.3010299\dots}$

$$= \textbf{2.322}, \text{correct to 4}$$

significant figures.

Problem 21. Solve the equation $2^{x+1} = 3^{2x-5}$ correct to 2 decimal places

Taking logarithms to base 10 of both sides gives

$$\log_{10} 2^{x+1} = \log_{10} 3^{2x-5}$$

i.e. $\qquad (x+1)\log_{10} 2 = (2x-5)\log_{10} 3$

$$x\log_{10} 2 + \log_{10} 2 = 2x\log_{10} 3 - 5\log_{10} 3$$

$$x(0.3010) + (0.3010) = 2x(0.4771) - 5(0.4771)$$

i.e. $\qquad 0.3010x + 0.3010 = 0.9542x - 2.3855$

Hence, $\quad 2.3855 + 0.3010 = 0.9542x - 0.3010x$

$$2.6865 = 0.6532x$$

from which $\qquad x = \dfrac{2.6865}{0.6532} = \textbf{4.11},$

correct to 2 decimal places.

Problem 22. Solve the equation $x^{2.7} = 34.68$, correct to 4 significant figures

Taking logarithms to base 10 of both sides gives

$$\log_{10} x^{2.7} = \log_{10} 34.68$$

$$2.7\log_{10} x = \log_{10} 34.68$$

Hence, $\qquad \log_{10} x = \dfrac{\log_{10} 34.68}{2.7} = 0.57040$

Thus, $\qquad x = \text{antilog } 0.57040 = 10^{0.57040}$

$$= \textbf{3.719},$$

correct to 4 significant figures.

Problem 23. The velocity v of a machine is given by the formula: $V = 15^{\left(\frac{40}{1.25b}\right)}$
(a) Transpose the formula to make b the subject.
(b) Evaluate b, correct to 3 significant figures, when v = 12

(a) Taking logarithms to the base of 10 of both sides of the equation gives:

$$\log_{10} v = \log_{10} 15^{\left(\frac{40}{1.25b}\right)}$$

From the third law of logarithms:

$$\log_{10} v = \left(\frac{40}{1.25b}\right)\log_{10} 15$$

$$= \left(\frac{40}{1.25b}\right)(1.1761) \text{ correct to 4 decimal places}$$

$$= \frac{47.044}{1.25b}$$

from which, $\quad 1.25b\log_{10} v = 47.044$

and $\qquad\qquad \textbf{b} = \dfrac{\textbf{47.044}}{\textbf{1.25}\log_{10}\textbf{v}}$

(b) When v = 12, $\quad b = \dfrac{47.044}{1.25\log_{10} 12} = 34.874$

i.e. $\qquad \textbf{b} = \textbf{34.9}$ correct to 3 significant figures.

Problem 24. A gas follows the polytropic law $PV^{1.25} = C$. Determine the new volume of the gas, given that its original pressure and volume are 101 kPa and 0.35 m³, respectively, and its final pressure is 1.18 MPa.

If $PV^{1.25} = C$ then $P_1 V_1^{1.25} = P_2 V_2^{1.25}$

$P_1 = 101$ kPa, $P_2 = 1.18$ MPa and $V_1 = 0.35\text{m}^3$

$$P_1 V_1^{1.25} = P_2 V_2^{1.25}$$

i.e. $\left(101 \times 10^3\right)(0.35)^{1.25} = (1.18 \times 10^6)V_2^{1.25}$

from which, $\qquad V_2^{1.25} = \dfrac{(101 \times 10^3)(0.35)^{1.25}}{(1.18 \times 10^6)}$

$$= 0.02304$$

Taking logarithms of both sides of the equation gives:

$$\log_{10} V_2^{1.25} = \log_{10} 0.02304$$

i.e. $\quad 1.25\log_{10} V_2 = \log_{10} 0.02304$ from the third law of logarithms

and
$$\log_{10} V_2 = \frac{\log_{10} 0.02304}{1.25} = -1.3100$$

from which, **volume $V_2 = 10^{-1.3100} = 0.049$ m^3**

Now try the following Practice Exercise

Practice Exercise 79   Indicial equations (answers on page 450)

In Problems 1 to 8, solve the indicial equations for x, each correct to 4 significant figures.

1. $3^x = 6.4$ 2. $2^x = 9$

3. $2^{x-1} = 3^{2x-1}$ 4. $x^{1.5} = 14.91$

5. $25.28 = 4.2^x$ 6. $4^{2x-1} = 5^{x+2}$

7. $x^{-0.25} = 0.792$ 8. $0.027^x = 3.26$

9. The decibel gain n of an amplifier is given by
$$n = 10\log_{10}\left(\frac{P_2}{P_1}\right),$$
where P_1 is the power input and P_2 is the power output. Find the power gain $\dfrac{P_2}{P_1}$ when $n = 25$ decibels.

10. A gas follows the polytropic law $PV^{1.26} = C$. Determine the new volume of the gas, given that its original pressure and volume are 101 kPa and 0.42 m^3, respectively, and its final pressure is 1.25 MPa.

15.4 Graphs of logarithmic functions

A graph of $y = \log_{10} x$ is shown in Fig. 15.1 and a graph of $y = \log_e x$ is shown in Fig. 15.2. Both can be seen to be of similar shape; in fact, the same general shape occurs for a logarithm to any base.

In general, with a logarithm to any base, a, it is noted that

(a) **$\log_a 1 = 0$**

Let $\log_a = x$ then $a^x = 1$ from the definition of the logarithm.

If $a^x = 1$ then $x = 0$ from the laws of logarithms.

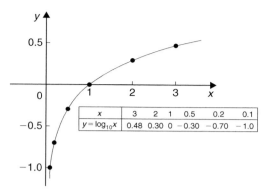

x	3	2	1	0.5	0.2	0.1
$y = \log_{10} x$	0.48	0.30	0	-0.30	-0.70	-1.0

Figure 15.1

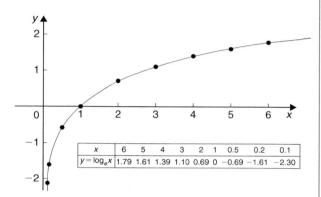

x	6	5	4	3	2	1	0.5	0.2	0.1
$y = \log_e x$	1.79	1.61	1.39	1.10	0.69	0	-0.69	-1.61	-2.30

Figure 15.2

Hence, **$\log_a 1 = 0$**. In the above graphs it is seen that $\log_{10} 1 = 0$ and $\log_e 1 = 0$

(b) **$\log_a a = 1$**

Let $\log_a a = x$ then $a^x = a$ from the definition of a logarithm.

If $a^x = a$ then $x = 1$

Hence, **$\log_a a = 1$**. (Check with a calculator that $\log_{10} 10 = 1$ and $\log_e e = 1$)

(c) **$\log_a 0 \to -\infty$**

Let $\log_a 0 = x$ then $a^x = 0$ from the definition of a logarithm.

If $a^x = 0$, and a is a positive real number, then x must approach minus infinity. (For example, check with a calculator, $2^{-2} = 0.25, 2^{-20} = 9.54 \times 10^{-7}, 2^{-200} = 6.22 \times 10^{-61}$, and so on.)

Hence, **$\log_a 0 \to -\infty$**

Practice Exercise 80   Multiple-choice questions on logarithms (answers on page 450)

Each question has only one correct answer

1. $\log_{16} 8$ is equal to:
    (a) $\frac{1}{2}$ (b) 144   (c) $\frac{3}{4}$ (d) 2

2. Simplifying $(\log 2 + \log 3)$ gives:
    (a) $\log 1$ (b) $\log 5$ (c) $\log 6$ (d) $\log \frac{2}{3}$

3. The value of $\log_{10} \frac{1}{100}$ is:
    (a) -5 (b) 2   (c) 5   (d) -2

4. Solving $(\log 30 - \log 5) = \log x$ for x gives:
    (a) 150   (b) 6   (c) 35   (d) 25

5. Solving $\log_x 81 = 4$ for x gives:
    (a) 3   (b) 9   (c) 20.25   (d) 324

6. $\frac{1}{4} \log 16 - 2 \log 5 + \frac{1}{3} \log 27$ is equivalent to:
    (a) $- \log 20$ (b) $\log 0.24$
    (c) $- \log 5$ (d) $\log 3$

7. Solving $\log_{10}(3x + 1) = 5$ for x gives:
    (a) 300   (b) 8   (c) 33333   (d) $\frac{4}{3}$

8. Solving $5^{2-x} = \frac{1}{125}$ for x gives:
    (a) $\frac{5}{3}$ (b) 5   (c) -1 (d) $\frac{7}{3}$

9. $\log_{10} 4 + \log_{10} 25$ is equal to:
    (a) 5   (b) 4   (c) 3   (d) 2

10. Simplifying $(2 \log 9)$ gives:
    (a) $\log 11$ (b) $\log 18$ (c) $\log 81$ (d) $\log \frac{2}{9}$

11. Solving the equation $\log(x + 2) - \log(x - 2) = \log 2$ for x gives:
    (a) 2   (b) 3   (c) 4   (d) 6

12. Writing $\left(3 \log 1 + \frac{1}{2} \log 16 \right)$ as a single logarithm gives:
    (a) $\log 24$ (b) $\log 18$ (c) $\log 5$ (d) $\log 4$

13. Solving $(3 \log 10) = \log x$ for x gives:
    (a) 1000   (b) 30   (c) 13   (d) 0.3

14. Solving $2^x = 6$ for x, correct to 1 decimal place, gives:
    (a) 2.6   (b) 1.6   (c) 12.0   (d) 5.2

15. The information I contained in a message is given by $I = \log_2 \left(\frac{1}{x} \right)$

    If $x = \frac{1}{8}$ then I is equal to:
    (a) 4   (b) 3   (c) 8   (d) 16

For fully worked solutions to each of the problems in Practice Exercises 77 to 79 in this chapter, go to the website:
www.routledge.com/cw/bird

Chapter 16

Exponential functions

Why it is important to understand: **Exponential functions**

Exponential functions are used in engineering, physics, biology and economics. There are many quantities that grow exponentially; some examples are population, compound interest and charge in a capacitor. With exponential growth, the rate of growth increases as time increases. We also have exponential decay; some examples are radioactive decay, atmospheric pressure, Newton's law of cooling and linear expansion. Understanding and using exponential functions is important in many branches of engineering.

At the end of this chapter you should be able to:

- evaluate exponential functions using a calculator
- state the exponential series for e^x
- plot graphs of exponential functions
- evaluate Napierian logarithms using a calculator
- solve equations involving Napierian logarithms
- appreciate the many examples of laws of growth and decay in engineering and science
- perform calculations involving the laws of growth and decay

16.1 Introduction to exponential functions

An exponential function is one which contains e^x, e being a constant called the exponent and having an approximate value of 2.7183. The exponent arises from the natural laws of growth and decay and is used as a base for natural or Napierian logarithms.

The most common method of evaluating an exponential function is by using a scientific notation **calculator**. Use your calculator to check the following values.

$$e^1 = 2.7182818, \text{ correct to 8 significant figures,}$$

$$e^{-1.618} = 0.1982949, \text{ correct to 7 significant figures,}$$

$$e^{0.12} = 1.1275, \text{ correct to 5 significant figures,}$$

$$e^{-1.47} = 0.22993, \text{ correct to 5 decimal places,}$$

$$e^{-0.431} = 0.6499, \text{ correct to 4 decimal places,}$$

$$e^{9.32} = 11159, \text{ correct to 5 significant figures,}$$

$$e^{-2.785} = 0.0617291, \text{ correct to 7 decimal places.}$$

Problem 1. Evaluate the following correct to 4 decimal places, using a calculator:

$$0.0256(e^{5.21} - e^{2.49})$$

$0.0256(e^{5.21} - e^{2.49})$

$= 0.0256(183.094058\ldots - 12.0612761\ldots)$

$= \mathbf{4.3784}$, correct to 4 decimal places.

Problem 2. Evaluate the following correct to 4 decimal places, using a calculator:

$$5\left(\frac{e^{0.25}-e^{-0.25}}{e^{0.25}+e^{-0.25}}\right)$$

$$5\left(\frac{e^{0.25}-e^{-0.25}}{e^{0.25}+e^{-0.25}}\right)$$

$$=5\left(\frac{1.28402541\ldots-0.77880078\ldots}{1.28402541\ldots+0.77880078\ldots}\right)$$

$$=5\left(\frac{0.5052246\ldots}{2.0628262\ldots}\right)$$

$$=\mathbf{1.2246},\text{ correct to 4 decimal places.}$$

Problem 3. The instantaneous voltage v in a capacitive circuit is related to time t by the equation $v = Ve^{-t/CR}$ where V, C and R are constants. Determine v, correct to 4 significant figures, when $t = 50$ ms, $C = 10\,\mu F$, $R = 47\,k\Omega$ and $V = 300$ volts

$$v = Ve^{-t/CR} = 300e^{(-50\times10^{-3})/(10\times10^{-6}\times47\times10^{3})}$$

Using a calculator, $\quad v = 300e^{-0.1063829\ldots}$

$$= 300(0.89908025\ldots)$$

$$= \mathbf{269.7\,volts.}$$

Now try the following Practice Exercise

Practice Exercise 81   Evaluating exponential functions (answers on page 450)

1. Evaluate the following, correct to 4 significant figures.
   (a) $e^{-1.8}$ (b) $e^{-0.78}$ (c) e^{10}

2. Evaluate the following, correct to 5 significant figures.
   (a) $e^{1.629}$ (b) $e^{-2.7483}$ (c) $0.62e^{4.178}$

In Problems 3 and 4, evaluate correct to 5 decimal places.

3. (a) $\frac{1}{7}e^{3.4629}$ (b) $8.52e^{-1.2651}$
   (c) $\frac{5e^{2.6921}}{3e^{1.1171}}$

4. (a) $\frac{5.6823}{e^{-2.1347}}$ (b) $\frac{e^{2.1127}-e^{-2.1127}}{2}$
   (c) $\frac{4\left(e^{-1.7295}-1\right)}{e^{3.6817}}$

5. The length of a bar, l, at a temperature, θ, is given by $l = l_0e^{\alpha\theta}$, where l_0 and α are constants. Evaluate l, correct to 4 significant figures, where $l_0 = 2.587, \theta = 321.7$ and $\alpha = 1.771\times10^{-4}$

6. When a chain of length $2L$ is suspended from two points, $2D$ metres apart on the same horizontal level, $D = k\left\{\ln\left(\frac{L+\sqrt{L^2+k^2}}{k}\right)\right\}$. Evaluate D when $k = 75$ m and $L = 180$ m.

16.2   The power series for e^x

The value of e^x can be calculated to any required degree of accuracy since it is defined in terms of the following **power series**:

$$e^x = 1+x+\frac{x^2}{2!}+\frac{x^3}{3!}+\frac{x^4}{4!}+\cdots \quad (1)$$

(where $3! = 3\times2\times1$ and is called '**factorial** 3').
The series is valid for all values of x.
The series is said to **converge**; i.e. if all the terms are added, an actual value for e^x (where x is a real number) is obtained. The more terms that are taken, the closer will be the value of e^x to its actual value. The value of the exponent e, correct to say 4 decimal places, may be determined by substituting $x = 1$ in the power series of equation (1). Thus,

$$e^1 = 1+1+\frac{(1)^2}{2!}+\frac{(1)^3}{3!}+\frac{(1)^4}{4!}+\frac{(1)^5}{5!}+\frac{(1)^6}{6!}$$
$$+\frac{(1)^7}{7!}+\frac{(1)^8}{8!}+\cdots$$

$$= 1+1+0.5+0.16667+0.04167+0.00833$$
$$+0.00139+0.00020+0.00002+\cdots$$

$$= 2.71828$$

i.e. $e = \mathbf{2.7183}$, correct to 4 decimal places.

The value of $e^{0.05}$, correct to say 8 significant figures, is found by substituting $x = 0.05$ in the power series for e^x. Thus,

$$e^{0.05} = 1 + 0.05 + \frac{(0.05)^2}{2!} + \frac{(0.05)^3}{3!} + \frac{(0.05)^4}{4!}$$
$$+ \frac{(0.05)^5}{5!} + \cdots$$
$$= 1 + 0.05 + 0.00125 + 0.000020833$$
$$+ 0.000000260 + 0.0000000026$$

i.e. $e^{0.05} = 1.0512711$, correct to 8 significant figures.

In this example, successive terms in the series grow smaller very rapidly and it is relatively easy to determine the value of $e^{0.05}$ to a high degree of accuracy. However, when x is nearer to unity or larger than unity, a very large number of terms are required for an accurate result.

If, in the series of equation (1), x is replaced by $-x$, then

$$e^{-x} = 1 + (-x) + \frac{(-x)^2}{2!} + \frac{(-x)^3}{3!} + \cdots$$

i.e. $$e^{-x} = 1 - x + \frac{x^2}{2!} - \frac{x^3}{3!} + \cdots$$

In a similar manner the power series for e^x may be used to evaluate any exponential function of the form ae^{kx}, where a and k are constants.

In the series of equation (1), let x be replaced by kx. Then

$$ae^{kx} = a\left\{1 + (kx) + \frac{(kx)^2}{2!} + \frac{(kx)^3}{3!} + \cdots\right\}$$

Thus, $$5e^{2x} = 5\left\{1 + (2x) + \frac{(2x)^2}{2!} + \frac{(2x)^3}{3!} + \cdots\right\}$$

$$= 5\left\{1 + 2x + \frac{4x^2}{2} + \frac{8x^3}{6} + \cdots\right\}$$

i.e. $$5e^{2x} = 5\left\{1 + 2x + 2x^2 + \frac{4}{3}x^3 + \cdots\right\}$$

Problem 4. Determine the value of $5e^{0.5}$, correct to 5 significant figures, by using the power series for e^x

From equation (1),

$$e^x = 1 + x + \frac{x^2}{2!} + \frac{x^3}{3!} + \frac{x^4}{4!} + \cdots$$

Hence, $$e^{0.5} = 1 + 0.5 + \frac{(0.5)^2}{(2)(1)} + \frac{(0.5)^3}{(3)(2)(1)}$$
$$+ \frac{(0.5)^4}{(4)(3)(2)(1)} + \frac{(0.5)^5}{(5)(4)(3)(2)(1)}$$
$$+ \frac{(0.5)^6}{(6)(5)(4)(3)(2)(1)}$$
$$= 1 + 0.5 + 0.125 + 0.020833$$
$$+ 0.0026042 + 0.0002604$$
$$+ 0.0000217$$

i.e. $e^{0.5} = 1.64872$, correct to 6 significant figures

Hence, $5e^{0.5} = 5(1.64872) = 8.2436$, correct to 5 significant figures.

Problem 5. Determine the value of $3e^{-1}$, correct to 4 decimal places, using the power series for e^x

Substituting $x = -1$ in the power series

$$e^x = 1 + x + \frac{x^2}{2!} + \frac{x^3}{3!} + \frac{x^4}{4!} + \cdots$$

gives $$e^{-1} = 1 + (-1) + \frac{(-1)^2}{2!} + \frac{(-1)^3}{3!}$$
$$+ \frac{(-1)^4}{4!} + \cdots$$
$$= 1 - 1 + 0.5 - 0.166667 + 0.041667$$
$$- 0.008333 + 0.001389$$
$$- 0.000198 + \cdots$$
$$= 0.367858 \text{ correct to 6 decimal places}$$

Hence, $3e^{-1} = (3)(0.367858) = 1.1036$, correct to 4 decimal places.

Problem 6. Expand $e^x(x^2 - 1)$ as far as the term in x^5

The power series for e^x is

$$e^x = 1 + x + \frac{x^2}{2!} + \frac{x^3}{3!} + \frac{x^4}{4!} + \frac{x^5}{5!} + \cdots$$

Hence,

$e^x(x^2 - 1)$

$$= \left(1 + x + \frac{x^2}{2!} + \frac{x^3}{3!} + \frac{x^4}{4!} + \frac{x^5}{5!} + \cdots\right)(x^2 - 1)$$

$$= \left(x^2 + x^3 + \frac{x^4}{2!} + \frac{x^5}{3!} + \cdots\right)$$

$$- \left(1 + x + \frac{x^2}{2!} + \frac{x^3}{3!} + \frac{x^4}{4!} + \frac{x^5}{5!} + \cdots\right)$$

Grouping like terms gives

$e^x(x^2 - 1)$

$$= -1 - x + \left(x^2 - \frac{x^2}{2!}\right) + \left(x^3 - \frac{x^3}{3!}\right)$$

$$+ \left(\frac{x^4}{2!} - \frac{x^4}{4!}\right) + \left(\frac{x^5}{3!} - \frac{x^5}{5!}\right) + \cdots$$

$$= -1 - x + \frac{1}{2}x^2 + \frac{5}{6}x^3 + \frac{11}{24}x^4 + \frac{19}{120}x^5$$

when expanded as far as the term in x^5.

Now try the following Practice Exercise

**Practice Exercise 82   Power series for e^x
(answers on page 451)**

1. Evaluate $5.6e^{-1}$, correct to 4 decimal places, using the power series for e^x.

2. Use the power series for e^x to determine, correct to 4 significant figures, (a) e^2 (b) $e^{-0.3}$ and check your results using a calculator.

3. Expand $(1 - 2x)e^{2x}$ as far as the term in x^4.

4. Expand $(2e^{x^2})(x^{1/2})$ to six terms.

16.3 Graphs of exponential functions

Values of e^x and e^{-x} obtained from a calculator, correct to 2 decimal places, over a range $x = -3$ to $x = 3$, are shown in Table 16.1.

Fig. 16.1 shows graphs of $y = e^x$ and $y = e^{-x}$

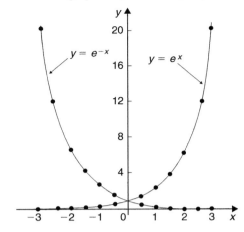

Figure 16.1

Problem 7.   Plot a graph of $y = 2e^{0.3x}$ over a range of $x = -3$ to $x = 3$. Then determine the value of y when $x = 2.2$ and the value of x when $y = 1.6$

A table of values is drawn up as shown below.

x	-3	-2	-1	0	1	2	3
$2e^{0.3x}$	0.81	1.10	1.48	2.00	2.70	3.64	4.92

A graph of $y = 2e^{0.3x}$ is shown plotted in Fig. 16.2. From the graph, **when $x = 2.2$, $y = 3.87$** and **when $y = 1.6$, $x = -0.74$**

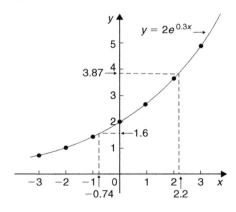

Figure 16.2

Table 16.1

x	-3.0	-2.5	-2.0	-1.5	-1.0	-0.5	0	0.5	1.0	1.5	2.0	2.5	3.0
e^x	0.05	0.08	0.14	0.22	0.37	0.61	1.00	1.65	2.72	4.48	7.39	12.18	20.09
e^{-x}	20.09	12.18	7.39	4.48	2.72	1.65	1.00	0.61	0.37	0.22	0.14	0.08	0.05

t	4	5	6
$e^{-t/3}$	0.2636	0.1889	0.1353
$v = 250e^{-t/3}$	65.90	47.22	33.83

The natural decay curve of $v = 250e^{-t/3}$ is shown in Fig. 16.4.

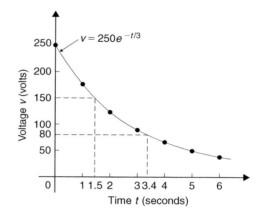

Figure 16.4

From the graph,

(a) when time $t = 3.4$ s, **voltage $v = 80$ V**, and

(b) when **voltage $v = 150$ V, time $t = 1.5$ s**.

Now try the following Practice Exercise

Practice Exercise 83   Exponential graphs (answers on page 451)

1. Plot a graph of $y = 3e^{0.2x}$ over the range $x = -3$ to $x = 3$. Hence determine the value of y when $x = 1.4$ and the value of x when $y = 4.5$

2. Plot a graph of $y = \dfrac{1}{2}e^{-1.5x}$ over a range $x = -1.5$ to $x = 1.5$ and then determine the value of y when $x = -0.8$ and the value of x when $y = 3.5$

3. In a chemical reaction the amount of starting material C cm^3 left after t minutes is given by $C = 40e^{-0.006t}$. Plot a graph of C against t and determine

    (a) the concentration C after 1 hour.
    (b) the time taken for the concentration to decrease by half.

Problem 8.   Plot a graph of $y = \dfrac{1}{3}e^{-2x}$ over the range $x = -1.5$ to $x = 1.5$. Determine from the graph the value of y when $x = -1.2$ and the value of x when $y = 1.4$

A table of values is drawn up as shown below.

x	-1.5	-1.0	-0.5	0	0.5	1.0	1.5
$\frac{1}{3}e^{-2x}$	6.70	2.46	0.91	0.33	0.12	0.05	0.02

A graph of $\dfrac{1}{3}e^{-2x}$ is shown in Fig. 16.3.
From the graph, **when $x = -1.2$, $y = 3.67$** and **when $y = 1.4$, $x = -0.72$**

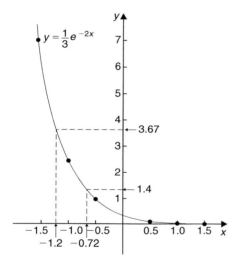

Figure 16.3

Problem 9.   The decay of voltage, v volts, across a capacitor at time t seconds is given by $v = 250e^{-t/3}$. Draw a graph showing the natural decay curve over the first 6 seconds. From the graph, find (a) the voltage after 3.4 s and (b) the time when the voltage is 150 V

A table of values is drawn up as shown below.

t	0	1	2	3
$e^{-t/3}$	1.00	0.7165	0.5134	0.3679
$v = 250e^{-t/3}$	250.0	179.1	128.4	91.97

4. The rate at which a body cools is given by $\theta = 250e^{-0.05t}$ where the excess of temperature of a body above its surroundings at time t minutes is $\theta\,°C$. Plot a graph showing the natural decay curve for the first hour of cooling. Then determine

(a) the temperature after 25 minutes.

(b) the time when the temperature is $195\,°C$.

16.4 Napierian logarithms

Logarithms having a base of e are called **hyperbolic**, **Napierian** or **natural logarithms** and the Napierian logarithm of x is written as $\log_e x$ or, more commonly, as $\ln x$. Logarithms were invented by **John Napier***, a Scotsman.

The most common method of evaluating a Napierian logarithm is by a scientific notation **calculator**. Use your calculator to check the following values:

$\ln 4.328 = 1.46510554\ldots = 1.4651$, correct to 4 decimal places

*Who was Napier? – **John Napier** of Merchiston (1550–4 April 1617) is best known as the discoverer of logarithms. The inventor of the so-called 'Napier's bones', Napier also made common the use of the decimal point in arithmetic and mathematics. To find out more go to **www.routledge.com/cw/bird**

$\ln 1.812 = 0.59443$, correct to 5 significant figures

$\ln 1 = 0$

$\ln 527 = 6.2672$, correct to 5 significant figures

$\ln 0.17 = -1.772$, correct to 4 significant figures

$\ln 0.00042 = -7.77526$, correct to 6 significant figures

$\ln e^3 = 3$

$\ln e^1 = 1$

From the last two examples we can conclude that

$$\log_e e^x = x$$

This is useful when **solving equations involving exponential functions**. For example, to solve $e^{3x} = 7$, take Napierian logarithms of both sides, which gives

$$\ln e^{3x} = \ln 7$$

i.e. $\qquad 3x = \ln 7$

from which $\qquad x = \dfrac{1}{3}\ln 7 = \mathbf{0.6486}$,

correct to 4 decimal places.

Problem 10. Evaluate the following, each correct to 5 significant figures: (a) $\dfrac{1}{2}\ln 4.7291$

(b) $\dfrac{\ln 7.8693}{7.8693}$ (c) $\dfrac{3.17\ln 24.07}{e^{-0.1762}}$

(a) $\dfrac{1}{2}\ln 4.7291 = \dfrac{1}{2}(1.5537349\ldots) = \mathbf{0.77687}$,

correct to 5 significant figures.

(b) $\dfrac{\ln 7.8693}{7.8693} = \dfrac{2.06296911\ldots}{7.8693} = \mathbf{0.26215}$, correct

to 5 significant figures.

(c) $\dfrac{3.17\ln 24.07}{e^{-0.1762}} = \dfrac{3.17(3.18096625\ldots)}{0.83845027\ldots} = \mathbf{12.027}$,

correct to 5 significant figures.

Problem 11. Evaluate the following: (a) $\dfrac{\ln e^{2.5}}{\lg 10^{0.5}}$

(b) $\dfrac{5e^{2.23}\lg 2.23}{\ln 2.23}$ (correct to 3 decimal places)

(a) $\dfrac{\ln e^{2.5}}{\lg 10^{0.5}} = \dfrac{2.5}{0.5} = \mathbf{5}$

(b) $\dfrac{5e^{2.23}\lg 2.23}{\ln 2.23}$

$$= \frac{5(9.29986607\ldots)(0.34830486\ldots)}{(0.80200158\ldots)}$$

$= \mathbf{20.194}$, correct to 3 decimal places.

Problem 12. Solve the equation $9 = 4e^{-3x}$ to find x, correct to 4 significant figures

Rearranging $9 = 4e^{-3x}$ gives $\dfrac{9}{4} = e^{-3x}$

Taking the reciprocal of both sides gives

$$\frac{4}{9} = \frac{1}{e^{-3x}} = e^{3x}$$

Taking Napierian logarithms of both sides gives

$$\ln\left(\frac{4}{9}\right) = \ln(e^{3x})$$

Since $\log_e e^{\alpha} = \alpha$, then $\ln\left(\dfrac{4}{9}\right) = 3x$

Hence, $x = \dfrac{1}{3}\ln\left(\dfrac{4}{9}\right) = \dfrac{1}{3}(-0.81093) = \mathbf{-0.2703}$, correct to 4 significant figures.

Problem 13. Given $32 = 70\left(1 - e^{-\frac{t}{2}}\right)$, determine the value of t, correct to 3 significant figures

Rearranging $32 = 70\left(1 - e^{-\frac{t}{2}}\right)$ gives

$$\frac{32}{70} = 1 - e^{-\frac{t}{2}}$$

and

$$e^{-\frac{t}{2}} = 1 - \frac{32}{70} = \frac{38}{70}$$

Taking the reciprocal of both sides gives

$$e^{\frac{t}{2}} = \frac{70}{38}$$

Taking Napierian logarithms of both sides gives

$$\ln e^{\frac{t}{2}} = \ln\left(\frac{70}{38}\right)$$

i.e. $\dfrac{t}{2} = \ln\left(\dfrac{70}{38}\right)$

from which, $t = 2\ln\left(\dfrac{70}{38}\right) = \mathbf{1.22}$, correct to 3 significant figures.

Problem 14. Solve the equation

$$2.68 = \ln\left(\frac{4.87}{x}\right) \text{ to find } x$$

From the definition of a logarithm, since $2.68 = \ln\left(\dfrac{4.87}{x}\right)$ then $e^{2.68} = \dfrac{4.87}{x}$

Rearranging gives $x = \dfrac{4.87}{e^{2.68}} = 4.87e^{-2.68}$

i.e. $\mathbf{x = 0.3339}$, correct to 4 significant figures.

Problem 15. Solve $\dfrac{7}{4} = e^{3x}$ correct to 4 significant figures

Taking natural logs of both sides gives

$$\ln\frac{7}{4} = \ln e^{3x}$$

$$\ln\frac{7}{4} = 3x\ln e$$

Since $\ln e = 1$, $\ln\dfrac{7}{4} = 3x$

i.e. $0.55962 = 3x$

i.e. $\mathbf{x = 0.1865}$, correct to 4 significant figures.

Problem 16. Solve $e^{x-1} = 2e^{3x-4}$ correct to 4 significant figures

Taking natural logarithms of both sides gives

$$\ln\left(e^{x-1}\right) = \ln\left(2e^{3x-4}\right)$$

and by the first law of logarithms,

$$\ln\left(e^{x-1}\right) = \ln 2 + \ln\left(e^{3x-4}\right)$$

i.e. $x - 1 = \ln 2 + 3x - 4$

Rearranging gives

$$4 - 1 - \ln 2 = 3x - x$$

i.e. $3 - \ln 2 = 2x$

from which, $x = \dfrac{3 - \ln 2}{2} = \mathbf{1.153}$

Problem 17. Solve, correct to 4 significant figures, $\ln(x-2)^2 = \ln(x-2) - \ln(x+3) + 1.6$

Rearranging gives

$$\ln(x-2)^2 - \ln(x-2) + \ln(x+3) = 1.6$$

and by the laws of logarithms,

$$\ln\left\{\frac{(x-2)^2(x+3)}{(x-2)}\right\} = 1.6$$

Cancelling gives

$$\ln\{(x-2)(x+3)\} = 1.6$$

and
$$(x-2)(x+3) = e^{1.6}$$

i.e.
$$x^2 + x - 6 = e^{1.6}$$

or
$$x^2 + x - 6 - e^{1.6} = 0$$

i.e.
$$x^2 + x - 10.953 = 0$$

Using a calculator or the quadratic formula,

$$x = 2.847 \text{ or } -3.8471$$

$x = -3.8471$ is not valid since the logarithm of a negative number has no real root.

Hence, **the solution of the equation is $x = 2.847$**

Now try the following Practice Exercise

Practice Exercise 84 Evaluating Napierian logarithms (answers on page 451)

In Problems 1 and 2, evaluate correct to 5 significant figures.

1. (a) $\frac{1}{3}\ln 5.2932$ (b) $\frac{\ln 82.473}{4.829}$

   (c) $\frac{5.62\ln 321.62}{e^{1.2942}}$

2. (a) $\frac{1.786\ln e^{1.76}}{\lg 10^{1.41}}$ (b) $\frac{5e^{-0.1629}}{2\ln 0.00165}$

   (c) $\frac{\ln 4.8629 - \ln 2.4711}{5.173}$

In Problems 3 to 16, solve the given equations, each correct to 4 significant figures.

3. $1.5 = 4e^{2t}$

4. $7.83 = 2.91e^{-1.7x}$

5. $16 = 24\left(1 - e^{-\frac{t}{2}}\right)$

6. $5.17 = \ln\left(\frac{x}{4.64}\right)$

7. $3.72\ln\left(\frac{1.59}{x}\right) = 2.43$

8. $\ln x = 2.40$

9. $24 + e^{2x} = 45$

10. $5 = e^{x+1} - 7$

11. $5 = 8\left(1 - e^{\frac{-x}{2}}\right)$

12. $\ln(x+3) - \ln x = \ln(x-1)$

13. $\ln(x-1)^2 - \ln 3 = \ln(x-1)$

14. $\ln(x+3) + 2 = 12 - \ln(x-2)$

15. $e^{(x+1)} = 3e^{(2x-5)}$

16. $\ln(x+1)^2 = 1.5 - \ln(x-2) + \ln(x+1)$

17. Transpose $b = \ln t - a\ln D$ to make t the subject.

18. If $\frac{P}{Q} = 10\log_{10}\left(\frac{R_1}{R_2}\right)$, find the value of R_1 when $P = 160$, $Q = 8$ and $R_2 = 5$

19. If $U_2 = U_1 e^{\left(\frac{W}{PV}\right)}$, make W the subject of the formula.

20. The velocity v_2 of a rocket is given by:
$$v_2 = v_1 + C\ln\left(\frac{m_1}{m_2}\right)$$ where v_1 is the initial rocket velocity, C is the velocity of the jet exhaust gases, m_1 is the mass of the rocket before the jet engine is fired, and m_2 is the mass of the rocket after the jet engine is switched off. Calculate the velocity of the rocket given $v_1 = 600$ m/s, $C = 3500$ m/s, $m_1 = 8.50 \times 10^4$ kg and $m_2 = 7.60 \times 10^4$ kg.

21. The work done in an isothermal expansion of a gas from pressure p_1 to p_2 is given by:
$$w = w_0 \ln\left(\frac{p_1}{p_2}\right)$$

If the initial pressure $p_1 = 7.0$ kPa, calculate the final pressure p_2 if $w = 3w_0$

22. A steel bar is cooled with running water. Its temperature, θ, in degrees Celsius, is given by: $\theta = 17 + 1250e^{-0.17t}$ where t is the time in minutes. Determine the time taken, correct to the nearest minute, for the temperature to fall to $35°C$.

16.5 Laws of growth and decay

Laws of exponential growth and decay are of the form $y = Ae^{-kx}$ and $y = A(1 - e^{-kx})$, where A and k are constants. When plotted, the form of these equations is as shown in Fig. 16.5.

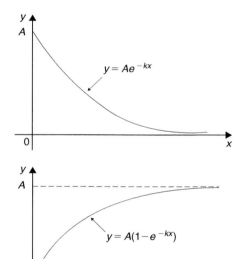

Figure 16.5

The laws occur frequently in engineering and science and examples of quantities related by a natural law include:

(a) Linear expansion $\qquad l = l_0 e^{\alpha\theta}$

(b) Change in electrical resistance with temperature $\qquad R_\theta = R_0 e^{\alpha\theta}$

(c) Tension in belts $\qquad T_1 = T_0 e^{\mu\theta}$

(d) Newton's law of cooling $\qquad \theta = \theta_0 e^{-kt}$

(e) Biological growth $\qquad y = y_0 e^{kt}$

(f) Discharge of a capacitor $\qquad q = Qe^{-t/CR}$

(g) Atmospheric pressure $\qquad p = p_0 e^{-h/c}$

(h) Radioactive decay $\qquad N = N_0 e^{-\lambda t}$

(i) Decay of current in an inductive circuit $\qquad i = Ie^{-Rt/L}$

(j) Growth of current in a capacitive circuit $\qquad i = I(1 - e^{-t/CR})$

Here are some worked problems to demonstrate the laws of growth and decay.

Problem 18. The resistance R of an electrical conductor at temperature $\theta°C$ is given by $R = R_0 e^{\alpha\theta}$, where α is a constant and $R_0 = 5\,k\Omega$. Determine the value of α correct to 4 significant figures when $R = 6\,k\Omega$ and $\theta = 1500°C$. Also, find the temperature, correct to the nearest degree, when the resistance R is $5.4\,k\Omega$

Transposing $R = R_0 e^{\alpha\theta}$ gives $\dfrac{R}{R_0} = e^{\alpha\theta}$

Taking Napierian logarithms of both sides gives

$$\ln \frac{R}{R_0} = \ln e^{\alpha\theta} = \alpha\theta$$

Hence, $\alpha = \dfrac{1}{\theta} \ln \dfrac{R}{R_0} = \dfrac{1}{1500} \ln \left(\dfrac{6 \times 10^3}{5 \times 10^3} \right)$

$$= \frac{1}{1500}(0.1823215\ldots)$$

$$= 1.215477\ldots \times 10^{-4}$$

Hence, $\alpha = \mathbf{1.215 \times 10^{-4}}$ correct to 4 significant figures.

From above, $\ln \dfrac{R}{R_0} = \alpha\theta$ hence $\theta = \dfrac{1}{\alpha} \ln \dfrac{R}{R_0}$

When $R = 5.4 \times 10^3$, $\alpha = 1.215477\ldots \times 10^{-4}$ and $R_0 = 5 \times 10^3$

$$\theta = \frac{1}{1.215477\ldots \times 10^{-4}} \ln \left(\frac{5.4 \times 10^3}{5 \times 10^3} \right)$$

$$= \frac{10^4}{1.215477\ldots}(7.696104\ldots \times 10^{-2})$$

$$= \mathbf{633°C}\ \text{correct to the nearest degree.}$$

Problem 19. In an experiment involving Newton's law of cooling*, the temperature $\theta(°C)$ is given by $\theta = \theta_0 e^{-kt}$. Find the value of constant k when $\theta_0 = 56.6°C$, $\theta = 16.5°C$ and $t = 79.0$ seconds

---

*Who was **Newton**? See page 65. To find out more go to www.routledge.com/cw/bird

Transposing $\theta = \theta_0 e^{-kt}$ gives $\dfrac{\theta}{\theta_0} = e^{-kt}$, from which

$$\frac{\theta_0}{\theta} = \frac{1}{e^{-kt}} = e^{kt}$$

Taking Napierian logarithms of both sides gives

$$\ln \frac{\theta_0}{\theta} = kt$$

from which,

$$k = \frac{1}{t}\ln \frac{\theta_0}{\theta} = \frac{1}{79.0}\ln\left(\frac{56.6}{16.5}\right)$$

$$= \frac{1}{79.0}(1.2326486\ldots)$$

Hence, $\quad k = 0.01560$ or 15.60×10^{-3}

Problem 20. The current i amperes flowing in a capacitor at time t seconds is given by $i = 8.0(1 - e^{-\frac{t}{CR}})$, where the circuit resistance R is $25\,\text{k}\Omega$ and capacitance C is $16\,\mu\text{F}$. Determine (a) the current i after 0.5 seconds and (b) the time, to the nearest millisecond, for the current to reach $6.0\,\text{A}$. Sketch the graph of current against time

---

(a) Current $i = 8.0\left(1 - e^{-\frac{t}{CR}}\right)$

$$= 8.0[1 - e^{-0.5/(16\times10^{-6})(25\times10^{3})}]$$

$$= 8.0(1 - e^{-1.25})$$

$$= 8.0(1 - 0.2865047\ldots)$$

$$= 8.0(0.7134952\ldots)$$

$$= \mathbf{5.71\,amperes}$$

(b) Transposing $i = 8.0\left(1 - e^{-\frac{t}{CR}}\right)$ gives

$$\frac{i}{8.0} = 1 - e^{-\frac{t}{CR}}$$

from which, $\quad e^{-\frac{t}{CR}} = 1 - \dfrac{i}{8.0} = \dfrac{8.0 - i}{8.0}$

Taking the reciprocal of both sides gives

$$e^{\frac{t}{CR}} = \frac{8.0}{8.0 - i}$$

Taking Napierian logarithms of both sides gives

$$\frac{t}{CR} = \ln\left(\frac{8.0}{8.0 - i}\right)$$

Hence,

$$t = CR\ln\left(\frac{8.0}{8.0 - i}\right)$$

When $i = 6.0\,\text{A}$,

$$t = (16 \times 10^{-6})(25 \times 10^{3})\ln\left(\frac{8.0}{8.0 - 6.0}\right)$$

i.e. $\quad t = \dfrac{400}{10^{3}}\ln\left(\dfrac{8.0}{2.0}\right) = 0.4\ln 4.0$

$$= 0.4(1.3862943\ldots)$$

$$= 0.5545\,\text{s}$$

$$= \mathbf{555\,ms} \text{ correct to the nearest ms.}$$

A graph of current against time is shown in Fig. 16.6.

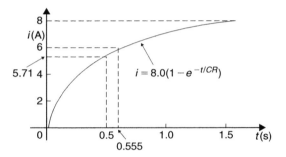

Figure 16.6

Problem 21. The temperature θ_2 of a winding which is being heated electrically at time t is given by $\theta_2 = \theta_1(1 - e^{-\frac{t}{\tau}})$, where θ_1 is the temperature (in degrees Celsius*) at time $t = 0$ and τ is a constant. Calculate

(a) θ_1, correct to the nearest degree, when θ_2 is $50°\text{C}$, t is $30\,\text{s}$ and τ is $60\,\text{s}$ and

(b) the time t, correct to 1 decimal place, for θ_2 to be half the value of θ_1

(a) Transposing the formula to make θ_1 the subject gives

$$\theta_1 = \frac{\theta_2}{\left(1 - e^{-t/\tau}\right)} = \frac{50}{1 - e^{-30/60}}$$

$$= \frac{50}{1 - e^{-0.5}} = \frac{50}{0.393469\ldots}$$

i.e. $\quad \boldsymbol{\theta_1 = 127°\text{C}}$ correct to the nearest degree.

---

*Who was Celsius? See page 84. To find out more go to www.routledge.com/cw/bird

(b) Transposing to make t the subject of the formula gives

$$\frac{\theta_2}{\theta_1} = 1 - e^{-\frac{t}{\tau}}$$

from which, $e^{-\frac{t}{\tau}} = 1 - \frac{\theta_2}{\theta_1}$

Hence, $\quad -\frac{t}{\tau} = \ln\left(1 - \frac{\theta_2}{\theta_1}\right)$

i.e. $\quad t = -\tau \ln\left(1 - \frac{\theta_2}{\theta_1}\right)$

Since $\theta_2 = \frac{1}{2}\theta_1$

$$t = -60\ln\left(1 - \frac{1}{2}\right) = -60\ln 0.5$$

$$= 41.59\,\text{s}$$

Hence, **the time for the temperature θ_2 to be one half of the value of θ_1 is 41.6 s, correct to 1 decimal place.**

Problem 22. The power, P, developed in a flat belt drive system is given by the equation: $P = Tv(1 - e^{-\mu\theta})$ where T is the belt tension, v is the linear velocity, θ is the angle of lap and μ is the coefficient of friction. (a) Transpose the equation for μ (b) Evaluate μ, correct to 3 significant figures, given that $P = 2750, T = 1300, v = 3$ and $\theta = 2.95$

(a) If $P = Tv(1 - e^{-\mu\theta})$ then $\frac{P}{Tv} = (1 - e^{-\mu\theta})$

and $\quad e^{-\mu\theta} = 1 - \frac{P}{Tv}$

Taking logs of both sides gives:

$$\ln(e^{-\mu\theta}) = \ln\left(1 - \frac{P}{Tv}\right)$$

from which, $\quad -\mu\theta = \ln\left(1 - \frac{P}{Tv}\right)$

and $\mu = \frac{1}{-\theta}\ln\left(1 - \frac{P}{Tv}\right)$ or $\mu = -\frac{1}{\theta}\ln\left(1 - \frac{P}{Tv}\right)$

(b) When $P = 2750$, $T = 1300$, $v = 3$ and $\theta = 2.95$, then

$$\mu = -\frac{1}{\theta}\ln\left(1 - \frac{P}{Tv}\right) = -\frac{1}{2.95}\ln\left(1 - \frac{2750}{1300 \times 3}\right)$$

i.e. $\quad$ **$\mu = 0.414$, correct to 3 significant figures.**

Now try the following Practice Exercise

Practice Exercise 85   Laws of growth and decay (answers on page 451)

1. The temperature, $T°C$, of a cooling object varies with time, t minutes, according to the equation $T = 150e^{-0.04t}$. Determine the temperature when (a) $t = 0$, (b) $t = 10$ minutes.

2. The pressure p pascals at height h metres above ground level is given by $p = p_0 e^{-h/C}$, where p_0 is the pressure at ground level and C is a constant. Find pressure p when $p_0 = 1.012 \times 10^5$ Pa, height $h = 1420$ m and $C = 71500$

3. The voltage drop, v volts, across an inductor L henrys at time t seconds is given by $v = 200e^{-\frac{Rt}{L}}$, where $R = 150\,\Omega$ and $L = 12.5 \times 10^{-3}$ H. Determine (a) the voltage when $t = 160 \times 10^{-6}$ s and (b) the time for the voltage to reach 85 V.

4. The length l metres of a metal bar at temperature $t°C$ is given by $l = l_0 e^{\alpha t}$, where l_0 and α are constants. Determine (a) the value of l when $l_0 = 1.894$, $\alpha = 2.038 \times 10^{-4}$ and $t = 250°C$ and (b) the value of l_0 when $l = 2.416, t = 310°C$ and $\alpha = 1.682 \times 10^{-4}$

5. The temperature $\theta_2°C$ of an electrical conductor at time t seconds is given by $\theta_2 = \theta_1(1 - e^{-t/T})$, where θ_1 is the initial temperature and T seconds is a constant. Determine (a) θ_2 when $\theta_1 = 159.9°C$, $t = 30$ s and $T = 80$ s and (b) the time t for θ_2 to fall to half the value of θ_1 if T remains at 80 s.

6. A belt is in contact with a pulley for a sector of $\theta = 1.12$ radians and the coefficient of friction between these two surfaces is $\mu = 0.26$. Determine the tension on the taut side of the belt, T newtons, when tension on the slack side is given by $T_0 = 22.7$ newtons, given that these quantities are related by the law $T = T_0 e^{\mu\theta}$

7. The instantaneous current i at time t is given by $i = 10e^{-t/CR}$ when a capacitor is being charged. The capacitance C is

7×10^{-6} farads and the resistance R is 0.3×10^6 ohms. Determine (a) the instantaneous current when t is 2.5 seconds and (b) the time for the instantaneous current to fall to 5 amperes. Sketch a curve of current against time from $t = 0$ to $t = 6$ seconds.

8. The amount of product x (in mol/cm^3) found in a chemical reaction starting with 2.5 mol/cm^3 of reactant is given by $x = 2.5(1 - e^{-4t})$ where t is the time, in minutes, to form product x. Plot a graph at 30 second intervals up to 2.5 minutes and determine x after 1 minute.

9. The current i flowing in a capacitor at time t is given by $i = 12.5(1 - e^{-t/CR})$, where resistance R is 30 kΩ and the capacitance C is 20 μF. Determine (a) the current flowing after 0.5 seconds and (b) the time for the current to reach 10 amperes.

10. The amount A after n years of a sum invested P is given by the compound interest law $A = Pe^{rn/100}$, when the per unit interest rate r is added continuously. Determine, correct to the nearest pound, the amount after 8 years for a sum of £1500 invested if the interest rate is 1.8% per annum.

11. The percentage concentration C of the starting material in a chemical reaction varies with time t according to the equation $C = 100e^{-0.004t}$. Determine the concentration when (a) $t = 0$, (b) $t = 100$ s, (c) $t = 1000$ s.

12. The current i flowing through a diode at room temperature is given by the equation $i = i_s\left(e^{40V} - 1\right)$ amperes. Calculate the current flowing in a silicon diode when the reverse saturation current $i_s = 50$ nA and the forward voltage $V = 0.27$ V.

13. A formula for chemical decomposition is given by: $C = A\left(1 - e^{-\frac{t}{10}}\right)$ where t is the time in seconds. Calculate the time, in milliseconds, for a compound to decompose to a value of $C = 0.12$ given $A = 8.5$.

14. The mass, m, of pollutant in a water reservoir decreases according to the law $m = m_0 e^{-0.1t}$ where t is the time in days and m_0 is the initial mass. Calculate the percentage decrease in the mass after 60 days, correct to 3 decimal places.

15. A metal bar is cooled with water. Its temperature, in $^\circ$C, is given by the equation $\theta = 15 + 1300e^{-0.2t}$ where t is the time in minutes. Calculate how long it will take for the temperature, θ, to decrease to 36°C, correct to the nearest second.

Practice Exercise 86   Multiple-choice questions on exponential functions (answers on page 451)

Each question has only one correct answer

1. The value of $\dfrac{\ln 2}{e^2 \lg 2}$, correct to 3 significant figures, is:
   (a) 0.0588   (b) 0.312   (c) 17.0   (d) 3.209

2. The value of $\dfrac{3.67 \ln 21.28}{e^{-0.189}}$, correct to 4 significant figures, is:
   (a) 9.289   (b) 13.56
   (c) 13.5566   (d) -3.844×10^9

3. A voltage, v, is given by $v = Ve^{-\frac{t}{CR}}$ volts. When $t = 30$ ms, $C = 150$ nF, $R = 50$ MΩ and $V = 250$ V, the value of v, correct to 3 significant figures, is:
   (a) 249 V   (b) 250 V   (c) 4.58 V   (d) 0 V

4. The value of $\log_x x^{2k}$ is:
   (a) k (b) k^2 (c) $2k$ (d) $2kx$

5. In the equation $5.0 = 3.0\ln\left(\dfrac{2.9}{x}\right)$, x has a value correct to 3 significant figures if:
   (a) 1.59   (b) 0.392   (c) 0.548   (d) 0.0625

6. Solving the equation $7 + e^{2k-4} = 8$ for k gives:
   (a) 2   (b) 3   (c) 4   (d) 5

7. Given that $2e^{x+1} = e^{2x-5}$, the value of x is:
   (a) 1   (b) 6.693   (c) 3   (d) 2.231

8. The current i amperes flowing in a capacitor at time t seconds is given by $i = 10(1 - e^{-t/CR})$, where resistance R is 25×10^3 ohms and capacitance C is 16×10^{-6} farads. When current i reaches 7 amperes, the time t is:
    (a) -0.48 s   (b) 0.14 s
    (c) 0.21 s   (d) 0.48 s

9. Solving $13 + 2e^{3x} = 31$ for x, correct to 4 significant figures, gives:
    (a) 0.4817 (b) 0.09060
    (c) 0.3181 (d) 0.7324

10. Chemical decomposition, C, is given by $C = A\left(1 - e^{-\frac{t}{15}}\right)$ where t is the time in seconds. If $A = 9$, the time, correct to 3 decimal places, for C to reach 0.10 is:
    (a) 34.388 s   (b) 0.007 s
    (c) 0.168 s   (d) 15.168 s

For fully worked solutions to each of the problems in Practice Exercises 81 to 85 in this chapter, go to the website:
www.routledge.com/cw/bird

This assignment covers the material contained in Chapters 14–16. *The marks available are shown in brackets at the end of each question.*

1. Solve the following equations by factorisation.
   (a) $x^2 - 9 = 0$ (b) $x^2 + 12x + 36 = 0$
   (c) $x^2 + 3x - 4 = 0$ (d) $3z^2 - z - 4 = 0$ (9)

2. Solve the following equations, correct to 3 decimal places.
   (a) $5x^2 + 7x - 3 = 0$ (b) $3a^2 + 4a - 5 = 0$ (8)

3. Solve the equation $3x^2 - x - 4 = 0$ by completing the square. (6)

4. Determine the quadratic equation in x whose roots are 1 and -3. (3)

5. The bending moment M at a point in a beam is given by $M = \dfrac{3x(20 - x)}{2}$ where x metres is the distance from the point of support. Determine the value of x when the bending moment is 50 Nm. (5)

6. The current i flowing through an electronic device is given by $i = (0.005v^2 + 0.014v)$ amperes where v is the voltage. Calculate the values of v when $i = 3 \times 10^{-3}$ (6)

7. Evaluate the following, correct to 4 significant figures.
   (a) $3.2 \ln 4.92 - 5 \lg 17.9$ (b) $\dfrac{5(1 - e^{-2.65})}{e^{1.73}}$ (4)

8. Solve the following equations.
   (a) $\lg x = 4$ (b) $\ln x = 2$
   (c) $\log_2 x = 6$ (d) $5^x = 2$
   (e) $3^{2t-1} = 7^{t+2}$ (f) $3e^{2x} = 4.2$ (18)

9. Evaluate $\log_{16}\left(\dfrac{1}{8}\right)$ (4)

10. Write the following as the logarithm of a single number.
    (a) $3 \log 2 + 2 \log 5 - \dfrac{1}{2} \log 16$

    (b) $3 \log 3 + \dfrac{1}{4} \log 16 - \dfrac{1}{3} \log 27$ (8)

11. Solve the equation
    $\log(x^2 + 8) - \log(2x) = \log 3$ (5)

12. Evaluate the following, each correct to 3 decimal places.
    (a) $\ln 462.9$

    (b) $\ln 0.0753$

    (c) $\dfrac{\ln 3.68 - \ln 2.91}{4.63}$ (3)

13. Expand xe^{3x} to six terms. (5)

14. Evaluate v given that $v = E\left(1 - e^{-\frac{t}{CR}}\right)$ volts when $E = 100$ V, $C = 15\,\mu$ F, $R = 50\,k\Omega$ and $t = 1.5$ s. Also, determine the time when the voltage is 60 V. (8)

15. Plot a graph of $y = \dfrac{1}{2}e^{-1.2x}$ over the range $x = -2$ to $x = +1$ and hence determine, correct to 1 decimal place,
    (a) the value of y when $x = -0.75$, and
    (b) the value of x when $y = 4.0$ (8)

For lecturers/instructors/teachers, fully worked solutions to each of the problems in Revision Test 6, together with a full marking scheme, are available at the website:
www.routledge.com/cw/bird

Chapter 17

Straight line graphs

Why it is important to understand: **Straight line graphs**

Graphs have a wide range of applications in engineering and in physical sciences because of their inherent simplicity. A graph can be used to represent almost any physical situation involving discrete objects and the relationship among them. If two quantities are directly proportional and one is plotted against the other, a straight line is produced. Examples include an applied force on the end of a spring plotted against spring extension, the speed of a flywheel plotted against time, and strain in a wire plotted against stress (Hooke's law). In engineering, the straight line graph is the most basic graph to draw and evaluate.

At the end of this chapter you should be able to:

- understand rectangular axes, scales and co-ordinates
- plot given co-ordinates and draw the best straight line graph
- determine the gradient of a straight line graph
- estimate the vertical axis intercept
- state the equation of a straight line graph
- plot straight line graphs involving practical engineering examples

17.1 Introduction to graphs

A graph is a visual representation of information, showing how one quantity varies with another related quantity.

We often see graphs in newspapers or in business reports, in travel brochures and government publications. For example, a graph of the share price (in pence) over a 6 month period for a drinks company, Fizzy Pops, is shown in Fig. 17.1.

Generally, we see that the share price increases to a high of 400 p in June, but dips down to around 280 p in August before recovering slightly in September.

A graph should convey information more quickly to the reader than if the same information was explained in words.

When this chapter is completed you should be able to draw up a table of values, plot co-ordinates, determine the gradient and state the equation of a straight line graph. Some typical practical examples are included in which straight lines are used.

17.2 Axes, scales and co-ordinates

We are probably all familiar with reading a map to locate a town, or a local map to locate a particular street. For example, a street map of central Portsmouth is shown in Fig. 17.2. Notice the squares drawn horizontally and vertically on the map; this is called a **grid** and enables us to locate a place of interest or a particular road. Most maps contain such a grid.

We locate places of interest on the map by stating a letter and a number – this is called the **grid reference**.

Figure 17.1

Figure 17.2 Reprinted with permission from AA Media Ltd.

For example, on the map, the Portsmouth & Southsea station is in square D2, King's Theatre is in square E5, HMS Warrior is in square A2, Gunwharf Quays is in square B3 and High Street is in square B4.

Portsmouth & Southsea station is located by moving horizontally along the bottom of the map until the square labelled D is reached and then moving vertically upwards until square 2 is met.

The letter/number, D2, is referred to as **co-ordinates**; i.e. co-ordinates are used to locate the position of a point on a map. If you are familiar with using a map in this way then you should have no difficulties with graphs, because similar co-ordinates are used with graphs.

As stated earlier, a **graph** is a visual representation of information, showing how one quantity varies with another related quantity. The most common method of showing a relationship between two sets of data is to use a pair of reference axes – these are two lines drawn at right angles to each other (often called **Cartesian**, named after Descartes*, or **rectangular axes**), as shown in Fig. 17.3.

The horizontal axis is labelled the x-axis and the vertical axis is labelled the y-axis. The point where x is 0 and y is 0 is called the **origin**.

x values have **scales** that are positive to the right of the origin and negative to the left. y values have scales that are positive up from the origin and negative down from the origin.

Co-ordinates are written with brackets and a comma in between two numbers. For example, point A is shown with co-ordinates (3, 2) and is located by starting at the

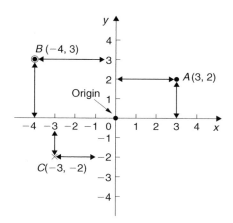

Figure 17.3

origin and moving 3 units in the positive x direction (i.e. to the right) and then 2 units in the positive y direction (i.e. up).

When co-ordinates are stated the first number is always the x value and the second number is always the y value. In Fig. 17.3, point B has co-ordinates $(-4, 3)$ and point C has co-ordinates $(-3, -2)$.

17.3   Straight line graphs

The distances travelled by a car in certain periods of time are shown in the following table of values.

Time (s)	10	20	30	40	50	60
Distance travelled (m)	50	100	150	200	250	300

We will plot time on the horizontal (or x) axis with a scale of 1 cm = 10 s.

We will plot distance on the vertical (or y) axis with a scale of 1 cm = 50 m.

(When choosing scales it is better to choose ones such as 1 cm = 1 unit, 1 cm = 2 units or 1 cm = 10 units because doing so makes reading values between these values easier.)

With the above data, the (x, y) co-ordinates become (time, distance) co-ordinates; i.e. the co-ordinates are (10, 50), (20, 100), (30, 150), and so on.

The co-ordinates are shown plotted in Fig. 17.4 using crosses. (Alternatively, a dot or a dot and circle may be used, as shown in Fig. 17.3.)

A straight line is drawn through the plotted co-ordinates as shown in Fig. 17.4.

*Who was Descartes? – **René Descartes** (31 March 1596–11 February 1650) was a French philosopher, mathematician and writer. He wrote many influential texts including *Meditations on First Philosophy*. Descartes is best known for the philosophical statement '*Cogito ergo sum*' (I think, therefore I am), found in part IV of *Discourse on the Method*. To find out more go to www.routledge.com/cw/bird

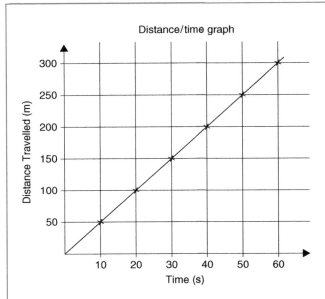

Distance/time graph

Figure 17.4

Student task

The following table gives the force F newtons which, when applied to a lifting machine, overcomes a corresponding load of L newtons.

F (newtons)	19	35	50	93	125	147
L (newtons)	40	120	230	410	540	680

1. Plot L horizontally and F vertically.

2. Scales are normally chosen such that the graph occupies as much space as possible on the graph paper. So in this case, the following scales are chosen.

   Horizontal axis (i.e. L): 1 cm = 50 N

   Vertical axis (i.e. F): 1 cm = 10 N

3. Draw the axes and label them L (newtons) for the horizontal axis and F (newtons) for the vertical axis.

4. Label the origin as 0

5. Write on the horizontal scaling at 100, 200, 300, and so on, every 2 cm.

6. Write on the vertical scaling at 10, 20, 30, and so on, every 1 cm.

7. Plot on the graph the co-ordinates (40, 19), (120, 35), (230, 50), (410, 93), (540, 125) and (680, 147), marking each with a cross or a dot.

8. Using a ruler, draw the best straight line through the points. You will notice that not all of the points lie exactly on a straight line. This is quite normal with experimental values. In a practical situation it would be surprising if all of the points lay exactly on a straight line.

9. Extend the straight line at each end.

10. From the graph, determine the force applied when the load is 325 N. It should be close to 75 N. This process of finding an equivalent value within the given data is called **interpolation**. Similarly, determine the load that a force of 45 N will overcome. It should be close to 170 N.

11. From the graph, determine the force needed to overcome a 750 N load. It should be close to 161 N. This process of finding an equivalent value outside the given data is called **extrapolation**. To extrapolate we need to have extended the straight line drawn. Similarly, determine the force applied when the load is zero. It should be close to 11 N. The point where the straight line crosses the vertical axis is called the **vertical axis intercept**. So, in this case, the vertical-axis intercept = 11 N at co-ordinates (0, 11).

The graph you have drawn should look something like Fig. 17.5 shown below.

Graph of F against L

Figure 17.5

In another example, let the relationship between two variables x and y be $y = 3x + 2$

When $x = 0$, $y = 0 + 2 = 2$
When $x = 1$, $y = 3 + 2 = 5$
When $x = 2$, $y = 6 + 2 = 8$, and so on.

The co-ordinates $(0, 2)$, $(1, 5)$ and $(2, 8)$ have been produced and are plotted, as shown in Fig. 17.6.
When the points are joined together **a straight line graph results**, i.e. $y = 3x + 2$ is a straight line graph.

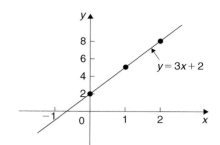

Figure 17.6 Graph of y/x

Summary of general rules to be applied when drawing graphs

(a) Give the graph a title clearly explaining what is being illustrated.

(b) Choose scales such that the graph occupies as much space as possible on the graph paper being used.

(c) Choose scales so that interpolation is made as easy as possible. Usually scales such as $1\,\text{cm} = 1\,\text{unit}$, $1\,\text{cm} = 2$ units or $1\,\text{cm} = 10$ units are used. Awkward scales such as $1\,\text{cm} = 3$ units or $1\,\text{cm} = 7$ units should not be used.

(d) The scales need not start at zero, particularly when starting at zero produces an accumulation of points within a small area of the graph paper.

(e) The co-ordinates, or points, should be clearly marked. This is achieved by a cross, or a dot and circle, or just by a dot (see Fig. 17.3).

(f) A statement should be made next to each axis explaining the numbers represented with their appropriate units.

(g) Sufficient numbers should be written next to each axis without cramping.

> **Problem 1.** Plot the graph $y = 4x + 3$ in the range $x = -3$ to $x = +4$. From the graph, find (a) the value of y when $x = 2.2$ and (b) the value of x when $y = -3$

Whenever an equation is given and a graph is required, a table giving corresponding values of the variable is necessary. The table is achieved as follows:

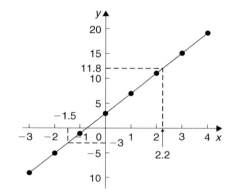

Figure 17.7

When $x = -3$, $y = 4x + 3 = 4(-3) + 3$
$= -12 + 3 = -9$

When $x = -2$, $y = 4(-2) + 3$
$= -8 + 3 = -5$, and so on.

Such a table is shown below.

x	-3	-2	-1	0	1	2	3	4
y	-9	-5	-1	3	7	11	15	19

The co-ordinates $(-3, -9)$, $(-2, -5)$, $(-1, -1)$, and so on, are plotted and joined together to produce the straight line shown in Fig. 17.7. (Note that the scales used on the x and y axes do not have to be the same.) From the graph:

(a) when $x = 2.2$, $y = \textbf{11.8}$, and

(b) when $y = -3$, $x = \textbf{-1.5}$

Now try the following Practice Exercise

Practice Exercise 87   Straight line graphs
(answers on page 451)

1. Assuming graph paper measuring 20 cm by 20 cm is available, suggest suitable scales for the following ranges of values.
   (a) Horizontal axis: 3 V to 55 V; vertical axis: 10 Ω to 180 Ω.

   (b) Horizontal axis: 7 m to 86 m; vertical axis: 0.3 V to 1.69 V.

   (c) Horizontal axis: 5 N to 150 N; vertical axis: 0.6 mm to 3.4 mm.

2. Corresponding values obtained experimentally for two quantities are

x	−5	−3	−1	0	2	4
y	−13	−9	−5	−3	1	5

   Plot a graph of y (vertically) against x (horizontally) to scales of 2 cm = 1 for the horizontal x-axis and 1 cm = 1 for the vertical y-axis. (This graph will need the whole of the graph paper with the origin somewhere in the centre of the paper.)
   From the graph, find
   (a) the value of y when $x = 1$

   (b) the value of y when $x = -2.5$

   (c) the value of x when $y = -6$

   (d) the value of x when $y = 5$

3. Corresponding values obtained experimentally for two quantities are

x	−2.0	−0.5	0	1.0	2.5	3.0	5.0
y	−13.0	−5.5	−3.0	2.0	9.5	12.0	22.0

   Use a horizontal scale for x of 1 cm = $\frac{1}{2}$ unit and a vertical scale for y of 1 cm = 2 units and draw a graph of x against y. Label the graph and each of its axes. By interpolation, find from the graph the value of y when x is 3.5

4. Draw a graph of $y - 3x + 5 = 0$ over a range of $x = -3$ to $x = 4$. Hence determine
   (a) the value of y when $x = 1.3$

   (b) the value of x when $y = -9.2$

5. The speed n rev/min of a motor changes when the voltage V across the armature is varied. The results are shown in the following table.

n (rev/min)	560	720	900	1010	1240	1410
V (volts)		80	100	120	140	160	180

   It is suspected that one of the readings taken of the speed is inaccurate. Plot a graph of speed (horizontally) against voltage (vertically) and find this value. Find also
   (a) the speed at a voltage of 132 V.

   (b) the voltage at a speed of 1300 rev/min.

17.4 Gradients, intercepts and equations of graphs

Gradients

The **gradient or slope** of a straight line is the ratio of the change in the value of y to the change in the value of x between any two points on the line. If, as x increases, ($\rightarrow$), y also increases, ($\uparrow$), then the gradient is positive. In Fig. 17.8(a), a straight line graph $y = 2x + 1$ is shown. To find the gradient of this straight line, choose two points on the straight line graph, such as A and C.

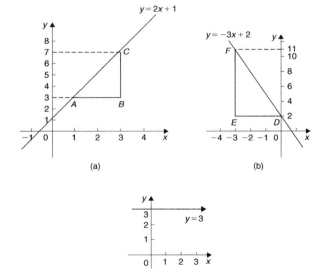

Figure 17.8

Then construct a right-angled triangle, such as ABC, where BC is vertical and AB is horizontal.

Then,   **gradient of** $AC = \dfrac{\text{change in } y}{\text{change in } x} = \dfrac{CB}{BA}$

$$= \dfrac{7-3}{3-1} = \dfrac{4}{2} = \mathbf{2}$$

In Fig. 17.8(b), a straight line graph $y = -3x + 2$ is shown. To find the gradient of this straight line, choose two points on the straight line graph, such as D and F. Then construct a right-angled triangle, such as DEF, where EF is vertical and DE is horizontal.

Then,   **gradient of** $DF = \dfrac{\text{change in } y}{\text{change in } x} = \dfrac{FE}{ED}$

$$= \dfrac{11-2}{-3-0} = \dfrac{9}{-3} = \mathbf{-3}$$

Fig. 17.8(c) shows a straight line graph $y = 3$. **Since the straight line is horizontal the gradient is zero.**

The y-axis intercept

The value of y when $x = 0$ is called the **y-axis intercept**. In Fig. 17.8(a) the y-axis intercept is 1 and in Fig. 17.8(b) the y-axis intercept is 2

The equation of a straight line graph

The general equation of a straight line graph is

$$y = mx + c$$

where m is the gradient and c is the y-axis intercept. Thus, as we have found in Fig. 17.8(a), $y = 2x + 1$ represents a straight line of gradient 2 and y-axis intercept 1. So, given the equation $y = 2x + 1$, we are able to state, on sight, that the gradient $= 2$ and the y-axis intercept $= 1$, without the need for any analysis.
Similarly, in Fig. 17.8(b), $y = -3x + 2$ represents a straight line of gradient -3 and y-axis intercept 2
In Fig. 17.8(c), $y = 3$ may be rewritten as $y = 0x + 3$ and therefore represents a straight line of gradient 0 and y-axis intercept 3
Here are some worked problems to help understanding of gradients, intercepts and equations of graphs.

> **Problem 2.** Plot the following graphs on the same axes in the range $x = -4$ to $x = +4$ and determine the gradient of each.

(a) $y = x$ (b) $y = x + 2$
(c) $y = x + 5$ (d) $y = x - 3$

A table of co-ordinates is produced for each graph.

(a) $y = x$

x	−4	−3	−2	−1	0	1	2	3	4
y	−4	−3	−2	−1	0	1	2	3	4

(b) $y = x + 2$

x	−4	−3	−2	−1	0	1	2	3	4
y	−2	−1	0	1	2	3	4	5	6

(c) $y = x + 5$

x	−4	−3	−2	−1	0	1	2	3	4
y	1	2	3	4	5	6	7	8	9

(d) $y = x - 3$

x	−4	−3	−2	−1	0	1	2	3	4
y	−7	−6	−5	−4	−3	−2	−1	0	1

The co-ordinates are plotted and joined for each graph. The results are shown in Fig. 17.9. Each of the straight lines produced is parallel to the others; i.e. the slope or gradient is the same for each.
To find the gradient of any straight line, say, $y = x - 3$, a horizontal and vertical component needs to be constructed. In Fig. 17.9, AB is constructed vertically at $x = 4$ and BC is constructed horizontally at $y = -3$.

The gradient of $AC = \dfrac{AB}{BC} = \dfrac{1-(-3)}{4-0} = \dfrac{4}{4} = 1$

i.e. the gradient of the straight line $y = x - 3$ is 1, which could have been deduced 'on sight' since $y = 1x - 3$ represents a straight line graph with gradient 1 and y-axis intercept of -3
The actual positioning of AB and BC is unimportant because the gradient is also given by

$$\dfrac{DE}{EF} = \dfrac{-1-(-2)}{2-1} = \dfrac{1}{1} = 1$$

The slope or gradient of each of the straight lines in Fig. 17.9 is thus 1 since they are parallel to each other.

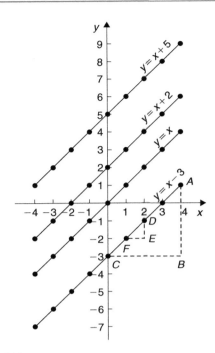

Figure 17.9

Problem 3. Plot the following graphs on the same axes between the values $x = -3$ to $x = +3$ and determine the gradient and y-axis intercept of each.

(a) $y = 3x$ (b) $y = 3x + 7$

(c) $y = -4x + 4$ (d) $y = -4x - 5$

A table of co-ordinates is drawn up for each equation.

(a) $y = 3x$

x	-3	-2	-1	0	1	2	3
y	-9	-6	-3	0	3	6	9

(b) $y = 3x + 7$

x	-3	-2	-1	0	1	2	3
y	-2	1	4	7	10	13	16

(c) $y = -4x + 4$

x	-3	-2	-1	0	1	2	3
y	16	12	8	4	0	-4	-8

(d) $y = -4x - 5$

x	-3	-2	-1	0	1	2	3
y	7	3	-1	-5	-9	-13	-17

Each of the graphs is plotted as shown in Fig. 17.10 and each is a straight line. $y = 3x$ and $y = 3x + 7$ are parallel to each other and thus have the same gradient. The gradient of AC is given by

$$\frac{CB}{BA} = \frac{16 - 7}{3 - 0} = \frac{9}{3} = 3$$

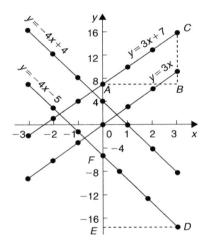

Figure 17.10

Hence, **the gradients of both $y = 3x$ and $y = 3x + 7$ are 3**, which could have been deduced 'on sight'.

$y = -4x + 4$ and $y = -4x - 5$ are parallel to each other and thus have the same gradient. The gradient of DF is given by

$$\frac{FE}{ED} = \frac{-5 - (-17)}{0 - 3} = \frac{12}{-3} = -4$$

Hence, **the gradient of both $y = -4x + 4$ and $y = -4x - 5$ is -4**, which, again, could have been deduced 'on sight'.

The y-axis intercept means the value of y where the straight line cuts the y-axis. From Fig. 17.10,

$y = 3x$ cuts the y-axis at $y = 0$
$y = 3x + 7$ cuts the y-axis at $y = +7$
$y = -4x + 4$ cuts the y-axis at $y = +4$
$y = -4x - 5$ cuts the y-axis at $y = -5$

Some general conclusions can be drawn from the graphs shown in Figs 17.9 and 17.10. When an equation is of the form $y = mx + c$, where m and c are constants, then

(a) a graph of y against x produces a straight line,

(b) m represents the slope or gradient of the line, and

(c) c represents the y-axis intercept.

Thus, given an equation such as $y = 3x + 7$, it may be deduced 'on sight' that its gradient is $+3$ and its y-axis intercept is $+7$, as shown in Fig. 17.10. Similarly, if $y = -4x - 5$, the gradient is -4 and the y-axis intercept is -5, as shown in Fig. 17.10.

When plotting a graph of the form $y = mx + c$, only two co-ordinates need be determined. When the co-ordinates are plotted a straight line is drawn between the two points. Normally, three co-ordinates are determined, the third one acting as a check.

Problem 4. Plot the graph $3x + y + 1 = 0$ and $2y - 5 = x$ on the same axes and find their point of intersection

---

Rearranging $3x + y + 1 = 0$ gives $y = -3x - 1$

Rearranging $2y - 5 = x$ gives $2y = x + 5$ and

$$y = \frac{1}{2}x + 2\frac{1}{2}$$

Since both equations are of the form $y = mx + c$, both are straight lines. Knowing an equation is a straight line means that only two co-ordinates need to be plotted and a straight line drawn through them. A third co-ordinate is usually determined to act as a check. A table of values is produced for each equation as shown below.

x	1	0	−1
$-3x - 1$	−4	−1	2

x	2	0	−3
$\frac{1}{2}x + 2\frac{1}{2}$	$3\frac{1}{2}$	$2\frac{1}{2}$	1

The graphs are plotted as shown in Fig. 17.11.
The **two straight lines are seen to intersect at $(-1, 2)$**

Problem 5. If graphs of y against x were to be plotted for each of the following, state (i) the gradient and (ii) the y-axis intercept.

(a) $y = 9x + 2$ (b) $y = -4x + 7$

(c) $y = 3x$ (d) $y = -5x - 3$

(e) $y = 6$ (f) $y = x$

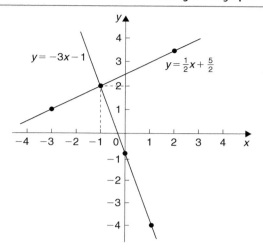

Figure 17.11

If $y = mx + c$ then $m = $ **gradient** and $c = y$-**axis intercept**.

(a) If $y = 9x + 2$, then   (i) **gradient $= 9$**

                                 (ii) y-**axis intercept $= 2$**

(b) If $y = -4x + 7$, then   (i) **gradient $= -4$**

                                 (ii) y-**axis intercept $= 7$**

(c) If $y = 3x$ i.e. $y = 3x + 0$,
    then   (i) **gradient $= 3$**

                                 (ii) y-**axis intercept $= 0$**

    i.e. the straight line passes through the origin.

(d) If $y = -5x - 3$, then   (i) **gradient $= -5$**

                                 (ii) y-**axis intercept $= -3$**

(e) If $y = 6$ i.e. $y = 0x + 6$,
    then   (i) **gradient $= 0$**

                                 (ii) y-**axis intercept $= 6$**

    i.e. $y = 6$ is a straight horizontal line.

(f) If $y = x$ i.e. $y = 1x + 0$,
    then   (i) **gradient $= 1$**

                                 (ii) y-**axis intercept $= 0$**

    Since $y = x$, as x increases, y increases by the same amount; i.e. **y is directly proportional to x**.

Problem 6. Without drawing graphs, determine the gradient and y-axis intercept for each of the following equations.

(a) $y + 4 = 3x$ (b) $2y + 8x = 6$ (c) $3x = 4y + 7$

If $y = mx + c$ then $m = $ **gradient** and $c = $ **y-axis intercept**.

(a) Transposing $y + 4 = 3x$ gives $y = 3x - 4$

Hence, **gradient $= 3$** and **y-axis intercept $= -4$**

(b) Transposing $2y + 8x = 6$ gives $2y = -8x + 6$

Dividing both sides by 2 gives $y = -4x + 3$

Hence, **gradient $= -4$** and **y-axis intercept $= 3$**

(c) Transposing $3x = 4y + 7$ gives $3x - 7 = 4y$

or $\qquad 4y = 3x - 7$

Dividing both sides by 4 gives $y = \dfrac{3}{4}x - \dfrac{7}{4}$

or $\qquad y = 0.75x - 1.75$

Hence, **gradient $= 0.75$** and **y-axis intercept $= -1.75$**

Problem 7. Without plotting graphs, determine the gradient and y-axis intercept values of the following equations.

(a) $y = 7x - 3$ $\qquad$ (b) $3y = -6x + 2$

(c) $y - 2 = 4x + 9$ $\qquad$ (d) $\dfrac{y}{3} = \dfrac{x}{2} - \dfrac{1}{5}$

(e) $2x + 9y + 1 = 0$

(a) $y = 7x - 3$ is of the form $y = mx + c$

Hence, **gradient, $m = 7$** and **y-axis intercept, $c = -3$**

(b) Rearranging $3y = -6x + 2$ gives $y = -\dfrac{6x}{3} + \dfrac{2}{3}$

i.e. $\qquad\qquad y = -2x + \dfrac{2}{3}$
which is of the form $y = mx + c$

Hence, **gradient $m = -2$** and **y-axis intercept, $c = \dfrac{2}{3}$**

(c) Rearranging $y - 2 = 4x + 9$ gives $y = 4x + 11$
Hence, **gradient $= 4$** and **y-axis intercept $= 11$**

(d) Rearranging $\dfrac{y}{3} = \dfrac{x}{2} - \dfrac{1}{5}$ gives $y = 3\left(\dfrac{x}{2} - \dfrac{1}{5}\right)$

$\qquad\qquad\qquad\qquad = \dfrac{3}{2}x - \dfrac{3}{5}$

Hence, **gradient $= \dfrac{3}{2}$** and **y-axis intercept $= -\dfrac{3}{5}$**

(e) Rearranging $2x + 9y + 1 = 0$ gives $9y = -2x - 1$,

i.e. $\quad y = -\dfrac{2}{9}x - \dfrac{1}{9}$

Hence, **gradient $= -\dfrac{2}{9}$** and **y-axis intercept $= -\dfrac{1}{9}$**

Problem 8. Determine for the straight line shown in Fig. 17.12 (a) the gradient and (b) the equation of the graph

Figure 17.12

(a) A right-angled triangle ABC is constructed on the graph as shown in Fig. 17.13.

$$\textbf{Gradient} = \frac{AC}{CB} = \frac{23 - 8}{4 - 1} = \frac{15}{3} = \textbf{5}$$

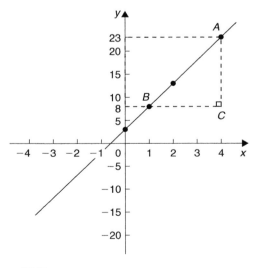

Figure 17.13

(b)   The *y*-axis intercept at $x = 0$ is seen to be at $y = 3$

$y = mx + c$ is a straight line graph where $m =$ gradient and $c = y$-axis intercept.

From above, $m = 5$ and $c = 3$

Hence, the equation of the graph is $y = 5x + 3$

Problem 9.   Determine the equation of the straight line shown in Fig. 17.14.

Figure 17.14

The triangle *DEF* is shown constructed in Fig. 17.14.

Gradient of $DE = \dfrac{DF}{FE} = \dfrac{3 - (-3)}{-1 - 2} = \dfrac{6}{-3} = -2$

and the *y*-axis intercept = 1
Hence, **the equation of the straight line is $y = mx + c$**
i.e. $y = -2x + 1$

Problem 10.   The velocity of a body was measured at various times and the results obtained were as follows:

Velocity v (m/s)	8	10.5	13	15.5	18	20.5	23
Time t (s)		1	2	3	4	5	6	7

Plot a graph of velocity (vertically) against time (horizontally) and determine the equation of the graph

Suitable scales are chosen and the co-ordinates (1, 8), (2, 10.5), (3, 13), and so on, are plotted as shown in Fig. 17.15.

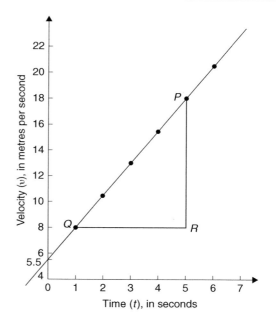

Figure 17.15

The right-angled triangle *PRQ* is constructed on the graph as shown in Fig. 17.15.

Gradient of $PQ = \dfrac{PR}{RQ} = \dfrac{18 - 8}{5 - 1} = \dfrac{10}{4} = 2.5$

The **vertical axis intercept** is at $v = 5.5$ m/s.
The equation of a straight line graph is $y = mx + c$. In this case, t corresponds to x and v corresponds to y. Hence, the equation of the graph shown in Fig. 17.15 is $v = mt + c$. But, from above, gradient, $m = 2.5$ and v-axis intercept, $c = 5.5$
Hence, **the equation of the graph is $v = 2.5t + 5.5$**

Problem 11.   Determine the gradient of the straight line graph passing through the co-ordinates (a) $(-2, 5)$ and $(3, 4)$, and (b) $(-2, -3)$ and $(-1, 3)$

From Fig. 17.16, a straight line graph passing through co-ordinates (x_1, y_1) and (x_2, y_2) has a gradient given by

$$m = \frac{y_2 - y_1}{x_2 - x_1}$$

(a)   A straight line passes through $(-2, 5)$ and $(3, 4)$, hence $x_1 = -2$, $y_1 = 5$, $x_2 = 3$ and $y_2 = 4$, hence,
gradient, $m = \dfrac{y_2 - y_1}{x_2 - x_1} = \dfrac{4 - 5}{3 - (-2)} = -\dfrac{1}{5}$

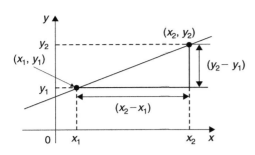

Figure 17.16

(b) A straight line passes through $(-2, -3)$ and $(-1, 3)$, hence $x_1 = -2, y_1 = -3$, $x_2 = -1$ and $y_2 = 3$, hence, **gradient,**

$$m = \frac{y_2 - y_1}{x_2 - x_1} = \frac{3 - (-3)}{-1 - (-2)} = \frac{3 + 3}{-1 + 2} = \frac{6}{1} = 6$$

Now try the following Practice Exercise

Practice Exercise 88   Gradients, intercepts and equations of graphs (answers on page 451)

1. The equation of a line is $4y = 2x + 5$. A table of corresponding values is produced and is shown below. Complete the table and plot a graph of y against x. Find the gradient of the graph.

x	-4	-3	-2	-1	0	1	2	3	4
y		-0.25			1.25				3.25

2. Determine the gradient and intercept on the y-axis for each of the following equations.
   (a) $y = 4x - 2$ (b) $y = -x$
   (c) $y = -3x - 4$ (d) $y = 4$

3. Find the gradient and intercept on the y-axis for each of the following equations.
   (a) $2y - 1 = 4x$ (b) $6x - 2y = 5$
   (c) $3(2y - 1) = \dfrac{x}{4}$

   Determine the gradient and y-axis intercept for each of the equations in Problems 4 and 5 and sketch the graphs.

4. (a) $y = 6x - 3$ (b) $y = -2x + 4$
   (c) $y = 3x$ (d) $y = 7$

5. (a) $2y + 1 = 4x$ (b) $2x + 3y + 5 = 0$
   (c) $3(2y - 4) = \dfrac{x}{3}$ (d) $5x - \dfrac{y}{2} - \dfrac{7}{3} = 0$

6. Determine the gradient of the straight line graphs passing through the co-ordinates:
   (a) $(2, 7)$ and $(-3, 4)$
   (b) $(-4, -1)$ and $(-5, 3)$
   (c) $\left(\dfrac{1}{4}, -\dfrac{3}{4}\right)$ and $\left(-\dfrac{1}{2}, \dfrac{5}{8}\right)$

7. State which of the following equations will produce graphs which are parallel to one another.
   (a) $y - 4 = 2x$ (b) $4x = -(y + 1)$
   (c) $x = \dfrac{1}{2}(y + 5)$ (d) $1 + \dfrac{1}{2}y = \dfrac{3}{2}x$
   (e) $2x = \dfrac{1}{2}(7 - y)$

8. Draw on the same axes the graphs of $y = 3x - 5$ and $3y + 2x = 7$. Find the co-ordinates of the point of intersection. Check the result obtained by solving the two simultaneous equations algebraically.

9. Plot the graphs $y = 2x + 3$ and $2y = 15 - 2x$ on the same axes and determine their point of intersection.

10. Draw on the same axes the graphs of $y = 3x - 1$ and $y + 2x = 4$. Find the co-ordinates of the point of intersection.

11. A piece of elastic is tied to a support so that it hangs vertically and a pan, on which weights can be placed, is attached to the free end. The length of the elastic is measured as various weights are added to the pan and the results obtained are as follows:

Load, W (N)	5	10	15	20	25
Length, l (cm)	60	72	84	96	108

Plot a graph of load (horizontally) against length (vertically) and determine
(a) the value of length when the load is 17 N

(b) the value of load when the length is 74 cm

(c) its gradient

(d) the equation of the graph.

12. The following table gives the effort P to lift a load W with a small lifting machine.

W (N)	10	20	30	40	50	60
P (N)	5.1	6.4	8.1	9.6	10.9	12.4

Plot W horizontally against P vertically and show that the values lie approximately on a straight line. Determine the probable relationship connecting P and W in the form $P = aW + b$.

13. In an experiment the speeds N rpm of a flywheel slowly coming to rest were recorded against the time t in minutes. Plot the results and show that N and t are connected by an equation of the form $N = at + b$. Find probable values of a and b.

t (min)	2	4	6	8	10	12	14
N (rev/min)	372	333	292	252	210	177	132

17.5 Practical problems involving straight line graphs

When a set of co-ordinate values are given or are obtained experimentally and it is believed that they follow a law of the form $y = mx + c$, if a straight line can be drawn reasonably close to most of the co-ordinate values when plotted, this verifies that a law of the form $y = mx + c$ exists. From the graph, constants m (i.e. gradient) and c (i.e. y-axis intercept) can be determined.

Here are some worked problems in which practical situations are featured.

Problem 12. The temperature in degrees Celsius* and the corresponding values in degrees

*Who was Celsius? See page 84. To find out more go to www.routledge.com/cw/bird

Fahrenheit are shown in the table below. Construct rectangular axes, choose suitable scales and plot a graph of degrees Celsius (on the horizontal axis) against degrees Fahrenheit (on the vertical scale).

°C	10	20	40	60	80	100
°F	50	68	104	140	176	212

From the graph find (a) the temperature in degrees Fahrenheit at 55°C, (b) the temperature in degrees Celsius at 167°F, (c) the Fahrenheit temperature at 0°C and (d) the Celsius temperature at 230°F

The co-ordinates (10, 50), (20, 68), (40, 104), and so on are plotted as shown in Fig. 17.17. When the co-ordinates are joined, a straight line is produced. Since a straight line results, there is a linear relationship between degrees Celsius and degrees Fahrenheit.

Figure 17.17

(a) To find the Fahrenheit temperature at 55°C, a vertical line AB is constructed from the horizontal axis to meet the straight line at B. The point where the horizontal line BD meets the vertical axis indicates the equivalent Fahrenheit temperature.

Hence, **55°C is equivalent to 131°F.**

This process of finding an equivalent value in between the given information in the above table is called **interpolation**.

(b) To find the Celsius temperature at 167°F, a horizontal line *EF* is constructed as shown in Fig. 17.17. The point where the vertical line *FG* cuts the horizontal axis indicates the equivalent Celsius temperature.

Hence, 167°F is equivalent to 75°C.

(c) If the graph is assumed to be linear even outside of the given data, the graph may be extended at both ends (shown by broken lines in Fig. 17.17).

From Fig. 17.17, **0°C corresponds to 32°F.**

(d) **230°F is seen to correspond to 110°C.**

The process of finding equivalent values outside of the given range is called **extrapolation**.

> ⚑ **Problem 13.** In an experiment on Charles's law,* the value of the volume of gas, $V\,\text{m}^3$, was measured for various temperatures $T°C$. The results are shown below.
>
> | $V\,\text{m}^3$ | 25.0 | 25.8 | 26.6 | 27.4 | 28.2 | 29.0 |
> |-----------------|------|------|------|------|------|------|
> | $T°C$ | 60 | 65 | 70 | 75 | 80 | 85 |
>
> Plot a graph of volume (vertical) against temperature (horizontal) and from it find (a) the temperature when the volume is $28.6\,\text{m}^3$ and (b) the volume when the temperature is 67°C

If a graph is plotted with both the scales starting at zero then the result is as shown in Fig. 17.18. All of the points lie in the top right-hand corner of the graph, making interpolation difficult. A more accurate graph is obtained if the temperature axis starts at 55°C and the volume axis starts at 24.5 m³. The axes corresponding to these values are shown by the broken lines in Fig. 17.18 and are called **false axes**, since the origin is now at zero. A magnified version of this relevant part of the graph is shown in Fig. 17.19. From the graph,

(a) When the volume is 28.6 m³, the equivalent temperature is **82.5°C.**

---

*Who was **Charles**? See page 51. To find out more go to www.routledge.com/cw/bird

(b) When the temperature is 67°C, the equivalent volume is **26.1 m³**.

Figure 17.18

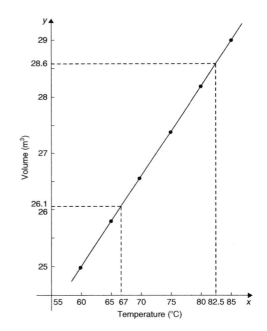

Figure 17.19

> ⚑ **Problem 14.** In an experiment demonstrating Hooke's law*, the strain in an aluminium wire was measured for various stresses. The results were:

---

*Who was **Hooke**? See page 51. To find out more go to www.routledge.com/cw/bird

Stress (N/mm²)	4.9	8.7	15.0
Strain	0.00007	0.00013	0.00021

Stress (N/mm²)	18.4	24.2	27.3
Strain	0.00027	0.00034	0.00039

Plot a graph of stress (vertically) against strain (horizontally). Find (a) Young's modulus of elasticity* for aluminium, which is given by the gradient of the graph, (b) the value of the strain at a stress of 20 N/mm² and (c) the value of the stress when the strain is 0.00020

The co-ordinates (0.00007, 4.9), (0.00013, 8.7), and so on, are plotted as shown in Fig. 17.20. The graph produced is the best straight line which can be drawn corresponding to these points. (With experimental results it is unlikely that all the points will lie exactly on a straight

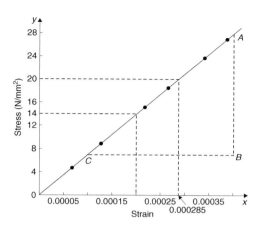

Figure 17.20

line.) The graph, and each of its axes, are labelled. Since the straight line passes through the origin, stress is directly proportional to strain for the given range of values.

(a) The gradient of the straight line AC is given by

$$\frac{AB}{BC} = \frac{28 - 7}{0.00040 - 0.00010} = \frac{21}{0.00030}$$

$$= \frac{21}{3 \times 10^{-4}} = \frac{7}{10^{-4}} = 7 \times 10^4$$

$$= 70\,000\,\text{N/mm}^2$$

Thus, **Young's modulus of elasticity for aluminium is 70 000 N/mm²**. Since $1\,\text{m}^2 = 10^6\,\text{mm}^2$, 70 000 N/mm² is equivalent to $70\,000 \times 10^6$ N/m², i.e. **70×10^9 N/m² (or pascals)**.

From Fig. 17.20,

(b) The value of the strain at a stress of 20 N/mm² is **0.000285**

(c) The value of the stress when the strain is 0.00020 is **14 N/mm²**

*Who was Young? **Thomas Young** (13 June 1773–10 May 1829) was an English polymath. He is famous for having partly deciphered Egyptian hieroglyphics (specifically the Rosetta Stone). Young made notable scientific contributions to the fields of vision, light, solid mechanics, energy, physiology, language, musical harmony and Egyptology. To find out more go to **www.routledge.com/cw/bird**

Problem 15. The following values of resistance R ohms and corresponding voltage V volts are obtained from a test on a filament lamp.

R ohms	30	48.5	73	107	128
V volts	16	29	52	76	94

Choose suitable scales and plot a graph with R representing the vertical axis and V the horizontal axis. Determine (a) the gradient of the graph, (b) the R-axis intercept value, (c) the equation of the graph, (d) the value of resistance when the voltage is 60 V and (e) the value of the voltage when the resistance is 40 ohms. (f) If the graph were to continue in the same manner, what value of resistance would be obtained at 110 V?

The co-ordinates (16, 30), (29, 48.5), and so on are shown plotted in Fig. 17.21, where the best straight line is drawn through the points.

(a) The slope or gradient of the straight line AC is given by

$$\frac{AB}{BC} = \frac{135 - 10}{100 - 0} = \frac{125}{100} = \mathbf{1.25}$$

(Note that the vertical line AB and the horizontal line BC may be constructed anywhere along the length of the straight line. However, calculations are made easier if the horizontal line BC is carefully chosen; in this case, 100)

(b) The R-axis intercept is at $R = \mathbf{10\,ohms}$ (by extrapolation).

(c) The equation of a straight line is $y = mx + c$, when y is plotted on the vertical axis and x on the horizontal axis. m represents the gradient and c the y-axis intercept. In this case, R corresponds to y, V

corresponds to x, $m = 1.25$ and $c = 10$. Hence, the equation of the graph is $R = \mathbf{(1.25\,V + 10)\,\Omega}$.

From Fig. 17.21,

(d) When the voltage is 60 V, the resistance is **85 Ω**.

(e) When the resistance is 40 ohms, the voltage is **24 V**.

(f) By extrapolation, when the voltage is 110 V, the resistance is **147 Ω**.

Problem 16. Experimental tests to determine the breaking stress σ of rolled copper at various temperatures t gave the following results.

Stress σ (N/cm^2)	8.46	8.04	7.78
Temperature t(°C)	70	200	280

Stress σ (N/cm^2)	7.37	7.08	6.63
Temperature t(°C)	410	500	640

Show that the values obey the law $\sigma = at + b$, where a and b are constants, and determine approximate values for a and b. Use the law to determine the stress at 250°C and the temperature when the stress is 7.54 N/cm^2

The co-ordinates (70, 8.46), (200, 8.04), and so on, are plotted as shown in Fig. 17.22. Since the graph is a straight line then the valuesobey the law $\sigma = at + b$, and

Figure 17.21

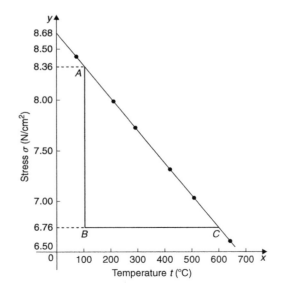

Figure 17.22

the gradient of the straight line is

$$a = \frac{AB}{BC} = \frac{8.36 - 6.76}{100 - 600} = \frac{1.60}{-500} = \mathbf{-0.0032}$$

Vertical axis intercept, $b = \mathbf{8.68}$

Hence, the law of the graph is $\boldsymbol{\sigma = -0.0032t + 8.68}$

When the temperature is 250°C, stress σ is given by

$$\sigma = -0.0032(250) + 8.68 = \mathbf{7.88\,N/cm^2}$$

Rearranging $\sigma = -0.0032t + 8.68$ gives

$$0.0032t = 8.68 - \sigma, \quad \text{i.e.} \quad t = \frac{8.68 - \sigma}{0.0032}$$

Hence, when the stress, $\sigma = 7.54\,\text{N/cm}^2$,

$$\mathbf{temperature,}\, t = \frac{8.68 - 7.54}{0.0032} = \mathbf{356.3°C}$$

Now try the following Practice Exercise

Practice Exercise 89   Practical problems involving straight line graphs (answers on page 451)

1. The resistance R ohms of a copper winding is measured at various temperatures $t°$C and the results are as follows:

R (ohms)	112	120	126	131	134
$t°$C	20	36	48	58	64

Plot a graph of R (vertically) against t (horizontally) and find from it (a) the temperature when the resistance is 122 Ω and (b) the resistance when the temperature is 52°C.

2. The speed of a motor varies with armature voltage as shown by the following experimental results.

n (rev/min)	285	517	615
V (volts)	60	95	110

n (rev/min)	750	917	1050
V (volts)	130	155	175

Plot a graph of speed (horizontally) against voltage (vertically) and draw the best straight line through the points. Find from the graph (a) the speed at a voltage of 145 V and (b) the voltage at a speed of 400 rev/min.

3. The following table gives the force F newtons which, when applied to a lifting machine, overcomes a corresponding load of L newtons.

Force F (newtons)	25	47	64
Load L (newtons)	50	140	210

Force F (newtons)	120	149	187
Load L (newtons)	430	550	700

Choose suitable scales and plot a graph of F (vertically) against L (horizontally). Draw the best straight line through the points. Determine from the graph

(a) the gradient,

(b) the F-axis intercept,

(c) the equation of the graph,

(d) the force applied when the load is 310 N, and

(e) the load that a force of 160 N will overcome.

(f) If the graph were to continue in the same manner, what value of force will be needed to overcome a 800 N load?

4. The following table gives the results of tests carried out to determine the breaking stress σ of rolled copper at various temperatures, t.

Stress σ (N/cm^2)	8.51	8.07	7.80
Temperature $t(°$C$)$	75	220	310

Stress σ (N/cm^2)	7.47	7.23	6.78
Temperature $t(°$C$)$	420	500	650

Plot a graph of stress (vertically) against temperature (horizontally). Draw the best straight line through the plotted co-ordinates. Determine the slope of the graph and the vertical axis intercept.

5. The velocity v of a body after varying time intervals t was measured as follows:

t (seconds)	2	5	8
v (m/s)	16.9	19.0	21.1

t (seconds)	11	15	18
v (m/s)	23.2	26.0	28.1

Plot v vertically and t horizontally and draw a graph of velocity against time. Determine from the graph (a) the velocity after 10 s, (b) the time at 20 m/s and (c) the equation of the graph.

6. The mass m of a steel joist varies with length L as follows:

mass, m (kg)	80	100	120	140	160
length, L (m)	3.00	3.74	4.48	5.23	5.97

Plot a graph of mass (vertically) against length (horizontally). Determine the equation of the graph.

7. The crushing strength of mortar varies with the percentage of water used in its preparation, as shown below.

Crushing strength, F (tonnes)	1.67	1.40	1.13
% of water used, w%	6	9	12

Crushing strength, F (tonnes)	0.86	0.59	0.32
% of water used, w%	15	18	21

Plot a graph of F (vertically) against w (horizontally).

(a) Interpolate and determine the crushing strength when 10% water is used.

(b) Assuming the graph continues in the same manner, extrapolate and determine the percentage of water used when the crushing strength is 0.15 tonnes.

(c) What is the equation of the graph?

8. In an experiment demonstrating Hooke's law, the strain in a copper wire was measured for various stresses. The results were

Stress (pascals)	10.6×10^6	18.2×10^6	24.0×10^6
Strain	0.00011	0.00019	0.00025

Stress (pascals)	30.7×10^6	39.4×10^6
Strain	0.00032	0.00041

Plot a graph of stress (vertically) against strain (horizontally). Determine
(a) Young's modulus of elasticity for copper, which is given by the gradient of the graph,

(b) the value of strain at a stress of 21×10^6 Pa,

(c) the value of stress when the strain is 0.00030

9. An experiment with a set of pulley blocks gave the following results.

Effort, E (newtons)	9.0	11.0	13.6
Load, L (newtons)	15	25	38

Effort, E (newtons)	17.4	20.8	23.6
Load, L (newtons)	57	74	88

Plot a graph of effort (vertically) against load (horizontally). Determine
(a) the gradient,
(b) the vertical axis intercept,
(c) the law of the graph,
(d) the effort when the load is 30 N,

(e) the load when the effort is 19 N.

10. The variation of pressure p in a vessel with temperature T is believed to follow a law of the form $p = aT + b$, where a and b are constants. Verify this law for the results given below and determine the approximate values of a and b. Hence, determine the pressures at temperatures of 285 K and 310 K and the temperature at a pressure of 250 kPa.

Pressure, p (kPa)	244	247	252
Temperature, T (K)	273	277	282

Pressure, p (kPa)	258	262	267
Temperature, T (K)	289	294	300

Practice Exercise 90   Practice Exercise 90 Multiple-choice questions on straight line graphs (answers on page 452)

Each question has only one correct answer

1. A graph of the line $x - y = 0$ passes through:
    (a) (2, 3)   (b) (3, 40)   (c) (5, 6)   (d) (0, 0)

2. A graph of the line $x + y = 5$ intersects the x-axis at the point:
    (a) (5, 0)   (b) (0, 5)   (c) (−5, 0)   (d) (0, −5)

3. The equation for the line with gradient $\frac{1}{2}$ and y-axis intercept 3 is:
    (a) $x + 2y = 6$ (b) $2x - y = 3$
    (c) $2y - x = 6$ (d) $\frac{1}{2}x - y = 3$

4. A graph of resistance against voltage for an electrical circuit is shown in Fig. 17.23. The equation relating resistance R and voltage V is:
    (a) $R = 1.45\,V + 40$ (b) $R = 0.8\,V + 20$
    (c) $R = 1.45\,V + 20$ (d) $R = 1.25\,V + 20$

5. Two points on rectangular axes are P at (−1, 6) and Q at (5, 2). The gradient of the straight line between points P and Q is:
    (a) 1   (b) −2/3   (c) −1   (d) 2/3

Figure 17.23

6. A graph of y against x, two engineering quantities, produces a straight line. A table of values is shown below:

x	2	−1	p
y	9	3	5

The value of p is:
(a) $-\frac{1}{2}$ (b) −2   (c) 3   (d) 0

7. Here are four equations in x and y. When x is plotted against y, in each case a straight line results.
    (i) $y + 3 = 3x$ (ii) $y + 3x = 3$
    (iii) $\frac{y}{2} - \frac{3}{2} = x$ (iv) $\frac{y}{3} = x + \frac{2}{3}$
    Which of these equations are parallel to each other?
    (a) (i) and (ii)   (b) (i) and (iv)
    (c) (ii) and (iii)   (d) (ii) and (iv)

8. Which of the straight lines shown in Fig. 17.24 has the equation $y + 4 = 2x$?
    (a) (iv)   (b) (ii)   (c) (iii)   (d) (i)

9. A graph relating effort E (plotted vertically) against load L (plotted horizontally) for a set of pulleys is given by $L + 30 = 6E$. The gradient of the graph is:
    (a) 6   (b) 5   (c) $\frac{1}{6}$ (d) $\frac{1}{5}$

10. The y-intercept of the equation $5x - 3y = 30$ is at:
    (a) 10   (b) 2   (c) −10   (d) 6

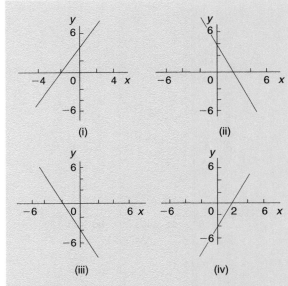

Figure 17.24

11. The graph of the linear equation $2x + 3y = 6$ intersects the y-axis at the point:
(a) (2, 0)   (b) (0, 3)   (c) (3, 0)   (d) (0, 2)

12. The mirror image about the y-axis of a point (3, 5) is:
(a) (3, 5)   (b) (−3, 5)
(c) (3, −5)   (d) (−3, −5)

13. The perpendicular distance of a point A at (5, 8) from the y-axis is:
(a) 5   (b) 8   (c) 3   (d) 13

14. The graph of the equation $3x + 5y = 7$ is a:
(a) vertical line
(b) straight line
(c) horizontal line
(d) line passing through the origin

15. In an experiment demonstrating Hooke's law, the strain in a copper wire was measured for various stresses. The results included

Stress (megapascals)	18.24	24.00	39.36
Strain	0.00019	0.00025	0.00041

When stress is plotted vertically against strain horizontally a straight-line graph results.

Young's modulus of elasticity for copper, which is given by the gradient of the graph, is:
(a) 96000 Pa   (b) 1.04×10^{-11} Pa
(c) 96 Pa   (d) 96×10^{9} Pa

Graphs reducing non-linear laws to linear form

Why it is important to understand: Graphs reducing non-linear laws to linear form

Graphs are important tools for analysing and displaying data between two experimental quantities. Many times situations occur in which the relationship between the variables is not linear. By manipulation, a straight line graph may be plotted to produce a law relating the two variables. Sometimes this involves using the laws of logarithms. The relationship between the resistance of wire and its diameter is not a linear one. Similarly, the periodic time of oscillations of a pendulum does not have a linear relationship with its length, and the head of pressure and the flow velocity are not linearly related. There are thus plenty of examples in engineering where determination of law is needed.

At the end of this chapter you should be able to:

- understand what is meant by determination of law
- prepare co-ordinates for a non-linear relationship between two variables
- plot prepared co-ordinates and draw a straight line graph
- determine the gradient and vertical axis intercept of a straight line graph
- state the equation of a straight line graph
- plot straight line graphs involving practical engineering examples
- determine straight line laws involving logarithms: $y = ax^n$, $y = ab^x$ and $y = ae^{bx}$
- plot straight line graphs involving logarithms

18.1 Introduction

In Chapter 17 we discovered that the equation of a straight line graph is of the form $y = mx + c$, where m is the gradient and c is the y-axis intercept. This chapter explains how the law of a graph can still be determined even when it is not of the linear form $y = mx + c$. The method used is called **determination of law** and is explained in the following sections.

18.2 Determination of law

Frequently, the relationship between two variables, say x and y, is not a linear one; i.e. when x is plotted against y a curve results. In such cases the non-linear equation may be modified to the linear form, $y = mx + c$, so that the constants, and thus the law relating the variables, can be determined. This technique is called '**determination of law**'.

Some examples of the reduction of equations to linear form include

(i) $y = ax^2 + b$ compares with $Y = mX + c$, where $m = a$, $c = b$ and $X = x^2$
Hence, y is plotted vertically against x^2 horizontally to produce a straight line graph of gradient a and y-axis intercept b.

(ii) $y = \dfrac{a}{x} + b$, i.e. $y = a\left(\dfrac{1}{x}\right) + b$

y is plotted vertically against $\dfrac{1}{x}$ horizontally to produce a straight line graph of gradient a and y-axis intercept b.

(iii) $y = ax^2 + bx$
Dividing both sides by x gives $\dfrac{y}{x} = ax + b$
Comparing with $Y = mX + c$ shows that $\dfrac{y}{x}$ is plotted vertically against x horizontally to produce a straight line graph of gradient a and $\dfrac{y}{x}$ axis intercept b.

Here are some worked problems to demonstrate determination of law.

Problem 1. Experimental values of x and y, shown below, are believed to be related by the law $y = ax^2 + b$. By plotting a suitable graph, verify this law and determine approximate values of a and b

x	1	2	3	4	5
y	9.8	15.2	24.2	36.5	53.0

If y is plotted against x a curve results and it is not possible to determine the values of constants a and b from the curve.
Comparing $y = ax^2 + b$ with $Y = mX + c$ shows that y is to be plotted vertically against x^2 horizontally. A table of values is drawn up as shown below.

x	1	2	3	4	5
x^2	1	4	9	16	25
y	9.8	15.2	24.2	36.5	53.0

A graph of y against x^2 is shown in Fig. 18.1, with the best straight line drawn through the points. **Since a straight line graph results, the law is verified**.

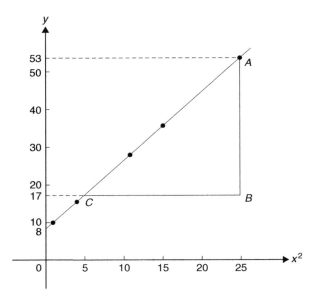

Figure 18.1

From the graph, **gradient, $a = \dfrac{AB}{BC} = \dfrac{53 - 17}{25 - 5}$**

$$= \dfrac{36}{20} = \mathbf{1.8}$$

and the **y-axis intercept, $b = 8.0$**

Hence, the law of the graph is $y = \mathbf{1.8}x^2 + \mathbf{8.0}$

Problem 2. Values of load L newtons and distance d metres obtained experimentally are shown in the following table.

Load, L(N)	32.3	29.6	27.0	23.2
Distance, d(m)	0.75	0.37	0.24	0.17

Load, L(N)	18.3	12.8	10.0	6.4
Distance, d(m)	0.12	0.09	0.08	0.07

(a) Verify that load and distance are related by a law of the form $L = \dfrac{a}{d} + b$ and determine approximate values of a and b

(b) Hence, calculate the load when the distance is 0.20 m and the distance when the load is 20 N

(a) Comparing $L = \dfrac{a}{d} + b$ i.e. $L = a\left(\dfrac{1}{d}\right) + b$ with $Y = mX + c$ shows that L is to be plotted vertically against $\dfrac{1}{d}$ horizontally. Another table of values is drawn up as shown below.

L	32.3	29.6	27.0	23.2	18.3	12.8	10.0	6.4
d	0.75	0.37	0.24	0.17	0.12	0.09	0.08	0.07
$\dfrac{1}{d}$	1.33	2.70	4.17	5.88	8.33	11.11	12.50	14.29

A graph of L against $\dfrac{1}{d}$ is shown in Fig. 18.2. **A straight line can be drawn through the points, which verifies that load and distance are related by a law of the form $L = \dfrac{a}{d} + b$**

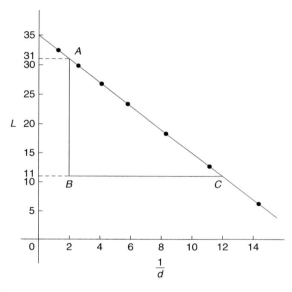

Figure 18.2

Gradient of straight line, $a = \dfrac{AB}{BC} = \dfrac{31 - 11}{2 - 12}$

$$= \dfrac{20}{-10} = -2$$

L-axis intercept, $b = 35$

Hence, the law of the graph is $L = -\dfrac{2}{d} + 35$

(b)   When the distance $d = 0.20\,$m,

load, $L = \dfrac{-2}{0.20} + 35 = \mathbf{25.0\,N}$

Rearranging $L = -\dfrac{2}{d} + 35$ gives

$\dfrac{2}{d} = 35 - L$ and $d = \dfrac{2}{35 - L}$

Hence, when the load $L = 20\,$N, **distance,**

$d = \dfrac{2}{35 - 20} = \dfrac{2}{15} = \mathbf{0.13\,m}$

Problem 3. The solubility s of potassium chlorate is shown by the following table.

$t\,°$C	10	20	30	40	50	60	80	100
s	4.9	7.6	11.1	15.4	20.4	26.4	40.6	58.0

The relationship between s and t is thought to be of the form $s = 3 + at + bt^2$. Plot a graph to test the supposition and use the graph to find approximate values of a and b. Hence, calculate the solubility of potassium chlorate at $70°$C

Rearranging $s = 3 + at + bt^2$ gives

$$s - 3 = at + bt^2$$

and $\quad \dfrac{s - 3}{t} = a + bt$

or $\quad \dfrac{s - 3}{t} = bt + a$

which is of the form $\quad Y = mX + c$

This shows that $\dfrac{s - 3}{t}$ is to be plotted vertically and t horizontally, with gradient b and vertical axis intercept a.

Another table of values is drawn up as shown below.

t	10	20	30	40	50	60	80	100
s	4.9	7.6	11.1	15.4	20.4	26.4	40.6	58.0
$\dfrac{s-3}{t}$	0.19	0.23	0.27	0.31	0.35	0.39	0.47	0.55

A graph of $\dfrac{s - 3}{t}$ against t is shown plotted in Fig. 18.3. **A straight line fits the points, which shows that s and t are related by $s = 3 + at + bt^2$**

Gradient of straight line, $b = \dfrac{AB}{BC} = \dfrac{0.39 - 0.19}{60 - 10}$

$$= \dfrac{0.20}{50} = \mathbf{0.004}$$

Vertical axis intercept, $a = \mathbf{0.15}$

Hence, the law of the graph is $s = \mathbf{3 + 0.15t + 0.004t^2}$

The solubility of potassium chlorate at $70°$C is given by

$$s = 3 + 0.15(70) + 0.004(70)^2$$

$$= 3 + 10.5 + 19.6 = \mathbf{33.1}$$

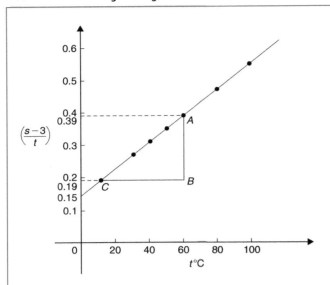

Figure 18.3

Now try the following Practice Exercise

Practice Exercise 91   Graphs reducing non-linear laws to linear form (answers on page 452)

In Problems 1 to 5, x and y are two related variables and all other letters denote constants. For the stated laws to be verified it is necessary to plot graphs of the variables in a modified form. State for each, (a) what should be plotted on the vertical axis, (b) what should be plotted on the horizontal axis, (c) the gradient and (d) the vertical axis intercept.

1. $y = d + cx^2$ 2. $y - a = b\sqrt{x}$

3. $y - e = \dfrac{f}{x}$ 4. $y - cx = bx^2$

5. $y = \dfrac{a}{x} + bx$

6.   In an experiment the resistance of wire is measured for wires of different diameters with the following results.

R (ohms)	1.64	1.14	0.89	0.76	0.63
d (mm)	1.10	1.42	1.75	2.04	2.56

It is thought that R is related to d by the law $R = \dfrac{a}{d^2} + b$, where a and b are constants. Verify this and find the approximate values for

a and b. Determine the cross-sectional area needed for a resistance reading of 0.50 ohms.

7.   Corresponding experimental values of two quantities x and y are given below.

x	1.5	3.0	4.5	6.0	7.5	9.0
y	11.5	25.0	47.5	79.0	119.5	169.0

By plotting a suitable graph, verify that y and x are connected by a law of the form $y = kx^2 + c$, where k and c are constants. Determine the law of the graph and hence find the value of x when y is 60.0

8.   Experimental results of the safe load L kN, applied to girders of varying spans, d m, are shown below.

Span, d (m)	2.0	2.8	3.6	4.2	4.8
Load, L (kN)	475	339	264	226	198

It is believed that the relationship between load and span is $L = c/d$, where c is a constant. Determine (a) the value of constant c and (b) the safe load for a span of 3.0 m.

9.   The following results give corresponding values of two quantities x and y which are believed to be related by a law of the form $y = ax^2 + bx$, where a and b are constants.

y	33.86	55.54	72.80	84.10	111.4	168.1
x	3.4	5.2	6.5	7.3	9.1	12.4

Verify the law and determine approximate values of a and b. Hence, determine (a) the value of y when x is 8.0 and (b) the value of x when y is 146.5

18.3   Revision of laws of logarithms

The laws of logarithms were stated in Chapter 15 as follows:

$$\log(A \times B) = \log A + \log B \qquad (1)$$

$$\log\left(\frac{A}{B}\right) = \log A - \log B \qquad (2)$$

$$\log A^n = n \times \log A \qquad (3)$$

Also, $\ln e = 1$ and if, say, $\lg x = 1.5$,

then $x = 10^{1.5} = 31.62$

Further, if $3^x = 7$ then $\lg 3^x = \lg 7$ and $x \lg 3 = \lg 7$, from which $x = \dfrac{\lg 7}{\lg 3} = 1.771$

These laws and techniques are used whenever non-linear laws of the form $y = ax^n, y = ab^x$ and $y = ae^{bx}$ are reduced to linear form with the values of a and b needing to be calculated. This is demonstrated in the following section.

18.4 Determination of laws involving logarithms

Examples of the reduction of equations to linear form involving logarithms include

(a) $y = ax^n$

Taking logarithms to **a base of 10** of both sides gives

$$\lg y = \lg(ax^n)$$

$$= \lg a + \lg x^n \quad \text{by law (1)}$$

i.e. $\lg y = n \lg x + \lg a$ by law (3)

which compares with $Y = mX + c$

and shows that **lg$y$ is plotted vertically against lg$x$ horizontally** to produce a straight line **graph of gradient n and lg$y$-axis intercept lg a.**
See worked Problems 4 and 5 to demonstrate how this law is determined.

(b) $y = ab^x$

Taking logarithms to **a base of 10** of both sides gives

$$\lg y = \lg(ab^x)$$

i.e. $\lg y = \lg a + \lg b^x$ by law (1)

$\lg y = \lg a + x \lg b$ by law (3)

i.e. $\lg y = x \lg b + \lg a$

or $\lg y = (\lg b)x + \lg a$

which compares with

$$Y = mX + c$$

and shows that **lg$y$ is plotted vertically against x horizontally** to produce **a straight line graph of gradient lg$b$ and lg$y$-axis intercept lg a.**
See worked Problem 6 to demonstrate how this law is determined.

(c) $y = ae^{bx}$

Taking logarithms to **a base of e** of both sides gives

$$\ln y = \ln(ae^{bx})$$

i.e. $\ln y = \ln a + \ln e^{bx}$ by law (1)

i.e. $\ln y = \ln a + bx \ln e$ by law (3)

i.e. $\ln y = bx + \ln a$ since $\ln e = 1$

which compares with

$$Y = mX + c$$

and shows that **ln$y$ is plotted vertically against x horizontally** to produce **a straight line graph of gradient b and ln y-axis intercept ln a.**

See worked Problem 7 to demonstrate how this law is determined.

> **Problem 4.** The current flowing in, and the power dissipated by, a resistor are measured experimentally for various values and the results are as shown below

Current, I (amperes)	2.2	3.6	4.1	5.6	6.8
Power, P (watts)	116	311	403	753	1110

Show that the law relating current and power is of the form $P = RI^n$, where R and n are constants, and determine the law

Taking logarithms to a base of 10 of both sides of $P = RI^n$ gives

$$\lg P = \lg(RI^n) = \lg R + \lg I^n = \lg R + n \lg I$$

by the laws of logarithms

i.e. $\lg P = n \lg I + \lg R$

which is of the form $Y = mX + c$, showing that $\lg P$ is to be plotted vertically against $\lg I$ horizontally.
A table of values for $\lg I$ and $\lg P$ is drawn up as shown below.

I	2.2	3.6	4.1	5.6	6.8
$\lg I$	0.342	0.556	0.613	0.748	0.833
P	116	311	403	753	1110
$\lg P$	2.064	2.493	2.605	2.877	3.045

A graph of $\lg P$ against $\lg I$ is shown in Fig. 18.4 and, **since a straight line results, the law $P = RI^n$ is verified**.

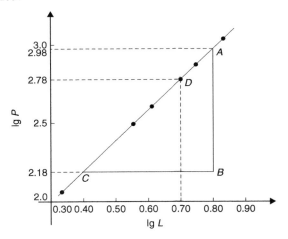

Figure 18.4

$$\text{Gradient of straight line, } n = \frac{AB}{BC} = \frac{2.98 - 2.18}{0.8 - 0.4}$$

$$= \frac{0.80}{0.4} = 2$$

It is not possible to determine the vertical axis intercept on sight since the horizontal axis scale does not start at zero. Selecting any point from the graph, say point D, where $\lg I = 0.70$ and $\lg P = 2.78$ and substituting values into

$$\lg P = n\lg I + \lg R$$

gives $\qquad 2.78 = (2)(0.70) + \lg R$

from which, $\quad \lg R = 2.78 - 1.40 = 1.38$

Hence, $\qquad\qquad R = \text{antilog } 1.38 = 10^{1.38} = \mathbf{24.0}$

Hence, **the law of the graph is $P = 24.0I^2$**

> **Problem 5.** The periodic time, T, of oscillation of a pendulum is believed to be related to its length, L, by a law of the form $T = kL^n$, where k and n are constants. Values of T were measured for various lengths of the pendulum and the results
>
> | Periodic time, T(s) | 1.0 | 1.3 | 1.5 | 1.8 | 2.0 | 2.3 |
> |---|---|---|---|---|---|---|
> | Length, L(m) | 0.25 | 0.42 | 0.56 | 0.81 | 1.0 | 1.32 |
>
> Show that the law is true and determine the approximate values of k and n. Hence find the periodic time when the length of the pendulum is 0.75 m

From para (a), page 189, if $\quad T = kL^n$

then $\qquad\qquad \lg T = n\lg L + \lg k$

and comparing with $\qquad Y = mX + c$

shows that $\lg T$ is plotted vertically against $\lg L$ horizontally, with gradient n and vertical axis intercept $\lg k$.

A table of values for $\lg T$ and $\lg L$ is drawn up as shown below.

T	1.0	1.3	1.5	1.8	2.0	2.3
$\lg T$	0	0.114	0.176	0.255	0.301	0.362
L	0.25	0.42	0.56	0.81	1.0	1.32
$\lg L$	−0.602	−0.377	−0.252	−0.092	0	0.121

A graph of $\lg T$ against $\lg L$ is shown in Fig. 18.5 and the **law $T = kL^n$ is true since a straight line results**.

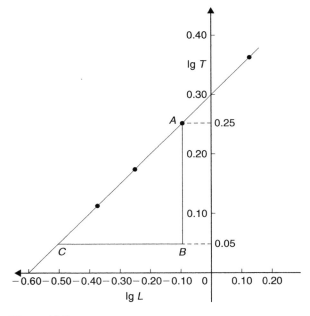

Figure 18.5

From the graph, gradient of straight line,

$$n = \frac{AB}{BC} = \frac{0.25 - 0.05}{-0.10 - (-0.50)} = \frac{0.20}{0.40} = \frac{1}{2}$$

Vertical axis intercept, $\lg k = 0.30$. Hence, $k = \text{antilog } 0.30 = 10^{0.30} = \mathbf{2.0}$

Hence, **the law of the graph is $T = 2.0L^{1/2}$ or $T = 2.0\sqrt{L}$.**

When length $L = 0.75$ m, $\mathbf{T = 2.0\sqrt{0.75} = 1.73\,s}$

Problem 6. Quantities x and y are believed to be related by a law of the form $y = ab^x$, where a and b are constants. The values of x and corresponding values of y are

x	0	0.6	1.2	1.8	2.4	3.0
y	5.0	9.67	18.7	36.1	69.8	135.0

Verify the law and determine the approximate values of a and b. Hence determine (a) the value of y when x is 2.1 and (b) the value of x when y is 100

From para (b), page 189, if $y = ab^x$

then $\qquad\qquad\qquad \lg y = (\lg b)x + \lg a$

and comparing with $\qquad Y = mX + c$

shows that $\lg y$ is plotted vertically and x horizontally, with gradient $\lg b$ and vertical axis intercept $\lg a$. Another table is drawn up as shown below.

x	0	0.6	1.2	1.8	2.4	3.0
y	5.0	9.67	18.7	36.1	69.8	135.0
$\lg y$	0.70	0.99	1.27	1.56	1.84	2.13

A graph of $\lg y$ against x is shown in Fig. 18.6 and, **since a straight line results, the law $y = ab^x$ is verified**.

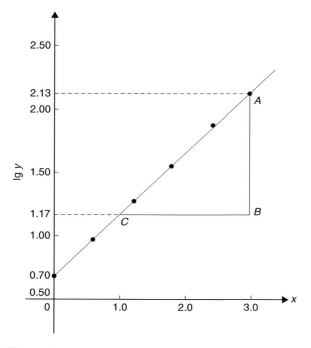

Figure 18.6

Gradient of straight line,

$$\mathbf{\lg b} = \frac{AB}{BC} = \frac{2.13 - 1.17}{3.0 - 1.0} = \frac{0.96}{2.0} = 0.48$$

Hence, $b = $ antilog $0.48 = 10^{0.48} = \mathbf{3.0}$, correct to 2 significant figures.

Vertical axis intercept, $\quad \lg a = 0.70$, from which

$$a = \text{antilog } 0.70$$
$$= 10^{0.70} = \mathbf{5.0}, \text{ correct to}$$
$$\text{2 significant figures.}$$

Hence, **the law of the graph is $y = 5.0(3.0)^x$**

(a) When $x = 2.1, y = 5.0(3.0)^{2.1} = \mathbf{50.2}$

(b) When $y = 100, 100 = 5.0(3.0)^x$, from which
$$100/5.0 = (3.0)^x$$
i.e. $\qquad 20 = (3.0)^x$

Taking logarithms of both sides gives
$$\lg 20 = \lg(3.0)^x = x\lg 3.0$$
Hence, $\quad x = \dfrac{\lg 20}{\lg 3.0} = \dfrac{1.3010}{0.4771} = \mathbf{2.73}$

Problem 7. The current i mA flowing in a capacitor which is being discharged varies with time t ms, as shown below.

i(mA)	203	61.14	22.49	6.13	2.49	0.615
t(ms)	100	160	210	275	320	390

Show that these results are related by a law of the form $i = Ie^{t/T}$, where I and T are constants. Determine the approximate values of I and T

Taking Napierian logarithms of both sides of
$$i = Ie^{t/T}$$
gives $\qquad \ln i = \ln(Ie^{t/T}) = \ln I + \ln e^{t/T}$
$$= \ln I + \frac{t}{T}\ln e$$
i.e. $\qquad \ln i = \ln I + \dfrac{t}{T}$ since $\ln e = 1$

or $\qquad \ln i = \left(\dfrac{1}{T}\right)t + \ln I$

which compares with $\quad y = mx + c$

showing that $\ln i$ is plotted vertically against t horizontally, with gradient $\dfrac{1}{T}$ and vertical axis intercept $\ln I$.

Another table of values is drawn up as shown below.

t	100	160	210	275	320	390
i	203	61.14	22.49	6.13	2.49	0.615
$\ln i$	5.31	4.11	3.11	1.81	0.91	−0.49

A graph of $\ln i$ against t is shown in Fig. 18.7 and, since a straight line results, **the law $i = Ie^{t/T}$ is verified**.

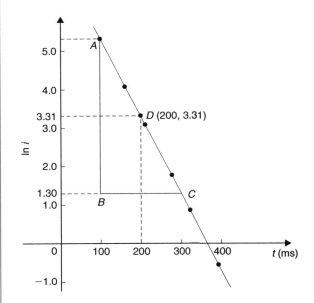

Figure 18.7

Gradient of straight line, $\dfrac{1}{T} = \dfrac{AB}{BC} = \dfrac{5.30 - 1.30}{100 - 300}$

$$= \dfrac{4.0}{-200} = -0.02$$

Hence, $$T = \dfrac{1}{-0.02} = -50$$

Selecting any point on the graph, say point D, where $t = 200$ and $\ln i = 3.31$, and substituting

into $$\ln i = \left(\dfrac{1}{T}\right) t + \ln I$$

gives $$3.31 = -\dfrac{1}{50}(200) + \ln I$$

from which, $\ln I = 3.31 + 4.0 = 7.31$

and $I = $ antilog $7.31 = e^{7.31} = 1495$ or **1500** correct to 3 significant figures.
Hence, **the law of the graph is $i = 1500e^{-t/50}$**

Now try the following Practice Exercise

Practice Exercise 92   Determination of laws involving logarithms (answers on page 452)

In Problems 1 to 3, x and y are two related variables and all other letters denote constants. For the stated laws to be verified it is necessary to plot graphs of the variables in a modified form. State for each, (a) what should be plotted on the vertical axis, (b) what should be plotted on the horizontal axis, (c) the gradient and (d) the vertical axis intercept.

1. $y = ba^x$ 2. $y = kx^L$ 3. $\dfrac{y}{m} = e^{nx}$

4. The luminosity I of a lamp varies with the applied voltage V, and the relationship between I and V is thought to be $I = kV^n$. Experimental results obtained are

I(candelas)	1.92	4.32	9.72
V(volts)	40	60	90

I(candelas)	15.87	23.52	30.72
V(volts)	115	140	160

Verify that the law is true and determine the law of the graph. Also determine the luminosity when 75 V is applied across the lamp.

5. The head of pressure h and the flow velocity v are measured and are believed to be connected by the law $v = ah^b$, where a and b are constants. The results are as shown below.

h	10.6	13.4	17.2	24.6	29.3
v	9.77	11.0	12.44	14.88	16.24

Verify that the law is true and determine values of a and b.

6. Experimental values of x and y are measured as follows.

x	0.4	0.9	1.2	2.3	3.8
y	8.35	13.47	17.94	51.32	215.20

The law relating x and y is believed to be of the form $y = ab^x$, where a and b are constants. Determine the approximate values of a and b.

Hence, find the value of y when x is 2.0 and the value of x when y is 100.

7. The activity of a mixture of radioactive isotopes is believed to vary according to the law $R = R_0 t^{-c}$, where R_0 and c are constants. Experimental results are shown below.

R	9.72	2.65	1.15	0.47	0.32	0.23
t	2	5	9	17	22	28

Verify that the law is true and determine approximate values of R_0 and c.

8. Determine the law of the form $y = ae^{kx}$ which relates the following values.

y	0.0306	0.285	0.841	5.21	173.2	1181
x	−4.0	5.3	9.8	17.4	32.0	40.0

9. The tension T in a belt passing round a pulley wheel and in contact with the pulley over an angle of θ radians is given by $T = T_0 e^{\mu\theta}$, where T_0 and μ are constants. Experimental results obtained are

T (newtons)	47.9	52.8	60.3	70.1	80.9
θ (radians)	1.12	1.48	1.97	2.53	3.06

Determine approximate values of T_0 and μ. Hence, find the tension when θ is 2.25 radians and the value of θ when the tension is 50.0 newtons.

Practice Exercise 93   Multiple-choice questions on graphs reducing non-linear laws to linear form (answers on page 452)

Each question has only one correct answer

Questions 1 to 4 relate to the following information:
x and y are two related engineering variables and p and q are constants. For the law $y - p = \dfrac{q}{x}$ to be verified it is necessary to plot a graph of the variables.

1. On the vertical axis is plotted:
(a) y (b) p (c) q (d) x

2. On the horizontal axis is plotted:
(a) x (b) $\dfrac{q}{x}$ (c) $\dfrac{1}{x}$ (d) p

3. The gradient of the graph is:
(a) y (b) p (c) x (d) q

4. The vertical axis intercept is:
(a) y (b) p (c) q (d) x

5. The relationship between two related engineering variables x and y is $y - cx = bx^2$ where b and c are constants. To produce a straight-line graph, it is necessary to plot:
(a) x vertically against y horizontally
(b) y vertically against x^2 horizontally
(c) $\dfrac{y}{x}$ vertically against x horizontally
(d) y vertically against x horizontally

6. The relationship between two related scientific variables x and y is $y = px^q$, where p and q are constants. To produce a straight-line graph, it is necessary to plot:
(a) x vertically against y horizontally
(b) lg y vertically against lg x horizontally
(c) lg y vertically against ln x^q horizontally
(d) y vertically against x horizontally

7. The relationship between two related engineering variables x and y is $y = cd^x$ where c and d are constants. To produce a straight-line graph, it is necessary to plot:
(a) x vertically against y horizontally
(b) ln y vertically against ln x horizontally
(c) y vertically against x horizontally
(d) lg y vertically against x horizontally

8. The relationship between two related scientific variables x and y is $\dfrac{y}{g} = e^{hx}$ where g and h are constants. To produce a straight-line graph, it is necessary to plot:
(a) ln y vertically against x horizontally
(b) ln y vertically against ln x horizontally
(c) x vertically against y horizontally
(d) y vertically against x horizontally

For fully worked solutions to each of the problems in Practice Exercises 91 and 92 in this chapter, go to the website:
www.routledge.com/cw/bird

Chapter 19

Graphical solution of equations

Why it is important to understand: Graphical solution of equations

It has been established in previous chapters that the solution of linear, quadratic, simultaneous and cubic equations occur often in engineering and science and may be solved using algebraic means. Being able to solve equations graphically provides another method to aid understanding and interpretation of equations. Engineers, including architects, surveyors and a variety of engineers in fields such as biomedical, chemical, electrical, mechanical and nuclear, all use equations which need solving by one means or another.

At the end of this chapter you should be able to:

- solve two simultaneous equations graphically
- solve a quadratic equation graphically
- solve a linear and quadratic equation simultaneously by graphical means
- solve a cubic equation graphically

19.1 Graphical solution of simultaneous equations

Linear simultaneous equations in two unknowns may be solved graphically by

(a) plotting the two straight lines on the same axes, and

(b) noting their point of intersection.

The co-ordinates of the point of intersection give the required solution.

Here are some worked problems to demonstrate the graphical solution of simultaneous equations.

Problem 1. Solve graphically the simultaneous equations

$$2x - y = 4$$
$$x + y = 5$$

Rearranging each equation into $y = mx + c$ form gives

$$y = 2x - 4$$
$$y = -x + 5$$

Only three co-ordinates need be calculated for each graph since both are straight lines.

x	0	1	2
$y = 2x - 4$	-4	-2	0

x	0	1	2
$y = -x + 5$	5	4	3

Each of the graphs is plotted as shown in Fig. 19.1. The point of intersection is at (3, 2) and since this is the only point which lies simultaneously on both lines then $x = 3, y = 2$ is the solution of the simultaneous equations.

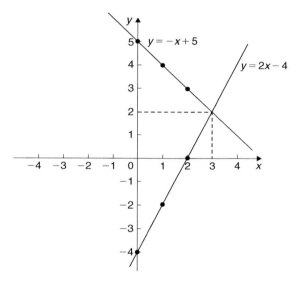

Figure 19.1

Problem 2. Solve graphically the equations

$$1.20x + y = 1.80$$
$$x - 5.0y = 8.50$$

Rearranging each equation into $y = mx + c$ form gives

$$y = -1.20x + 1.80 \qquad (1)$$

$$y = \frac{x}{5.0} - \frac{8.5}{5.0}$$

i.e. $\qquad y = 0.20x - 1.70 \qquad (2)$

Three co-ordinates are calculated for each equation as shown below.

x	0	1	2
$y = -1.20x + 1.80$	1.80	0.60	-0.60

x	0	1	2
$y = 0.20x - 1.70$	-1.70	-1.50	-1.30

The two sets of co-ordinates are plotted as shown in Fig. 19.2. The point of intersection is (2.50, −1.20). Hence, the solution of the simultaneous equations is $x = 2.50, y = -1.20$

(It is sometimes useful to initially sketch the two straight lines to determine the region where the point of intersection is. Then, for greater accuracy, a graph having a smaller range of values can be drawn to 'magnify' the point of intersection.)

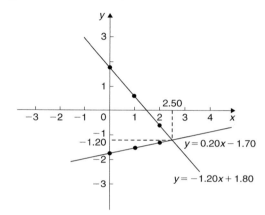

Figure 19.2

Now try the following Practice Exercise

Practice Exercise 94   Graphical solution of simultaneous equations (Answers on page 452)

In Problems 1 to 6, solve the simultaneous equations graphically.

1. $y = 3x - 2$
    $y = -x + 6$

2. $x + y = 2$
    $3y - 2x = 1$

3. $y = 5 - x$
    $x - y = 2$

4. $3x + 4y = 5$
    $2x - 5y + 12 = 0$

5. $1.4x - 7.06 = 3.2y$
    $2.1x - 6.7y = 12.87$

6. $3x - 2y = 0$
    $4x + y + 11 = 0$

7. The friction force F newtons and load L newtons are connected by a law of the form $F = aL + b$, where a and b are constants. When $F = 4\,\text{N}, L = 6\,\text{N}$ and when $F = 2.4\,\text{N}$, $L = 2\,\text{N}$. Determine graphically the values of a and b.

19.2 Graphical solution of quadratic equations

A general **quadratic equation** is of the form $y = ax^2 + bx + c$, where a, b and c are constants and a is not equal to zero.

A graph of a quadratic equation always produces a shape called a **parabola**.

The gradients of the curves between 0 and A and between B and C in Fig. 19.3 are positive, whilst the gradient between A and B is negative. Points such as A and B are called **turning points**. At A the gradient is zero and, as x increases, the gradient of the curve changes from positive just before A to negative just after. Such a point is called a **maximum value**. At B the gradient is also zero and, as x increases, the gradient of the curve changes from negative just before B to positive just after. Such a point is called a **minimum value**.

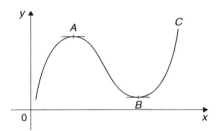

Figure 19.3

Following are three examples of **quadratic graphs.**

(a) $y = ax^2$

Graphs of $y = x^2$, $y = 3x^2$ and $y = \frac{1}{2}x^2$ are shown in Fig. 19.4. All have minimum values at the origin $(0, 0)$

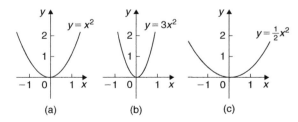

Figure 19.4

Graphs of $y = -x^2$, $y = -3x^2$ and $y = -\frac{1}{2}x^2$ are shown in Fig. 19.5. All have maximum values at the origin $(0, 0)$

When $y = ax^2$,

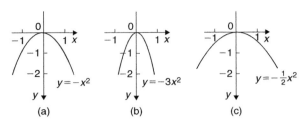

Figure 19.5

(i)   curves are symmetrical about the y-axis,
(ii) the magnitude of a affects the gradient of the curve, and
(iii) the sign of a determines whether it has a maximum or minimum value.

(b) $y = ax^2 + c$

Graphs of $y = x^2 + 3$, $y = x^2 - 2$, $y = -x^2 + 2$ and $y = -2x^2 - 1$ are shown in Fig. 19.6.

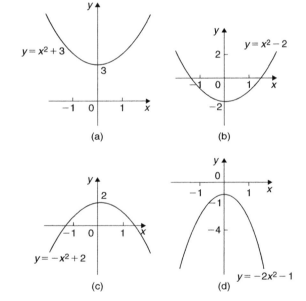

Figure 19.6

When $y = ax^2 + c$,

(i)   curves are symmetrical about the y-axis,
(ii) the magnitude of a affects the gradient of the curve, and
(iii) the constant c is the y-axis intercept.

(c) $y = ax^2 + bx + c$

Whenever b has a value other than zero the curve is displaced to the right or left of the y-axis.

When b/a is positive, the curve is displaced $b/2a$ to the left of the y-axis, as shown in Fig. 19.7(a).

When b/a is negative, the curve is displaced $b/2a$ to the right of the y-axis, as shown in Fig. 19.7(b).

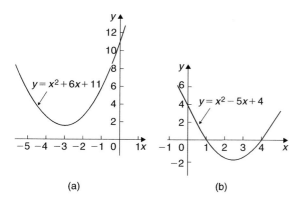

(a)   (b)

Figure 19.7

Quadratic equations of the form $ax^2 + bx + c = 0$ may be solved graphically by

(a) plotting the graph $y = ax^2 + bx + c$, and

(b) noting the points of intersection on the x-axis (i.e. where $y = 0$)

The x values of the points of intersection give the required solutions since at these points both $y = 0$ and $ax^2 + bx + c = 0$

The number of solutions, or roots, of a quadratic equation depends on how many times the curve cuts the x-axis. There can be no real roots, as in Fig. 19.7(a), one root, as in Figs 19.4 and 19.5, or two roots, as in Fig. 19.7(b).

Here are some worked problems to demonstrate the graphical solution of quadratic equations.

Problem 3. Solve the quadratic equation $4x^2 + 4x - 15 = 0$ graphically, given that the solutions lie in the range $x = -3$ to $x = 2$. Determine also the co-ordinates and nature of the turning point of the curve

---

Let $y = 4x^2 + 4x - 15$. A table of values is drawn up as shown below.

x			-3	-2	-1	0	1	2
$y = 4x^2 + 4x - 15$			9	-7	-15	-15	-7	9

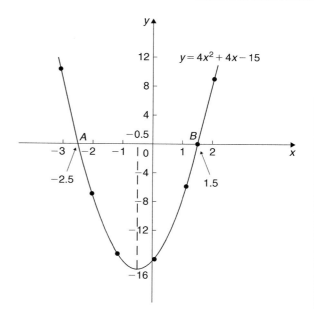

Figure 19.8

A graph of $y = 4x^2 + 4x - 15$ is shown in Fig. 19.8. The only points where $y = 4x^2 + 4x - 15$ and $y = 0$ are the points marked A and B. This occurs at $x = -2.5$ and $x = 1.5$ and these are the solutions of the quadratic equation $4x^2 + 4x - 15 = 0$

By substituting $x = -2.5$ and $x = 1.5$ into the original equation the solutions may be checked.

The curve has a turning point at $(-0.5, -16)$ and the nature of the point is a minimum.

An alternative graphical method of solving $4x^2 + 4x - 15 = 0$ is to rearrange the equation as $4x^2 = -4x + 15$ and then plot two separate graphs – in this case, $y = 4x^2$ and $y = -4x + 15$. Their points of intersection give the roots of the equation $4x^2 = -4x + 15$, i.e. $4x^2 + 4x - 15 = 0$. This is shown in Fig. 19.9, where the roots are $x = -2.5$ and $x = 1.5$, as before.

Problem 4. Solve graphically the quadratic equation $-5x^2 + 9x + 7.2 = 0$ given that the solutions lie between $x = -1$ and $x = 3$. Determine also the co-ordinates of the turning point and state its nature

---

Let $y = -5x^2 + 9x + 7.2$. A table of values is drawn up as shown below.

x	-1	-0.5	0	1
$y = -5x^2 + 9x + 7.2$	-6.8	1.45	7.2	11.2

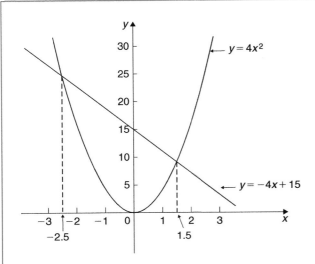

Figure 19.9

x		2	2.5	3
$y = -5x^2 + 9x + 7.2$		5.2	-1.55	-10.8

A graph of $y = -5x^2 + 9x + 7.2$ is shown plotted in Fig. 19.10. The graph crosses the x-axis (i.e. where $y = 0$) at $x = -0.6$ and $x = 2.4$ and these are the solutions of the quadratic equation $-5x^2 + 9x + 7.2 = 0$

The turning point is a **maximum**, having co-ordinates **(0.9, 11.25)**

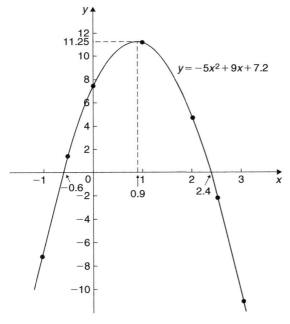

Figure 19.10

A graph of $y = 2x^2$ is shown in Fig. 19.11.

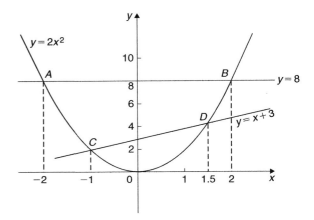

Figure 19.11

(a) Rearranging $2x^2 - 8 = 0$ gives $2x^2 = 8$ and the solution of this equation is obtained from the points of intersection of $y = 2x^2$ and $y = 8$; i.e. at co-ordinates $(-2, 8)$ and $(2, 8)$, shown as A and B, respectively, in Fig. 19.11. Hence, the solutions of $2x^2 - 8 = 0$ are $x = -2$ and $x = +2$

(b) Rearranging $2x^2 - x - 3 = 0$ gives $2x^2 = x + 3$ and the solution of this equation is obtained from the points of intersection of $y = 2x^2$ and $y = x + 3$; i.e. at C and D in Fig. 19.11. Hence, the solutions of $2x^2 - x - 3 = 0$ are $x = -1$ and $x = 1.5$

Problem 6. Plot the graph of $y = -2x^2 + 3x + 6$ for values of x from $x = -2$ to $x = 4$. Use the graph to find the roots of the following equations.
(a) $-2x^2 + 3x + 6 = 0$ (b) $-2x^2 + 3x + 2 = 0$
(c) $-2x^2 + 3x + 9 = 0$ (d) $-2x^2 + x + 5 = 0$

A table of values for $y = -2x^2 + 3x + 6$ is drawn up as shown below.

x	-2	-1	0	1	2	3	4
y	-8	1	6	7	4	-3	-14

A graph of $y = -2x^2 + 3x + 6$ is shown in Fig. 19.12.

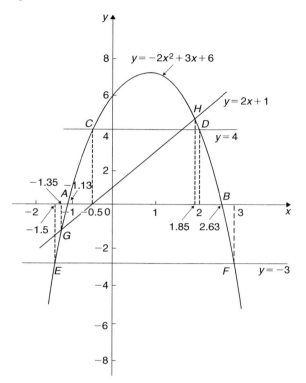

Figure 19.12

(a) The parabola $y = -2x^2 + 3x + 6$ and the straight line $y = 0$ intersect at A and B, where $x = -1.13$ and $x = 2.63$, and these are the roots of the equation $-2x^2 + 3x + 6 = 0$

(b) Comparing $y = -2x^2 + 3x + 6$ (1)

with $0 = -2x^2 + 3x + 2$ (2)

shows that, if 4 is added to both sides of equation (2), the RHS of both equations will be the same. Hence, $4 = -2x^2 + 3x + 6$. The solution of this equation is found from the points of intersection of the line $y = 4$ and the parabola $y = -2x^2 + 3x + 6$; i.e. points C and D in Fig. 19.12. Hence, the roots of $-2x^2 + 3x + 2 = 0$ are $x = -0.5$ and $x = 2$

(c) $-2x^2 + 3x + 9 = 0$ may be rearranged as $-2x^2 + 3x + 6 = -3$ and the solution of this equation is obtained from the points of intersection of the line $y = -3$ and the parabola $y = -2x^2 + 3x + 6$; i.e. at points E and F in Fig. 19.12. Hence, the roots of $-2x^2 + 3x + 9 = 0$ are $x = -1.5$ and $x = 3$

(d) Comparing $y = -2x^2 + 3x + 6$ (3)

with $0 = -2x^2 + x + 5$ (4)

shows that, if $2x + 1$ is added to both sides of equation (4), the RHS of both equations will be the same. Hence, equation (4) may be written as $2x + 1 = -2x^2 + 3x + 6$. The solution of this equation is found from the points of intersection of the line $y = 2x + 1$ and the parabola $y = -2x^2 + 3x + 6$; i.e. points G and H in Fig. 19.12. Hence, the roots of $-2x^2 + x + 5 = 0$ are $x = -1.35$ and $x = 1.85$

Now try the following Practice Exercise

Practice Exercise 95   Solving quadratic equations graphically (answers on page 452)

1. Sketch the following graphs and state the nature and co-ordinates of their respective turning points.
   (a) $y = 4x^2$ (b) $y = 2x^2 - 1$
   (c) $y = -x^2 + 3$ (d) $y = -\frac{1}{2}x^2 - 1$

Solve graphically the quadratic equations in Problems 2 to 5 by plotting the curves between the given limits. Give answers correct to 1 decimal place.

2. $4x^2 - x - 1 = 0$; $x = -1$ to $x = 1$

3. $x^2 - 3x = 27$; $x = -5$ to $x = 8$

4. $2x^2 - 6x - 9 = 0$; $x = -2$ to $x = 5$

5. $2x(5x - 2) = 39.6$; $x = -2$ to $x = 3$

6. Solve the quadratic equation $2x^2 + 7x + 6 = 0$ graphically, given that the solutions lie in the range $x = -3$ to $x = 1$. Determine also the nature and co-ordinates of its turning point.

7. Solve graphically the quadratic equation $10x^2 - 9x - 11.2 = 0$, given that the roots lie between $x = -1$ and $x = 2$

8. Plot a graph of $y = 3x^2$ and hence solve the following equations.
   (a) $3x^2 - 8 = 0$ (b) $3x^2 - 2x - 1 = 0$

9. Plot the graphs $y = 2x^2$ and $y = 3 - 4x$ on the same axes and find the co-ordinates of the points of intersection. Hence, determine the roots of the equation $2x^2 + 4x - 3 = 0$

10. Plot a graph of $y = 10x^2 - 13x - 30$ for values of x between $x = -2$ and $x = 3$. Solve the equation $10x^2 - 13x - 30 = 0$ and from the graph determine
    (a) the value of y when x is 1.3
    (b) the value of x when y is 10
    (c) the roots of the equation $10x^2 - 15x - 18 = 0$

19.3 Graphical solution of linear and quadratic equations simultaneously

The solution of **linear and quadratic equations simultaneously** may be achieved graphically by

(a) plotting the straight line and parabola on the same axes, and

(b) noting the points of intersection.

The co-ordinates of the points of intersection give the required solutions.
Here is a worked problem to demonstrate the simultaneous solution of a linear and quadratic equation.

Problem 7. Determine graphically the values of x and y which simultaneously satisfy the equations $y = 2x^2 - 3x - 4$ and $y = 2 - 4x$

$y = 2x^2 - 3x - 4$ is a parabola and a table of values is drawn up as shown below.

x	-2	-1	0	1	2	3
y	10	1	-4	-5	-2	5

$y = 2 - 4x$ is a straight line and only three co-ordinates need be calculated:

x	0	1	2
y	2	-2	-6

The two graphs are plotted in Fig. 19.13 and the points of intersection, shown as A and B, are at co-ordinates $(-2, 10)$ and $(1.5, -4)$. Hence, the simultaneous solutions occur when $x = -2$, $y = 10$ and when $x = 1.5$, $y = -4$
These solutions may be checked by substituting into each of the original equations.

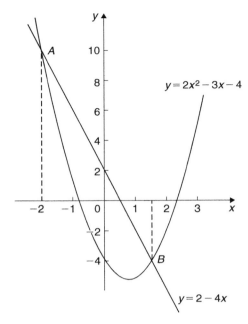

Figure 19.13

Now try the following Practice Exercise

Practice Exercise 96   Solving linear and quadratic equations simultaneously (answers on page 452)

1. Determine graphically the values of x and y which simultaneously satisfy the equations $y = 2(x^2 - 2x - 4)$ and $y + 4 = 3x$

2. Plot the graph of $y = 4x^2 - 8x - 21$ for values of x from -2 to $+4$. Use the graph to find the roots of the following equations.
   (a) $4x^2 - 8x - 21 = 0$
   (b) $4x^2 - 8x - 16 = 0$
   (c) $4x^2 - 6x - 18 = 0$

19.4 Graphical solution of cubic equations

A **cubic equation** of the form $ax^3 + bx^2 + cx + d = 0$ may be solved graphically by

(a) plotting the graph $y = ax^3 + bx^2 + cx + d$, and

(b) noting the points of intersection on the x-axis (i.e. where $y = 0$)

The x-values of the points of intersection give the required solution since at these points both $y = 0$ and $ax^3 + bx^2 + cx + d = 0$.

The number of solutions, or roots, of a cubic equation depends on how many times the curve cuts the x-axis and there can be one, two or three possible roots, as shown in Fig. 19.14.

(a)   (b)   (c)

Figure 19.14

Here are some worked problems to demonstrate the graphical solution of cubic equations.

Problem 8. Solve graphically the cubic equation $4x^3 - 8x^2 - 15x + 9 = 0$, given that the roots lie between $x = -2$ and $x = 3$. Determine also the co-ordinates of the turning points and distinguish between them

Let $y = 4x^3 - 8x^2 - 15x + 9$. A table of values is drawn up as shown below.

x	-2	-1	0	1	2	3
y	-25	12	9	-10	-21	0

A graph of $y = 4x^3 - 8x^2 - 15x + 9$ is shown in Fig. 19.15.

The graph crosses the x-axis (where $y = 0$) at $x = -1.5$, $x = 0.5$ and $x = 3$ and these are the solutions to the cubic equation $4x^3 - 8x^2 - 15x + 9 = 0$

The turning points occur at $(-0.6, 14.2)$, which is a **maximum**, and $(2, -21)$, which is a **minimum**.

Problem 9. Plot the graph of $y = 2x^3 - 7x^2 + 4x + 4$ for values of x between $x = -1$ and $x = 3$. Hence, determine the roots of the equation $2x^3 - 7x^2 + 4x + 4 = 0$

A table of values is drawn up as shown below.

x	-1	0	1	2	3
y	-9	4	3	0	7

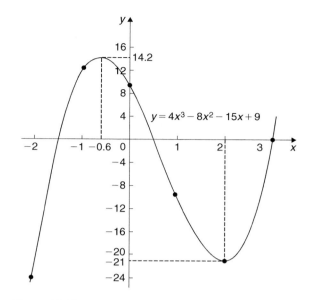

Figure 19.15

A graph of $y = 2x^3 - 7x^2 + 4x + 4$ is shown in Fig. 19.16. The graph crosses the x-axis at $x = -0.5$ and touches the x-axis at $x = 2$

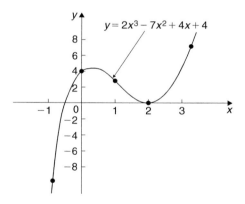

Figure 19.16

Hence the solutions of the equation $2x^3 - 7x^2 + 4x + 4 = 0$ are $x = -0.5$ and $x = 2$

Now try the following Practice Exercise

Practice Exercise 97   Solving cubic equations (answers on page 452)

1. Plot the graph $y = 4x^3 + 4x^2 - 11x - 6$ between $x = -3$ and $x = 2$ and use

the graph to solve the cubic equation $4x^3 + 4x^2 - 11x - 6 = 0$

2. By plotting a graph of $y = x^3 - 2x^2 - 5x + 6$ between $x = -3$ and $x = 4$, solve the equation $x^3 - 2x^2 - 5x + 6 = 0$. Determine also the co-ordinates of the turning points and distinguish between them.

In Problems 3 to 6, solve graphically the cubic equations given, each correct to 2 significant figures.

3. $x^3 - 1 = 0$

4. $x^3 - x^2 - 5x + 2 = 0$

5. $x^3 - 2x^2 = 2x - 2$

6. $2x^3 - x^2 - 9.08x + 8.28 = 0$

7. Show that the cubic equation $8x^3 + 36x^2 + 54x + 27 = 0$ has only one real root and determine its value.

Practice Exercise 98   Multiple-choice questions on graphical solution of equations (answers on page 452)

Each question has only one correct answer

1. The equation of the graph shown in Fig. 19.17 is:

   (a) $x(x+1) = \dfrac{15}{4}$ (b) $4x^2 - 4x - 15 = 0$

   (c) $x^2 - 4x - 5 = 0$ (d) $4x^2 + 4x - 15 = 0$

Figure 19.17

2. The numbers of real roots to the equation $ax^2 + bx + c = 0$ may be as many as:

   (a) 1   (b) 2   (c) 3   (d) 4

3. The numbers of real roots to the equation $ax^3 + bx^2 + cx + d = 0$ may be as many as:

   (a) 1   (b) 2   (c) 3   (d) 4

4. A parabola cuts the x-axis at $x = -3$ and at $x = 2$. The equation of the graph is:

   (a) $y = x^2 - x + 6$ (b) $y = x^2 + x + 6$
   (c) $y = x^2 - x - 6$ (d) $y = x^2 + x - 6$

5. Graphs of $y = 2x^2 - 3x + 1$ and $y = 2x - 1$ are drawn on the same axes. The two graphs intersect where:
   (a) $x = 0.5$ and $x = 2$ (b) $x = 1$ and $x = 2$
   (c) $x = 0.5$ and $x = 1$ (d) $x = 1$ and $x = 3$

For fully worked solutions to each of the problems in Practice Exercises 94 to 97 in this chapter, go to the website:
www.routledge.com/cw/bird

Chapter 20

Graphs with logarithmic scales

Why it is important to understand: Graphs with logarithmic scales

As mentioned in previous chapters, graphs are important tools for analysing and displaying data between two experimental quantities and that many times situations occur in which the relationship between the variables is not linear. By manipulation, a straight line graph may be plotted to produce a law relating the two variables. Knowledge of logarithms may be used to simplify plotting the relation between one variable and another. In particular, we consider those situations in which one of the variables requires scaling because the range of its data values is very large in comparison to the range of the other variable. Log-log and log-linear graph paper is available to make the plotting process easier.

At the end of this chapter you should be able to:

- understand logarithmic scales
- understand log-log and log-linear graph paper
- plot a graph of the form $y = ax^n$ using log-log graph paper and determine constants 'a' and 'n'
- plot a graph of the form $y = ab^x$ using log-linear graph paper and determine constants 'a' and 'b'
- plot a graph of the form $y = ae^{kx}$ using log-linear graph paper and determine constants 'a' and 'k'

20.1 Logarithmic scales and logarithmic graph paper

Graph paper is available where the scale markings along the horizontal and vertical axes are proportional to the logarithms of the numbers. Such graph paper is called **log-log graph paper**.

A **logarithmic scale** is shown in Fig. 20.1 where the distance between, say 1 and 2, is proportional to $\lg 2 - \lg 1$, i.e. 0.3010 of the total distance from 1 to 10. Similarly, the distance between 7 and 8 is proportional to $\lg 8 - \lg 7$, i.e. 0.05799 of the total distance from 1 to 10. Thus the distance between markings progressively decreases as the numbers increase from 1 to 10.

Figure 20.1

With log-log graph paper the scale markings are from 1 to 9, and this pattern can be repeated several times. The number of times the pattern of markings is repeated on an axis signifies the number of **cycles**. When the vertical axis has, say, 3 sets of values from 1 to 9, and the horizontal axis has, say, 2 sets of values from 1 to 9, then this log-log graph paper is called 'log 3 cycle × 2 cycle' (see Fig. 20.2). Many different arrangements are available ranging from 'log 1 cycle × 1 cycle' through to 'log 5 cycle × 5 cycle'.

To depict a set of values, say, from 0.4 to 161, on an axis of log-log graph paper, 4 cycles are required, from 0.1 to 1, 1 to 10, 10 to 100 and 100 to 1000.

20.2 Graphs of the form $y = ax^n$

Taking logarithms to a base of 10 of both sides of $y = ax^n$

gives: $\lg y = \lg(ax^n) = \lg a + \lg x^n$

i.e. $\lg y = n\lg x + \lg a$

which compares with $Y = mX + c$

Thus, by plotting $\lg y$ vertically against $\lg x$ horizontally, a straight line results, i.e. the equation $y = ax^n$ is reduced to linear form. With log-log graph paper available x and y may be plotted directly, without having first to determine their logarithms, as shown in Chapter 18.

> **Problem 1.** Experimental values of two related quantities x and y are shown below:
>
> | x | 0.41 | 0.63 | 0.92 | 1.36 | 2.17 | 3.95 |
> |---|---|---|---|---|---|---|
> | y | 0.45 | 1.21 | 2.89 | 7.10 | 20.79 | 82.46 |
>
> The law relating x and y is believed to be $y = ax^b$, where 'a' and 'b' are constants. Verify that this law is true and determine the approximate values of 'a' and 'b'

If $y = ax^b$ then $\lg y = b\lg x + \lg a$, from above, which is of the form $Y = mX + c$, showing that to produce a straight line graph $\lg y$ is plotted vertically against $\lg x$ horizontally. x and y may be plotted directly on to log-log graph paper as shown in Fig. 20.2. The values of y range from 0.45 to 82.46 and 3 cycles are needed (i.e. 0.1 to 1, 1 to 10 and 10 to 100). The values of x range from 0.41 to 3.95 and 2 cycles are needed (i.e. 0.1 to 1 and 1 to 10). Hence 'log 3 cycle × 2 cycle' is used as shown in Fig. 20.2 where the axes are marked and the points plotted. **Since the points lie on a straight line the law $y = ax^b$ is verified.**

To evaluate constants 'a' and 'b':

Method 1. Any two points on the straight line, say points A and C, are selected, and AB and BC are

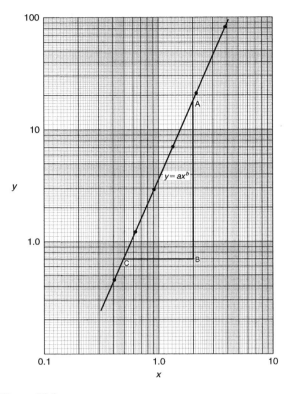

Figure 20.2

measured (say in centimetres). Then,

$$\text{gradient, } b = \frac{AB}{BC} = \frac{11.5\,\text{units}}{5\,\text{units}} = 2.3$$

Since $\lg y = b\lg x + \lg a$, when $x = 1$, $\lg x = 0$ and $\lg y = \lg a$.

The straight line crosses the ordinate $x = 1.0$ at $y = 3.5$.

Hence, $\lg a = \lg 3.5$, i.e. $a = 3.5$

Method 2. Any two points on the straight line, say points A and C, are selected. A has co-ordinates (2, 17.25) and C has co-ordinates (0.5, 0.7).

Since $y = ax^b$ then $17.25 = a(2)^b$ (1)

and $0.7 = a(0.5)^b$ (2)

i.e. two simultaneous equations are produced and may be solved for 'a' and 'b'.

Dividing equation (1) by equation (2) to eliminate 'a' gives:

$$\frac{17.25}{0.7} = \frac{(2)^b}{(0.5)^b} = \left(\frac{2}{0.5}\right)^b$$

i.e. $24.643 = (4)^b$

Taking logarithms of both sides gives:

$\lg 24.643 = b \lg 4$,

i.e. $b = \dfrac{\lg 24.643}{\lg 4}$

$= \mathbf{2.3}$, correct to 2 significant figures.

Substituting $b = 2.3$ in equation (1) gives:
$17.25 = a(2)^{2.3}$

i.e. $a = \dfrac{17.25}{(2)^{2.3}} = \dfrac{17.25}{4.925}$

$= \mathbf{3.5}$, correct to 2 significant figures

Hence, the law of the graph is: $y = 3.5x^{2.3}$

⚑ **Problem 2.** The power dissipated by a resistor was measured for varying values of current flowing in the resistor and the results are as shown:

Current, I amperes	1.4	4.7	6.8	9.1	11.2	13.1
Power, P watts	49	552	1156	2070	3136	4290

Prove that the law relating current and power is of the form $P = RI^n$, where R and n are constants, and determine the law. Hence calculate the power when the current is 12 A and the current when the power is 1000 W.

Since $P = RI^n$ then $\lg P = n \lg I + \lg R$, which is of the form $Y = mX + c$, showing that to produce a straight line graph $\lg P$ is plotted vertically against $\lg I$ horizontally. Power values range from 49 to 4290, hence 3 cycles of log-log graph paper are needed (10 to 100, 100 to 1000 and 1000 to 10 000). Current values range from 1.4 to 13.1, hence 2 cycles of log-log graph paper are needed (1 to 10 and 10 to 100). Thus 'log 3 cycles × 2 cycles' is used as shown in Fig. 20.3 (or, if not available, graph paper having a larger number of cycles per axis can be used).

The co-ordinates are plotted and **a straight line results which proves that the law relating current and power is of the form $P = RI^n$**

Gradient of straight line, $n = \dfrac{AB}{BC} = \dfrac{14\,\text{units}}{7\,\text{units}} = \mathbf{2}$

At point $C, I = 2$ and $P = 100$. Substituting these values into $P = RI^n$ gives:

$$100 = R(2)^2$$

Hence, $R = \dfrac{100}{2^2} = 25$

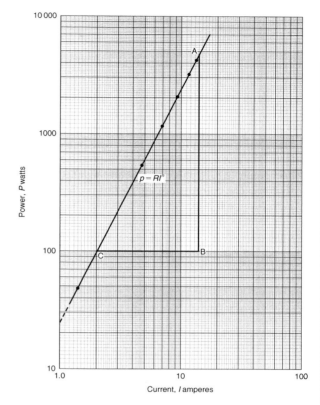

Figure 20.3

which may have been found from the intercept on the $I = 1.0$ axis in Fig. 20.3.

Hence, the law of the graph is: $P = 25I^2$
When current $I = 12$,
power $P = 25(12)^2 = \mathbf{3600\ watts}$ (which may be read from the graph).

When power $P = 1000, 1000 = 25I^2$

Hence $I^2 = \dfrac{1000}{25} = 40$

from which,   **current $I = \sqrt{40} = \mathbf{6.32A}$**

⚑ **Problem 3.** The pressure p and volume v of a gas are believed to be related by a law of the form $p = cv^n$, where c and n are constants. Experimental values of p and corresponding values of v obtained in a laboratory are:

p pascals	2.28×10^5	8.04×10^5	2.03×10^6
v m^3	3.2×10^{-2}	1.3×10^{-2}	6.7×10^{-3}

p pascals	5.05×10^6	1.82×10^7
v m^3	3.5×10^{-3}	1.4×10^{-3}

Verify that the law is true and determine approximate values of c and n.

Since $p = cv^n$, then $\lg p = n \lg v + \lg c$, which is of the form $Y = mX + c$, showing that to produce a straight line graph $\lg p$ is plotted vertically against $\lg v$ horizontally. The co-ordinates are plotted on 'log 3 cycle $\times$ 2 cycle' graph paper as shown in Fig. 20.4. With the data expressed in standard form, the axes are marked in standard form also. Since a straight line results the law $p = cv^n$ is verified.

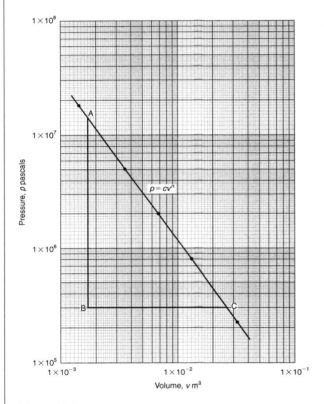

Figure 20.4

The straight line has a negative gradient and the value of the gradient is given by:

$$\frac{AB}{BC} = \frac{14 \, \text{units}}{10 \, \text{units}} = 1.4, \text{ hence } n = -\mathbf{1.4}$$

Selecting any point on the straight line, say point C, having co-ordinates $(2.63 \times 10^{-2}, 3 \times 10^5)$, and substituting these values in $p = cv^n$

gives: $\qquad 3 \times 10^5 = c(2.63 \times 10^{-2})^{-1.4}$

Hence, $\quad c = \dfrac{3 \times 10^5}{(2.63 \times 10^{-2})^{-1.4}} = \dfrac{3 \times 10^5}{(0.0263)^{-1.4}}$

$$= \frac{3 \times 10^5}{1.63 \times 10^2}$$

$$= \mathbf{1840}, \text{ correct to 3}$$
significant figures.

Hence, the law of the graph is: $p = 1840 v^{-1.4}$ or $pv^{1.4} = 1840$

Now try the following Practice Exercise

1. Quantities x and y are believed to be related by a law of the form $y = ax^n$, where a and n are constants. Experimental values of x and corresponding values of y are:

x	0.8	2.3	5.4	11.5	21.6	42.9
y	8	54	250	974	3028	10410

   Show that the law is true and determine the values of 'a' and 'n'. Hence determine the value of y when x is 7.5 and the value of x when y is 5000.

2. Show from the following results of voltage V and admittance Y of an electrical circuit that the law connecting the quantities is of the form $V = kY^n$, and determine the values of k and n.

Voltage, V volts	2.88	2.05	1.60	1.22	0.96
Admittance, Y siemens	0.52	0.73	0.94	1.23	1.57

3. Quantities x and y are believed to be related by a law of the form $y = mx^n$. The values of x and corresponding values of y are:

x	0.5	1.0	1.5	2.0	2.5	3.0
y	0.53	3.0	8.27	16.97	29.69	46.77

   Verify the law and find the values of m and n.

20.3   Graphs of the form $y = ab^x$

Taking logarithms to a base of 10 of both sides of $y = ab^x$

gives:
$$\lg y = \lg (ab^x) = \lg a + \lg b^x$$
$$= \lg a + x\lg b$$

i.e.
$$\lg y = (\lg b)x + \lg a$$

which compares with
$$Y = mX + c$$

Thus, by plotting $\lg y$ vertically against x horizontally a straight line results, i.e. the graph $y = ab^x$ is reduced to linear form. In this case, graph paper having a linear horizontal scale and a logarithmic vertical scale may be used. This type of graph paper is called **log-linear graph paper**, and is specified by the number of cycles on the logarithmic scale. For example, graph paper having 3 cycles on the logarithmic scale is called 'log 3 cycle × linear' graph paper.

> **Problem 4.**   Experimental values of quantities x and y are believed to be related by a law of the form $y = ab^x$, where 'a' and 'b' are constants. The values of x and corresponding values of y are:
>
> | x | 0.7 | 1.4 | 2.1 | 2.9 | 3.7 | 4.3 |
> |---|---|---|---|---|---|---|
> | y | 18.4 | 45.1 | 111 | 308 | 858 | 1850 |
>
> Verify the law and determine the approximate values of 'a' and 'b'. Hence evaluate (i) the value of y when x is 2.5, and (ii) the value of x when y is 1200.

Since $y = ab^x$ then $\lg y = (\lg b)x + \lg a$ (from above), which is of the form $Y = mX + c$, showing that to produce a straight line graph $\lg y$ is plotted vertically against x horizontally. Using log-linear graph paper, values of x are marked on the horizontal scale to cover the range 0.7 to 4.3. Values of y range from 18.4 to 1850 and 3 cycles are needed (i.e. 10 to 100, 100 to 1000 and 1000 to 10000). Thus, using 'log 3 cycles × linear' graph paper the points are plotted as shown in Fig. 20.5. **A straight line is drawn through the co-ordinates, hence the law $y = ab^x$ is verified.**
Gradient of straight line, $\lg b = AB/BC$. Direct measurement (say in centimetres) is not made with log-linear graph paper since the vertical scale is logarithmic and the horizontal scale is linear.

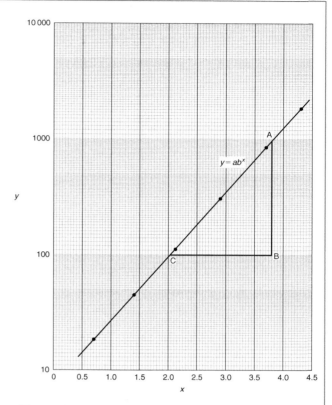

Figure 20.5

Hence,
$$\lg b = \frac{AB}{BC} = \frac{\lg 1000 - \lg 100}{3.82 - 2.02}$$
$$= \frac{3 - 2}{1.80} = \frac{1}{1.80} = 0.5556$$

Hence,
$$b = \text{antilog } 0.5556$$
$$= 10^{0.5556} = \mathbf{3.6}, \text{ correct to 2 significant figures.}$$

Point A has co-ordinates $(3.82, 1000)$.
Substituting these values into $y = ab^x$

gives:
$$1000 = a(3.6)^{3.82}$$

i.e.
$$a = \frac{1000}{(3.6)^{3.82}} = \mathbf{7.5}, \text{ correct to 2 significant figures.}$$

Hence, the law of the graph is: $y = 7.5(3.6)^x$

(i)   When $x = 2.5, y = 7.5(3.6)^{2.5} = \mathbf{184}$

(ii)   When $y = 1200, 1200 = 7.5(3.6)^x$,

hence $(3.6)^x = \dfrac{1200}{7.5} = 160$

Taking logarithms gives:

$$x \lg 3.6 = \lg 160$$

i.e. $$x = \frac{\lg 160}{\lg 3.6} = \frac{2.2041}{0.5563} = \mathbf{3.96}$$

Now try the following Practice Exercise

1. Experimental values of p and corresponding
    values of q are shown below.

p	−13.2	−27.9	−62.2	−383.2	−1581	−2931
q	0.30	0.75	1.23	2.32	3.17	3.54

    Show that the law relating p and q is
    $p = ab^q$, where 'a' and 'b' are constants.
    Determine (i) values of 'a' and 'b', and state
    the law, (ii) the value of p when q is 2.0, and
    (iii) the value of q when p is −2000

20.4 Graphs of the form $y = ae^{kx}$

Taking logarithms to a base of e of both sides of $y = ae^{kx}$

gives: $$\ln y = \ln(ae^{kx}) = \ln a + \ln e^{kx}$$
$$= \ln a + kx \ln e$$

i.e. $$\ln y = kx + \ln a \quad \text{since} \quad \ln e = 1$$

which compares with $Y = mX + c$

Thus, by plotting $\ln y$ vertically against x horizontally, a
straight line results, i.e. the equation $y = ae^{kx}$ is reduced
to linear form. In this case, graph paper having a linear
horizontal scale and a logarithmic vertical scale may be
used.

Problem 5. The data given below is believed to
be related by a law of the form $y = ae^{kx}$, where 'a'
and 'k' are constants. Verify that the law is true and
determine approximate values of 'a' and 'k'. Also,
determine the value of y when x is 3.8 and the value
of x when y is 85.

x	−1.2	0.38	1.2	2.5	3.4	4.2	5.3
y	9.3	22.2	34.8	71.2	117	181	332

Since $y = ae^{kx}$ then $\ln y = kx + \ln a$ (from above), which
is of the form $Y = mX + c$, showing that to produce a
straight line graph $\ln y$ is plotted vertically against x horizontally. The value of y ranges from 9.3 to 332 hence
'log 3 cycle × linear' graph paper is used. The plotted
co-ordinates are shown in Fig. 20.6 and **since a straight
line passes through the points the law** $y = ae^{kx}$ **is
verified**.

Gradient of straight line,

$$k = \frac{AB}{BC} = \frac{\ln 100 - \ln 10}{3.12 - (-1.08)} = \frac{2.3026}{4.20}$$

$$= \mathbf{0.55}, \text{ correct to 2 significant figures.}$$

Since $\ln y = kx + \ln a$, when $x = 0, \ln y = \ln a$ i.e.
$y = a$.
The vertical axis intercept value at $x = 0$ is 18,
hence, $a = \mathbf{18}$

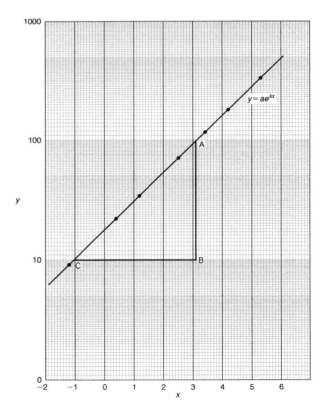

Figure 20.6

The law of the graph is thus: $y = 18e^{0.55x}$

When x is 3.8,

$$y = 18e^{0.55(3.8)} = 18e^{2.09} = 18(8.0849) = \mathbf{146}$$

When y is 85, $85 = 18e^{0.55x}$

Hence, $e^{0.55x} = \dfrac{85}{18} = 4.7222$

and $0.55x = \ln 4.7222 = 1.5523$

Hence, $x = \dfrac{1.5523}{0.55} = \mathbf{2.82}$

⚑ **Problem 6.**   The voltage, v volts, across an inductor is believed to be related to time, t ms, by the law $v = Ve^{t/T}$, where V and T are constants. Experimental results obtained are:

v volts	883	347	90	55.5	18.6	5.2
t ms	10.4	21.6	37.8	43.6	56.7	72.0

Show that the law relating voltage and time is as stated, and determine the approximate values of V and T. Find also the value of voltage after 25 ms and the time when the voltage is 30.0 V

Since $v = Ve^{t/T}$ then: $\ln v = \ln(Ve^{t/T})$

$$= \ln V + \ln e^{t/T} = \ln V + \frac{t}{T}\ln e = \frac{t}{T} + \ln V$$

i.e. $\ln v = \dfrac{1}{T}t + \ln V$ which is of the form $Y = mX + c$

Using 'log 3 cycle × linear' graph paper, the points are plotted as shown in Fig. 20.7.
Since the points are joined by a straight line the law $v = Ve^{t/T}$ is verified.

Gradient of straight line,

$$\frac{1}{T} = \frac{AB}{BC} = \frac{\ln 100 - \ln 10}{36.5 - 64.2} = \frac{2.3026}{-27.7}$$

Hence, $T = \dfrac{-27.7}{2.3026} = \mathbf{-12.0}$, correct to 3 significant figures.

Since the straight line does not cross the vertical axis at $t = 0$ in Fig. 20.7, the value of V is determined by selecting any point, say A, having co-ordinates (36.5, 100) and substituting these values into $v = Ve^{t/T}$

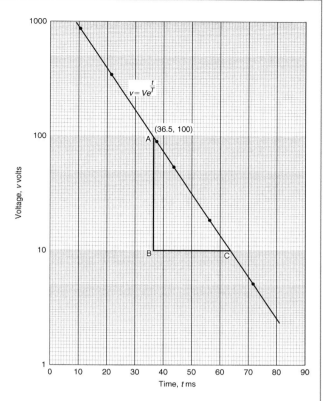

Figure 20.7

Thus, $100 = Ve^{36.5/-12.0}$

i.e. $V = \dfrac{100}{e^{-36.5/12.0}}$

$= \mathbf{2090\ volts}$, correct to 3 significant figures.

Hence, the law of the graph is: $v = 2090e^{-t/12.0}$

When time $t = 25$ ms, voltage $v = 2090e^{-25/12.0}$

$$= \mathbf{260\,V}$$

When the voltage is 30.0 volts, $30.0 = 2090e^{-t/12.0}$

Hence, $e^{-t/12.0} = \dfrac{30.0}{2090}$ and $e^{t/12.0} = \dfrac{2090}{30.0} = 69.67$

Taking Napierian logarithms gives:

$$\frac{t}{12.0} = \ln 69.67 = 4.2438$$

from which,   time $t = (12.0)(4.2438) = \mathbf{50.9\,ms}$

Now try the following Practice Exercise

Practice Exercise 101   Logarithmic graphs of the form $y = ae^{kx}$ **(answers on page 453)**

1. Atmospheric pressure p is measured at varying altitudes h and the results are as shown below:

Altitude, h m	500	1500	3000	5000	8000
pressure, p kPa	73.39	68.42	61.60	53.56	43.41

Show that the quantities are related by the law $p = ae^{kh}$, where 'a' and 'k' are constants. Determine the values of 'a' and 'k' and state the law. Find also the atmospheric pressure at 10 000 m.

2. At particular times, t minutes, measurements are made of the temperature, $\theta°$C, of a cooling liquid and the following results are obtained:

Temperature $\theta°$C	92.2	55.9	33.9	20.6	12.5
Time t minutes	10	20	30	40	50

Prove that the quantities follow a law of the form $\theta = \theta_0 e^{kt}$, where θ_0 and k are constants, and determine the approximate values of θ_0 and k.

Practice Exercise 102   Multiple-choice questions on graphs with logarithmic scales (answers on page 453)

Each question has only one correct answer

Questions 1 to 3 relate to the following information.
A straight-line graph is plotted for the equation $y = ax^n$, where y and x are the variables and a and n are constants.

1. On the vertical axis is plotted:
(a) y (b) x (c) lg y (d) a

2. On the horizontal axis is plotted:
(a) lg x (b) x (c) x^n (d) a

3. The gradient of the graph is given by:
(a) y (b) a (c) x (d) n

4. To depict a set of values from 0.05 to 275, the minimum number of cycles required on logarithmic graph paper is:
(a) 5 (b) 3 (c) 4 (d) 2

Questions 5 to 7 relate to the following information.

A straight-line graph is plotted for the equation $y = ae^{bx}$, where y and x are the variables and a and b are constants.

5. On the vertical axis is plotted:
(a) y (b) x (c) ln y (d) a

6. The gradient of the graph is given by:
(a) y (b) a (c) x (d) b

7. The vertical axis intercept is given by:
(a) b (b) ln a (c) x (d) ln y

Questions 8 to 10 relate to the following information.

A straight-line graph is plotted for the equation $y = ab^x$, where y and x are the variables and a and b are constants.

8. On the horizontal axis is plotted:
(a) lg x (b) x (c) b^x (d) a

9. The gradient of the graph is given by:
(a) y (b) a (c) b (d) lg b

10. The vertical axis intercept is given by:
(a) b (b) lg a (c) x (d) lg y

For fully worked solutions to each of the problems in Practice Exercises 99 to 101 in this chapter, go to the website:
www.routledge.com/cw/bird

This assignment covers the material contained in Chapters 17–20. *The marks available are shown in brackets at the end of each question.*

1. Determine the value of P in the following table of values.

x	0	1	4
$y = 3x - 5$	-5	-2	P

   (2)

2. Assuming graph paper measuring 20 cm by 20 cm is available, suggest suitable scales for the following ranges of values.

   Horizontal axis: 5 N to 70 N; vertical axis: 20 mm to 190 mm. (2)

3. Corresponding values obtained experimentally for two quantities are:

x	-5	-3	-1	0	2	4
y	-17	-11	-5	-2	4	10

   Plot a graph of y (vertically) against x (horizontally) to scales of 1 cm = 1 for the horizontal x-axis and 1 cm = 2 for the vertical y-axis. From the graph, find
   (a) the value of y when $x = 3$
   (b) the value of y when $x = -4$
   (c) the value of x when $y = 1$
   (d) the value of x when $y = -20$ (8)

4. If graphs of y against x were to be plotted for each of the following, state (i) the gradient, and (ii) the y-axis intercept.
   (a) $y = -5x + 3$ (b) $y = 7x$
   (c) $2y + 4 = 5x$ (d) $5x + 2y = 6$
   (e) $2x - \dfrac{y}{3} = \dfrac{7}{6}$ (10)

5. The resistance R ohms of a copper winding is measured at various temperatures $t°C$ and the results are as follows.

$R\ (\Omega)$	38	47	55	62	72
$t\ (°C)$	16	34	50	64	84

   Plot a graph of R (vertically) against t (horizontally) and find from it
   (a) the temperature when the resistance is 50 Ω
   (b) the resistance when the temperature is 72°C
   (c) the gradient
   (d) the equation of the graph. (10)

6. x and y are two related variables and all other letters denote constants. For the stated laws to be verified it is necessary to plot graphs of the variables in a modified form. State for each
   (a) what should be plotted on the vertical axis,
   (b) what should be plotted on the horizontal axis,
   (c) the gradient,
   (d) the vertical axis intercept.
   (i) $y = p + rx^2$ (ii) $y = \dfrac{a}{x} + bx$ (4)

7. The following results give corresponding values of two quantities x and y which are believed to be related by a law of the form $y = ax^2 + bx$ where a and b are constants.

y	33.9	55.5	72.8	84.1	111.4	168.1
x	3.4	5.2	6.5	7.3	9.1	12.4

   Verify the law and determine approximate values of a and b.
   Hence determine (i) the value of y when x is 8.0 and (ii) the value of x when y is 146.5 (18)

8. By taking logarithms of both sides of $y = kx^n$, show that $\lg y$ needs to be plotted vertically and $\lg x$ needs to be plotted horizontally to produce a straight line graph. Also, state the gradient and vertical axis intercept. (6)

9. By taking logarithms of both sides of $y = ae^{kx}$ show that $\ln y$ needs to be plotted vertically and x needs to be plotted horizontally to produce a straight line graph. Also, state the gradient and vertical axis intercept. (6)

10. Show from the following results of voltage V and admittance Y of an electrical circuit that the law connecting the quantities is of the form $V = kY^n$ and determine the values of k and n.

Voltage, V (volts)	2.88	2.05	1.60	1.22	0.96
Admittance, Y (siemens)	0.52	0.73	0.94	1.23	1.57

    (12)

11. The current i flowing in a discharging capacitor varies with time t as shown.

i (mA)	50.0	17.0	5.8	1.7	0.58	0.24
t (ms)	200	255	310	375	425	475

Show that these results are connected by the law of the form $i = Ie^{\frac{t}{T}}$ where I is the initial current flowing and T is a constant. Determine approximate values of constants I and T. (15)

12. Solve, correct to 1 decimal place, the quadratic equation $2x^2 - 6x - 9 = 0$ by plotting values of x from $x = -2$ to $x = 5$ (8)

13. Plot the graph of $y = x^3 + 4x^2 + x - 6$ for values of x between $x = -4$ and $x = 2$. Hence determine the roots of the equation $x^3 + 4x^2 + x - 6 = 0$ (9)

14. Plot a graph of $y = 2x^2$ from $x = -3$ to $x = +3$ and hence solve the following equations.
    (a) $2x^2 - 8 = 0$
    (b) $2x^2 - 4x - 6 = 0$ (9)

15. State the minimum number of cycles on logarithmic graph paper needed to plot a set of values ranging from 0.065 to 480 (2)

16. The current i flowing in a discharging capacitor varies with time as shown below:

i (mA)	50.0	17.0	5.8	1.7	0.58	0.24
t (ms)	200	255	310	375	425	475

Using logarithmic graph paper, show that these results are connected by the law of the form $i = Ie^{\frac{t}{T}}$ where I, the initial current flowing, and T are constants. Determine the approximate values of I and T. (14)

Chapter 21

Angles and triangles

Why it is important to understand: **Angles and triangles**

Knowledge of angles and triangles is very important in engineering. Trigonometry is needed in surveying and architecture, for building structures/systems, designing bridges and solving scientific problems. Trigonometry is also used in electrical engineering; the functions that relate angles and side lengths in right-angled triangles are useful in expressing how a.c. electric current varies with time. Engineers use triangles to determine how much force it will take to move along an incline, GPS satellite receivers use triangles to determine exactly where they are in relation to satellites orbiting hundreds of miles away. Whether you want to build a skateboard ramp, a stairway or a bridge, you can't escape trigonometry.

At the end of this chapter you should be able to:

- define an angle: acute, right, obtuse, reflex, complementary, supplementary
- define parallel lines, transversal, vertically opposite angles, corresponding angles, alternate angles, interior angles
- define degrees, minutes, seconds, radians
- add and subtract angles
- state types of triangles – acute, right, obtuse, equilateral, isosceles, scalene
- define hypotenuse, adjacent and opposite sides with reference to an angle in a right-angled triangle
- recognise congruent triangles
- recognise similar triangles
- construct triangles, given certain sides and angles

21.1 Introduction

Trigonometry is a subject that involves the measurement of sides and angles of triangles and their relationship to each other. This chapter involves the measurement of angles and introduces types of triangle.

21.2 Angular measurement

An **angle** is the amount of rotation between two straight lines. Angles may be measured either in **degrees** or in **radians**.

If a circle is divided into 360 equal parts, then each part is called **1 degree** and is written as **1°**

i.e.   **1 revolution = 360°**

or   **1 degree is $\frac{1}{360}$ th of a revolution**

Some angles are given **special names**.

- Any angle between 0° and 90° is called an **acute angle**.
- An angle equal to 90° is called a **right angle**.
- Any angle between 90° and 180° is called an **obtuse angle**.

- Any angle greater than 180° and less than 360° is called a **reflex angle**.

- An angle of 180° lies on a **straight line**.

- If two angles add up to 90° they are called **complementary angles**.

- If two angles add up to 180° they are called **supplementary angles**.

- **Parallel lines** are straight lines which are in the same plane and never meet. Such lines are denoted by arrows, as in Fig. 21.1.

- A straight line which crosses two parallel lines is called a **transversal** (see *MN* in Fig. 21.1).

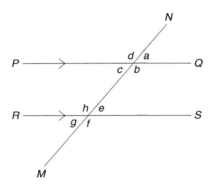

Figure 21.1

With reference to Fig. 21.1,

(a) $a = c, b = d, e = g$ and $f = h$. Such pairs of angles are called **vertically opposite angles**.

(b) $a = e, b = f, c = g$ and $d = h$. Such pairs of angles are called **corresponding angles**.

(c) $c = e$ and $b = h$. Such pairs of angles are called **alternate angles**.

(d) $b + e = 180°$ and $c + h = 180°$. Such pairs of angles are called **interior angles**.

21.2.1 Minutes and seconds

One degree may be subdivided into 60 parts, called **minutes**.

i.e. **1 degree = 60 minutes**

which is written as **1° = 60′**

41 degrees and 29 minutes is written as 41°29′. 41°29′ is equivalent to $41\dfrac{29°}{60} = 41.483°$ as a decimal, correct to 3 decimal places by calculator.

1 minute further subdivides into 60 seconds,

i.e. **1 minute = 60 seconds**

which is written as **1′ = 60″**

(Notice that for minutes, 1 dash is used and for seconds, 2 dashes are used.)

For example, 56 degrees, 36 minutes and 13 seconds is written as 56°36′13″

21.2.2 Radians and degrees

One radian is defined as the angle subtended at the centre of a circle by an arc equal in length to the radius. (For more on circles, see Chapter 27.)

With reference to Fig. 21.2, for arc length s,

θ radians $= \dfrac{s}{r}$

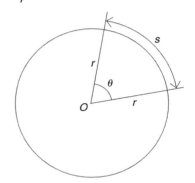

Figure 21.2

When s is the whole circumference, i.e. when $s = 2\pi r$,

$$\theta = \frac{s}{r} = \frac{2\pi r}{r} = 2\pi$$

In one revolution, $\theta = 360°$. Hence, the relationship between **degrees and radians** is

$$360° = 2\pi \text{ radians or } \mathbf{180° = \pi \, rad}$$

i.e. $\mathbf{1 \, rad = \dfrac{180°}{\pi} \approx 57.30°}$

Since π rad $= 180°$, then $\dfrac{\pi}{2}$rad $= 90°$, $\dfrac{\pi}{4}$rad $= 45°$, $\dfrac{\pi}{3}$rad $= 60°$ and $\dfrac{\pi}{6}$rad $= 30°$

Here are some worked examples on angular measurement.

Problem 1. Evaluate 43°29′ + 27°43′

$$
\begin{array}{r}
43° \ 29' \\
+ \quad 27° \ 43' \\
\hline
71° \ 12' \\
1° \\
\end{array}
$$

(i) $29' + 43' = 72'$

(ii)   Since $60' = 1°, 72' = 1°12'$

(iii)   The $12'$ is placed in the minutes column and $1°$ is carried in the degrees column.

(iv) $43° + 27° + 1°$ (carried) $= 71°$. Place $71°$ in the degrees column.

This answer can be obtained using the **calculator** as follows.

1. Enter 43   2. Press ° ′ ′ ′   3. Enter 29
4. Press ° ′ ′ ′   5. Press $+$ 6. Enter 27
7. Press ° ′ ′ ′   8. Enter 43   9. Press ° ′ ′ ′
10. Press $=$ Answer $= \mathbf{71°12'}$

Thus, $\mathbf{43°29' + 27°43' = 71°12'}$

> **Problem 2.**   Evaluate $84°13' - 56°39'$

$$\begin{array}{r} 84°\ 13' \\ - \underline{56°\ 39'} \\ 27°\ 34' \end{array}$$

(i) $13' - 39'$ cannot be done.

(ii) $1°$ or $60'$ is 'borrowed' from the degrees column, which leaves $83°$ in that column.

(iii) $(60' + 13') - 39' = 34'$, which is placed in the minutes column.

(iv) $83° - 56° = 27°$, which is placed in the degrees column.

This answer can be obtained using the **calculator** as follows.

1. Enter 84   2. Press ° ′ ′ ′   3. Enter 13
4. Press ° ′ ′ ′   5. Press $-$ 6. Enter 56
7. Press ° ′ ′ ′   8. Enter 39   9. Press ° ′ ′ ′
10. Press $=$ Answer $= \mathbf{27°34'}$

Thus, $\mathbf{84°13' - 56°39' = 27°34'}$

> **Problem 3.**   Evaluate $19°\,51'47'' + 63°27'34''$

$$\begin{array}{r} 19°\ 51'\ 47'' \\ + \underline{63°\ 27'\ 34''} \\ 83°\ 19'\ 21'' \\ 1°\ \ 1' \end{array}$$

(i) $47'' + 34'' = 81''$

(ii)   Since $60'' = 1', 81'' = 1'21''$

(iii)   The $21''$ is placed in the seconds column and $1'$ is carried in the minutes column.

(iv) $51' + 27' + 1' = 79'$

(v)   Since $60' = 1°, 79' = 1°19'$

(vi)   The $19'$ is placed in the minutes column and $1°$ is carried in the degrees column.

(vii) $19' + 63° + 1°$ (carried) $= 83°$. Place $83°$ in the degrees column.

This answer can be obtained using the **calculator** as follows.

1. Enter 19   2. Press ° ′ ′ ′   3. Enter 51
4. Press ° ′ ′ ′   5. Enter 47   6. Press ° ′ ′ ′
7. Press $+$ 8. Enter 63   9. Press ° ′ ′ ′
10. Enter 27   11. Press ° ′ ′ ′
12. Enter 34   13. Press ° ′ ′ ′
14. Press $=$ Answer $= \mathbf{83°19'21''}$

Thus, $\mathbf{19°51'47'' + 63°27'34'' = 83°19'21''}$

> **Problem 4.**   Convert $39°\,27'$ to degrees in decimal form

$$39°27' = 39\frac{27°}{60}$$

$$\frac{27°}{60} = 0.45° \quad \text{by calculator}$$

Hence, $\mathbf{39°27' = 39\dfrac{27°}{60} = 39.45°}$

This answer can be obtained using the **calculator** as follows.

1. Enter 39   2. Press ° ′ ′ ′   3. Enter 27
4. Press ° ′ ′ ′   5. Press $=$ 6. Press ° ′ ′ ′
Answer $= \mathbf{39.45°}$

> **Problem 5.**   Convert $63°\,26'\,51''$ to degrees in decimal form, correct to 3 decimal places

$$63°26'51'' = 63°26\frac{51'}{60} = 63°26.85'$$

$$63°26.85' = 63\frac{26.85°}{60} = 63.4475°$$

Hence, $\mathbf{63°26'51'' = 63.448°}$ correct to 3 decimal places.

This answer can be obtained using the **calculator** as follows.

1. Enter 63   2. Press ° ′ ′ ′   3. Enter 26
4. Press ° ′ ′ ′   5. Enter 51   6. Press ° ′ ′ ′
7. Press $=$ 8. Press ° ′ ′ ′   Answer $= \mathbf{63.4475°}$

Problem 6. Convert 53.753° to degrees, minutes and seconds

$$0.753° = 0.753 × 60' = 45.18'$$

$$0.18' = 0.18 × 60'' = 11''$$
to the nearest second

Hence, **53.753° = 53°45′11″**

This answer can be obtained using the **calculator** as follows.

1. Enter 53.753   2. Press =
3. Press ° ′ ″   Answer = **53°45′10.8″**

Now try the following Practice Exercise

Practice Exercise 103   Angular measurement (answers on page 453)

1. Evaluate $52°39' + 29°48'$

2. Evaluate $76°31' − 48°37'$

3. Evaluate $77°22' + 41°36' − 67°47'$

4. Evaluate $41°37'16'' + 58°29'36''$

5. Evaluate $54°37'42'' − 38°53'25''$

6. Evaluate $79°26'19'' − 45°58'56'' + 53°21'38''$

7. Convert $72°33'$ to degrees in decimal form.

8. Convert $27°45'15''$ to degrees correct to 3 decimal places.

9. Convert $37.952°$ to degrees and minutes.

10. Convert $58.381°$ to degrees, minutes and seconds.

Here are some further worked examples on angular measurement.

Problem 7. State the general name given to the following angles: (a) 157° (b) 49° (c) 90° (d) 245°

(a) Any angle between 90° and 180° is called an obtuse angle.

Thus, **157° is an obtuse angle**.

(b) Any angle between 0° and 90° is called an acute angle.

Thus, **49° is an acute angle**.

(c) An angle equal to 90° is called a **right angle**.

(d) Any angle greater than 180° and less than 360° is called a reflex angle.

Thus, **245° is a reflex angle**.

Problem 8. Find the angle complementary to 48° 39′

If two angles add up to 90° they are called **complementary angles**. Hence, **the angle complementary to 48°39′ is**

$$90° − 48°39' = 41°21'$$

Problem 9. Find the angle supplementary to 74°25′

If two angles add up to 180° they are called **supplementary angles**. Hence, **the angle supplementary to 74°25′ is**

$$180° − 74° 25' = 105° 35'$$

Problem 10. Evaluate angle $θ$ in each of the diagrams shown in Fig. 21.3

Figure 21.3

(a) The symbol shown in Fig. 21.4 is called a **right angle** and it equals 90°

Figure 21.4

Hence, from Fig. 21.3(a),

$$θ + 41° = 90°$$
from which, $$θ = 90° − 41° = 49°$$

(b) An angle of 180° lies on a straight line.
Hence, from Fig. 21.3(b),

$$180° = 53° + \theta + 44°$$

from which, $\theta = 180° - 53° - 44° = \mathbf{83°}$

Problem 11.   Evaluate angle θ in the diagram shown in Fig. 21.5

Figure 21.5

There are 360° in a complete revolution of a circle.
Thus, $360° = 58° + 108° + 64° + 39° + \theta$
from which, $\theta = 360° - 58° - 108° - 64° - 39° = \mathbf{91°}$

Problem 12.   Two straight lines AB and CD intersect at 0. If $\angle AOC$ is 43°, find $\angle AOD$, $\angle DOB$ and $\angle BOC$

From Fig. 21.6, $\angle AOD$ is supplementary to $\angle AOC$.
Hence, $\angle \mathbf{AOD} = 180° - 43° = \mathbf{137°}$
When two straight lines intersect, the vertically opposite angles are equal.
Hence, $\angle \mathbf{DOB} = \mathbf{43°}$ and $\angle \mathbf{BOC}\ \mathbf{137°}$

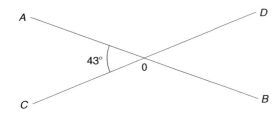

Figure 21.6

Problem 13.   Determine angle β in Fig. 21.7

Figure 21.7

$\alpha = 180° - 133° = 47°$ (i.e. supplementary angles).
$\alpha = \beta = \mathbf{47°}$ (corresponding angles between parallel lines).

Problem 14.   Determine the value of angle θ in Fig. 21.8

Figure 21.8

Let a straight line FG be drawn through E such that FG is parallel to AB and CD.
$\angle BAE = \angle AEF$ (alternate angles between parallel lines AB and FG), hence $\angle AEF = 23°37'$
$\angle ECD = \angle FEC$ (alternate angles between parallel lines FG and CD), hence $\angle FEC = 35°49'$

$$\text{Angle } \theta = \angle AEF + \angle FEC = 23°37' + 35°49'$$

$$= \mathbf{59°26'}$$

Problem 15.   Determine angles c and d in Fig. 21.9

Figure 21.9

$a = b = 46°$ (corresponding angles between parallel lines).

Also, $b + c + 90° = 180°$ (angles on a straight line).

Hence, $46° + c + 90° = 180°$, from which, $c = \mathbf{44°}$

b and d are supplementary, hence $d = 180° - 46°$
$$= \mathbf{134°}$$

Alternatively, $90° + c = d$ (vertically opposite angles).

> **Problem 16.** Convert the following angles to radians, correct to 3 decimal places.
> (a) 73° (b) 25°37′

Although we may be more familiar with degrees, radians is the SI unit of angular measurement in engineering (1 radian $\approx 57.3°$).

(a) Since $180° = \pi$ rad then $1° = \dfrac{\pi}{180}$rad.

Hence, $\mathbf{73°} = 73 \times \dfrac{\pi}{180}$rad $= \mathbf{1.274\ rad}$.

(b) $25°37′ = 25\dfrac{37°}{60} = 25.616666\ldots$

Hence, $\mathbf{25°37′} = 25.616666\ldots°$
$$= 25.616666\ldots \times \dfrac{\pi}{180}\text{rad}$$
$$= \mathbf{0.447\ rad}.$$

> **Problem 17.** Convert 0.743 rad to degrees and minutes

Since $180° = \pi$ rad then 1 rad $= \dfrac{180°}{\pi}$

Hence, $\mathbf{0.743\ rad} = 0.743 \times \dfrac{180°}{\pi} = 42.57076\ldots°$
$$= \mathbf{42°34′}$$

Now try the following Practice Exercise

1. State the general name given to an angle of 197°

2. State the general name given to an angle of 136°

3. State the general name given to an angle of 49°

4. State the general name given to an angle of 90°

5. Determine the angles complementary to the following.
   (a) 69° (b) 27°37′ (c) 41°3′43″

6. Determine the angles supplementary to
   (a) 78° (b) 15° (c) 169°41′11″

7. Find the values of angle θ in diagrams (a) to (i) of Fig. 21.10.

Figure 21.10

8. With reference to Fig. 21.11, what is the name given to the line *XY*? Give examples of each of the following.

Figure 21.11

(a) vertically opposite angles
(b) supplementary angles
(c) corresponding angles
(d) alternate angles

9. In Fig. 21.12, find angle α.

Figure 21.12

10. In Fig. 21.13, find angles a, b and c.

Figure 21.13

11. Find angle β in Fig. 21.14.

Figure 21.14

12. Convert 76° to radians, correct to 3 decimal places.

13. Convert 34°40′ to radians, correct to 3 decimal places.

14. Convert 0.714 rad to degrees and minutes.

21.3 Triangles

A **triangle** is a figure enclosed by three straight lines. **The sum of the three angles of a triangle is equal to 180°**

21.3.1 Types of triangle

An **acute-angled triangle** is one in which all the angles are acute; i.e. all the angles are less than 90°. An example is shown in triangle *ABC* in Fig. 21.15(a).
A **right-angled triangle** is one which contains a right angle; i.e. one in which one of the angles is 90°. An example is shown in triangle *DEF* in Fig. 21.15(b).

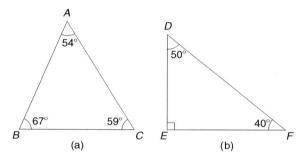

Figure 21.15

An **obtuse-angled triangle** is one which contains an obtuse angle; i.e. one angle which lies between 90° and 180°. An example is shown in triangle *PQR* in Fig. 21.16(a).
An **equilateral triangle** is one in which all the sides and all the angles are equal; i.e. each is 60°. An example is shown in triangle *ABC* in Fig. 21.16(b).

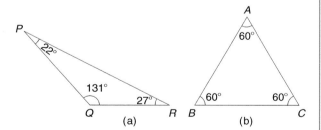

Figure 21.16

An **isosceles triangle** is one in which two angles and two sides are equal. An example is shown in triangle *EFG* in Fig. 21.17(a).
A **scalene triangle** is one with unequal angles and therefore unequal sides. An example of an acute-angled scalene triangle is shown in triangle *ABC* in Fig. 21.17(b).

Figure 21.17

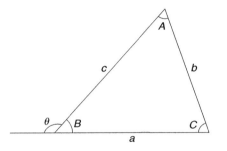

Figure 21.18

With reference to Fig. 21.18,

(a) Angles A, B and C are called **interior angles** of the triangle.

(b) Angle θ is called an **exterior angle** of the triangle and is equal to the sum of the two opposite interior angles; i.e. $\theta = A + C$.

(c) $a + b + c$ is called the **perimeter** of the triangle.

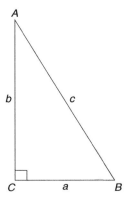

Figure 21.19

A right-angled triangle ABC is shown in Fig. 21.19. The point of intersection of two lines is called a vertex (plural **vertices**); the three vertices of the triangle are labelled as A, B and C, respectively. The right angle is angle C. The side opposite the right angle is given the special name of the **hypotenuse**. The hypotenuse, length AB in Fig. 21.19, is always the longest side of a right-angled triangle. With reference to angle B, AC is the **opposite** side and BC is called the **adjacent** side. With reference to angle A, BC is the **opposite** side and AC is the **adjacent** side.

Often sides of a triangle are labelled with lower case letters, a being the side opposite angle A, b being the side opposite angle B and c being the side opposite angle C. So, in the triangle ABC, length $AB = c$, length $BC = a$ and length $AC = b$. Thus, c is the hypotenuse in the triangle ABC.

$\angle$ is the symbol used for 'angle'. For example, in the triangle shown, $\angle C = 90°$. Another way of indicating an angle is to use all three letters. For example, $\angle ABC$ actually means $\angle B$; i.e. we take the middle letter as the angle. Similarly, $\angle BAC$ means $\angle A$ and $\angle ACB$ means $\angle C$.

Here are some worked examples to help us understand more about triangles.

Problem 18. Name the types of triangle shown in Fig. 21.20

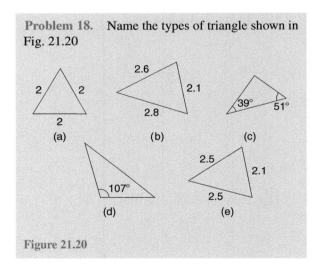

Figure 21.20

(a) **Equilateral triangle** (since all three sides are equal).

(b) **Acute-angled scalene triangle** (since all the angles are less than $90°$).

(c) **Right-angled triangle** ($39° + 51° = 90°$; hence, the third angle must be $90°$, since there are $180°$ in a triangle).

(d) **Obtuse-angled scalene triangle** (since one of the angles lies between 90° and 180°).

(e) **Isosceles triangle** (since two sides are equal).

Problem 19. In the triangle *ABC* shown in Fig. 21.21, with reference to angle θ, which side is the adjacent?

Figure 21.21

The triangle is right-angled; thus, side *AC* is the hypotenuse. With reference to angle θ, the opposite side is *BC*. The remaining side, **AB**, **is the adjacent side**.

Problem 20. In the triangle shown in Fig. 21.22, determine angle θ

Figure 21.22

The sum of the three angles of a triangle is equal to 180°

The triangle is right-angled. Hence,

$90° + 56° + \angle\theta = 180°$

from which, $\angle\theta = 180° - 90° - 56° = \mathbf{34°}$

Problem 21. Determine the value of θ and α in Fig. 21.23

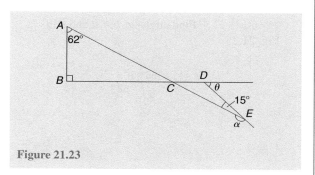

Figure 21.23

In triangle *ABC*, $\angle A + \angle B + \angle C = 180°$ (the angles in a triangle add up to 180°)
Hence, $\angle C = 180° - 90° - 62° = 28°$. Thus,
$\angle DCE = 28°$ (vertically opposite angles).
$\theta = \angle DCE + \angle DEC$ (the exterior angle of a triangle is equal to the sum of the two opposite interior angles).
Hence, $\angle\theta = 28° + 15° = \mathbf{43°}$
$\angle\alpha$ and $\angle DEC$ are supplementary; thus,
$\alpha = 180° - 15° = \mathbf{165°}$

Problem 22. *ABC* is an isosceles triangle in which the unequal angle *BAC* is 56°. *AB* is extended to *D* as shown in Fig. 21.24. Find, for the triangle, $\angle ABC$ and $\angle ACB$. Also, calculate $\angle DBC$

Figure 21.24

Since triangle *ABC* is isosceles, two sides – i.e. *AB* and *AC* – are equal and two angles – i.e. $\angle ABC$ and $\angle ACB$ – are equal.
The sum of the three angles of a triangle is equal to 180°.
Hence, $\angle ABC + \angle ACB = 180° - 56° = 124°$
Since $\angle ABC = \angle ACB$ then
$\angle ABC = \angle ACB = \dfrac{124°}{2} = \mathbf{62°}$

An angle of 180° lies on a straight line; hence,
$\angle ABC + \angle DBC = 180°$ from which,
$\angle DBC = 180° - \angle ABC = 180° - 62° = \mathbf{118°}$
Alternatively, $\angle DBC = \angle A + \angle C$ (exterior angle equals sum of two interior opposite angles),
i.e. $\angle DBC = 56° + 62° = \mathbf{118°}$

Problem 23. Find angles a, b, c, d and e in Fig. 21.25

Figure 21.25

$a = 62°$ and $c = 55°$ (alternate angles between parallel lines).
$55° + b + 62° = 180°$ (angles in a triangle add up to $180°$); hence, $b = 180° - 55° - 62° = \mathbf{63°}$
$b = d = \mathbf{63°}$ (alternate angles between parallel lines).
$e + 55° + 63° = 180°$ (angles in a triangle add up to $180°$); hence, $e = 180° - 55° - 63° = \mathbf{62°}$
Check: $e = a = 62°$ (corresponding angles between parallel lines).

Now try the following Practice Exercise

Practice Exercise 105   Triangles (answers on page 453)

1. Name the types of triangle shown in diagrams (a) to (f) in Fig. 21.26.

Figure 21.26

2. Find the angles a to f in Fig. 21.27.

Figure 21.27

3. In the triangle DEF of Fig. 21.28, which side is the hypotenuse? With reference to angle D, which side is the adjacent?

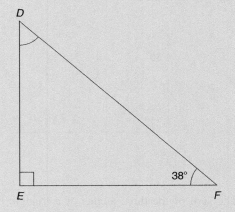

Figure 21.28

4. In triangle DEF of Fig. 21.28, determine angle D.

5. MNO is an isosceles triangle in which the unequal angle is $65°$ as shown in Fig. 21.29. Calculate angle θ.

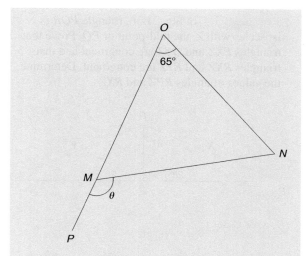

Figure 21.29

6.   Determine $\angle\phi$ and $\angle x$ in Fig. 21.30.

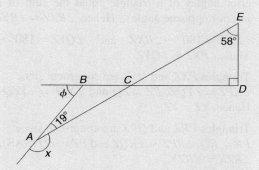

Figure 21.30

7.   In Fig. 21.31(a) and (b), find angles w, x, y and z. What is the name given to the types of triangle shown in (a) and (b)?

Figure 21.31

8.   Find the values of angles a to g in Fig. 21.32(a) and (b).

Figure 21.32

9.   Find the unknown angles a to k in Fig. 21.33.

Figure 21.33

10.   Triangle ABC has a right angle at B and $\angle BAC$ is $34°$. BC is produced to D. If the bisectors of $\angle ABC$ and $\angle ACD$ meet at E, determine $\angle BEC$.

11.   If in Fig. 21.34 triangle BCD is equilateral, find the interior angles of triangle ABE.

Figure 21.34

21.4   Congruent triangles

Two triangles are said to be **congruent** if they are equal in all respects; i.e. three angles and three sides in one triangle are equal to three angles and three sides in the other triangle. Two triangles are congruent if

(a)   the three sides of one are equal to the three sides of the other (SSS),

(b)   two sides of one are equal to two sides of the other and the angles included by these sides are equal (SAS),

(c)   two angles of the one are equal to two angles of the other and any side of the first is equal to the corresponding side of the other (ASA), or

(d)   their hypotenuses are equal and one other side of one is equal to the corresponding side of the other (RHS).

Problem 24.   State which of the pairs of triangles shown in Fig. 21.35 are congruent and name their sequence

Figure 21.35

(a)   Congruent *ABC*, *FDE* (angle, side, angle; i.e. ASA).

(b)   Congruent *GIH*, *JLK* (side, angle, side; i.e. SAS).

(c)   Congruent *MNO*, *RQP* (right angle, hypotenuse, side; i.e. RHS).

(d)   Not necessarily congruent. It is not indicated that any side coincides.

(e)   Congruent *ABC*, *FED* (side, side, side; i.e. SSS).

Problem 25.   In Fig. 21.36, triangle *PQR* is isosceles with *Z*, the mid-point of *PQ*. Prove that triangles *PXZ* and *QYZ* are congruent and that triangles *RXZ* and *RYZ* are congruent. Determine the values of angles *RPZ* and *RXZ*

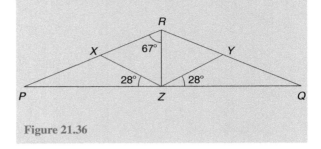

Figure 21.36

Since triangle *PQR* is isosceles, $PR = RQ$ and thus $\angle QPR = \angle RQP$.

$\angle RXZ = \angle QPR + 28°$ and $\angle RYZ = \angle RQP + 28°$ (exterior angles of a triangle equal the sum of the two interior opposite angles). Hence, $\angle RXZ = \angle RYZ$.

$\angle PXZ = 180° - \angle RXZ$ and $\angle QYZ = 180° - \angle RYZ$. Thus, $\angle PXZ = \angle QYZ$.

Triangles *PXZ* and *QYZ* are congruent since $\angle XPZ = \angle YQZ$, $PZ = ZQ$ and $\angle XZP = \angle YZQ$ (ASA). Hence, $XZ = YZ$.

Triangles *PRZ* and *QRZ* are congruent since $PR = RQ$, $\angle RPZ = \angle RQZ$ and $PZ = ZQ$ (SAS). Hence, $\angle RZX = \angle RZY$.

Triangles *RXZ* and *RYZ* are congruent since $\angle RXZ = \angle RYZ$, $XZ = YZ$ and $\angle RZX = \angle RZY$ (ASA).

$\angle QRZ = 67°$ and thus $\angle PRQ = 67° + 67° = 134°$

Hence, $\angle \mathbf{RPZ} = \angle RQZ = \dfrac{180° - 134°}{2} = \mathbf{23°}$

$\angle \mathbf{RXZ} = 23° + 28° = \mathbf{51°}$ (external angle of a triangle equals the sum of the two interior opposite angles).

Now try the following Practice Exercise

Practice Exercise 106   Congruent triangles (answers on page 453)

1.   State which of the pairs of triangles in Fig. 21.37 are congruent and name their sequences.

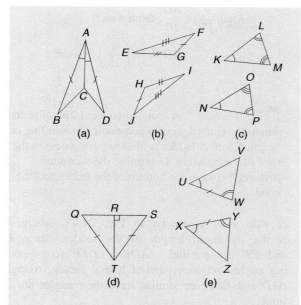

Figure 21.37

2. In a triangle ABC, $AB = BC$ and D and E are points on AB and BC, respectively, such that $AD = CE$. Show that triangles AEB and CDB are congruent.

21.5 Similar triangles

Two triangles are said to be **similar** if the angles of one triangle are equal to the angles of the other triangle. With reference to Fig. 21.38, triangles ABC and PQR are similar and the corresponding sides are in proportion to each other,

i.e.
$$\frac{p}{a} = \frac{q}{b} = \frac{r}{c}$$

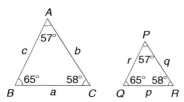

Figure 21.38

Problem 26.   In Fig. 21.39, find the length of side a

Figure 21.39

In triangle ABC, $50° + 70° + \angle C = 180°$, from which $\angle C = 60°$

In triangle DEF, $\angle E = 180° - 50° - 60° = 70°$ Hence, triangles ABC and DEF are similar, since their angles are the same. Since corresponding sides are in proportion to each other,

$$\frac{a}{d} = \frac{c}{f} \quad \text{i.e.} \quad \frac{a}{4.42} = \frac{12.0}{5.0}$$

Hence, **side, $a = \dfrac{12.0}{5.0}(4.42) = $ 10.61 cm**.

Problem 27.   In Fig. 21.40, find the dimensions marked r and p

Figure 21.40

In triangle PQR, $\angle Q = 180° - 90° - 35° = 55°$
In triangle XYZ, $\angle X = 180° - 90° - 55° = 35°$
Hence, triangles PQR and ZYX are similar since their angles are the same. The triangles may he redrawn as shown in Fig. 21.41.

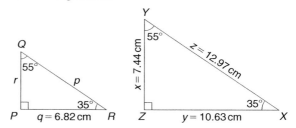

Figure 21.41

By proportion: $\dfrac{p}{z} = \dfrac{r}{x} = \dfrac{q}{y}$

i.e. $\dfrac{p}{12.97} = \dfrac{r}{7.44} = \dfrac{6.82}{10.63}$

from which, $r = 7.44 \left(\dfrac{6.82}{10.63} \right) = \textbf{4.77\,cm}$

By proportion: $\dfrac{p}{z} = \dfrac{q}{y}$ i.e. $\dfrac{p}{12.97} = \dfrac{6.82}{10.63}$

Hence, $p = 12.97 \left(\dfrac{6.82}{10.63} \right) = \textbf{8.32\,cm}$

Problem 28. In Fig. 21.42, show that triangles *CBD* and *CAE* are similar and hence find the length of *CD* and *BD*. The dimensions are in centimetres.

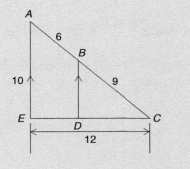

Figure 21.42

Since *BD* is parallel to *AE* then $\angle CBD = \angle CAE$ and $\angle CDB = \angle CEA$ (corresponding angles between parallel lines). Also, $\angle C$ is common to triangles *CBD* and *CAE*.

Since the angles in triangle *CBD* are the same as in triangle *CAE*, the triangles are similar. Hence,

by proportion: $\dfrac{CB}{CA} = \dfrac{CD}{CE} \left(= \dfrac{BD}{AE} \right)$

i.e. $\dfrac{9}{6+9} = \dfrac{CD}{12}$, from which

$$CD = 12 \left(\dfrac{9}{15} \right) = \textbf{7.2\,cm}$$

Also, $\dfrac{9}{15} = \dfrac{BD}{10}$, from which

$$BD = 10 \left(\dfrac{9}{15} \right) = \textbf{6\,cm}$$

Problem 29. A rectangular shed 2 m wide and 3 m high stands against a perpendicular building of height 5.5 m. A ladder is used to gain access to the roof of the building. Determine the minimum distance between the bottom of the ladder and the shed

A side view is shown in Fig. 21.43, where *AF* is the minimum length of the ladder. Since *BD* and *CF* are parallel, $\angle ADB = \angle DFE$ (corresponding angles between parallel lines). Hence, triangles *BAD* and *EDF* are similar since their angles are the same.

$$AB = AC - BC = AC - DE = 5.5 - 3 = 2.5\,\text{m}$$

By proportion: $\dfrac{AB}{DE} = \dfrac{BD}{EF}$ i.e. $\dfrac{2.5}{3} = \dfrac{2}{EF}$

Hence, $EF = 2 \left(\dfrac{3}{2.5} \right) = \textbf{2.4\,m = minimum distance}$ **from bottom of ladder to the shed**.

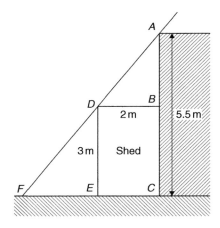

Figure 21.43

Now try the following Practice Exercise

Practice Exercise 107   Similar triangles
(answers on page 453)

1. In Fig. 21.44, find the lengths x and y.

Figure 21.44

2. PQR is an equilateral triangle of side 4 cm. When PQ and PR are produced to S and T, respectively, ST is found to be parallel with QR. If PS is 9 cm, find the length of ST. X is a point on ST between S and T such that the line PX is the bisector of $\angle SPT$. Find the length of PX.

3. In Fig. 21.45, find
   (a) the length of BC when $AB = 6$ cm, $DE = 8$ cm and $DC = 3$ cm,
   (b) the length of DE when $EC = 2$ cm, $AC = 5$ cm and $AB = 10$ cm.

Figure 21.45

4. In Fig. 21.46, $AF = 8$ m, $AB = 5$ m and $BC = 3$ m. Find the length of BD.

Figure 21.46

21.6 Construction of triangles

To construct any triangle, the following drawing instruments are needed:

(a) ruler and/or straight edge

(b) compass

(c) protractor

(d) pencil.

Here are some worked problems to demonstrate triangle construction.

Problem 30.   Construct a triangle whose sides are 6 cm, 5 cm and 3 cm

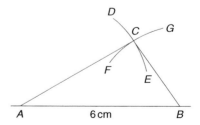

Figure 21.47

With reference to Fig. 21.47:

(i)   Draw a straight line of any length and, with a pair of compasses, mark out 6cm length and label it *AB*.

(ii)   Set compass to 5cm and with centre at *A* describe arc *DE*.

(iii)   Set compass to 3cm and with centre at *B* describe arc *FG*.

(iv)   The intersection of the two curves at *C* is the vertex of the required triangle. Join *AC* and *BC* by straight lines.

It may be proved by measurement that the ratio of the angles of a triangle is not equal to the ratio of the sides (i.e. in this problem, the angle opposite the 3cm side is not equal to half the angle opposite the 6cm side).

Problem 31.   Construct a triangle *ABC* such that $a = 6$cm, $b = 3$cm and $\angle C = 60°$

With reference to Fig. 21.48:

Figure 21.48

(i)   Draw a line *BC*, 6cm long.

(ii)   Using a protractor centred at *C*, make an angle of 60° to *BC*.

(iii)   From *C* measure a length of 3cm and label *A*.

(iv)   Join *B* to *A* by a straight line.

Problem 32.   Construct a triangle *PQR* given that $QR = 5$cm, $\angle Q = 70°$ and $\angle R = 44°$

With reference to Fig. 21.49:

(i)   Draw a straight line 5cm long and label it *QR*.

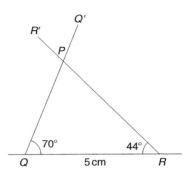

Figure 21.49

(ii)   Use a protractor centred at *Q* and make an angle of 70°. Draw *QQ'*.

(iii)   Use a protractor centred at *R* and make an angle of 44°. Draw *RR'*.

(iv)   The intersection of *QQ'* and *RR'* forms the vertex *P* of the triangle.

Problem 33.   Construct a triangle *XYZ* given that $XY = 5$cm, the hypotenuse $YZ = 6.5$cm and $\angle X = 90°$

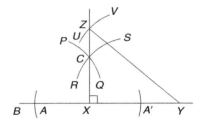

Figure 21.50

With reference to Fig. 21.50:

(i)   Draw a straight line 5cm long and label it *XY*.

(ii)   Produce *XY* any distance to *B*. With compass centred at *X* make an arc at *A* and *A'*. (The length *XA* and *XA'* is arbitrary.) With compass centred at *A* draw the arc *PQ*. With the same compass setting and centred at *A'*, draw the arc *RS*. Join the intersection of the arcs, *C* to *X*, and a right angle to *XY* is produced at *X*. (Alternatively, a protractor can be used to construct a 90° angle.)

(iii)   The hypotenuse is always opposite the right angle. Thus, *YZ* is opposite $\angle X$. Using a compass centred at *Y* and set to 6.5cm, describe the arc *UV*.

(iv) The intersection of the arc UV with XC produced, forms the vertex Z of the required triangle. Join YZ with a straight line.

Now try the following Practice Exercise

Practice Exercise 108   Construction of triangles (answers on page 453)

In the following, construct the triangles ABC for the given sides/angles.

1. $a = 8\,\text{cm}, b = 6\,\text{cm}$ and $c = 5\,\text{cm}$.

2. $a = 40\,\text{mm}, b = 60\,\text{mm}$ and $C = 60°$.

3. $a = 6\,\text{cm}, C = 45°$ and $B = 75°$.

4. $c = 4\,\text{cm}, A = 130°$ and $C = 15°$.

5. $a = 90\,\text{mm}, B = 90°, \text{hypotenuse} = 105\,\text{mm}$.

Practice Exercise 109   Multiple-choice questions on angles and triangles (answers on page 453)

Each question has only one correct answer

1. $\dfrac{3\pi}{4}$ radians is equivalent to:
    (a) 135°   (b) 270°   (c) 45°   (d) 67.5°

2. An angle of 213° is called:
    (a) an acute angle
    (b) a reflex angle
    (c) an obtuse angle
    (d) a complementary angle

3. An angle of 154° is called:
    (a) an acute angle
    (b) an obtuse angle
    (c) a reflex angle
    (d) a complementary angle

4. An angle of 90° is called:
    (a) an acute angle   (b) an obtuse angle
    (c) a reflex angle   (d) a right angle

5. An angle of 77° is called:
    (a) an acute angle   (b) an obtuse angle
    (c) a reflex angle   (d) a right angle

6. 60° is equivalent to:
    (a) $\dfrac{\pi}{4}$ rad   (b) $\dfrac{\pi}{2}$ rad
    (c) $\dfrac{\pi}{3}$ rad   (d) $\dfrac{\pi}{6}$ rad

7. A triangle which has two angles the same is called:
    (a) a right-angled triangle
    (b) an equilateral triangle
    (c) an isosceles triangle
    (d) an obtuse-angled triangle

8. The angle that is supplementary to 70° is:
    (a) 20°   (b) 110°   (c) 290°   (d) 200°

9. The angle that is complementary to 20° is:
    (a) 70°   (b) 160°   (c) 250°   (d) 340°

10. 2.13 radians is equivalent to:
    (a) 61.02°   (b) 84.51°
    (c) 383.40°   (d) 122.04°

For fully worked solutions to each of the problems in Practice Exercises 103 to 108 in this chapter, go to the website:
www.routledge.com/cw/bird

Chapter 22

Introduction to trigonometry

Why it is important to understand: Introduction to trigonometry

There are an enormous number of uses of trigonometry and trigonometric functions. Fields that use trigonometry or trigonometric functions include astronomy (especially for locating apparent positions of celestial objects, in which spherical trigonometry is essential) and hence navigation (on the oceans, in aircraft and in space), music theory, acoustics, optics, analysis of financial markets, electronics, probability theory, statistics, biology, medical imaging (CAT scans and ultrasound), pharmacy, chemistry, number theory (and hence cryptology), seismology, meteorology, oceanography, many physical sciences, land surveying and geodesy (a branch of earth sciences), architecture, phonetics, economics, electrical engineering, mechanical engineering, civil engineering, computer graphics, cartography, crystallography and game development. It is clear that a good knowledge of trigonometry is essential in most fields of engineering.

At the end of this chapter you should be able to:

- state the theorem of Pythagoras and use it to find the unknown side of a right-angled triangle
- define sine, cosine and tangent of an angle in a right-angled triangle
- evaluate trigonometric ratios of angles
- solve right-angled triangles
- understand angles of elevation and depression

22.1 Introduction

Trigonometry is a subject that involves the measurement of sides and angles of triangles and their relationship to each other.

The theorem of Pythagoras and trigonometric ratios are used with right-angled triangles only. However, there are many practical examples in engineering where knowledge of right-angled triangles is very important.

In this chapter, three trigonometric ratios – i.e. sine, cosine and tangent – are defined and then evaluated using a calculator. Finally, solving right-angled triangle problems using Pythagoras and trigonometric ratios is demonstrated, together with some practical examples involving angles of elevation and depression.

22.2 The theorem of Pythagoras

The **theorem of Pythagoras*** states:
In any right-angled triangle, the square of the hypotenuse is equal to the sum of the squares of the other two sides.

---

*Who was **Pythagoras**? See page 111. To find out more go to **www.routledge.com/cw/bird**

In the right-angled triangle *ABC* shown in Fig. 22.1, this means

$$b^2 = a^2 + c^2 \qquad (1)$$

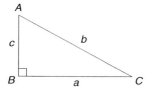

Figure 22.1

If the lengths of any two sides of a right-angled triangle are known, the length of the third side may be calculated by Pythagoras' theorem.

From equation (1): $b = \sqrt{a^2 + c^2}$

Transposing equation (1) for a gives $a^2 = b^2 - c^2$, from which $a = \sqrt{b^2 - c^2}$
Transposing equation (1) for c gives $c^2 = b^2 - a^2$, from which $c = \sqrt{b^2 - a^2}$
Here are some worked problems to demonstrate the theorem of Pythagoras.

Problem 1.   In Fig. 22.2, find the length of *BC*

Figure 22.2

From Pythagoras, $a^2 = b^2 + c^2$

i.e. $a^2 = 4^2 + 3^2$

$= 16 + 9 = 25$

Hence, $a = \sqrt{25} = 5\,\text{cm}.$

$\sqrt{25} = \pm 5$ but in a practical example like this an answer of $a = -5\,\text{cm}$ has no meaning, so we take only the positive answer.

Thus $a = BC = 5\,\text{cm}.$

ABC is a 3, 4, 5 triangle. There are not many right-angled triangles which have integer values (i.e. whole numbers) for all three sides.

Problem 2.   In Fig. 22.3, find the length of *EF*

Figure 22.3

By Pythagoras' theorem, $e^2 = d^2 + f^2$

Hence, $13^2 = d^2 + 5^2$

$169 = d^2 + 25$

$d^2 = 169 - 25 = 144$

Thus, $d = \sqrt{144} = 12\,\text{cm}$

i.e. $d = EF = 12\,\text{cm}$

DEF is a 5, 12, 13 triangle, another right-angled triangle which has integer values for all three sides.

Problem 3.   Two aircraft leave an airfield at the same time. One travels due north at an average speed of 300 km/h and the other due west at an average speed of 220 km/h. Calculate their distance apart after 4 hours

After 4 hours, the first aircraft has travelled

$4 \times 300 = 1200\,\text{km}$ due north

and the second aircraft has travelled

$4 \times 220 = 880\,\text{km}$ due west,

as shown in Fig. 22.4. The distance apart after 4 hours $= BC$.

Figure 22.4

From Pythagoras' theorem,

$$BC^2 = 1200^2 + 880^2$$
$$= 1440000 + 774400 = 2214400$$

and $\quad BC = \sqrt{2214400} = 1488 \, \text{km}.$

Hence, distance apart after 4 hours = 1488 km.

Now try the following Practice Exercise

Practice Exercise 110   Theorem of Pythagoras (answers on page 453)

1. Find the length of side x in Fig. 22.5.

Figure 22.5

2. Find the length of side x in Fig. 22.6(a).

3. Find the length of side x in Fig. 22.6(b), correct to 3 significant figures.

(a)

(b)

Figure 22.6

4. In a triangle ABC, $AB = 17$ cm, $BC = 12$ cm and $\angle ABC = 90°$. Determine the length of AC, correct to 2 decimal places.

5. A tent peg is 4.0 m away from a 6.0 m high tent. What length of rope, correct to the nearest centimetre, runs from the top of the tent to the peg?

6. In a triangle ABC, $\angle B$ is a right angle, $AB = 6.92$ cm and $BC = 8.78$ cm. Find the length of the hypotenuse.

7. In a triangle CDE, $D = 90°$, $CD = 14.83$ mm and $CE = 28.31$ mm. Determine the length of DE.

8. Show that if a triangle has sides of 8, 15 and 17 cm it is right-angled.

9. Triangle PQR is isosceles, Q being a right angle. If the hypotenuse is 38.46 cm find (a) the lengths of sides PQ and QR and (b) the value of $\angle QPR$.

10. A man cycles 24 km due south and then 20 km due east. Another man, starting at the same time and position as the first man, cycles 32 km due east and then 7 km due south. Find the distance between the two men.

11. A ladder 3.5 m long is placed against a perpendicular wall with its foot 1.0 m from the wall. How far up the wall (to the nearest centimetre) does the ladder reach? If the foot of the ladder is now moved 30 cm further away from the wall, how far does the top of the ladder fall?

12. Two ships leave a port at the same time. One travels due west at 18.4 knots and the other due south at 27.6 knots. If 1 knot = 1 nautical mile per hour, calculate how far apart the two ships are after 4 hours.

13. Fig. 22.7 shows a bolt rounded off at one end. Determine the dimension h.

Figure 22.7

14. Fig. 22.8 shows a cross-section of a component that is to be made from a round bar. If the diameter of the bar is 74 mm, calculate the dimension x.

Figure 22.8

22.3 Sines, cosines and tangents

With reference to angle θ in the right-angled triangle ABC shown in Fig. 22.9,

$$\textbf{sine}\,\theta = \frac{\textbf{opposite side}}{\textbf{hypotenuse}}$$

'Sine' is abbreviated to 'sin', thus $\sin\theta = \dfrac{BC}{AC}$

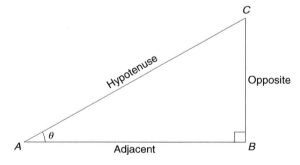

Figure 22.9

Also, $\textbf{cosine}\,\theta = \dfrac{\textbf{adjacent side}}{\textbf{hypotenuse}}$

'Cosine' is abbreviated to 'cos', thus $\cos\theta = \dfrac{AB}{AC}$

Finally, $\textbf{tangent}\,\theta = \dfrac{\textbf{opposite side}}{\textbf{adjacent side}}$

'Tangent' is abbreviated to 'tan', thus $\tan\theta = \dfrac{BC}{AB}$

These three trigonometric ratios only apply to right-angled triangles. Remembering these three equations is very important and the mnemonic '**SOH CAH TOA**' is one way of remembering them.

SOH indicates <u>s</u>in $=$ <u>o</u>pposite $\div$ <u>h</u>ypotenuse
CAH indicates <u>c</u>os $=$ <u>a</u>djacent $\div$ <u>h</u>ypotenuse
TOA indicates <u>t</u>an $=$ <u>o</u>pposite $\div$ <u>a</u>djacent

Here are some worked problems to help familiarise ourselves with trigonometric ratios.

Problem 4. In triangle PQR shown in Fig. 22.10, determine $\sin\theta$, $\cos\theta$ and $\tan\theta$

Figure 22.10

$$\sin\theta = \frac{\text{opposite side}}{\text{hypotenuse}} = \frac{PQ}{PR} = \frac{5}{13} = \textbf{0.3846}$$

$$\cos\theta = \frac{\text{adjacent side}}{\text{hypotenuse}} = \frac{QR}{PR} = \frac{12}{13} = \textbf{0.9231}$$

$$\tan\theta = \frac{\text{opposite side}}{\text{adjacent side}} = \frac{PQ}{QR} = \frac{5}{12} = \textbf{0.4167}$$

Problem 5. In triangle ABC of Fig. 22.11, determine length $AC, \sin C, \cos C, \tan C, \sin A, \cos A$ and $\tan A$

Figure 22.11

By Pythagoras, $\quad AC^2 = AB^2 + BC^2$

i.e. $\quad AC^2 = 3.47^2 + 4.62^2$

from which $\quad AC = \sqrt{3.47^2 + 4.62^2} = \textbf{5.778 cm}$

$$\sin C = \frac{\text{opposite side}}{\text{hypotenuse}} = \frac{AB}{AC} = \frac{3.47}{5.778} = \mathbf{0.6006}$$

$$\cos C = \frac{\text{adjacent side}}{\text{hypotenuse}} = \frac{BC}{AC} = \frac{4.62}{5.778} = \mathbf{0.7996}$$

$$\tan C = \frac{\text{opposite side}}{\text{adjacent side}} = \frac{AB}{BC} = \frac{3.47}{4.62} = \mathbf{0.7511}$$

$$\sin A = \frac{\text{opposite side}}{\text{hypotenuse}} = \frac{BC}{AC} = \frac{4.62}{5.778} = \mathbf{0.7996}$$

$$\cos A = \frac{\text{adjacent side}}{\text{hypotenuse}} = \frac{AB}{AC} = \frac{3.47}{5.778} = \mathbf{0.6006}$$

$$\tan A = \frac{\text{opposite side}}{\text{adjacent side}} = \frac{BC}{AB} = \frac{4.62}{3.47} = \mathbf{1.3314}$$

Problem 6.   If $\tan B = \dfrac{8}{15}$, determine the value of $\sin B, \cos B, \sin A$ and $\tan A$

A right-angled triangle ABC is shown in Fig. 22.12. If $\tan B = \dfrac{8}{15}$, then $AC = 8$ and $BC = 15$

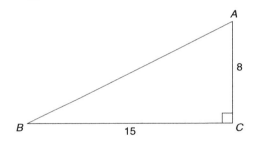

Figure 22.12

By Pythagoras, $AB^2 = AC^2 + BC^2$

i.e. $\qquad\qquad AB^2 = 8^2 + 15^2$

from which $\qquad AB = \sqrt{8^2 + 15^2} = 17$

$$\sin B = \frac{AC}{AB} = \frac{8}{17} \text{ or } \mathbf{0.4706}$$

$$\cos B = \frac{BC}{AB} = \frac{15}{17} \text{ or } \mathbf{0.8824}$$

$$\sin A = \frac{BC}{AB} = \frac{15}{17} \text{ or } \mathbf{0.8824}$$

$$\tan A = \frac{BC}{AC} = \frac{15}{8} \text{ or } \mathbf{1.8750}$$

Problem 7.   Point A lies at co-ordinate (2, 3) and point B at (8, 7). Determine (a) the distance AB and (b) the gradient of the straight line AB

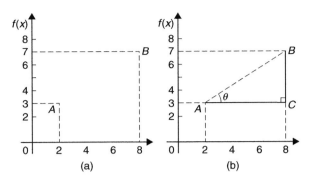

Figure 22.13

(a) Points A and B are shown in Fig. 22.13(a).

In Fig. 22.13(b), the horizontal and vertical lines AC and BC are constructed. Since ABC is a right-angled triangle, and $AC = (8 - 2) = 6$ and $BC = (7 - 3) = 4$, by Pythagoras' theorem,

$$AB^2 = AC^2 + BC^2 = 6^2 + 4^2$$

and $\quad \boldsymbol{AB} = \sqrt{6^2 + 4^2} = \sqrt{52}$

$$= \mathbf{7.211} \text{ correct to 3 decimal places.}$$

(b) The gradient of AB is given by $\tan\theta$, i.e.

$$\textbf{gradient} = \tan\theta = \frac{BC}{AC} = \frac{4}{6} = \frac{\mathbf{2}}{\mathbf{3}}$$

Now try the following Practice Exercise

Practice Exercise 111   Trigonometric ratios (answers on page 453)

1. Sketch a triangle XYZ such that $\angle Y = 90°, XY = 9$ cm and $YZ = 40$ cm. Determine $\sin Z, \cos Z, \tan X$ and $\cos X$.

2. In triangle ABC shown in Fig. 22.14, find $\sin A, \cos A, \tan A, \sin B, \cos B$ and $\tan B$.

Figure 22.14

3. If $\cos A = \dfrac{15}{17}$, find $\sin A$ and $\tan A$, in fraction form.

4. If $\tan X = \dfrac{15}{112}$, find $\sin X$ and $\cos X$, in fraction form.

5. For the right-angled triangle shown in Fig. 22.15, find (a) $\sin \alpha$ (b) $\cos \theta$ (c) $\tan \theta$.

Figure 22.15

6. If $\tan \theta = \dfrac{7}{24}$, find (a) $\sin \theta$ and (b) $\cos \theta$ in fraction form.

7. Point P lies at co-ordinate $(-3, 1)$ and point Q at $(5, -4)$. Determine
   (a) the distance PQ
   (b) the gradient of the straight line PQ

22.4   Evaluating trigonometric ratios of acute angles

The easiest way to evaluate trigonometric ratios of any angle is to use a calculator. Use a calculator to check the following (each correct to 4 decimal places).

$$\sin 29° = 0.4848 \quad \sin 53.62° = 0.8051$$

$$\cos 67° = 0.3907 \quad \cos 83.57° = 0.1120$$

$$\tan 34° = 0.6745 \quad \tan 67.83° = 2.4541$$

$$\sin 67°43' = \sin 67\dfrac{43°}{60} = \sin 67.7166666\ldots° = 0.9253$$

$$\cos 13°28' = \cos 13\dfrac{28°}{60} = \cos 13.466666\ldots° = 0.9725$$

$$\tan 56°54' = \tan 56\dfrac{54°}{60} = \tan 56.90° = 1.5340$$

If we know the value of a trigonometric ratio and need to find the angle we use the **inverse function** on our calculators. For example, using shift and sin on our calculator gives $\sin^{-1}($

If, for example, we know the sine of an angle is 0.5 then the value of the angle is given by

$$\sin^{-1} 0.5 = 30° \text{ (Check that } \sin 30° = 0.5)$$

Similarly, if

$$\cos \theta = 0.4371 \text{ then } \theta = \cos^{-1} 0.4371 = 64.08°$$

and if

$$\tan A = 3.5984 \text{ then } A = \tan^{-1} 3.5984 = 74.47°$$

each correct to 2 decimal places.
Use your calculator to check the following worked problems.

Problem 8. Determine, correct to 4 decimal places, $\sin 43°39'$

$$\sin 43°39' = \sin 43\dfrac{39°}{60} = \sin 43.65° = 0.6903$$

This answer can be obtained using the **calculator** as follows:

1. Press sin   2. Enter 43   3. Press° '''
4. Enter 39   5. Press° '''   6. Press)
7. Press =   Answer = **0.6902512**...

Problem 9. Determine, correct to 3 decimal places, $6 \cos 62°12'$

$$6 \cos 62°12' = 6 \cos 62\dfrac{12°}{60} = 6 \cos 62.20° = 2.798$$

This answer can be obtained using the **calculator** as follows:

1. Enter 6   2. Press cos   3. Enter 62
4. Press° '''   5. Enter 12   6. Press° '''
7. Press)
8. Press =   Answer = **2.798319**...

Problem 10. Evaluate $\sin 1.481$, correct to 4 significant figures

$\sin 1.481$ means the sine of 1.481 **radians**. (If there is no degrees sign, i.e. ° , then radians are assumed). Therefore a calculator needs to be on the radian function. Hence, $\sin 1.481 = 0.9960$

Problem 11. Evaluate $\cos(3\pi/5)$, correct to 4 significant figures

As in Problem 10, $3\pi/5$ is in radians.
Hence, $\cos(3\pi/5) = \cos 1.884955\ldots = -0.3090$
Since, from page 204, π radians $= 180°$,

$$3\pi/5 \text{ rad} = \dfrac{3}{5} \times 180° = 108°$$

i.e. $3\pi/5$ rad $= 108°$. Check with your calculator that $\cos 108° = -0.3090$

> **Problem 12.** Evaluate tan 2.93, correct to 4 significant figures

Again, since there is no degrees sign, 2.93 means 2.93 radians.
Hence, **tan 2.93 = −0.2148**
It is important to know when to have your calculator on either degrees mode or radian mode. A lot of mistakes can arise from this if we are not careful.

> **Problem 13.** Find the acute angle $\sin^{-1} 0.4128$ in degrees, correct to 2 decimal places

$\sin^{-1} 0.4128$ means 'the angle whose sine is 0.4128'. Using a calculator,

1. Press shift   2. Press sin   3. Enter 0.4128
4. Press)   5. Press =

The answer 24.380848... is displayed.
Hence, $\sin^{-1} 0.4128 = 24.38°$ correct to 2 decimal places.

> **Problem 14.** Find the acute angle $\cos^{-1} 0.2437$ in degrees and minutes

$\cos^{-1} 0.2437$ means 'the angle whose cosine is 0.2437'. Using a calculator,

1. Press shift   2. Press cos   3. Enter 0.2437
4. Press)   5. Press =

The answer 75.894979... is displayed.

6. Press ° '" and 75° 53' 41.93" is displayed.

Hence, $\cos^{-1} 0.2437 = 75.89° = 77°54'$ correct to the nearest minute.

> **Problem 15.** Find the acute angle $\tan^{-1} 7.4523$ in degrees and minutes

$\tan^{-1} 7.4523$ means 'the angle whose tangent is 7.4523'. Using a calculator,

1. Press shift   2. Press tan   3. Enter 7.4523
4. Press)   5. Press =

The answer 82.357318... is displayed.

6. Press ° '" and 82° 21' 26.35" is displayed.

Hence, $\tan^{-1} 7.4523 = 82.36° = 82°21'$ correct to the nearest minute.

> **Problem 16.** In triangle *EFG* in Fig. 22.16, calculate angle *G*

Figure 22.16

With reference to ∠*G*, the two sides of the triangle given are the opposite side *EF* and the hypotenuse *EG*; hence, sine is used,

i.e. $\sin G = \dfrac{2.30}{8.71} = 0.26406429\ldots$

from which, $G = \sin^{-1} 0.26406429\ldots$

i.e. $G = 15.311360°$

Hence, $\angle G = 15.31°$ or $15°19'$

> **Problem 17.** A locomotive moves around a curve of radius, $r = 500$ m. The angle of banking, θ, is given by: $\theta = \tan^{-1}\left(\dfrac{v^2}{rg}\right)$ where $g = 9.81$ m/s² and v is the speed in m/s. Calculate the angle of banking when the speed of the locomotive is 30 km/h.

Angle of banking, $\theta = \tan^{-1}\left(\dfrac{v^2}{rg}\right)$

$$v = 30\,\text{km/h} = 30\frac{\text{km}}{\text{h}} \times \frac{1\,\text{h}}{3600\,\text{s}} \times \frac{1000\,\text{m}}{1\,\text{km}}$$

$$= \frac{30}{3.6} = 8.3333\,\text{m/s}$$

Hence,

$$\theta = \tan^{-1}\left(\frac{8.3333^2\,\text{m}^2/\text{s}^2}{500\,\text{m} \times 9.81\,\text{m/s}^2}\right) = \tan^{-1}(0.014158)$$

i.e.   **angle of banking, $\theta = 0.811°$**

Now try the following Practice Exercise

Practice Exercise 112   Evaluating trigonometric ratios (answers on page 454)

1. Determine, correct to 4 decimal places, $3 \sin 66°41'$

2. Determine, correct to 3 decimal places, $5 \cos 14°15'$

3. Determine, correct to 4 significant figures, $7 \tan 79°9'$

4. Determine

   (a) sine $\dfrac{2\pi}{3}$ (b) $\cos 1.681$ (c) $\tan 3.672$

5. Find the acute angle $\sin^{-1} 0.6734$ in degrees, correct to 2 decimal places.

6. Find the acute angle $\cos^{-1} 0.9648$ in degrees, correct to 2 decimal places.

7. Find the acute angle $\tan^{-1} 3.4385$ in degrees, correct to 2 decimal places.

8. Find the acute angle $\sin^{-1} 0.1381$ in degrees and minutes.

9. Find the acute angle $\cos^{-1} 0.8539$ in degrees and minutes.

10. Find the acute angle $\tan^{-1} 0.8971$ in degrees and minutes.

11. In the triangle shown in Fig. 22.17, determine angle θ, correct to 2 decimal places.

Figure 22.17

12. In the triangle shown in Fig. 22.18, determine angle θ in degrees and minutes.

Figure 22.18

13. Evaluate, correct to 4 decimal places,
$$\frac{4.5 \cos 67°34' - \sin 90°}{2 \tan 45°}$$

14. Evaluate, correct to 4 significant figures,
$$\frac{(3 \sin 37.83°)(2.5 \tan 57.48°)}{4.1 \cos 12.56°}$$

15. For the supported beam AB shown in Fig. 22.19, determine (a) the angle the supporting stay CD makes with the beam, i.e. θ, correct to the nearest degree, (b) the length of the stay, CD, correct to the nearest centimetre.

Figure 22.19

16. A cantilever is subjected to a vertically applied downward load at its free end, the position, β, of the maximum bending stress is given by: $\beta = \tan^{-1}\left(-\dfrac{I_{UU}\tan\theta}{I_{YY}}\right)$
Evaluate angle β, in degrees, correct to 2 decimal places, when $I_{UU} = 1.381 \times 10^{-5}$ m^4, $\theta = 17.64°$ and $I_{YY} = 1.741 \times 10^{-6}$ m^4.

22.5 Solving right-angled triangles

'Solving a right-angled triangle' means 'finding the unknown sides and angles'. This is achieved using

(a) the theorem of Pythagoras and/or

(b) trigonometric ratios.

Six pieces of information describe a triangle completely; i.e. three sides and three angles. As long as at least three facts are known, the other three can usually be calculated.

Here are some worked problems to demonstrate the solution of right-angled triangles.

> **Problem 18.** In the triangle ABC shown in Fig. 22.20, find the lengths AC and AB

Figure 22.20

There is usually more than one way to solve such a triangle.

In triangle ABC,

$$\tan 42° = \frac{AC}{BC} = \frac{AC}{6.2}$$

(Remember SOH CAH TOA)

Transposing gives
$$AC = 6.2 \tan 42° = \textbf{5.583 mm}$$

$$\cos 42° = \frac{BC}{AB} = \frac{6.2}{AB}, \text{ from which}$$

$$AB = \frac{6.2}{\cos 42°} = \textbf{8.343 mm}$$

Alternatively, by Pythagoras, $AB^2 = AC^2 + BC^2$
from which $AB = \sqrt{AC^2 + BC^2} = \sqrt{5.583^2 + 6.2^2}$
$$= \sqrt{69.609889} = \textbf{8.343 mm}.$$

> **Problem 19.** Sketch a right-angled triangle ABC such that $B = 90°$, $AB = 5$ cm and $BC = 12$ cm. Determine the length of AC and hence evaluate $\sin A, \cos C$ and $\tan A$

Triangle ABC is shown in Fig. 22.21.

Figure 22.21

By Pythagoras' theorem, $AC = \sqrt{5^2 + 12^2} = 13$

By definition: $\sin A = \dfrac{\text{opposite side}}{\text{hypotenuse}} = \dfrac{12}{13}$ or **0.9231**

(Remember SOH CAH TOA)

$$\cos C = \frac{\text{adjacent side}}{\text{hypotenuse}} = \frac{12}{13} \text{ or } \textbf{0.9231}$$

and $\tan A = \dfrac{\text{opposite side}}{\text{adjacent side}} = \dfrac{12}{5}$ or **2.400**

> **Problem 20.** In triangle PQR shown in Fig. 22.22, find the lengths of PQ and PR

Figure 22.22

$$\tan 38° = \frac{PQ}{QR} = \frac{PQ}{7.5}, \text{ hence}$$

$$PQ = 7.5 \tan 38° = 7.5(0.7813) = \textbf{5.860 cm}$$

$$\cos 38° = \frac{QR}{PR} = \frac{7.5}{PR}, \text{ hence}$$

$$PR = \frac{7.5}{\cos 38°} = \frac{7.5}{0.7880} = \textbf{9.518 cm}$$

Check: using Pythagoras' theorem,
$(7.5)^2 + (5.860)^2 = 90.59 = (9.518)^2$

Problem 21. Solve the triangle *ABC* shown in Fig. 22.23

Figure 22.23

To 'solve the triangle *ABC*' means 'to find the length *AC* and angles *B* and *C*'

$$\sin C = \frac{35}{37} = 0.94595, \text{ hence}$$

$$C = \sin^{-1} 0.94595 = \mathbf{71.08°}$$

$B = 180° - 90° - 71.08° = \mathbf{18.92°}$ (since the angles in a triangle add up to $180°$)

$$\sin B = \frac{AC}{37}, \text{ hence}$$

$$AC = 37 \sin 18.92° = 37(0.3242) = \mathbf{12.0\,mm}$$

or, using Pythagoras' theorem, $37^2 = 35^2 + AC^2$, from which $AC = \sqrt{(37^2 - 35^2)} = \mathbf{12.0\,mm}$.

Problem 22. Solve triangle *XYZ* given $\angle X = 90°$, $\angle Y = 23°17'$ and $YZ = 20.0\,mm$

It is always advisable to make a reasonably accurate sketch so as to visualise the expected magnitudes of unknown sides and angles. Such a sketch is shown in Fig. 22.24.

$$\angle Z = 180° - 90° - 23°17' = \mathbf{66°43'}$$

Figure 22.24

$$\sin 23°17' = \frac{XZ}{20.0}, \text{ hence } XZ = 20.0 \sin 23°17'$$

$$= 20.0(0.3953) = \mathbf{7.906\,mm}$$

$$\cos 23°17' = \frac{XY}{20.0}, \text{ hence } XY = 20.0 \cos 23°17'$$

$$= 20.0(0.9186) = \mathbf{18.37\,mm}$$

Check: using Pythagoras' theorem,
$(18.37)^2 + (7.906)^2 = 400.0 = (20.0)^2$

Now try the following Practice Exercise

Practice Exercise 113   Solving right-angled triangles (answers on page 454)

1. Calculate the dimensions shown as x in Figs 22.25(a) to (f), each correct to 4 significant figures.

Figure 22.25

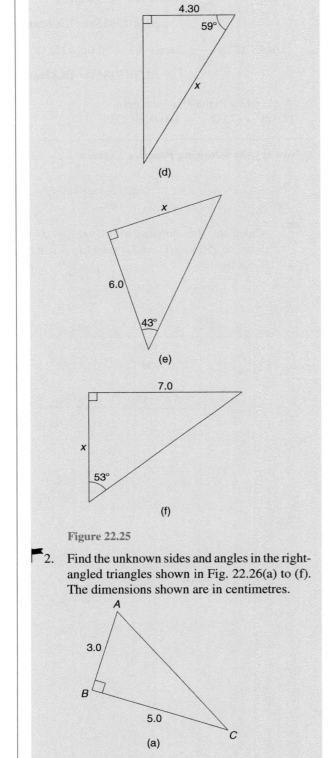

Figure 22.25

2. Find the unknown sides and angles in the right-angled triangles shown in Fig. 22.26(a) to (f). The dimensions shown are in centimetres.

Figure 22.26

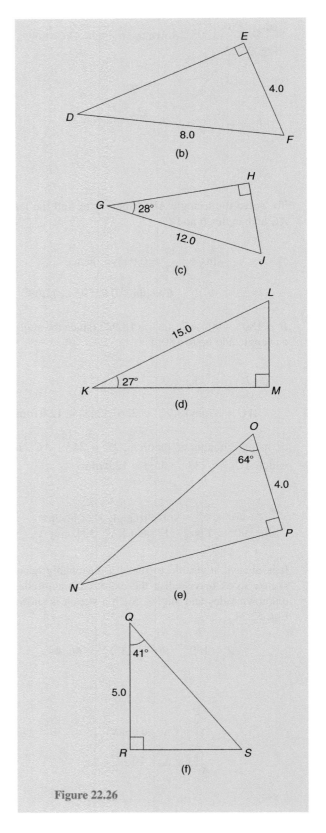

Figure 22.26

3. A ladder rests against the top of the perpendicular wall of a building and makes an angle of 73° with the ground. If the foot of the ladder is 2 m from the wall, calculate the height of the building.

4. Determine the length x in Fig. 22.27.

Figure 22.27

5. A symmetrical part of a bridge lattice is shown in Fig. 22.28. If $AB = 6$ m, angle $BAD = 56°$ and E is the mid-point of $ABCD$, determine the height h, correct to the nearest centimetre.

Figure 22.28

22.6 Angles of elevation and depression

If, in Fig. 22.29, BC represents horizontal ground and AB a vertical flagpole, the **angle of elevation** of the top of the flagpole, A, from the point C is the angle that the imaginary straight line AC must be raised (or elevated) from the horizontal CB; i.e. angle θ. In practice, **surveyors** measure this angle using an instrument called a **theodolite**.

Figure 22.29

Figure 22.30

If, in Fig. 22.30, PQ represents a vertical cliff and R a ship at sea, the **angle of depression** of the ship from point P is the angle through which the imaginary straight line PR must be lowered (or depressed) from the horizontal to the ship; i.e. angle ϕ. (Note, $\angle PRQ$ is also ϕ − **alternate angles** between parallel lines from page 214).

> **Problem 23.** An electricity pylon stands on horizontal ground. At a point 80 m from the base of the pylon, the angle of elevation of the top of the pylon is 23°. Calculate the height of the pylon to the nearest metre

Fig. 22.31 shows the pylon AB and the angle of elevation of A from point C is 23°

Figure 22.31

$$\tan 23° = \frac{AB}{BC} = \frac{AB}{80}$$

Hence, height of pylon $AB = 80 \tan 23°$
$$= 80(0.4245) = 33.96 \text{ m}$$
$$= \textbf{34 m to the nearest metre}.$$

> **Problem 24.** A surveyor measures the angle of elevation of the top of a perpendicular building as 19°. He moves 120 m nearer to the building and finds the angle of elevation is now 47°. Determine the height of the building

The building PQ and the angles of elevation are shown in Fig. 22.32.

Figure 22.32

In triangle PQS, $\tan 19° = \dfrac{h}{x + 120}$

Hence, $h = \tan 19°(x + 120)$

i.e. $h = 0.3443(x + 120)$ (1)

In triangle PQR, $\tan 47° = \dfrac{h}{x}$

Hence, $h = \tan 47°(x)$ i.e. $h = 1.0724x$ (2)

Equating equations (1) and (2) gives

$$0.3443(x + 120) = 1.0724x$$
$$0.3443x + (0.3443)(120) = 1.0724x$$
$$(0.3443)(120) = (1.0724 - 0.3443)x$$
$$41.316 = 0.7281x$$
$$x = \frac{41.316}{0.7281} = 56.74\,\text{m}$$

From equation (2), **height of building**, $h = 1.0724x = 1.0724(56.74) = \mathbf{60.85\,m}$.

⚑ Problem 25. The angle of depression of a ship viewed at a particular instant from the top of a 75 m vertical cliff is 30°. Find the distance of the ship from the base of the cliff at this instant. The ship is sailing away from the cliff at constant speed and 1 minute later its angle of depression from the top of the cliff is 20°. Determine the speed of the ship in km/h

Fig. 22.33 shows the cliff AB, the initial position of the ship at C and the final position at D. Since the angle of depression is initially 30°, $\angle ACB = 30°$ (alternate angles between parallel lines).

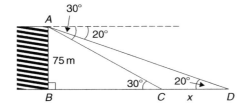

Figure 22.33

$\tan 30° = \dfrac{AB}{BC} = \dfrac{75}{BC}$ hence,

$$BC = \frac{75}{\tan 30°}$$
$$= \mathbf{129.9\,m} = \textbf{initial position}$$
$$\textbf{of ship from base of cliff}$$

In triangle ABD,

$$\tan 20° = \frac{AB}{BD} = \frac{75}{BC + CD} = \frac{75}{129.9 + x}$$

Hence, $129.9 + x = \dfrac{75}{\tan 20°} = 206.06\,\text{m}$

from which $x = 206.06 - 129.9 = 76.16\,\text{m}$

Thus, the ship sails 76.16 m in 1 minute; i.e. 60 s,

Hence, **speed of ship** $= \dfrac{\text{distance}}{\text{time}} = \dfrac{76.16}{60}\,\text{m/s}$

$$= \frac{76.16 \times 60 \times 60}{60 \times 1000}\,\text{km/h} = \mathbf{4.57\,km/h}.$$

Now try the following Practice Exercise

Practice Exercise 114   Angles of elevation and depression (answers on page 454)

⚑1. A vertical tower stands on level ground. At a point 105 m from the foot of the tower the angle of elevation of the top is 19°. Find the height of the tower.

⚑2. If the angle of elevation of the top of a vertical 30 m high aerial is 32°, how far is it to the aerial?

⚑3. From the top of a vertical cliff 90.0 m high the angle of depression of a boat is 19°50′. Determine the distance of the boat from the cliff.

⚑4. From the top of a vertical cliff 80.0 m high the angles of depression of two buoys lying due west of the cliff are 23° and 15°, respectively. How far apart are the buoys?

⚑5. From a point on horizontal ground a surveyor measures the angle of elevation of the top of a flagpole as 18°40′. He moves 50 m nearer to the flagpole and measures the angle of elevation as 26°22′. Determine the height of the flagpole.

⚑6. A flagpole stands on the edge of the top of a building. At a point 200 m from the building the angles of elevation of the top and bottom of the pole are 32° and 30° respectively. Calculate the height of the flagpole.

⚑7. From a ship at sea, the angles of elevation of the top and bottom of a vertical lighthouse standing on the edge of a vertical cliff are 31° and 26°, respectively. If the lighthouse is 25.0 m high, calculate the height of the cliff.

⚑8. From a window 4.2 m above horizontal ground the angle of depression of the foot of a building across the road is 24° and the angle of elevation of the top of the building is 34°. Determine, correct to the nearest centimetre, the width of the road and the height of the building.

9. The elevation of a tower from two points, one due west of the tower and the other due east of it are 20° and 24°, respectively, and the two points of observation are 300 m apart. Find the height of the tower to the nearest metre.

Practice Exercise 115   Multiple-choice questions on an introduction to trigonometry (answers on page 454)

Each question has only one correct answer

1. In the right-angled triangle ABC shown in Figure 22.34, sine A is given by:
   (a) b/a   (b) c/b   (c) b/c   (d) a/b

Figure 22.34

2. In the right-angled triangle ABC shown in Figure 22.34, cosine C is given by:
   (a) a/b   (b) c/b   (c) a/c   (d) b/a

3. In the right-angled triangle shown in Figure 22.34, tangent A is given by:
   (a) b/c   (b) a/c   (c) a/b   (d) c/a

4. In the triangular template ABC shown in Figure 22.35, the length AC is:
   (a) 6.17 cm   (b) 11.17 cm
   (c) 9.22 cm   (d) 12.40 cm

Figure 22.35

5. Correct to 3 decimal places, sin(−2.6 rad) is:
   (a) 0.516   (b) −0.045
   (c) −0.516   (d) 0.045

6. For the right-angled triangle PQR shown in Figure 22.36, angle R is equal to:
   (a) 41.41°   (b) 48.59°
   (c) 36.87°   (d) 53.13°

Figure 22.36

7. If $\cos A = \dfrac{12}{13}$, then sin A is equal to:
   (a) $\dfrac{5}{13}$ (b) $\dfrac{13}{12}$ (c) $\dfrac{5}{12}$ (d) $\dfrac{12}{5}$

8. The area of triangle XYZ in Figure 22.37 is:
   (a) 24.22 cm^2 (b) 19.35 cm^2
   (c) 38.72 cm^2 (d) 32.16 cm^2

Figure 22.37

9. The length of side XZ in Figure 22.37 is:
   (a) 4.313 cm   (b) 7.166 cm
   (c) 8.973 cm   (d) 6.762 cm

10. The value, correct to 3 decimal places, of $\cos\left(\dfrac{-3\pi}{4}\right)$ is:
    (a) 0.999   (b) 0.707
    (c) −0.999   (d) −0.707

11. Tan 60° is equivalent to:

   (a) $\dfrac{1}{\sqrt{3}}$ (b) $\dfrac{\sqrt{3}}{2}$ (c) $\dfrac{1}{2}$ (d) $\sqrt{3}$

12. A vertical tower stands on level ground. At a point 100 m from the foot of the tower the angle of elevation of the top is 20°. The height of the tower is:

   (a) 274.7 m   (b) 36.4 m
   (c) 34.3 m   (d) 94.0 m

13. Cos 30° is equivalent to:

   (a) $\dfrac{1}{2}$ (b) $\dfrac{2}{\sqrt{3}}$ (c) $\dfrac{\sqrt{3}}{2}$ (d) $\dfrac{1}{\sqrt{3}}$

14. A 10 m long ladder is placed 2.5 m from the perpendicular wall of a building. How high will the top of the ladder reach on the building?

   (a) 9.68 m   (b) 7.5 m
   (c) 10.31 m   (d) 2.74 m

15. A surveyor, standing 120 m from the base of a tower, looks up 60° to see the top of the structure. The height of the tower, correct to the nearest metre is:

   (a) 60 m   (b) 69 m   (c) 104 m   (d) 208 m

For fully worked solutions to each of the problems in Practice Exercises 110 to 114 in this chapter, go to the website:

www.routledge.com/cw/bird

Revision Test 8: Angles, triangles and trigonometry

This assignment covers the material contained in Chapters 21 and 22. *The marks available are shown in brackets at the end of each question.*

1. Determine $38°48' + 17°23'$ (2)

2. Determine $47°43'12'' - 58°35'53'' + 26°17'29''$ (3)

3. Change $42.683°$ to degrees and minutes. (2)

4. Convert $77°42'34''$ to degrees correct to 3 decimal places. (3)

5. Determine angle θ in Fig. RT8.1. (3)

Figure RT8.1

6. Determine angle θ in the triangle in Fig. RT8.2. (2)

Figure RT8.2

7. Determine angle θ in the triangle in Fig. RT8.3. (2)

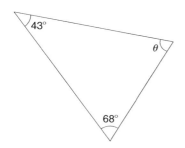

Figure RT8.3

8. In Fig. RT8.4, if triangle ABC is equilateral, determine $\angle CDE$. (3)

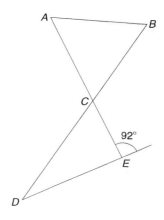

Figure RT8.4

9. Find angle J in Fig. RT8.5. (2)

Figure RT8.5

10. Determine angle θ in Fig. RT8.6. (3)

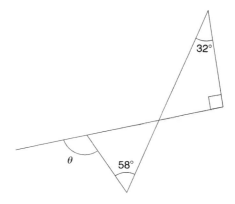

Figure RT8.6

11. State the angle (a) supplementary to $49°$ (b) complementary to $49°$ (2)

12. In Fig. RT8.7, determine angles x, y and z. (3)

Figure RT8.7

13. In Fig. RT8.8, determine angles a to e. (5)

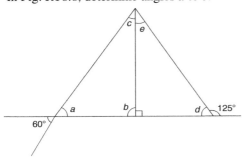

Figure RT8.8

14. In Fig. RT8.9, determine the length of AC. (4)

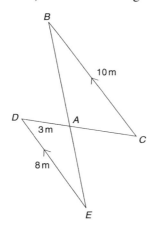

Figure RT8.9

15. In triangle JKL in Fig. RT8.10, find
    (a) the length KJ correct to 3 significant figures.
    (b) $\sin L$ and $\tan K$, each correct to 3 decimal places. (4)

16. Two ships leave a port at the same time. Ship X travels due west at 30 km/h and ship Y travels due

Figure RT8.10

north. After 4 hours the two ships are 130 km apart. Calculate the velocity of ship Y. (4)

17. If $\sin A = \dfrac{12}{37}$, find $\tan A$ in fraction form. (3)

18. Evaluate $5 \tan 62°11'$ correct to 3 significant figures. (2)

19. Determine the acute angle $\cos^{-1} 0.3649$ in degrees and minutes. (2)

20. In triangle PQR in Fig. RT8.11, find angle P in decimal form, correct to 2 decimal places. (2)

Figure RT8.11

21. Evaluate, correct to 3 significant figures, $3 \tan 81.27° - 5 \cos 7.32° - 6 \sin 54.81°$ (2)

22. In triangle ABC in Fig. RT8.12, find lengths AB and AC, correct to 2 decimal places. (4)

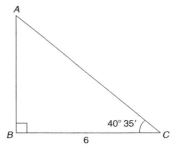

Figure RT8.12

23. From a point P, the angle of elevation of a 40 m high electricity pylon is $20°$. How far is point P from the base of the pylon, correct to the nearest metre? (3)

For lecturers/instructors/teachers, fully worked solutions to each of the problems in Revision Test 8, together with a full marking scheme, are available at the website:

www.routledge.com/cw/bird

Chapter 23

Trigonometric waveforms

Why it is important to understand: Trigonometric waveforms

Trigonometric graphs are commonly used in all areas of science and engineering for modelling many different natural and mechanical phenomena such as waves, engines, acoustics, electronics, populations, UV intensity, growth of plants and animals, and so on. Periodic trigonometric graphs mean that the shape repeats itself exactly after a certain amount of time. Anything that has a regular cycle, like the tides, temperatures, rotation of the earth, and so on, can be modelled using a sine or cosine curve. The most common periodic signal waveform that is used in electrical and electronic engineering is the sinusoidal waveform. However, an alternating a.c. waveform may not always take the shape of a smooth shape based around the sine and cosine function; a.c. waveforms can also take the shape of square or triangular waves, i.e. complex waves. In engineering, it is therefore important to have some clear understanding of trigonometric waveforms.

At the end of this chapter you should be able to:

- sketch sine, cosine and tangent waveforms
- determine angles of any magnitude
- understand cycle, amplitude, period, periodic time, frequency, lagging/leading angles with reference to sine and cosine waves
- perform calculations involving sinusoidal form $A \sin(\omega t \pm \alpha)$

23.1 Graphs of trigonometric functions

By drawing up tables of values from $0°$ to $360°$, graphs of $y = \sin A, y = \cos A$ and $y = \tan A$ may be plotted. Values obtained with a calculator (correct to 3 decimal places – which is more than sufficient for plotting graphs), using $30°$ intervals, are shown below, with the respective graphs shown in Fig. 23.1.

(a) $y = \sin A$

A	0	30°	60°	90°	120°	150°	180°
sin A	0	0.500	0.866	1.000	0.866	0.500	0

A	210°	240°	270°	300°	330°	360°
sin A	−0.500	−0.866	−1.000	−0.866	−0.500	0

(b) $y = \cos A$

A	0	30°	60°	90°	120°	150°	180°
cos A	1.000	0.866	0.500	0	−0.500	−0.866	−1.000

A	210°	240°	270°	300°	330°	360°
cos A	−0.866	−0.500	0	0.500	0.866	1.000

(c) $y = \tan A$

A	0	30°	60°	90°	120°	150°	180°
tan A	0	0.577	1.732	∞	−1.732	−0.577	0

A	210°	240°	270°	300°	330°	360°
tan A	0.577	1.732	∞	−1.732	−0.577	0

Figure 23.1

From Fig. 23.1 it is seen that

(a)   Sine and cosine graphs oscillate between peak values of ±1

(b)   The cosine curve is the same shape as the sine curve but displaced by 90°

(c)   The sine and cosine curves are continuous and they repeat at intervals of 360°, and the tangent curve appears to be discontinuous and repeats at intervals of 180°

23.2   Angles of any magnitude

Fig. 23.2 shows rectangular axes XX' and YY' intersecting at origin 0. As with graphical work, measurements made to the right and above 0 are positive, while those to the left and downwards are negative.

Let a phasor $0A$ be free to rotate about 0. By convention, when $0A$ moves anticlockwise angular measurement is considered positive, and vice versa.

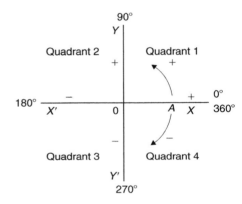

Figure 23.2

Let $0A$ be rotated anticlockwise so that θ_1 is any angle in the first quadrant and let perpendicular AB be constructed to form the right-angled triangle $0AB$ in Fig. 23.3. Since all three sides of the triangle are positive, the trigonometric ratios sine, cosine and tangent will all be positive in the first quadrant. (Note: $0A$ is always positive since it is the radius of a circle.)

Let $0A$ be further rotated so that θ_2 is any angle in the second quadrant and let AC be constructed to form the right-angled triangle $0AC$. Then,

$$\sin\theta_2 = \frac{+}{+} = + \qquad \cos\theta_2 = \frac{-}{+} = -$$

$$\tan\theta_2 = \frac{+}{-} = -$$

Let $0A$ be further rotated so that θ_3 is any angle in the third quadrant and let AD be constructed to form the

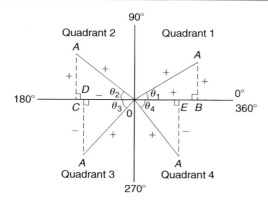

Figure 23.3

right-angled triangle $0AD$. Then,

$$\sin\theta_3 = \frac{-}{+} = - \qquad \cos\theta_3 = \frac{-}{+} = -$$

$$\tan\theta_3 = \frac{-}{-} = +$$

Let $0A$ be further rotated so that θ_4 is any angle in the fourth quadrant and let AE be constructed to form the right-angled triangle $0AE$. Then,

$$\sin\theta_4 = \frac{-}{+} = - \qquad \cos\theta_4 = \frac{+}{+} = +$$

$$\tan\theta_4 = \frac{-}{+} = -$$

The above results are summarised in Fig. 23.4, in which all three trigonometric ratios are positive in the first quadrant, only sine is positive in the second quadrant, only tangent is positive in the third quadrant and only cosine is positive in the fourth quadrant.

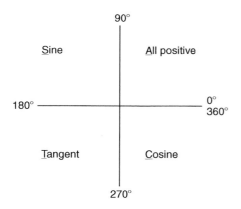

Figure 23.4

The underlined letters in Fig. 23.4 spell the word CAST when starting in the fourth quadrant and moving in an anticlockwise direction.

It is seen that, in the first quadrant of Fig. 23.1, all of the curves have positive values; in the second only sine is positive; in the third only tangent is positive; and in the fourth only cosine is positive – exactly as summarised in Fig. 23.4.

A knowledge of angles of any magnitude is needed when finding, for example, all the angles between $0°$ and $360°$ whose sine is, say, 0.3261. If 0.3261 is entered into a calculator and then the inverse sine key pressed (or $\sin^{-1}$ key) the answer $19.03°$ appears. However, there is a second angle between $0°$ and $360°$ which the calculator does not give. Sine is also positive in the second quadrant (either from CAST or from Fig. 23.1(a)). The other angle is shown in Fig. 23.5 as angle θ, where $\theta = 180° - 19.03° = 160.97°$. Thus, $19.03°$ **and** $160.97°$ are the angles between $0°$ and $360°$ whose sine is 0.3261 (check that $\sin 160.97° = 0.3261$ on your calculator).

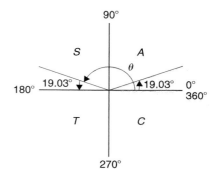

Figure 23.5

Be careful! Your calculator only gives you one of these answers. The second answer needs to be deduced from a knowledge of angles of any magnitude, as shown in the following worked problems.

⚑ Problem 1. Determine all of the angles between $0°$ and $360°$ whose sine is -0.4638

The angles whose sine is -0.4638 occur in the third and fourth quadrants since sine is negative in these quadrants – see Fig. 23.6.
From Fig. 23.7, $\theta = \sin^{-1} 0.4638 = 27.63°$. Measured from $0°$, the two angles between $0°$ and $360°$ whose sine is -0.4638 are $180° + 27.63°$ i.e. **207.63°** and $360° - 27.63°$, i.e. **332.37°**.
(Note that if a calculator is used to determine $\sin^{-1}(-0.4638)$ it only gives one answer: $-27.632588°$)

⚑ Problem 2. Determine all of the angles between $0°$ and $360°$ whose tangent is 1.7629

Figure 23.6

Figure 23.7

A tangent is positive in the first and third quadrants – see Fig. 23.8.

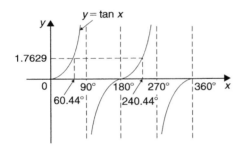

Figure 23.8

From Fig. 23.9, $\theta = \tan^{-1} 1.7629 = 60.44°$. Measured from $0°$, the two angles between $0°$ and $360°$ whose tangent is 1.7629 are **60.44°** and $180° + 60.44°$, i.e. **240.44°**

Figure 23.9

▶ **Problem 3.** Solve the equation $\cos^{-1}(-0.2348) = \alpha$ for angles of α between $0°$ and $360°$

Cosine is positive in the first and fourth quadrants and thus negative in the second and third quadrants – see Fig. 23.10 or from Fig. 23.1(b).

In Fig. 23.10, angle $\theta = \cos^{-1}(0.2348) = 76.42°$. Measured from $0°$, the two angles whose cosine is -0.2348 are $\alpha = 180° - 76.42°$, i.e. **103.58°** and $\alpha = 180° + 76.42°$, i.e. **256.42°**

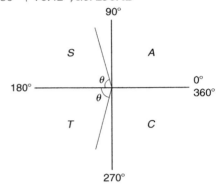

Figure 23.10

Now try the following Practice Exercise

Practice Exercise 116   Angles of any magnitude (answers on page 454)

▶ 1. Determine all of the angles between $0°$ and $360°$ whose sine is
    (a)   0.6792   (b)   −0.1483

▶ 2. Solve the following equations for values of x between $0°$ and $360°$.
    (a) $x = \cos^{-1} 0.8739$

    (b) $x = \cos^{-1}(-0.5572)$

▶ 3. Find the angles between $0°$ to $360°$ whose tangent is
    (a)   0.9728   (b)   −2.3420

In Problems 4 to 6, solve the given equations in the range $0°$ to $360°$, giving the answers in degrees and minutes.

▶ 4. $\cos^{-1}(-0.5316) = t$

▶ 5. $\sin^{-1}(-0.6250) = \alpha$

▶ 6. $\tan^{-1} 0.8314 = \theta$

23.3 The production of sine and cosine waves

In Fig. 23.11, let OR be a vector 1 unit long and free to rotate anticlockwise about 0. In one revolution a circle is produced and is shown with $15°$ sectors. Each radius arm has a vertical and a horizontal component. For example, at $30°$, the vertical component is TS and the horizontal component is OS.

From triangle OST,

$$\sin 30° = \frac{TS}{TO} = \frac{TS}{1} \quad \text{i.e.} \quad TS = \sin 30°$$

$$\text{and} \quad \cos 30° = \frac{OS}{TO} = \frac{OS}{1} \quad \text{i.e.} \quad OS = \cos 30°$$

Sine waves

The vertical component TS may be projected across to $T'S'$, which is the corresponding value of $30°$ on the graph of y against angle $x°$. If all such vertical components as TS are projected on to the graph, a **sine wave** is produced as shown in Fig. 23.11.

Cosine waves

If all horizontal components such as OS are projected on to a graph of y against angle $x°$, a **cosine wave** is produced. It is easier to visualise these projections by redrawing the circle with the radius arm OR initially in a vertical position as shown in Fig. 23.12.

It is seen from Figs 23.11 and 23.12 that a cosine curve is of the same form as the sine curve but is displaced by $90°$ (or $\pi/2$ radians). Both sine and cosine waves repeat every $360°$

23.4 Terminology involved with sine and cosine waves

Sine waves are extremely important in engineering, with examples occurring with alternating currents and voltages – the mains supply is a sine wave – and with simple harmonic motion.

Cycle

When a sine wave has passed through a complete series of values, both positive and negative, it is said to have completed one **cycle**. One cycle of a sine wave is shown in Fig. 23.1(a) on page 237 and in Fig. 23.11.

Amplitude

The amplitude is the maximum value reached in a half cycle by a sine wave. Another name for **amplitude** is **peak value** or **maximum value**.

A sine wave $y = 5 \sin x$ has an amplitude of 5, a sine wave $v = 200 \sin 314t$ has an amplitude of 200 and the sine wave $y = \sin x$ shown in Fig. 23.11 has an amplitude of 1.

Period

The waveforms $y = \sin x$ and $y = \cos x$ repeat themselves every $360°$. Thus, for each, the **period** is $360°$. A waveform of $y = \tan x$ has a period of $180°$ (from Fig. 23.1(c)).

A graph of $y = 3 \sin 2A$, as shown in Fig. 23.13, has an **amplitude of 3** and **period 180°**

A graph of $y = \sin 3A$, as shown in Fig. 23.14, has an **amplitude of 1** and **period of 120°**

Figure 23.11

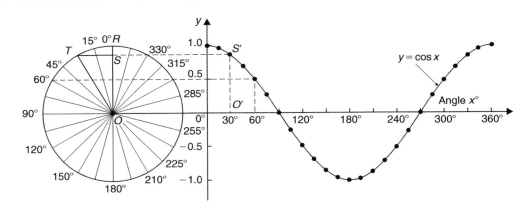

Figure 23.12

A graph of $y = 4\cos 2x$, as shown in Fig. 23.15, has an amplitude of 4 and a period of 180°

In general, **if $y = A\sin px$ or $y = A\cos px$,**

$$\textbf{amplitude} = A \textbf{ and period} = \frac{360°}{p}$$

Figure 23.13

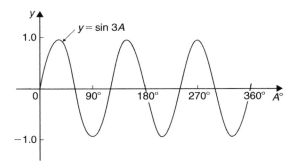

Figure 23.14

> **Problem 4.** Sketch $y = 2\sin\dfrac{3}{5}A$ over one cycle

$$\textbf{Amplitude} = 2; \textbf{ period} = \frac{360°}{\frac{3}{5}} = \frac{360° \times 5}{3} = \textbf{600°}$$

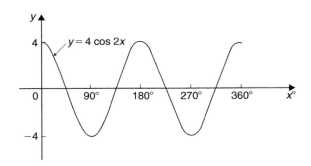

Figure 23.15

A sketch of $y = 2\sin\dfrac{3}{5}A$ is shown in Fig. 23.16.

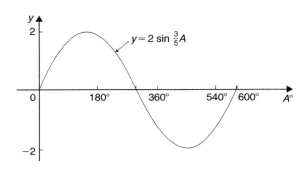

Figure 23.16

Periodic time

In practice, the horizontal axis of a sine wave will be time. The time taken for a sine wave to complete one cycle is called the **periodic time, T**.

In the sine wave of voltage v (volts) against time t (milliseconds) shown in Fig. 23.17, the amplitude is 10 V and the periodic time is 20 ms; i.e. $T = 20$ ms.

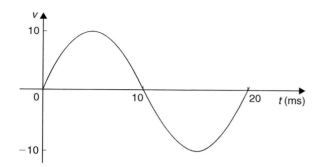

Figure 23.17

Frequency

The number of cycles completed in one second is called the **frequency** f and is measured in **hertz, Hz**.

$$f = \frac{1}{T} \text{ or } T = \frac{1}{f}$$

> **Problem 5.** Determine the frequency of the sine wave shown in Fig. 23.17

In the sine wave shown in Fig. 23.17, $T = 20$ ms, hence

$$\textbf{frequency, } f = \frac{1}{T} = \frac{1}{20 \times 10^{-3}} = \textbf{50 Hz}$$

> **Problem 6.** If a waveform has a frequency of 200 kHz, determine the periodic time

If a waveform has a frequency of 200 kHz, the periodic time T is given by

$$\textbf{periodic time, } T = \frac{1}{f} = \frac{1}{200 \times 10^3}$$

$$= 5 \times 10^{-6}\text{s} = \textbf{5}\,\boldsymbol{\mu}\textbf{s}$$

Lagging and leading angles

A sine or cosine curve may not always start at $0°$. To show this, a periodic function is represented by $y = A\sin(x \pm \alpha)$ where α is a phase displacement compared with $y = A\sin x$. For example, $y = \sin A$ is shown by the broken line in Fig. 23.18 and, on the same axes, $y = \sin(A - 60°)$ is shown. **The graph** $y = \sin(A - 60°)$ **is said to lag** $y = \sin A$ **by** $60°$
In another example, $y = \cos A$ is shown by the broken line in Fig. 23.19 and, on the same axes, $y = \cos(A + 45°)$ is shown. **The graph** $y = \cos(A + 45°)$ **is said to lead** $y = \cos A$ **by** $45°$

Figure 23.18

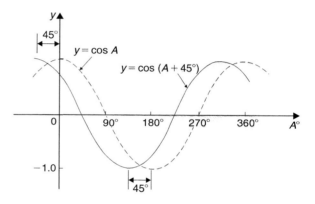

Figure 23.19

> **Problem 7.** Sketch $y = 5\sin(A + 30°)$ from $A = 0°$ to $A = 360°$

Amplitude = 5 and **period** $= 360°/1 = \textbf{360}°$

$5\sin(A + 30°)$ **leads** $5\sin A$ by $30°$ (i.e. starts $30°$ earlier).
A sketch of $y = 5\sin(A + 30°)$ is shown in Fig. 23.20.

> **Problem 8.** Sketch $y = 7\sin(2A - \pi/3)$ in the range $0 \leq A \leq 360°$

Amplitude = 7 and **period** $= 2\pi/2 = \pi$ **radians**

In general, $y = \sin(pt - \alpha)$ **lags** $y = \sin pt$ **by** α/p, hence $7\sin(2A - \pi/3)$ lags $7\sin 2A$ by $(\pi/3)/2$, i.e. $\pi/6$ rad or $30°$
A sketch of $y = 7\sin(2A - \pi/3)$ is shown in Fig. 23.21.

Figure 23.20

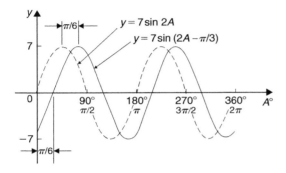

Figure 23.21

Problem 9.   Sketch $y = 2\cos(\omega t - 3\pi/10)$ over one cycle

Amplitude = 2 and **period = $2\pi/\omega$ rad**

$2\cos(\omega t - 3\pi/10)$ **lags** $2\cos\omega t$ by $3\pi/10\omega$ seconds. A sketch of $y = 2\cos(\omega t - 3\pi/10)$ is shown in Fig. 23.22.

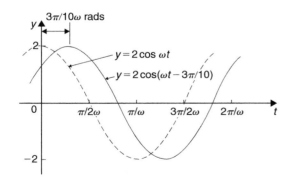

Figure 23.22

Now try the following Practice Exercise

Practice Exercise 117   Trigonometric waveforms (answers on page 454)

1. A sine wave is given by $y = 5\sin 3x$. State its peak value.

2. A sine wave is given by $y = 4\sin 2x$. State its period in degrees.

3. A periodic function is given by $y = 30\cos 5x$. State its maximum value.

4. A periodic function is given by $y = 25\cos 3x$. State its period in degrees.

In Problems 5 to 11, state the amplitude and period of the waveform and sketch the curve between $0°$ and $360°$.

5. $y = \cos 3A$ 6. $y = 2\sin\dfrac{5x}{2}$

7. $y = 3\sin 4t$ 8. $y = 5\cos\dfrac{\theta}{2}$

9. $y = \dfrac{7}{2}\sin\dfrac{3x}{8}$ 10. $y = 6\sin(t - 45°)$

11. $y = 4\cos(2\theta + 30°)$

12. The frequency of a sine wave is $200\,\text{Hz}$. Calculate the periodic time.

13. Calculate the frequency of a sine wave that has a periodic time of $25\,\text{ms}$.

14. Calculate the periodic time for a sine wave having a frequency of $10\,\text{kHz}$.

15. An alternating current completes 15 cycles in $24\,\text{ms}$. Determine its frequency.

16. Graphs of $y_1 = 2\sin x$ and $y_2 = 3\sin(x + 50°)$ are drawn on the same axes. Is y_2 lagging or leading y_1?

17. Graphs of $y_1 = 6\sin x$ and $y_2 = 5\sin(x - 70°)$ are drawn on the same axes. Is y_1 lagging or leading y_2?

23.5   Sinusoidal form: $A\sin(\omega t \pm \alpha)$

If a sine wave is expressed in the form $y = A\sin(\omega t \pm \alpha)$ then

(a) A = amplitude

(b) ω = angular velocity = $2\pi f\,\text{rad/s}$

(c)   frequency, $f = \dfrac{\omega}{2\pi}$ hertz

(d)   periodic time, $T = \dfrac{2\pi}{\omega}$ seconds $\left(\text{i.e. } T = \dfrac{1}{f}\right)$

(e) $\alpha =$ angle of lead or lag (compared with $y = A\sin\omega t$)

Here are some worked problems involving the sinusoidal form $A\sin(\omega t \pm \alpha)$

> **Problem 10.** An alternating current is given by $i = 30\sin(100\pi t + 0.35)$ amperes. Find the (a) amplitude, (b) frequency, (c) periodic time and (d) phase angle (in degrees and minutes)

(a) $i = 30\sin(100\pi t + 0.35)A$; hence, **amplitude = 30 A**

(b)   Angular velocity, $\omega = 100\pi$, rad/s, hence

$$\textbf{frequency, } f = \frac{\omega}{2\pi} = \frac{100\pi}{2\pi} = \textbf{50 Hz}$$

(c)   **Periodic time, $T = \dfrac{1}{f} = \dfrac{1}{50} = $ 0.02 s or 20 ms.**

(d)   0.35 is the angle in **radians**. The relationship between radians and degrees is

$$360° = 2\pi \text{ radians or } \mathbf{180° = \pi \, radians}$$

from which,

$$\mathbf{1° = \frac{\pi}{180} \, rad} \text{ and } \mathbf{1 \, rad = \frac{180°}{\pi}} \, (\approx 57.30°)$$

Hence, **phase angle**, $\alpha = 0.35$ rad

$$= \left(0.35 \times \frac{180}{\pi}\right)^\circ = \mathbf{20.05° \text{ or } 20°3' \text{ leading}}$$

$$i = \mathbf{30\sin(100\pi t)}$$

> **Problem 11.** An oscillating mechanism has a maximum displacement of 2.5 m and a frequency of 60 Hz. At time $t = 0$ the displacement is 90 cm. Express the displacement in the general form $A\sin(\omega t \pm \alpha)$

Amplitude = maximum displacement = 2.5 m.

Angular velocity, $\omega = 2\pi f = 2\pi(60) = 120\pi$ rad/s.

Hence,   **displacement = $2.5\sin(120\pi t + \alpha)$ m.**

When $t = 0$,   displacement = 90 cm = 0.90 m

Hence, $0.90 = 2.5\sin(0 + \alpha)$

i.e. $\sin\alpha = \dfrac{0.90}{2.5} = 0.36$

Hence, $\alpha = \sin^{-1} 0.36 = 21.10°$

$$= 21°6' = 0.368 \text{ rad.}$$

Thus, **displacement = $2.5\sin(120\pi t + 0.368)$ m**

> **Problem 12.** The instantaneous value of voltage in an a.c. circuit at any time t seconds is given by $v = 340\sin(50\pi t - 0.541)$ volts. Determine the (a) amplitude, frequency, periodic time and phase angle (in degrees), (b) value of the voltage when $t = 0$, (c) value of the voltage when $t = 10$ ms, (d) time when the voltage first reaches 200 V and (e) time when the voltage is a maximum. Also, (f) sketch one cycle of the waveform

(a)   **Amplitude = 340 V**

Angular velocity, $\omega = 50\pi$

$$\textbf{Frequency, } f = \frac{\omega}{2\pi} = \frac{50\pi}{2\pi} = \textbf{25 Hz}$$

$$\textbf{Periodic time, } T = \frac{1}{f} = \frac{1}{25} = \textbf{0.04 s or 40 ms}$$

$$\textbf{Phase angle} = 0.541 \text{ rad} = \left(0.541 \times \frac{180}{\pi}\right)^\circ$$

$$= \mathbf{31° \text{ lagging } v = 340\sin(50\pi t)}$$

(b)   **When $t = 0$,**

$$v = 340\sin(0 - 0.541)$$

$$= 340\sin(-31°) = \mathbf{-175.1 \, V}$$

(c)   **When $t = 10$ ms,**

$$v = 340\sin(50\pi \times 10 \times 10^{-3} - 0.541)$$

$$= 340\sin(1.0298)$$

$$= 340\sin 59° = \mathbf{291.4 \text{ volts}}$$

(d)   When $v = 200$ volts,

$$200 = 340\sin(50\pi t - 0.541)$$

$$\frac{200}{340} = \sin(50\pi t - 0.541)$$

Hence, $(50\pi t - 0.541) = \sin^{-1}\dfrac{200}{340}$

$$= 36.03° \text{ or } 0.628875 \text{ rad}$$

$$50\pi t = 0.628875 + 0.541$$

$$= 1.169875$$

Hence, when $v = 200$ V,

time, $t = \dfrac{1.169875}{50\pi} = \textbf{7.448 ms}$

(e)   When the voltage is a maximum, $v = 340$ V.

Hence, $340 = 340\sin(50\pi t - 0.541)$

$$1 = \sin(50\pi t - 0.541)$$

$$50\pi t - 0.541 = \sin^{-1} 1 = 90° \text{ or } 1.5708 \text{ rad}$$

$$50\pi t = 1.5708 + 0.541 = 2.1118$$

Hence, **time**, $t = \dfrac{2.1118}{50\pi} = \textbf{13.44 ms}$

(f)   A sketch of $v = 340\sin(50\pi t - 0.541)$ volts is shown in Fig. 23.23.

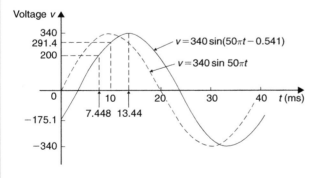

Figure 23.23

Now try the following Practice Exercise

Practice Exercise 118   Sinusoidal form
$A\sin(\omega t \pm \alpha)$ **(answers on page 454)**

In Problems 1 to 3 find the (a) amplitude, (b) frequency, (c) periodic time and (d) phase angle (stating whether it is leading or lagging $\sin\omega t$) of the alternating quantities given.

1. $i = 40\sin(50\pi t + 0.29)$ mA

2. $y = 75\sin(40t - 0.54)$ cm

3. $v = 300\sin(200\pi t - 0.412)$ V

4.   A sinusoidal voltage has a maximum value of 120 V and a frequency of 50 Hz. At time $t = 0$, the voltage is (a) zero and (b) 50 V. Express the instantaneous voltage v in the form $v = A\sin(\omega t \pm \alpha)$.

5.   An alternating current has a periodic time of 25 ms and a maximum value of 20 A. When time $= 0$, current $i = -10$ amperes. Express the current i in the form $i = A\sin(\omega t \pm \alpha)$.

6.   An oscillating mechanism has a maximum displacement of 3.2 m and a frequency of 50 Hz. At time $t = 0$ the displacement is 150 cm. Express the displacement in the general form $A\sin(\omega t \pm \alpha)$.

7.   The current in an a.c. circuit at any time t seconds is given by

$$i = 5\sin(100\pi t - 0.432) \text{ amperes}$$

Determine the

(a)   amplitude, frequency, periodic time and phase angle (in degrees),

(b)   value of current at $t = 0$,

(c)   value of current at $t = 8$ ms,

(d)   time when the current is first a maximum,

(e)   time when the current first reaches 3A.

Also,

(f)   sketch one cycle of the waveform showing relevant points.

Practice Exercise 119   Multiple-choice questions on trigonometric waveforms (answers on page 454)

Each question has only one correct answer.

1. An alternating current is given by: $i = 15\sin(100\pi t - 0.25)$ amperes. When time $t = 5$ ms, the current i has a value of:

    (a) 0.35 A   (b) 14.53 A
    (c) 15 A   (d) 0.41 A

2. The displacement x metres of a mass from a fixed point about which it is oscillating is given by $x = 3\cos\omega t - 4\sin\omega t$, where t is the time in seconds. x may be expressed as:

    (a) $5\sin(\omega t + 2.50)$ metres
    (b) $7\sin(\omega t - 36.87°)$ metres
    (c) $5\sin\omega t$ metres
    (d) $-\sin(\omega t - 2.50)$ metres

3. A sinusoidal current is given by: $i = R\sin(\omega t + \alpha)$. Which of the following statements is incorrect?

    (a) R is the average value of the current
    (b) frequency $= \dfrac{\omega}{2\pi}$ Hz
    (c) $\omega =$ angular velocity
    (d) periodic time $= \dfrac{2\pi}{\omega}$ s

4. An alternating voltage v is given by $v = 100\sin\left(100\pi t + \dfrac{\pi}{4}\right)$ volts. When $v = 50$ volts, the time t is equal to:

    (a) 0.093 s   (b) −0.908 ms
    (c) −0.833 ms   (d) −0.162 s

5. The values of θ that are true for the equation $5\sin\theta + 2 = 0$ in the range $\theta = 0°$ to $\theta = 360°$ are:

    (a) 23.58° and 336.42°
    (b) 23.58° and 203.58°
    (c) 156.42° and 336.42°
    (d) 203.58° and 336.42°

For fully worked solutions to each of the problems in Practice Exercises 116 to 118 in this chapter, go to the website:
www.routledge.com/cw/bird

Chapter 24

Non-right-angled triangles and some practical applications

Why it is important to understand: Non-right-angled triangles and some practical applications

As was mentioned earlier, fields that use trigonometry include astronomy, navigation, music theory, acoustics, optics, electronics, probability theory, statistics, biology, medical imaging (CAT scans and ultrasound), pharmacy, chemistry, seismology, meteorology, oceanography, many physical sciences, land surveying, architecture, economics, electrical engineering, mechanical engineering, civil engineering, computer graphics, cartography and crystallography. There are so many examples where triangles are involved in engineering, and the ability to solve such triangles is of great importance.

At the end of this chapter you should be able to:

- state and use the sine rule
- state and use the cosine rule
- use various formulae to determine the area of any triangle
- apply the sine and cosine rules to solving practical trigonometric problems

24.1 The sine and cosine rules

To '**solve a triangle**' means 'to find the values of unknown sides and angles'. If a triangle is **right-angled**, trigonometric ratios and the theorem of Pythagoras* may be used for its solution, as shown in Chapter 22. However, for a **non-right-angled triangle**,

---

*Who was **Pythagoras**? See page 111. To find out more go to **www.routledge.com/cw/bird**

trigonometric ratios and Pythagoras' theorem cannot be used. Instead, two rules, called the **sine rule** and the **cosine rule**, are used.

The sine rule

With reference to triangle *ABC* of Fig. 24.1, the **sine rule** states

$$\frac{a}{\sin A} = \frac{b}{\sin B} = \frac{c}{\sin C}$$

The rule may be used only when

Figure 24.1

(a) 1 side and any 2 angles are initially given, or
(b) 2 sides and an angle (not the included angle) are initially given.

The cosine rule

With reference to triangle ABC of Fig. 24.1, the **cosine rule** states

$$a^2 = b^2 + c^2 - 2bc\cos A$$

or

$$b^2 = a^2 + c^2 - 2ac\cos B$$

$$-2pt]$$

or

$$c^2 = a^2 + b^2 - 2ab\cos C$$

The rule may be used only when

(a) 2 sides and the included angle are initially given, or

(b) 3 sides are initially given.

24.2 Area of any triangle

The **area of any triangle** such as ABC of Fig. 24.1 is given by

(a) $\dfrac{1}{2} \times$ **base** $\times$ **perpendicular height**

or (b) $\dfrac{1}{2}ab\sin C$ or $\dfrac{1}{2}ac\sin B$ or $\dfrac{1}{2}bc\sin A$

or (c) $\sqrt{[s(s-a)(s-b)(s-c)]}$ where $s = \dfrac{a+b+c}{2}$

This latter formula is called **Heron's formula** (or sometimes **Hero's formula**).

24.3 Worked problems on the solution of triangles and their areas

Problem 1. In a triangle XYZ, $\angle X = 51°$, $\angle Y = 67°$ and $YZ = 15.2$ cm. Solve the triangle and find its area

The triangle XYZ is shown in Fig. 24.2. Solving the triangle means finding $\angle Z$ and sides XZ and XY.
Since the angles in a triangle add up to $180°$, $Z = 180° - 51° - 67° = \mathbf{62°}$

Figure 24.2

Applying the sine rule, $\dfrac{15.2}{\sin 51°} = \dfrac{y}{\sin 67°}$

$$= \dfrac{z}{\sin 62°}$$

Using $\dfrac{15.2}{\sin 51°} = \dfrac{y}{\sin 67°}$

and transposing gives $y = \dfrac{15.2\sin 67°}{\sin 51°}$

$$= \mathbf{18.00\,cm} = \mathbf{\mathit{XZ}}$$

Using $\dfrac{15.2}{\sin 51°} = \dfrac{z}{\sin 62°}$

and transposing gives $z = \dfrac{15.2\sin 62°}{\sin 51°}$

$$= \mathbf{17.27\,cm} = \mathbf{\mathit{XY}}$$

Area of triangle $XYZ = \dfrac{1}{2}xy\sin Z$

$$= \dfrac{1}{2}(15.2)(18.00)\sin 62° = \mathbf{120.8\,cm^2}$$

(or area $= \dfrac{1}{2}xz\sin Y = \dfrac{1}{2}(15.2)(17.27)\sin 67°$

$$= \mathbf{120.8\,cm^2})$$

It is always worth checking with triangle problems that the longest side is opposite the largest angle and vice versa. In this problem, Y is the largest angle and XZ is the longest of the three sides.

Problem 2. Solve the triangle ABC given $B = 78°51'$, $AC = 22.31$ mm and $AB = 17.92$ mm. Also find its area

Triangle ABC is shown in Fig. 24.3. Solving the triangle means finding angles A and C and side BC.

Applying the sine rule, $\dfrac{22.31}{\sin 78°51'} = \dfrac{17.92}{\sin C}$

from which $\sin C = \dfrac{17.92\sin 78°51'}{22.31} = 0.7881$

Hence, $C = \sin^{-1} 0.7881 = 52°0'$ or $128°0'$

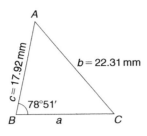

Figure 24.3

Since $B = 78°51'$, C cannot be $128°0'$, since $128°0' + 78°51'$ is greater than $180°$. Thus, only $C = 52°0'$ is valid.

Angle $A = 180° - 78°51' - 52°0' = $ **49°9'**

Applying the sine rule, $\dfrac{a}{\sin 49°9'} = \dfrac{22.31}{\sin 78°51'}$

from which $a = \dfrac{22.31 \sin 49°9'}{\sin 78°51'} = $ **17.20 mm**

Hence, $A = $ **49°9'**, $C = $ **52°0'** and $BC = $ **17.20 mm**.

Area of triangle $ABC = \dfrac{1}{2} ac \sin B$

$$= \dfrac{1}{2}(17.20)(17.92) \sin 78°51'$$

$$= \textbf{151.2 mm}^2$$

> **Problem 3.** Solve the triangle PQR and find its area given that $QR = 36.5$ mm, $PR = 29.6$ mm and $\angle Q = 36°$

Triangle PQR is shown in Fig. 24.4.

Figure 24.4

Applying the sine rule, $\dfrac{29.6}{\sin 36°} = \dfrac{36.5}{\sin P}$

from which $\sin P = \dfrac{36.5 \sin 36°}{29.6} = 0.7248$

Hence, $P = \sin^{-1} 0.7248 = 46.45°$ or $133.55°$

When $P = 46.45°$ and $Q = 36°$ then
$R = 180° - 46.45° - 36° = 97.55°$
When $P = 133.55°$ and $Q = 36°$ then
$R = 180° - 133.55° - 36° = 10.45°$

Thus, in this problem, there are **two** separate sets of results and both are feasible solutions. Such a situation is called the **ambiguous case**.

Case 1. $P = 46.45°, Q = 36°, R = 97.55°$, $p = 36.5$ mm and $q = 29.6$ mm

From the sine rule, $\dfrac{r}{\sin 97.55°} = \dfrac{29.6}{\sin 36°}$

from which $r = \dfrac{29.6 \sin 97.55°}{\sin 36°} = $ **49.92 mm** $= PQ$

Area of $PQR = \dfrac{1}{2} pq \sin R = \dfrac{1}{2}(36.5)(29.6) \sin 97.55°$

$$= \textbf{535.5 mm}^2$$

Case 2. $P = 133.55°, Q = 36°, R = 10.45°$, $p = 36.5$ mm and $q = 29.6$ mm

From the sine rule, $\dfrac{r}{\sin 10.45°} = \dfrac{29.6}{\sin 36°}$

from which $r = \dfrac{29.6 \sin 10.45°}{\sin 36°} = $ **9.134 mm** $= PQ$

Area of $PQR = \dfrac{1}{2} pq \sin R = \dfrac{1}{2}(36.5)(29.6) \sin 10.45°$

$$= \textbf{97.98 mm}^2$$

The triangle PQR for case 2 is shown in Fig. 24.5.

Figure 24.5

Now try the following Practice Exercise

> **Practice Exercise 120 Solution of triangles and their areas (answers on page 454)**
>
> In Problems 1 and 2, use the sine rule to solve the triangles ABC and find their areas.
> 1. $A = 29°, B = 68°, b = 27$ mm
>
> 2. $B = 71°26', C = 56°32', b = 8.60$ cm
>
> In Problems 3 and 4, use the sine rule to solve the triangles DEF and find their areas.
>
> 3. $d = 17$ cm, $f = 22$ cm, $F = 26°$
>
> 4. $d = 32.6$ mm, $e = 25.4$ mm, $D = 104°22'$
>
> In Problems 5 and 6, use the sine rule to solve the triangles JKL and find their areas.

5. $j = 3.85\,\text{cm}, k = 3.23\,\text{cm}, K = 36°$

6. $k = 46\,\text{mm}, l = 36\,\text{mm}, L = 35°$

24.4 Further worked problems on the solution of triangles and their areas

Problem 4. Solve triangle DEF and find its area given that $EF = 35.0\,\text{mm}, DE = 25.0\,\text{mm}$ and $\angle E = 64°$

Triangle DEF is shown in Fig. 24.6. Solving the triangle means finding angles D and F and side DF. Since two sides and the angle in between the two sides are given, the cosine rule needs to be used.

Figure 24.6

Applying the cosine rule, $e^2 = d^2 + f^2 - 2df\cos E$

i.e. $e^2 = (35.0)^2 + (25.0)^2 - [2(35.0)(25.0)\cos 64°]$

$= 1225 + 625 - 767.15$

$= 1082.85$

from which $e = \sqrt{1082.85}$

$= \mathbf{32.91\,mm} = DF$

Applying the sine rule, $\dfrac{32.91}{\sin 64°} = \dfrac{25.0}{\sin F}$

from which $\sin F = \dfrac{25.0\sin 64°}{32.91} = 0.6828$

Thus, $\angle F = \sin^{-1} 0.6828 = 43°4'$ or $136°56'$

$F = 136°56'$ is not possible in this case since $136°56' + 64°$ is greater than $180°$. Thus, only $\boldsymbol{F = 43°4'}$ is valid. Then $\boldsymbol{\angle D} = 180° - 64° - 43°4' = \mathbf{72°56'}$

Area of triangle DEF $= \dfrac{1}{2}\,df\sin E$

$= \dfrac{1}{2}(35.0)(25.0)\sin 64° = \mathbf{393.2\,mm^2}$

Problem 5. A triangle ABC has sides $a = 9.0\,\text{cm}, b = 7.5\,\text{cm}$ and $c = 6.5\,\text{cm}$. Determine its three angles and its area

Triangle ABC is shown in Fig. 24.7. It is usual first to calculate the largest angle to determine whether the triangle is acute or obtuse. In this case the largest angle is A (i.e. opposite the longest side).

Figure 24.7

Applying the cosine rule, $a^2 = b^2 + c^2 - 2bc\cos A$

from which $2bc\cos A = b^2 + c^2 - a^2$

and $\cos A = \dfrac{b^2 + c^2 - a^2}{2bc} = \dfrac{7.5^2 + 6.5^2 - 9.0^2}{2(7.5)(6.5)}$

$= 0.1795$

Hence, $A = \cos^{-1} 0.1795 = \mathbf{79.67°}$
(or $280.33°$, which is clearly impossible)

The triangle is thus acute angled since $\cos A$ is positive. (If $\cos A$ had been negative, angle A would be obtuse; i.e. would lie between $90°$ and $180°$)

Applying the sine rule, $\dfrac{9.0}{\sin 79.67°} = \dfrac{7.5}{\sin B}$

from which $\sin B = \dfrac{7.5\sin 79.67°}{9.0} = 0.8198$

Hence, $B = \sin^{-1} 0.8198 = \mathbf{55.07°}$

and $C = 180° - 79.67° - 55.07° = \mathbf{45.26°}$

Area $= \sqrt{[s(s-a)(s-b)(s-c)]}$, where

$s = \dfrac{a+b+c}{2} = \dfrac{9.0 + 7.5 + 6.5}{2} = 11.5\,\text{cm}$

Hence,

$\mathbf{area} = \sqrt{[11.5(11.5 - 9.0)(11.5 - 7.5)(11.5 - 6.5)]}$

$= \sqrt{[11.5(2.5)(4.0)(5.0)]} = \mathbf{23.98\,cm^2}$

Alternatively, **area** $= \dfrac{1}{2}\,ac\sin B$

$= \dfrac{1}{2}(9.0)(6.5)\sin 55.07° = \mathbf{23.98\,cm^2}$

Problem 6. Solve triangle XYZ, shown in Fig. 24.8, and find its area given that $Y = 128°, XY = 7.2$ cm and $YZ = 4.5$ cm

Figure 24.8

Applying the cosine rule,

$$y^2 = x^2 + z^2 - 2xz \cos Y$$

$$= 4.5^2 + 7.2^2 - [2(4.5)(7.2)\cos 128°]$$

$$= 20.25 + 51.84 - [-39.89]$$

$$= 20.25 + 51.84 + 39.89 = 112.0$$

$$y = \sqrt{112.0} = \mathbf{10.58\,cm} = XZ$$

Applying the sine rule, $\dfrac{10.58}{\sin 128°} = \dfrac{7.2}{\sin Z}$

from which $\sin Z = \dfrac{7.2 \sin 128°}{10.58} = 0.5363$

Hence, $Z = \sin^{-1} 0.5363 = \mathbf{32.43°}$
(or $147.57°$ which is not possible)

Thus, $X = 180° - 128° - 32.43° = \mathbf{19.57°}$

$$\textbf{Area of } XYZ = \frac{1}{2}xz \sin Y = \frac{1}{2}(4.5)(7.2)\sin 128°$$

$$= \mathbf{12.77\,cm^2}$$

Now try the following Practice Exercise

Practice Exercise 121   Solution of triangles and their areas (answers on page 455)

In Problems 1 and 2, use the cosine and sine rules to solve the triangles PQR and find their areas.

1. $q = 12$ cm, $r = 16$ cm, $P = 54°$

2. $q = 3.25$ m, $r = 4.42$ m, $P = 105°$

In Problems 3 and 4, use the cosine and sine rules to solve the triangles XYZ and find their areas.

3. $x = 10.0$ cm, $y = 8.0$ cm, $z = 7.0$ cm

4. $x = 21$ mm, $y = 34$ mm, $z = 42$ mm

24.5   Practical situations involving trigonometry

There are a number of **practical situations** in which the use of trigonometry is needed to find unknown sides and angles of triangles. This is demonstrated in the following worked problems.

Problem 7. A room 8.0 m wide has a span roof which slopes at 33° on one side and 40° on the other. Find the length of the roof slopes, correct to the nearest centimetre

A section of the roof is shown in Fig. 24.9.

Figure 24.9

Angle at ridge, $B = 180° - 33° - 40° = 107°$

From the sine rule, $\dfrac{8.0}{\sin 107°} = \dfrac{a}{\sin 33°}$

from which $a = \dfrac{8.0 \sin 33°}{\sin 107°} = \mathbf{4.556\,m} = BC$

Also from the sine rule, $\dfrac{8.0}{\sin 107°} = \dfrac{c}{\sin 40°}$

from which $c = \dfrac{8.0 \sin 40°}{\sin 107°} = \mathbf{5.377\,m} = AB$

Hence, **the roof slopes are 4.56 m and 5.38 m**, correct to the nearest centimetre.

Problem 8. A man leaves a point walking at 6.5 km/h in a direction E 20° N (i.e. a bearing of 70°). A cyclist leaves the same point at the same time in a direction E 40° S (i.e. a bearing of 130°) travelling at a constant speed. Find the average speed of the cyclist if the walker and cyclist are 80 km apart after 5 hours

After 5 hours the walker has travelled $5 \times 6.5 = 32.5$ km (shown as AB in Fig. 24.10). If AC is the distance the cyclist travels in 5 hours then $BC = 80$ km.

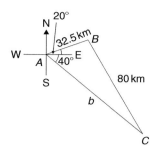

Figure 24.10

Applying the sine rule, $\dfrac{80}{\sin 60°} = \dfrac{32.5}{\sin C}$

from which $\sin C = \dfrac{32.5 \sin 60°}{80} = 0.3518$

Hence, $\quad C = \sin^{-1} 0.3518 = 20.60°$

$\qquad$ (or $159.40°$, which is not possible)

and $\qquad B = 180° - 60° - 20.60° = 99.40°$

Applying the sine rule again, $\dfrac{80}{\sin 60°} = \dfrac{b}{\sin 99.40°}$

from which $\qquad b = \dfrac{80 \sin 99.40°}{\sin 60°} = 91.14 \, \text{km}$

Since the cyclist travels 91.14 km in 5 hours,

$\quad$ **average speed** $= \dfrac{\text{distance}}{\text{time}} = \dfrac{91.14}{5} = \textbf{18.23 km/h}$

�⬛ **Problem 9.** Two voltage phasors are shown in Fig. 24.11. If $V_1 = 40$ V and $V_2 = 100$ V, determine the value of their resultant (i.e. length OA) and the angle the resultant makes with V_1

Figure 24.11

Angle $OBA = 180° - 45° = 135°$

Applying the cosine rule,

$$OA^2 = V_1^2 + V_2^2 - 2V_1 V_2 \cos OBA$$
$$= 40^2 + 100^2 - \{2(40)(100) \cos 135°\}$$
$$= 1600 + 10000 - \{-5657\}$$
$$= 1600 + 10000 + 5657 = 17257$$

Thus, $\quad$ **resultant**, $OA = \sqrt{17257} = \textbf{131.4 V}$

Applying the sine rule $\qquad \dfrac{131.4}{\sin 135°} = \dfrac{100}{\sin AOB}$

from which $\quad \sin AOB = \dfrac{100 \sin 135°}{131.4} = 0.5381$

Hence, angle $AOB = \sin^{-1} 0.5381 = 32.55°$ (or $147.45°$, which is not possible)

Hence, **the resultant voltage is 131.4 volts at 32.55° to V_1**

▬

▬ **Problem 10.** In Fig. 24.12, PR represents the inclined jib of a crane and is 10.0 m long. PQ is 4.0 m long. Determine the inclination of the jib to the vertical and the length of tie QR.

Figure 24.12

Applying the sine rule, $\qquad \dfrac{PR}{\sin 120°} = \dfrac{PQ}{\sin R}$

from which $\quad \sin R = \dfrac{PQ \sin 120°}{PR} = \dfrac{(4.0) \sin 120°}{10.0}$

$$= 0.3464$$

Hence, $\quad \angle R = \sin^{-1} 0.3464 = 20.27°$ (or $159.73°$, which is not possible)

$\angle P = 180° - 120° - 20.27° = \textbf{39.73°}$, **which is the inclination of the jib to the vertical**

Applying the sine rule, $\qquad \dfrac{10.0}{\sin 120°} = \dfrac{QR}{\sin 39.73°}$

from which $\quad$ **length of tie,** $QR = \dfrac{10.0 \sin 39.73°}{\sin 120°}$

$$= \textbf{7.38 m}$$

Now try the following Practice Exercise

1. A ship P sails at a steady speed of 45 km/h in a direction of W 32° N (i.e. a bearing of 302°) from a port. At the same time another ship Q leaves the port at a steady speed of 35 km/h in a direction N 15° E (i.e. a bearing of 015°). Determine their distance apart after 4 hours.

2. Two sides of a triangular plot of land are 52.0 m and 34.0 m, respectively. If the area of the plot is 620 m², find (a) the length of fencing required to enclose the plot and (b) the angles of the triangular plot.

3. A jib crane is shown in Fig. 24.13. If the tie rod PR is 8.0 m long and PQ is 4.5 m long, determine (a) the length of jib RQ and (b) the angle between the jib and the tie rod.

Figure 24.13

4. A building site is in the form of a quadrilateral, as shown in Fig. 24.14, and its area is 1510 m². Determine the length of the perimeter of the site.

Figure 24.14

5. Determine the length of members BF and EB in the roof truss shown in Fig. 24.15.

Figure 24.15

6. A laboratory 9.0 m wide has a span roof which slopes at 36° on one side and 44° on the other. Determine the lengths of the roof slopes.

7. PQ and QR are the phasors representing the alternating currents in two branches of a circuit. Phasor PQ is 20.0 A and is horizontal. Phasor QR (which is joined to the end of PQ to form triangle PQR) is 14.0 A and is at an angle of 35° to the horizontal. Determine the resultant phasor PR and the angle it makes with phasor PQ.

24.6 Further practical situations involving trigonometry

Problem 11. A vertical aerial stands on horizontal ground. A surveyor positioned due east of the aerial measures the elevation of the top as 48°. He moves due south 30.0 m and measures the elevation as 44°. Determine the height of the aerial

In Fig. 24.16, DC represents the aerial, A is the initial position of the surveyor and B his final position.

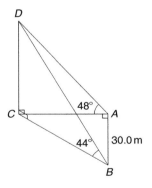

Figure 24.16

From triangle ACD, $\tan 48° = \dfrac{DC}{AC}$ from which

$$AC = \dfrac{DC}{\tan 48°}$$

Similarly, from triangle BCD, $BC = \dfrac{DC}{\tan 44°}$

For triangle ABC, using Pythagoras' theorem,

$$BC^2 = AB^2 + AC^2$$

$$\left(\frac{DC}{\tan 44°}\right)^2 = (30.0)^2 + \left(\frac{DC}{\tan 48°}\right)^2$$

$$DC^2\left(\frac{1}{\tan^2 44°} - \frac{1}{\tan^2 48°}\right) = 30.0^2$$

$$DC^2(1.072323 - 0.810727) = 30.0^2$$

$$DC^2 = \frac{30.0^2}{0.261596} = 3440.4$$

Hence, **height of aerial, $DC = \sqrt{3340.4} = 58.65$ m**.

Problem 12. A crank mechanism of a petrol engine is shown in Fig. 24.17. Arm OA is 10.0 cm long and rotates clockwise about O. The connecting rod AB is 30.0 cm long and end B is constrained to move horizontally.

(a) For the position shown in Fig. 24.17, determine the angle between the connecting rod AB and the horizontal, and the length of OB.

(b) How far does B move when angle AOB changes from 50° to 120°?

Figure 24.17

(a) Applying the sine rule, $\dfrac{AB}{\sin 50°} = \dfrac{AO}{\sin B}$

from which $\sin B = \dfrac{AO \sin 50°}{AB} = \dfrac{10.0 \sin 50°}{30.0}$

$$= 0.2553$$

Hence, $B = \sin^{-1} 0.2553 = 14.79°$ (or 165.21°, which is not possible)

Hence, **the connecting rod AB makes an angle of 14.79° with the horizontal**.

Angle $OAB = 180° - 50° - 14.79° = 115.21°$

Applying the sine rule: $\dfrac{30.0}{\sin 50°} = \dfrac{OB}{\sin 115.21°}$

from which

$$OB = \frac{30.0 \sin 115.21°}{\sin 50°}$$

$$= \mathbf{35.43\,cm}$$

(b) Fig. 24.18 shows the initial and final positions of the crank mechanism.

Figure 24.18

In triangle $OA'B'$, applying the sine rule,

$$\frac{30.0}{\sin 120°} = \frac{10.0}{\sin A'B'O}$$

from which $\sin A'B'O = \dfrac{10.0 \sin 120°}{30.0} = 0.28868$

Hence, $A'B'O = \sin^{-1} 0.28868 = 16.78°$ (or 163.22°, which is not possible)

Angle $OA'B' = 180° - 120° - 16.78° = 43.22°$

Applying the sine rule, $\dfrac{30.0}{\sin 120°} = \dfrac{OB'}{\sin 43.22°}$

from which $OB' = \dfrac{30.0 \sin 43.22°}{\sin 120°} = 23.72$ cm

Since $OB = 35.43$ cm and $OB' = 23.72$ cm, $BB' = 35.43 - 23.72 = 11.71$ cm

Hence, **B moves 11.71 cm when angle AOB changes from 50° to 120°**

Problem 13. The area of a field is in the form of a quadrilateral $ABCD$ as shown in Fig. 24.19. Determine its area

Figure 24.19

A diagonal drawn from B to D divides the quadrilateral into two triangles.

Area of quadrilateral $ABCD$

$$= \text{area of triangle } ABD$$
$$\qquad + \text{area of triangle } BCD$$
$$= \frac{1}{2}(39.8)(21.4)\sin 114°$$
$$\qquad + \frac{1}{2}(42.5)(62.3)\sin 56°$$
$$= 389.04 + 1097.5$$
$$= \mathbf{1487\,m^2}$$

Now try the following Practice Exercise

Practice Exercise 123   More practical situations involving trigonometry (answers on page 455)

1. Three forces acting on a fixed point are represented by the sides of a triangle of dimensions 7.2 cm, 9.6 cm and 11.0 cm. Determine the angles between the lines of action of the three forces.

2. A vertical aerial AB, 9.60 m high, stands on ground which is inclined 12° to the horizontal. A stay connects the top of the aerial A to a point C on the ground 10.0 m downhill from B, the foot of the aerial. Determine (a) the length of the stay and (b) the angle the stay makes with the ground.

3. A reciprocating engine mechanism is shown in Fig. 24.20. The crank AB is 12.0 cm long and the connecting rod BC is 32.0 cm long. For the position shown determine the length of AC and the angle between the crank and the connecting rod.

Figure 24.20

4. From Fig. 24.20, determine how far C moves, correct to the nearest millimetre, when angle CAB changes from 40° to 160°, B moving in an anticlockwise direction.

5. A surveyor standing W 25°S of a tower measures the angle of elevation of the top of the tower as 46°30'. From a position E 23°S from the tower the elevation of the top is 37°15'. Determine the height of the tower if the distance between the two observations is 75 m.

6. Calculate, correct to 3 significant figures, the co-ordinates x and y to locate the hole centre at P shown in Fig. 24.21.

Figure 24.21

7. An idler gear, 30 mm in diameter, has to be fitted between a 70 mm diameter driving gear and a 90 mm diameter driven gear, as shown in Fig. 24.22. Determine the value of angle θ between the centre lines.

Figure 24.22

8. 16 holes are equally spaced on a pitch circle of 70 mm diameter. Determine the length of the chord joining the centres of two adjacent holes.

Practice Exercise 124   Multiple-choice questions on non-right-angled triangles and some practical applications (Answers on page 455)

Each question has only one correct answer

1. In a non-right-angled triangular plate $XYZ, XY = 12$ cm, $YZ = 18$ cm and $\angle XZY = 30°$. The value of $\sin X$ is equal to:
    (a) $\dfrac{1}{3}$ (b) $\dfrac{1}{2}$ (c) $\dfrac{2}{3}$ (d) $\dfrac{3}{4}$

2. In the triangular template DEF shown in Figure 24.23, angle F is equal to:
    (a) 43.5°   (b) 28.6°
    (c) 116.4°   (d) 101.5°

Figure 24.23

3. The length of side DE in Figure 24.23 is:
    (a) 51.25 mm   (b) 46.85 mm
    (c) 50.55 mm   (d) 73.70 mm

4. The area of the triangular template DEF shown in Figure 24.23 is:
    (a) 529.2 mm^2 (b) 258.5 mm^2
    (c) 483.7 mm^2 (d) 371.7 mm^2

5. A triangle has sides $a = 9.0$ cm, $b = 8.0$ cm and $c = 6.0$ cm. Angle A is equal to:
    (a) 82.42°   (b) 56.49°
    (c) 78.58°   (d) 79.87°

6. The area of a triangle PQR is given by:
    (a) $\dfrac{1}{2}$ pr cos Q
    (b) $\sqrt{[(s-p)(s-q)(s-r)]}$
       where $s = \dfrac{p+q+r}{2}$
    (c) $\dfrac{1}{2}$ rq sin P   (d) $\dfrac{1}{2}$ pq sin Q

7. In triangle ABC in Figure 24.24, length AC is:
    (a) 18.15 cm   (b) 14.90 cm
    (c) 13.16 cm   (d) 14.04 cm

Figure 24.24

8. The area of triangle ABC in Figure 24.24 is:
    (a) 70.0 cm^2 (b) 18.51 cm^2
    (c) 67.51 cm^2 (d) 71.85 cm^2

9. In triangle ABC in Figure 24.25, the length AC is:
    (a) 18.79 cm   (b) 70.89 cm
    (c) 22.89 cm   (d) 16.10 cm

Figure 24.25

10. The area of triangle ABC in Figure 24.25 is:
    (a) 53.07 cm^2 (b) 66.47 cm^2
    (c) 11.72 cm^2 (d) 138.78 cm^2

For fully worked solutions to each of the problems in Practice Exercises 120 to 123 in this chapter, go to the website:
www.routledge.com/cw/bird

Chapter 25

Cartesian and polar co-ordinates

Why it is important to understand: **Cartesian and polar co-ordinates**

Applications where polar co-ordinates would be used include terrestrial navigation with sonar-like devices, and those in engineering and science involving energy radiation patterns. Applications where Cartesian co-ordinates would be used include any navigation on a grid and anything involving raster graphics (i.e. bitmap – a dot matrix data structure representing a generally rectangular grid of pixels). The ability to change from Cartesian to polar co-ordinates is vitally important when using complex numbers and their use in a.c. electrical circuit theory and with vector geometry.

At the end of this chapter you should be able to:

- change from Cartesian to polar co-ordinates
- change from polar to Cartesian co-ordinates
- use a scientific notation calculator to change from Cartesian to polar co-ordinates and vice versa

25.1 Introduction

There are two ways in which the position of a point in a plane can be represented. These are

(a) Cartesian co-ordinates, (named after Descartes*), i.e. (x, y)

(b) Polar co-ordinates, i.e. (r, θ), where r is a radius from a fixed point and θ is an angle from a fixed point.

---

*Who was Descartes? See page 167. To find out more go to www.routledge.com/cw/bird

25.2 Changing from Cartesian to polar co-ordinates

In Fig. 25.1, if lengths x and y are known then the length of r can be obtained from Pythagoras' theorem (see Chapter 22) since OPQ is a right-angled triangle.

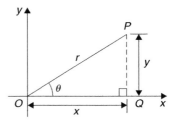

Figure 25.1

Hence, $r^2 = (x^2 + y^2)$, from which $\mathbf{r = \sqrt{x^2 + y^2}}$

From trigonometric ratios (see Chapter 22), $\tan\theta = \dfrac{y}{x}$

from which $\boldsymbol{\theta = \tan^{-1}\dfrac{y}{x}}$

$r = \sqrt{x^2 + y^2}$ and $\theta = \tan^{-1}\dfrac{y}{x}$ are the two formulae we need to change from Cartesian to polar co-ordinates. The angle θ, which may be expressed in degrees or radians, must **always** be measured from the positive x-axis; i.e. measured from the line OQ in Fig. 25.1. It is suggested that when changing from Cartesian to polar co-ordinates a diagram should always be sketched.

Problem 1.   Change the Cartesian co-ordinates (3, 4) into polar co-ordinates

---

A diagram representing the point (3, 4) is shown in Fig. 25.2.

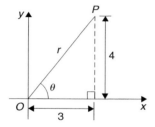

Figure 25.2

From Pythagoras' theorem, $r = \sqrt{3^2 + 4^2} = 5$ (note that -5 has no meaning in this context).

By trigonometric ratios, $\theta = \tan^{-1}\dfrac{4}{3} = 53.13°$ or 0.927 rad.

Note that $53.13° = 53.13 \times \dfrac{\pi}{180}$ rad $= 0.927$ rad.

Hence, **(3, 4) in Cartesian co-ordinates corresponds to (5, 53.13°) or (5, 0.927 rad) in polar co-ordinates**.

Problem 2.   Express in polar co-ordinates the position $(-4, 3)$

A diagram representing the point using the Cartesian co-ordinates $(-4, 3)$ is shown in Fig. 25.3.

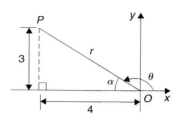

Figure 25.3

From Pythagoras' theorem, $r = \sqrt{4^2 + 3^2} = 5$

By trigonometric ratios, $\alpha = \tan^{-1}\dfrac{3}{4} = 36.87°$ or 0.644 rad

Hence, $\theta = 180° - 36.87° = 143.13°$

or $\theta = \pi - 0.644 = 2.498$ rad

Hence, **the position of point P in polar co-ordinate form is (5, 143.13°) or (5, 2.498 rad)**.

Problem 3.   Express $(-5, -12)$ in polar co-ordinates

A sketch showing the position $(-5, -12)$ is shown in Fig. 25.4.

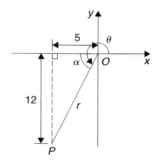

Figure 25.4

$r = \sqrt{5^2 + 12^2} = 13$ and $\alpha = \tan^{-1}\dfrac{12}{5} = 67.38°$ or 1.176 rad

Hence, $\theta = 180° + 67.38° = 247.38°$

or $\theta = \pi + 1.176 = 4.318$ rad.

Thus, **$(-5, -12)$ in Cartesian co-ordinates corresponds to (13, 247.38°) or (13, 4.318 rad) in polar co-ordinates**.

Problem 4.   Express $(2, -5)$ in polar co-ordinates

A sketch showing the position $(2, -5)$ is shown in Fig. 25.5.

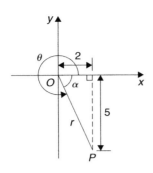

Figure 25.5

$r = \sqrt{2^2 + 5^2} = \sqrt{29} = 5.385$, correct to 3 decimal places

$\alpha = \tan^{-1}\dfrac{5}{2} = 68.20°$ or 1.190 rad

Hence, $\theta = 360° - 68.20° = 291.80°$
or $\theta = \quad 2\pi - 1.190 = 5.093$ rad.

Thus, **(2, −5) in Cartesian co-ordinates corresponds to (5.385, 291.80°) or (5.385, 5.093 rad) in polar co-ordinates**.

Now try the following Practice Exercise

Practice Exercise 125   Changing from Cartesian to polar co-ordinates (answers on page 455)

In Problems 1 to 8, express the given Cartesian co-ordinates as polar co-ordinates, correct to 2 decimal places, in both degrees and radians.

1. (3, 5)   2. (6.18, 2.35)

3. (−2, 4)   4. (−5.4, 3.7)

5. (−7, −3)   6. (−2.4, −3.6)

7. (5, −3)   8. (9.6, −12.4)

25.3 Changing from polar to Cartesian co-ordinates

From the right-angled triangle OPQ in Fig. 25.6,

$$\cos\theta = \frac{x}{r} \text{ and } \sin\theta = \frac{y}{r} \text{ from trigonometric ratios}$$

Hence, $x = r\cos\theta$ and $y = r\sin\theta$

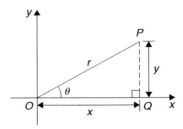

Figure 25.6

If lengths r and angle θ are known then $x = r\cos\theta$ and $y = r\sin\theta$ are the two formulae we need to change from polar to Cartesian co-ordinates.

Problem 5.   Change (4, 32°) into Cartesian co-ordinates

A sketch showing the position (4, 32°) is shown in Fig. 25.7.

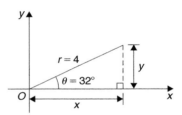

Figure 25.7

Now $x = r\cos\theta = 4\cos 32° = 3.39$

and $y = r\sin\theta = 4\sin 32° = 2.12$

Hence, **(4, 32°) in polar co-ordinates corresponds to (3.39, 2.12) in Cartesian co-ordinates**.

Problem 6.   Express (6, 137°) in Cartesian co-ordinates

A sketch showing the position (6, 137°) is shown in Fig. 25.8.

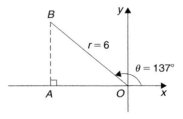

Figure 25.8

$$x = r\cos\theta = 6\cos 137° = -4.388$$

which corresponds to length OA in Fig. 25.8.

$$y = r\sin\theta = 6\sin 137° = 4.092$$

which corresponds to length AB in Fig. 25.8.

Thus, **(6, 137°) in polar co-ordinates corresponds to (−4.388, 4.092) in Cartesian co-ordinates**.

(Note that when changing from polar to Cartesian co-ordinates it is not quite so essential to draw a sketch. Use of $x = r\cos\theta$ and $y = r\sin\theta$ automatically produces the correct values and signs.)

> **Problem 7.**   Express (4.5, 5.16 rad) in Cartesian co-ordinates

A sketch showing the position (4.5, 5.16 rad) is shown in Fig. 25.9.

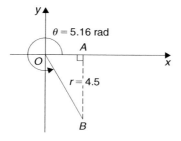

Figure 25.9

$$x = r\cos\theta = 4.5\cos 5.16 = 1.948$$

which corresponds to length OA in Fig. 25.9.

$$y = r\sin\theta = 4.5\sin 5.16 = -4.057$$

which corresponds to length AB in Fig. 25.9.

Thus, **(1.948, −4.057) in Cartesian co-ordinates corresponds to (4.5, 5.16 rad) in polar co-ordinates**.

Now try the following Practice Exercise

> **Practice Exercise 126   Changing polar to Cartesian co-ordinates (answers on page 455)**
>
> In Problems 1 to 8, express the given polar co-ordinates as Cartesian co-ordinates, correct to 3 decimal places.

1. (5, 75°)

2. (4.4, 1.12 rad)

3. (7, 140°)

4. (3.6, 2.5 rad)

5. (10.8, 210°)

6. (4, 4 rad)

7. (1.5, 300°)

8. (6, 5.5 rad)

9. Fig. 25.10 shows 5 equally spaced holes on an 80 mm pitch circle diameter. Calculate their co-ordinates relative to axes Ox and Oy in (a) polar form, (b) Cartesian form.

10. In Fig. 25.10, calculate the shortest distance between the centres of two adjacent holes.

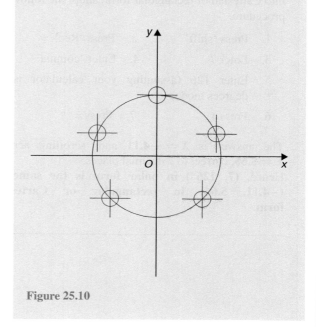

Figure 25.10

25.4   Use of Pol/Rec functions on calculators

Another name for Cartesian co-ordinates is **rectangular** co-ordinates. Many scientific notation calculators have **Pol** and **Rec** functions. 'Rec' is an abbreviation of 'rectangular' (i.e. Cartesian) and 'Pol' is an abbreviation of 'polar'. Check the operation manual for your particular calculator to determine how to use these two functions. They make changing from Cartesian to polar co-ordinates, and vice versa, so much quicker and easier.

For example, with the Casio fx-991ES PLUS calculator, or similar, to change the Cartesian number (3, 4) into polar form, the following procedure is adopted.

1. Press 'shift'   2. Press 'Pol'   3. Enter 3
4. Enter 'comma' (obtained by 'shift' then))
5. Enter 4   6. Press)   7. Press =

The answer is $r = 5, \theta = 53.13°$

Hence, **(3, 4) in Cartesian form is the same as (5, 53.13°) in polar form**.

If the angle is required in **radians**, then before repeating the above procedure press 'shift', 'mode' and then 4 to change your calculator to radian mode.

Similarly, to change the polar form number (7, 126°) into Cartesian or rectangular form, adopt the following procedure.

1. Press 'shift'   2. Press 'Rec'
3. Enter 7   4. Enter 'comma'
5. Enter 126 (assuming your calculator is in degrees mode)
6. Press)   7. Press =

The answer is $X = -4.11$ and, scrolling across, $Y = 5.66$, correct to 2 decimal places.

Hence, **(7, 126°) in polar form is the same as (-4.11, 5.66) in rectangular or Cartesian form**.

Now return to Practice Exercises 125 and 126 in this chapter and use your calculator to determine the answers, and see how much more quickly they may be obtained.

Practice Exercise 127   Multiple-choice questions on Cartesian and polar co-ordinates (answers on page 455)

Each question has only one correct answer

1. $(-4, 3)$ in polar co-ordinates is:
    (a) (5, 2.498 rad)   (b) (7, 36.87°)
    (c) (5, 36.87°)   (d) (5, 323.13°)

2. $(7, 141°)$ in Cartesian co-ordinates is:
    (a) (5.44, -4.41)   (b) (-5.44, -4.41)
    (c) (5.44, 4.41)   (d) (-5.44, 4.41)

3. $(-3, -7)$ in polar co-ordinates is:
    (a) (-7.62, -113.20°)   (b) (7.62, 246.80°)
    (c) (7.62, 23.20°)   (d) (7.62, 203.20°)

4. (6.3, 5.23 rad) in Cartesian co-ordinates is:
    (a) (3.12, -5.47)   (b) (6.27, 0.57)
    (c) (-5.47, 3.12)   (d) (-3.12, 5.47)

5. $(-12, -5)$ in polar co-ordinates is:
    (a) (13, 157.38°)   (b) (13, 2.75 rad)
    (c) (13, -2.75 rad)   (d) (13, 22.62°)

For fully worked solutions to each of the problems in Practice Exercises 125 and 126 in this chapter, go to the website:
www.routledge.com/cw/bird

Revision Test 9: Trigonometric waveforms, non-right-angled triangles, and Cartesian and polar co-ordinates

This assignment covers the material contained in Chapters 23–25. *The marks available are shown in brackets at the end of each question.*

1. A sine wave is given by $y = 8\sin 4x$. State its peak value and its period, in degrees. (2)

2. A periodic function is given by $y = 15\tan 2x$. State its period in degrees. (2)

3. The frequency of a sine wave is $800\,Hz$. Calculate the periodic time in milliseconds. (2)

4. Calculate the frequency of a sine wave that has a periodic time of $40\,\mu s$. (2)

5. Calculate the periodic time for a sine wave having a frequency of $20\,kHz$. (2)

6. An alternating current completes 12 cycles in $16\,ms$. What is its frequency? (3)

7. A sinusoidal voltage is given by $e = 150\sin(500\pi t - 0.25)$ volts. Determine the
   (a) amplitude,
   (b) frequency,
   (c) periodic time,
   (d) phase angle (stating whether it is leading or lagging $150\sin 500\pi t$). (4)

8. Determine the acute angles in degrees, degrees and minutes, and radians.
   (a) $\sin^{-1} 0.4721$ (b) $\cos^{-1} 0.8457$
   (c) $\tan^{-1} 1.3472$ (9)

9. Sketch the following curves, labelling relevant points.
   (a) $y = 4\cos(\theta + 45°)$ (b) $y = 5\sin(2t - 60°)$ (8)

10. The current in an alternating current circuit at any time t seconds is given by $i = 120\sin(100\pi t + 0.274)$ amperes. Determine
    (a) the amplitude, frequency, periodic time and phase angle (with reference to $120\sin 100\pi t$),
    (b) the value of current when $t = 0$,
    (c) the value of current when $t = 6\,ms$.
    Sketch one cycle of the oscillation. (16)

11. A triangular plot of land ABC is shown in Fig. RT9.1. Solve the triangle and determine its area. (10)

Figure RT9.1

12. A car is travelling $20\,m$ above sea level. It then travels $500\,m$ up a steady slope of $17°$. Determine, correct to the nearest metre, how high the car is now above sea level. (3)

13. Fig. RT9.2 shows a roof truss PQR with rafter $PQ = 3\,m$. Calculate the length of
    (a) the roof rise PP',
    (b) rafter PR,
    (c) the roof span QR.
    Find also (d) the cross-sectional area of the roof truss. (11)

Figure RT9.2

14. Solve triangle ABC given $b = 10\,cm, c = 15\,cm$ and $\angle A = 60°$. (10)

15. Change the following Cartesian co-ordinates into polar co-ordinates, correct to 2 decimal places, in both degrees and in radians.
    (a) $(-2.3, 5.4)$ (b) $(7.6, -9.2)$ (10)

16. Change the following polar co-ordinates into Cartesian co-ordinates, correct to 3 decimal places.
    (a) $(6.5, 132°)$ (b) $(3, 3\,rad)$ (6)

Chapter 26

Areas of common shapes

Why it is important to understand: Areas of common shapes

To paint, wallpaper or panel a wall, you must know the total area of the wall so you can buy the appropriate amount of finish. When designing a new building, or seeking planning permission, it is often necessary to specify the total floor area of the building. In construction, calculating the area of a gable end of a building is important when determining the number of bricks and amount of mortar to order. When using a bolt, the most important thing is that it is long enough for your particular application and it may also be necessary to calculate the shear area of the bolt connection. Ridge vents allow a home to properly vent, while disallowing rain or other forms of precipitation to leak into the attic or crawlspace underneath the roof. Equal amounts of cool air and warm air flowing through the vents is paramount for proper heat exchange. Calculating how much surface area is available on the roof aids in determining how long the ridge vent should run. Arches are everywhere, from sculptures and monuments to pieces of architecture and strings on musical instruments; finding the height of an arch or its cross-sectional area is often required. Determining the cross-sectional areas of beam structures is vitally important in design engineering. There are thus a large number of situations in engineering where determining area is important.

At the end of this chapter you should be able to:

- state the SI unit of area
- identify common polygons – triangle, quadrilateral, pentagon, hexagon, heptagon and octagon
- identify common quadrilaterals – rectangle, square, parallelogram, rhombus and trapezium
- calculate areas of quadrilaterals and circles
- appreciate that areas of similar shapes are proportional to the squares of the corresponding linear dimensions

26.1 Introduction

Area is a measure of the size or extent of a plane surface. Area is measured in **square units** such as mm^2, cm^2 and m^2. This chapter deals with finding the areas of common shapes.

In engineering it is often important to be able to calculate simple areas of various shapes. In everyday life its important to be able to measure area to, say, lay a carpet, order sufficient paint for a decorating job or order sufficient bricks for a new wall.

On completing this chapter you will be able to recognise common shapes and be able to find the areas of rectangles, squares, parallelograms, triangles, trapeziums and circles.

26.2 Common shapes

Polygons

A polygon is a closed plane figure bounded by straight lines. A polygon which has

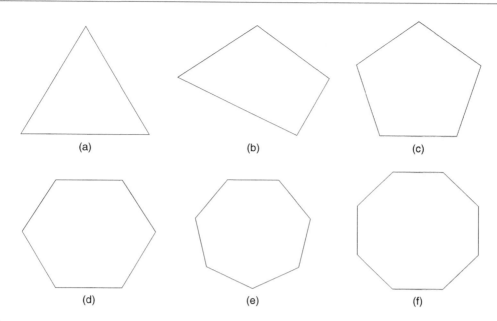

(a)   (b)   (c)

(d)   (e)   (f)

Figure 26.1

3 sides is called a **triangle** – see Fig. 26.1(a)
4 sides is called a **quadrilateral** – see Fig. 26.1(b)
5 sides is called a **pentagon** – see Fig. 26.1(c)
6 sides is called a **hexagon** – see Fig. 26.1(d)
7 sides is called a **heptagon** – see Fig. 26.1(e)
8 sides is called an **octagon** – see Fig. 26.1(f)

Quadrilaterals

There are five types of quadrilateral, these being rectangle, square, parallelogram, rhombus and trapezium.
If the opposite corners of any quadrilateral are joined by a straight line, two triangles are produced. Since the sum of the angles of a triangle is 180°, the sum of the angles of a quadrilateral is 360°

Rectangle

In the rectangle *ABCD* shown in Fig. 26.2,

(a)   all four angles are right angles,

(b)   the opposite sides are parallel and equal in length, and

(c)   diagonals *AC* and *BD* are equal in length and bisect one another.

Square

In the square *PQRS* shown in Fig. 26.3,

(a)   all four angles are right angles,

(b)   the opposite sides are parallel,

(c)   all four sides are equal in length, and

(d)   diagonals *PR* and *QS* are equal in length and bisect one another at right angles.

Parallelogram

In the parallelogram *WXYZ* shown in Fig. 26.4,

(a)   opposite angles are equal,

(b)   opposite sides are parallel and equal in length, and

(c)   diagonals *WY* and *XZ* bisect one another.

Rhombus

In the rhombus *ABCD* shown in Fig. 26.5,

(a)   opposite angles are equal,

(b)   opposite angles are bisected by a diagonal,

(c)   opposite sides are parallel,

(d)   all four sides are equal in length, and

(e)   diagonals *AC* and *BD* bisect one another at right angles.

Trapezium

In the trapezium *EFGH* shown in Fig. 26.6,

(a)   only one pair of sides is parallel.

Figure 26.2   **Figure 26.3**

Figure 26.4

Figure 26.5   **Figure 26.6**

Problem 1. State the types of quadrilateral shown in Fig. 26.7 and determine the angles marked a to l

Figure 26.7

(a) **$ABCD$ is a square**

The diagonals of a square bisect each of the right angles, hence

$$a = \frac{90°}{2} = 45°$$

(b) **$EFGH$ is a rectangle**

In triangle FGH, $40° + 90° + b = 180°$, since the angles in a triangle add up to $180°$, from which $b = 50°$. Also, $c = 40°$ (alternate angles between parallel lines EF and HG). (Alternatively, b and c are complementary; i.e. add up to $90°$)

$d = 90° + c$ (external angle of a triangle equals the sum of the interior opposite angles), hence $d = 90° + 40° = 130°$ (or $\angle EFH = 50°$ and $d = 180° - 50° = 130°$)

(c) **$JKLM$ is a rhombus**

The diagonals of a rhombus bisect the interior angles and the opposite internal angles are equal. Thus, $\angle JKM = \angle MKL = \angle JMK = \angle LMK = 30°$, hence, $e = 30°$

In triangle KLM, $30° + \angle KLM + 30° = 180°$ (the angles in a triangle add up to $180°$), hence, $\angle KLM = 120°$. The diagonal JL bisects $\angle KLM$, hence, $f = \dfrac{120°}{2} = 60°$

(d) **$NOPQ$ is a parallelogram**

$g = 52°$ since the opposite interior angles of a parallelogram are equal.

In triangle NOQ, $g + h + 65° = 180°$ (the angles in a triangle add up to $180°$), from which $h = 180° - 65° - 52° = 63°$

$i = 65°$ (alternate angles between parallel lines NQ and OP).

$j = 52° + i = 52° + 65° = 117°$ (the external angle of a triangle equals the sum of the interior opposite angles). (Alternatively, $\angle PQO = h = 63°$; hence, $j = 180° - 63° = 117°$)

(e) **$RSTU$ is a trapezium**

$35° + k = 75°$ (external angle of a triangle equals the sum of the interior opposite angles), hence, $k = 40°$

$\angle STR = 35°$ (alternate angles between parallel lines RU and ST). $l + 35° = 115°$ (external angle of a triangle equals the sum of the interior opposite angles), hence, $l = 115° - 35° = 80°$

Now try the following Practice Exercise

Practice Exercise 128   Common shapes (answers on page 455)

1. Find the angles p and q in Fig. 26.8(a).

2. Find the angles r and s in Fig. 26.8(b).

3. Find the angle t in Fig. 26.8(c).

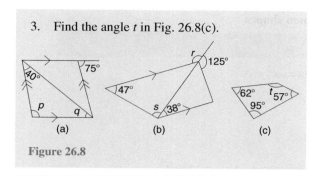

Figure 26.8

26.3   Areas of common shapes

The formulae for the areas of common shapes are shown in Table 26.1, on page 278.

Here are some worked problems to demonstrate how the formulae are used to determine the area of common shapes.

Problem 2. Calculate the area and length of the perimeter of the metal square shown in Fig. 26.9

Figure 26.9

Area of square $= x^2 = (4.0)^2 = 4.0\,\text{cm} \times 4.0\,\text{cm}$

$$= \mathbf{16.0\,cm^2}$$

(Note the unit of area is cm × cm = cm^2; i.e. square centimetres or centimetres squared.)

Perimeter of square $= 4.0\,\text{cm} + 4.0\,\text{cm} + 4.0\,\text{cm}$

$$+ 4.0\,\text{cm} = \mathbf{16.0\,cm}$$

Problem 3. Calculate the area and length of the perimeter of the rectangular plate shown in Fig. 26.10

Figure 26.10

Area of rectangle $= l \times b = 7.0 \times 4.5$

$$= \mathbf{31.5\,cm^2}$$

Perimeter of rectangle $= 7.0\,\text{cm} + 4.5\,\text{cm}$

$$+ 7.0\,\text{cm} + 4.5\,\text{cm}$$

$$= \mathbf{23.0\,cm}$$

Problem 4. Calculate the area of the aluminium sheet in the form of a parallelogram shown in Fig. 26.11

Figure 26.11

Area of a parallelogram = base × perpendicular height

The perpendicular height h is not shown in Fig. 26.11 but may be found using Pythagoras' theorem (see Chapter 22).

From Fig. 26.12, $9^2 = 5^2 + h^2$, from which $h^2 = 9^2 - 5^2 = 81 - 25 = 56$

Hence, perpendicular height,

$$h = \sqrt{56} = 7.48\,\text{mm}.$$

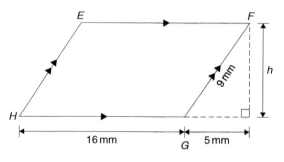

Figure 26.12

Hence, **area of parallelogram EFGH**

$$= 16\,\text{mm} \times 7.48\,\text{mm}$$

$$= \mathbf{120\,mm^2}$$

Table 26.1 Formulae for the areas of common shapes

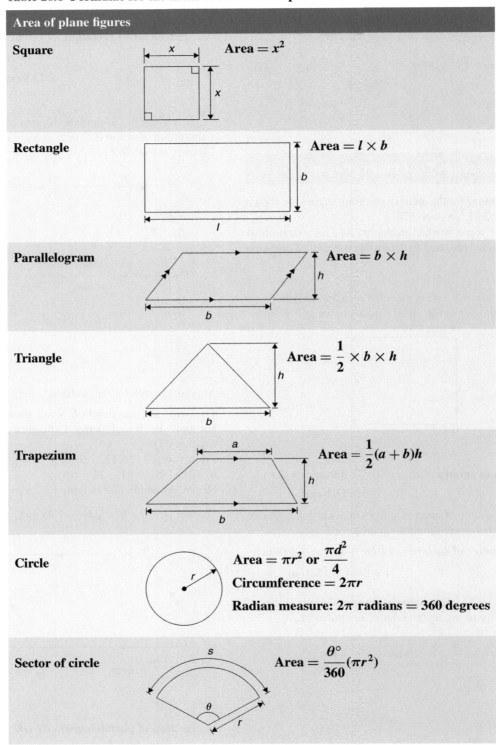

Area of plane figures		
Square		$\text{Area} = x^2$
Rectangle		$\text{Area} = l \times b$
Parallelogram		$\text{Area} = b \times h$
Triangle		$\text{Area} = \frac{1}{2} \times b \times h$
Trapezium		$\text{Area} = \frac{1}{2}(a+b)h$
Circle		$\text{Area} = \pi r^2$ or $\dfrac{\pi d^2}{4}$ $\text{Circumference} = 2\pi r$ Radian measure: 2π radians $= 360$ degrees
Sector of circle		$\text{Area} = \dfrac{\theta^\circ}{360}(\pi r^2)$

Problem 5. Calculate the area of the triangular plastic template shown in Fig. 26.13

Figure 26.13

Area of triangle $IJK = \dfrac{1}{2} \times$ base $\times$ perpendicular height

$$= \dfrac{1}{2} \times IJ \times JK$$

To find JK, Pythagoras' theorem is used; i.e.

$$5.68^2 = 1.92^2 + JK^2, \text{ from which}$$

$$JK = \sqrt{5.68^2 - 1.92^2} = 5.346 \text{cm}$$

Hence, **area of triangle IJK** $= \dfrac{1}{2} \times 1.92 \times 5.346$

$$= \mathbf{5.132\,cm^2}$$

Problem 6. Calculate the area of the trapezium shown in Fig. 26.14

Figure 26.14

Area of a trapezium $= \dfrac{1}{2} \times$ (sum of parallel sides)

$$\times \text{(perpendicular distance between the parallel sides)}$$

Hence, **area of trapezium $LMNO$**

$$= \dfrac{1}{2} \times (27.4 + 8.6) \times 5.5$$

$$= \dfrac{1}{2} \times 36 \times 5.5 = \mathbf{99\,mm^2}$$

Problem 7. A rectangular tray is 820 mm long and 400 mm wide. Find its area in (a) mm² (b) cm² (c) m²

(a) **Area of tray** $=$ length $\times$ width $= 820 \times 400$

$$= \mathbf{328000\,mm^2}$$

(b) Since $1\,\text{cm} = 10\,\text{mm}, 1\,\text{cm}^2 = 1\,\text{cm} \times 1\,\text{cm}$

$$= 10\,\text{mm} \times 10\,\text{mm} = 100\,\text{mm}^2, \text{ or}$$

$$1\,\text{mm}^2 = \dfrac{1}{100}\text{cm}^2 = 0.01\,\text{cm}^2$$

Hence, $\mathbf{328000\,mm^2} = 328000 \times 0.01\,\text{cm}^2$

$$= \mathbf{3280\,cm^2}$$

(c) Since $1\,\text{m} = 100\,\text{cm}, 1\,\text{m}^2 = 1\,\text{m} \times 1\,\text{m}$

$$= 100\,\text{cm} \times 100\,\text{cm} = 10000\,\text{cm}^2, \text{ or}$$

$$1\,\text{cm}^2 = \dfrac{1}{10000}\text{m}^2 = 0.0001\text{m}^2$$

Hence, $\mathbf{3280\,cm^2} = 3280 \times 0.0001\,\text{m}^2$

$$= \mathbf{0.3280\,m^2}$$

Problem 8. The outside measurements of a picture frame are 100 cm by 50 cm. If the frame is 4 cm wide, find the area of the wood used to make the frame

A sketch of the frame is shown shaded in Fig. 26.15.

Figure 26.15

Area of wood $=$ area of large rectangle $-$ area of small rectangle

$$= (100 \times 50) - (92 \times 42)$$

$$= 5000 - 3864$$

$$= \mathbf{1136\,cm^2}$$

Problem 9. Find the cross-sectional area of the girder shown in Fig. 26.16

Figure 26.16

The girder may be divided into three separate rectangles, as shown.

$$\text{Area of rectangle } A = 50 \times 5 = 250\,\text{mm}^2$$

$$\text{Area of rectangle } B = (75 - 8 - 5) \times 6$$

$$= 62 \times 6 = 372\,\text{mm}^2$$

$$\text{Area of rectangle } C = 70 \times 8 = 560\,\text{mm}^2$$

Total area of girder $= 250 + 372 + 560$

$$= \mathbf{1182\,mm^2 \text{ or } 11.82\,cm^2}$$

Problem 10. Fig. 26.17 shows the gable end of a building. Determine the area of brickwork in the gable end

Figure 26.17

The shape is that of a rectangle and a triangle.

$$\text{Area of rectangle } = 6 \times 8 = 48\,\text{m}^2$$

$$\text{Area of triangle } = \frac{1}{2} \times \text{base} \times \text{height}$$

$CD = 4\,$m and $AD = 5\,$m, hence $AC = 3\,$m (since it is a 3, 4, 5 triangle – or by Pythagoras).

Hence, area of triangle $ABD = \frac{1}{2} \times 8 \times 3 = 12\,\text{m}^2$

Total area of brickwork $= 48 + 12$

$$= \mathbf{60\,m^2}$$

Now try the following Practice Exercise

Practice Exercise 129   Areas of common shapes (answers on page 455)

1. Name the types of quadrilateral shown in Fig. 26.18(i) to (iv) and determine for each (a) the area and (b) the perimeter.

Figure 26.18

2. A rectangular plate is 85 mm long and 42 mm wide. Find its area in square centimetres.

3. A rectangular field has an area of 1.2 hectares and a length of 150 m. If 1 hectare $= 10000\,\text{m}^2$, find (a) the field's width and (b) the length of a diagonal.

4. Find the area of a triangular template whose base is 8.5 cm and perpendicular height is 6.4 cm.

5. A square box lid has an area of 162 cm². Determine the length of a diagonal.

6. A rectangular picture has an area of 0.96 m². If one of the sides has a length of 800 mm, calculate, in millimetres, the length of the other side.

7. Determine the area of each of the angle iron sections shown in Fig. 26.19.

Figure 26.19

8. Fig. 26.20 shows a 4 m wide path around the outside of a 41 m by 37 m garden. Calculate the area of the path.

Figure 26.20

9. The area of a trapezium is 13.5 cm² and the perpendicular distance between its parallel sides is 3 cm. If the length of one of the parallel sides is 5.6 cm, find the length of the other parallel side.

10. Calculate the area of the steel plate shown in Fig. 26.21.

Figure 26.21

11. Determine the area of an equilateral triangle of side 10.0 cm.

12. If paving slabs are produced in 250 mm by 250 mm squares, determine the number of slabs required to cover an area of 2 m².

13. Fig. 26.22 shows a plan view of an office block to be built. The walls will have a height of 8 m, and it is necessary to make an estimate of the number of bricks required to build the walls. Assuming that any doors and windows are ignored in the calculation and that 48 bricks are required to build 1 m² of wall, calculate the number of external bricks required.

Figure 26.22

Here are some further worked problems on finding the areas of common shapes, using the formulae in Table 26.1, page 270.

Problem 11. Find the area of a circular gold medallion having a radius of 5 cm

$$\text{Area of circle} = \pi r^2 = \pi(5)^2 = 25\pi = \mathbf{78.54\,cm^2}$$

Problem 12. Find the area of a circular metal plate having a diameter of 15 mm

$$\textbf{Area of circle} = \frac{\pi d^2}{4} = \frac{\pi(15)^2}{4} = \frac{225\pi}{4} = \mathbf{176.7\,mm^2}$$

Problem 13. Find the area of a circular disc having a circumference of 70 mm

Circumference, $c = 2\pi r$, hence

$$\text{radius}, r = \frac{c}{2\pi} = \frac{70}{2\pi} = \frac{35}{\pi}\,\text{mm}$$

$$\textbf{Area of circle} = \pi r^2 = \pi\left(\frac{35}{\pi}\right)^2 = \frac{35^2}{\pi}$$

$$= \mathbf{389.9\,mm^2 \text{ or } 3.899\,cm^2}$$

Problem 14. Calculate the area of the sector of a circle having radius 6 cm with angle subtended at centre 50°

$$\textbf{Area of sector} = \frac{\theta}{360}(\pi r^2) = \frac{50}{360}(\pi 6^2)$$

$$= \frac{50 \times \pi \times 36}{360} = \textbf{15.71 cm}^2$$

Problem 15. Calculate the area of the sector of a circle having diameter 80 mm with angle subtended at centre 107°42′

If diameter = 80 mm then radius, $r = 40$ mm, and

$$\text{area of sector} = \frac{107°42′}{360}(\pi 40^2) = \frac{107\frac{42}{60}}{360}(\pi 40^2)$$

$$= \frac{107.7}{360}(\pi 40^2)$$

$$= \textbf{1504 mm}^2 \text{ or } \textbf{15.04 cm}^2$$

Problem 16. A hollow shaft has an outside diameter of 5.45 cm and an inside diameter of 2.25 cm. Calculate the cross-sectional area of the shaft

The cross-sectional area of the shaft is shown by the shaded part in Fig. 26.23 (often called an **annulus**).

Figure 26.23

Area of shaded part = area of large circle − area of
small circle

$$= \frac{\pi D^2}{4} - \frac{\pi d^2}{4} = \frac{\pi}{4}(D^2 - d^2)$$

$$= \frac{\pi}{4}(5.45^2 - 2.25^2)$$

$$= \textbf{19.35 cm}^2$$

Now try the following Practice Exercise

Practice Exercise 130   Areas of common shapes (answers on page 455)

1. A rectangular garden measures 40 m by 15 m. A 1 m flower border is made round the two shorter sides and one long side. A circular swimming pool of diameter 8 m is constructed in the middle of the garden. Find, correct to the nearest square metre, the area remaining.

2. Determine the area of a circular template having (a) a radius of 4 cm (b) a diameter of 30 mm (c) a circumference of 200 mm.

3. A washer in the form of an annulus has an outside diameter of 60 mm and an inside diameter of 20 mm. Determine its area.

4. If the area of a circle is 320 mm², find (a) its diameter and (b) its circumference.

5. Calculate the areas of the following sectors of circles.

   (a) radius 9 cm, angle subtended at centre 75°
   (b) diameter 35 mm, angle subtended at centre 48°37′

6. Determine the shaded area of the template shown in Fig. 26.24.

Figure 26.24

7. An archway consists of a rectangular opening topped by a semicircular arch, as shown in Fig. 26.25. Determine the area of the opening if the width is 1 m and the greatest height is 2 m.

Figure 26.25

8. A base plate is in the form of a quadrant of a circle of radius 0.5 m. Calculate the area and perimeter of the plate.

9. A rectangular gasket 350 mm by 200 mm has four holes cut in it, each of diameter 60 mm. Calculate the area of the gasket in square centimetres.

Here are some further worked problems of common shapes.

Problem 17. Calculate the area of a regular octagonal metal feature if each side is 5 cm and the width across the flats is 12 cm

An octagon is an 8-sided polygon. If radii are drawn from the centre of the polygon to the vertices then 8 equal triangles are produced, as shown in Fig. 26.26.

Figure 26.26

$$\text{Area of one triangle} = \frac{1}{2} \times \text{base} \times \text{height}$$
$$= \frac{1}{2} \times 5 \times \frac{12}{2} = 15 \text{cm}^2$$
Area of octagon $= 8 \times 15 = \mathbf{120\,cm^2}$

Problem 18. Determine the area of a regular hexagonal sheet of metal which has sides 8 cm long

A hexagon is a 6-sided polygon which may be divided into 6 equal triangles as shown in Fig. 26.27. The angle subtended at the centre of each triangle is $360° \div 6 = 60°$. The other two angles in the triangle add up to 120° and are equal to each other. Hence, each of the triangles is equilateral with each angle 60° and each side 8 cm.

Figure 26.27

$$\text{Area of one triangle} = \frac{1}{2} \times \text{base} \times \text{height}$$
$$= \frac{1}{2} \times 8 \times h$$

h is calculated using Pythagoras' theorem:

$$8^2 = h^2 + 4^2$$
from which $h = \sqrt{8^2 - 4^2} = 6.928 \text{cm}$

Hence,

$$\text{Area of one triangle} = \frac{1}{2} \times 8 \times 6.928 = 27.71 \text{cm}^2$$
Area of hexagon $= 6 \times 27.71$
$$= \mathbf{166.3\,cm^2}$$

Problem 19. Fig. 26.28 shows a plan of a floor of a building which is to be carpeted. Calculate the area of the floor in square metres. Calculate the cost, correct to the nearest pound, of carpeting the floor with carpet costing £21.50 per m², assuming

30% extra carpet is required due to wastage in fitting

Figure 26.28

Area of floor plan

= area of triangle ABC + area of semicircle

+ area of rectangle $CGLM$

+ area of rectangle $CDEF$

− area of trapezium $HIJK$

Triangle ABC is equilateral since $AB = BC = 3$ m and, hence, angle $B'CB = 60°$

$$\sin B'CB = BB'/3$$

i.e. $$BB' = 3\sin 60° = 2.598\,\text{m}.$$

$$\text{Area of triangle } ABC = \frac{1}{2}(AC)(BB')$$

$$= \frac{1}{2}(3)(2.598) = 3.897\,\text{m}^2$$

$$\text{Area of semicircle} = \frac{1}{2}\pi r^2 = \frac{1}{2}\pi(2.5)^2$$

$$= 9.817\,\text{m}^2$$

$$\text{Area of } CGLM = 5 \times 7 = 35\,\text{m}^2$$

$$\text{Area of } CDEF = 0.8 \times 3 = 2.4\,\text{m}^2$$

$$\text{Area of } HIJK = \frac{1}{2}(KH + IJ)(0.8)$$

Since $MC = 7$ m then $LG = 7$ m, hence $IJ = 7 - 5.2 = 1.8$ m. Hence,

$$\text{Area of } HIJK = \frac{1}{2}(3 + 1.8)(0.8) = 1.92\,\text{m}^2$$

Total floor area $= 3.897 + 9.817 + 35 + 2.4 - 1.92$

$$= 49.194\,\text{m}^2$$

To allow for 30% wastage, amount of carpet required $= 1.3 \times 49.194 = 63.95\,\text{m}^2$

Cost of carpet at £21.50 per m^2
$= 63.95 \times 21.50 = $ **£1375**, correct to the nearest pound.

Now try the following Practice Exercise

Practice Exercise 131　Areas of common shapes (answers on page 455)

1. Calculate the area of a regular octagonal metal template if each side is 20 mm and the width across the flats is 48.3 mm.

2. Determine the area of a regular hexagonal wooden feature which has sides 25 mm.

3. A plot of land is in the shape shown in Fig. 26.29. Determine

Figure 26.29

   (a) its area in hectares (1 ha $= 10^4\,\text{m}^2$)

   (b) the length of fencing required, to the nearest metre, to completely enclose the plot of land.

4. Calculate the number of turns (to the nearest whole number) on a solenoid which is made by winding 25 m of fine copper wire around a cylindrical former of diameter 26 mm.

5. Determine the area of the largest regular hexagon that can be cut from a circular sheet of aluminium of radius 400 mm. Give the answer in square metres, correct to 3 significant figures.

26.4 Areas of similar shapes

Fig. 26.30 shows two squares, one of which has sides three times as long as the other.

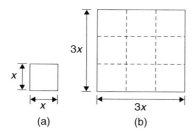

(a)   (b)

Figure 26.30

$$\text{Area of Fig. 26.30(a)} = (x)(x) = x^2$$

$$\text{Area of Fig. 26.30(b)} = (3x)(3x) = 9x^2$$

Hence, Fig. 26.30(b) has an area $(3)^2$; i.e. 9 times the area of Fig. 26.30(a).

In summary, **the areas of similar shapes are proportional to the squares of corresponding linear dimensions**.

Problem 20. A rectangular garage is shown on a building plan having dimensions 10 mm by 20 mm. If the plan is drawn to a scale of 1 to 250, determine the true area of the garage in square metres

$$\text{Area of garage on the plan} = 10\,\text{mm} \times 20\,\text{mm}$$

$$= 200\,\text{mm}^2$$

Since the areas of similar shapes are proportional to the squares of corresponding dimensions,

$$\text{True area of garage} = 200 \times (250)^2$$

$$= 12.5 \times 10^6\,\text{mm}^2$$

$$= \frac{12.5 \times 10^6}{10^6}\,\text{m}^2$$

$$\text{since } 1\,\text{m}^2 = 10^6\,\text{mm}^2$$

$$= \mathbf{12.5\,m^2}$$

Now try the following Practice Exercise

Practice Exercise 132   Areas of similar shapes (answers on page 456)

1. The area of a park on a map is $500\,\text{mm}^2$. If the scale of the map is 1 to 40 000, determine the true area of the park in hectares (1 hectare $= 10^4\,\text{m}^2$).

2. A model of a boiler is made having an overall height of 75 mm corresponding to an overall height of the actual boiler of 6 m. If the area of metal required for the model is $12\,500\,\text{mm}^2$, determine, in square metres, the area of metal required for the actual boiler.

3. The scale of an Ordnance Survey map is $1 : 2500$. A circular sports field has a diameter of 8 cm on the map. Calculate its area in hectares, giving your answer correct to 3 significant figures. (1 hectare $= 10^4\,\text{m}^2$)

Practice Exercise 133   Multiple-choice questions on areas of common shapes (answers on page 456)

Each question has only one correct answer

1. If the circumference of a circle is 100 mm, its area is:
    (a) 314.2 cm^2 (b) 7.96 cm^2
    (c) 31.83 mm^2 (d) 78.54 cm^2

2. The outside measurements of a rectangular picture frame are 80 cm by 30 cm. If the frame is 3 cm wide, the area of the metal used to make the frame is:
    (a) 624 cm^2 (b) 2079 cm^2
    (c) 660 cm^2 (d) 588 cm^2

3. A square piece of metal plate has sides x cm. A square of side $(x-3)$ is machined out of the centre. In terms of x, the remaining area of metal in square centimetres is:
    (a) $2x^2 - 9$ (b) $6x - 9$
    (c) $2x^2 - 6x - 9$ (d) $6x + 9$

4. The interior angle of a regular hexagon is:
    (a) 60°   (b) 360°   (c) 150°   (d) 120°

5. The length of each side of an equilateral triangle is 2 cm. The area of the triangle is:

    (a) $2\,\text{cm}^2$ (b) $\sqrt{3}\,\text{cm}^2$
    (c) $4\,\text{cm}^2$ (d) $\sqrt{5}\,\text{cm}^2$

6. The area of the path shown shaded in Figure 26.31 is:

    (a) $300\,\text{m}^2$ (b) $234\,\text{m}^2$
    (c) $124\,\text{m}^2$ (d) $66\,\text{m}^2$

Figure 26.31

7. The area of a semi-circular faceplate with a diameter of 10 cm is:

    (a) $12.5\pi\,\text{cm}^2$ (b) $25\pi\,\text{cm}^2$
    (c) $50\pi\,\text{cm}^2$ (d) $100\pi\,\text{cm}^2$

8. A steel shape is a square of area $120\,\text{cm}^2$. The length of the diagonal of the square is:

    (a) $4.68\,\text{cm}$ (b) $10.95\,\text{cm}$
    (c) $15.49\,\text{cm}$ (d) $169.71\,\text{cm}$

9. A hollow shaft has an outside diameter of 8 cm and an inside diameter of 4 cm. The cross-sectional area of the shaft is:

    (a) $20\pi\,\text{cm}^2$ (b) $48\pi\,\text{cm}^2$
    (c) $12\pi\,\text{cm}^2$ (d) $24\pi\,\text{cm}^2$

10. A rectangular building is shown on a building plan having dimensions 20 mm by 10 mm. If the plan is drawn to a scale of 1 to 300, the true area of the building in m^2 is:

    (a) $60000\,\text{m}^2$ (b) $1800\,\text{m}^2$
    (c) $0.06\,\text{m}^2$ (d) $18\,\text{m}^2$

For fully worked solutions to each of the problems in Practice Exercises 128 to 132 in this chapter, go to the website:
www.routledge.com/cw/bird

Chapter 27

The circle and its properties

Why it is important to understand: **The circle and its properties**

A circle is one of the fundamental shapes of geometry; it consists of all the points that are equidistant from a central point. Knowledge of calculations involving circles is needed with crank mechanisms, with determinations of latitude and longitude, with pendulums, and even in the design of paper clips. The floodlit area at a football ground, the area an automatic garden sprayer sprays and the angle of lap of a belt drive all rely on calculations involving the arc of a circle. The ability to handle calculations involving circles and their properties is clearly essential in several branches of engineering design.

At the end of this chapter you should be able to:

- define a circle
- state some properties of a circle – including radius, circumference, diameter, semicircle, quadrant, tangent, sector, chord, segment and arc
- appreciate the angle in a semicircle is a right angle
- define a radian, and change radians to degrees, and vice versa
- determine arc length, area of a circle and area of a sector of a circle
- state the equation of a circle
- sketch a circle given its equation

27.1 Introduction

A **circle** is a plane figure enclosed by a curved line, every point on which is equidistant from a point within, called the **centre**.

In Chapter 26, worked problems on the areas of circles and sectors were demonstrated. In this chapter, properties of circles are listed and arc lengths are calculated, together with more practical examples on the areas of sectors of circles. Finally, the equation of a circle is explained.

27.2 Properties of circles

(a) The distance from the centre to the curve is called the **radius**, *r*, of the circle (see *OP* in Fig. 27.1).

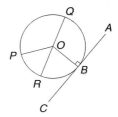

Figure 27.1

(b) The boundary of a circle is called the **circumference**, c

(c) Any straight line passing through the centre and touching the circumference at each end is called the **diameter**, *d* (see *QR* in Fig. 27.1). Thus, *d = 2r*

(d) The ratio $\dfrac{\text{circumference}}{\text{diameter}}$ is a constant for any circle. This constant is denoted by the Greek letter π (pronounced 'pie'), where $\pi = 3.14159$, correct to 5 decimal places (check with your calculator). Hence, $\dfrac{c}{d} = \pi$ or $c = \pi d$ or $c = 2\pi r$

(e) A **semicircle** is one half of a whole circle.

(f) A **quadrant** is one quarter of a whole circle.

(g) A **tangent** to a circle is a straight line which meets the circle at one point only and does not cut the circle when produced. *AC* in Fig. 27.1 is a tangent to the circle since it touches the curve at point *B* only. If radius *OB* is drawn, **angle *ABO* is a right angle**.

(h) The **sector** of a circle is the part of a circle between radii (for example, the portion *OXY* of Fig. 27.2 is a sector). If a sector is less than a semicircle it is called a **minor sector**; if greater than a semicircle it is called a **major sector**.

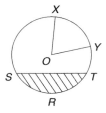

Figure 27.2

(i) The **chord** of a circle is any straight line which divides the circle into two parts and is terminated at each end by the circumference. *ST*, in Fig. 27.2, is a chord.

(j) **Segment** is the name given to the parts into which a circle is divided by a chord. If the segment is less than a semicircle it is called a **minor segment** (see shaded area in Fig. 27.2). If the segment is greater than a semicircle it is called a **major segment** (see the un-shaded area in Fig. 27.2).

(k) An **arc** is a portion of the circumference of a circle. The distance *SRT* in Fig. 27.2 is called a **minor arc** and the distance *SXYT* is called a **major arc**.

(l) The angle at the centre of a circle, subtended by an arc, is double the angle at the circumference subtended by the same arc. With reference to Fig. 27.3,

Angle *AOC* = 2 × angle *ABC*

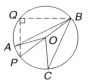

Figure 27.3

(m) The angle in a semicircle is a right angle (see angle *BQP* in Fig. 27.3).

⚑ **Problem 1.** Find the circumference of a circular template of radius 12.0 cm

Circumference, $c = 2 \times \pi \times \text{radius} = 2\pi r = 2\pi(12.0)$

$$= \mathbf{75.40\,cm}$$

⚑ **Problem 2.** If the diameter of a circular disc is 75 mm, find its circumference

Circumference, $c = \pi \times \text{diameter} = \pi d = \pi(75)$

$$= \mathbf{235.6\,mm}$$

⚑ **Problem 3.** Determine the radius of a circular pond if its perimeter is 112 m

Perimeter = circumference, $c = 2\pi r$

Hence, **radius of pond,** $r = \dfrac{c}{2\pi} = \dfrac{112}{2\pi} = \mathbf{17.83\,cm}$

Problem 4. In Fig. 27.4, *AB* is a tangent to the circle at *B*. If the circle radius is 40 mm and *AB* = 150 mm, calculate the length *AO*

Figure 27.4

A tangent to a circle is at right angles to a radius drawn from the point of contact; i.e. $\mathbf{ABO = 90°}$. Hence, using Pythagoras' theorem,

$$AO^2 = AB^2 + OB^2$$

from which, $$AO = \sqrt{AB^2 + OB^2}$$

$$= \sqrt{150^2 + 40^2} = \mathbf{155.2\,mm}$$

Now try the following Practice Exercise

Practice Exercise 134   Properties of a circle
(answers on page 456)

1. Calculate the length of the circumference of a circular template of radius 7.2 cm.

2. If the diameter of a circular disc is 82.6 mm, calculate the circumference of the disc.

3. Determine the radius of a circular badge whose circumference is 16.52 cm.

4. Find the diameter of a circular metal plate whose perimeter is 149.8 cm.

5. A crank mechanism is shown in Fig. 27.5, where XY is a tangent to the circle at point X. If the circle radius OX is 10 cm and length OY is 40 cm, determine the length of the connecting rod XY.

Figure 27.5

6. If the circumference of the earth is 40 000 km at the equator, calculate its diameter.

7. Calculate the length of wire in the paper clip shown in Fig. 27.6. The dimensions are in millimetres.

Figure 27.6

27.3   Radians and degrees

One **radian** is defined as the angle subtended at the centre of a circle by an arc equal in length to the radius. With reference to Fig. 27.7, for arc length s,

$$\theta \text{ radians } = \frac{s}{r}$$

Figure 27.7

When $s = $ whole circumference $(= 2\pi r)$ then

$$\theta = \frac{s}{r} = \frac{2\pi r}{r} = 2\pi$$

i.e. $\mathbf{2\pi}$ **radians** $\mathbf{= 360°}$ or $\mathbf{\pi}$ **radians** $\mathbf{= 180°}$

Thus, $\mathbf{1\,rad} = \dfrac{\mathbf{180°}}{\pi} = \mathbf{57.30°}$, correct to 2 decimal places.

Since $\pi\,\text{rad} = 180°$, then $\dfrac{\pi}{2} = 90°, \dfrac{\pi}{3} = 60°, \dfrac{\pi}{4} = 45°$, and so on.

Problem 5.   Convert to radians: (a) 125°
(b) 69°47′

(a)   Since $180° = \pi$ rad, $1° = \dfrac{\pi}{180}$ rad, therefore

$$125° = 125\left(\dfrac{\pi}{180}\right) \text{rad} = \textbf{2.182 radians}.$$

(b) $69°47' = 69\dfrac{47°}{60} = 69.783°$ (or, with your calculator, enter $69°47'$ using $°\;'\;''$ function, press $=$ and press $°\;'\;''$ again).

and $69.783° = 69.783\left(\dfrac{\pi}{180}\right)$ rad

$$= \textbf{1.218 radians}.$$

> **Problem 6.**   Convert to degrees and minutes:
> (a) 0.749 radians (b) $3\pi/4$ radians

(a)   Since π rad $= 180°$, 1 rad $= \dfrac{180°}{\pi}$

therefore 0.749 rad $= 0.749\left(\dfrac{180}{\pi}\right)^{°} = 42.915°$

$0.915° = (0.915 \times 60)' = 55'$, correct to the nearest minute.

Hence, **0.749 radians $= 42°55'$**

(b)   Since 1 rad $= \left(\dfrac{180}{\pi}\right)^{°}$ then

$$\dfrac{3\pi}{4}\text{ rad} = \dfrac{3\pi}{4}\left(\dfrac{180}{\pi}\right)^{°} = \dfrac{3}{4}(180)° = \textbf{135°}$$

> **Problem 7.**   Express in radians, in terms of π,
> (a) 150° (b) 270° (c) 37.5°

Since $180° = \pi$ rad, $1° = \dfrac{\pi}{180}$ rad

(a) $150° = 150\left(\dfrac{\pi}{180}\right)$ rad $= \dfrac{5\pi}{6}$ **rad**

(b) $270° = 270\left(\dfrac{\pi}{180}\right)$ rad $= \dfrac{3\pi}{2}$ **rad**

(c) $37.5° = 37.5\left(\dfrac{\pi}{180}\right)$ rad $= \dfrac{75\pi}{360}$ rad $= \dfrac{5\pi}{24}$ **rad**

Now try the following Practice Exercise

Practice Exercise 135   Radians and degrees (answers on page 456)

> 1.   Convert to radians in terms of π:
>   (a)   30°   (b)   75°   (c)   225°
>
> 2.   Convert to radians, correct to 3 decimal places:
>   (a)   48°   (b)   84°51'   (c)   232°15'

> 3.   Convert to degrees:
>   (a) $\dfrac{7\pi}{6}$ rad   (b) $\dfrac{4\pi}{9}$ rad   (c) $\dfrac{7\pi}{12}$ rad
>
> 4.   Convert to degrees and minutes:
>   (a)   0.0125 rad   (b)   2.69 rad
>   (c)   7.241 rad
>
> 5.   A car engine speed is 1000 rev/min. Convert this speed into rad/s.

27.4   Arc length and area of circles and sectors

Arc length

From the definition of the radian in the previous section and Fig. 27.7,

$$\textbf{arc length}, s = r\theta \qquad \text{where } \theta \text{ is in radians}$$

Area of a circle

From Chapter 26, for any circle, area $= \pi \times (\text{radius})^2$

i.e. $$\textbf{area} = \pi r^2$$

Since $r = \dfrac{d}{2}$, $$\textbf{area} = \pi r^2 \text{ or } \dfrac{\pi d^2}{4}$$

Area of a sector

$$\textbf{Area of a sector} = \dfrac{\theta}{360}(\pi r^2) \text{ when } \theta \text{ is in degrees}$$

$$= \dfrac{\theta}{2\pi}(\pi r^2)$$

$$= \dfrac{1}{2}r^2\theta \qquad \text{when } \theta \text{ is in radians}$$

> **Problem 8.**   A hockey pitch has a semicircle of radius 14.63 m around each goal net. Find the area enclosed by the semicircle, correct to the nearest square metre

Area of a semicircle $= \dfrac{1}{2}\pi r^2$

When $r = 14.63$ m,   area $= \dfrac{1}{2}\pi\left(14.63\right)^2$

i.e.   **area of semicircle $= 336$ m^2**

Problem 9. Find the area of a circular metal plate having a diameter of 35.0 mm, correct to the nearest square millimetre

Area of a circle $= \pi r^2 = \dfrac{\pi d^2}{4}$

When $d = 35.0$ mm,   area $= \dfrac{\pi \left(35.0\right)^2}{4}$

i.e.   **area of circular plate = 962 mm²**

Problem 10. Find the area of a circular copper disc having a circumference of 60.0 mm

Circumference, $c = 2\pi r$

from which radius, $r = \dfrac{c}{2\pi} = \dfrac{60.0}{2\pi} = \dfrac{30.0}{\pi}$

Area of a circular disc $= \pi r^2$

i.e.   **area** $= \pi \left(\dfrac{30.0}{\pi}\right)^2 = \mathbf{286.5\,mm^2}$

Problem 11. Find the length of the arc of a circle of radius 5.5 cm when the angle subtended at the centre is 1.20 radians

Length of arc, $s = r\theta$, where θ is in radians.
Hence, **arc length, $s = (5.5)(1.20) = \mathbf{6.60\,cm}$**.

Problem 12. Determine the diameter and circumference of a circular metal plate if an arc of length 4.75 cm subtends an angle of 0.91 radians

Since arc length, $s = r\theta$ then
radius, $r = \dfrac{s}{\theta} = \dfrac{4.75}{0.91} = 5.22$ cm
Diameter $= 2 \times$ radius $= 2 \times 5.22 = \mathbf{10.44\,cm}$
Circumference, $c = \pi d = \pi(10.44) = \mathbf{32.80\,cm}$

Problem 13. If an angle of 125° is subtended by an arc of a circle of radius 8.4 cm, find the length of (a) the minor arc and (b) the major arc, correct to 3 significant figures

Since $180° = \pi$ rad then $1° = \left(\dfrac{\pi}{180}\right)$ rad and
$125° = 125\left(\dfrac{\pi}{180}\right)$ rad

(a)   Length of minor arc,

$$s = r\theta = (8.4)(125)\left(\dfrac{\pi}{180}\right) = \mathbf{18.3\,cm},$$

correct to 3 significant figures

(b)   Length of major arc = (circumference − minor arc) $= 2\pi(8.4) - 18.3 = \mathbf{34.5\,cm}$, correct to 3 significant figures.
(Alternatively, major arc $= r\theta$
$$= 8.4(360 - 125)\left(\dfrac{\pi}{180}\right) = \mathbf{34.5\,cm})$$

Problem 14. Determine the angle, in degrees and minutes, subtended at the centre of a circular disc of diameter 42 mm by an arc of length 36 mm. Calculate also the area of the minor sector formed

Since length of arc, $s = r\theta$ then $\theta = \dfrac{s}{r}$

Radius, $r = \dfrac{\text{diameter}}{2} = \dfrac{42}{2} = 21$ mm,

hence $\theta = \dfrac{s}{r} = \dfrac{36}{21} = 1.7143$ radians.

$1.7143\,\text{rad} = 1.7143 \times \left(\dfrac{180}{\pi}\right)^{\circ} = 98.22° = \mathbf{98°13'} =$ angle subtended at centre of circle.
From page 281,

$$\textbf{area of sector} = \dfrac{1}{2}r^2\theta = \dfrac{1}{2}(21)^2(1.7143)$$
$$= \mathbf{378\,mm^2}$$

Problem 15. A football stadium floodlight can spread its illumination over an angle of 45° to a distance of 55 m. Determine the maximum area that is floodlit.

Floodlit area = area of sector $= \dfrac{1}{2}r^2\theta$

$$= \dfrac{1}{2}(55)^2 \left(45 \times \dfrac{\pi}{180}\right)$$
$$= \mathbf{1188\,m^2}$$

Problem 16. An automatic garden sprayer produces spray to a distance of 1.8 m and revolves through an angle α which may be varied. If the desired spray catchment area is to be 2.5 m², to what should angle α be set, correct to the nearest degree?

Area of sector $= \dfrac{1}{2}r^2\theta$, hence $2.5 = \dfrac{1}{2}\left(1.8\right)^2\alpha$

from which,

$$\alpha = \frac{2.5 \times 2}{1.8^2} = 1.5432 \text{ radians}$$

$$1.5432 \text{ rad} = \left(1.5432 \times \frac{180}{\pi}\right)^\circ = 88.42^\circ$$

Hence, **angle $\alpha = 88^\circ$**, correct to the nearest degree.

Problem 17. The angle of a tapered groove is checked using a 20 mm diameter roller as shown in Fig. 27.8. If the roller lies 2.12 mm below the top of the groove, determine the value of angle θ

Figure 27.8

In Fig. 27.9, triangle ABC is right-angled at C (see property (g) in Section 27.2).

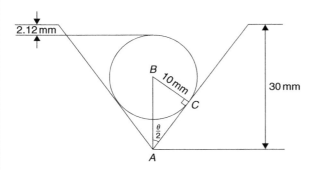

Figure 27.9

Length $BC = 10$ mm (i.e. the radius of the circle), and $AB = 30 - 10 - 2.12 = 17.88$ mm, from Fig. 27.9.

Hence, $\sin\frac{\theta}{2} = \frac{10}{17.88}$ and $\frac{\theta}{2} = \sin^{-1}\left(\frac{10}{17.88}\right) = 34^\circ$

and **angle $\theta = 68^\circ$**.

Now try the following Practice Exercise

Practice Exercise 136   Arc length and area of circles and sectors (answers on page 456)

1. Calculate the area of a circular DVD of radius 6.0 cm, correct to the nearest square centimetre.

2. The diameter of a circular medallion is 55.0 mm. Determine its area, correct to the nearest square millimetre.

3. The perimeter of a circular coin is 150 mm. Find its area, correct to the nearest square millimetre.

4. Find the area of the sector, correct to the nearest square millimetre, of a circle having a radius of 35 mm with angle subtended at centre of 75°

5. An annulus-shaped washer has an outside diameter of 49.0 mm and an inside diameter of 15.0 mm. Find its area correct to 4 significant figures.

6. Find the area, correct to the nearest square metre, of a 2 m wide path surrounding a circular plot of land 200 m in diameter.

7. A rectangular park measures 50 m by 40 m. A 3 m flower bed is made round the two longer sides and one short side. A circular fish pond of diameter 8.0 m is constructed in the centre of the park. It is planned to grass the remaining area. Find, correct to the nearest square metre, the area of grass.

8. With reference to Fig. 27.10, determine (a) the perimeter and (b) the area.

Figure 27.10

9. Find the area of the shaded portion of Fig. 27.11.

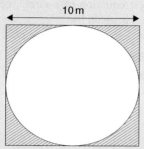

Figure 27.11

10. Find the length of an arc of a circle of radius 8.32 cm when the angle subtended at the centre is 2.14 radians. Calculate also the area of the minor sector formed.

11. If the angle subtended at the centre of a circle of diameter 82 mm is 1.46 rad, find the lengths of the (a) minor arc and (b) major arc.

12. A pendulum of length 1.5 m swings through an angle of 10° in a single swing. Find, in centimetres, the length of the arc traced by the pendulum bob.

13. Determine the shaded area of the section shown in Fig. 27.12

Figure 27.12

14. Determine the length of the radius and circumference of a circle if an arc length of 32.6 cm subtends an angle of 3.76 radians.

15. Determine the angle of lap, in degrees and minutes, if 180 mm of a belt drive are in contact with a pulley of diameter 250 mm.

16. Determine the number of complete revolutions a motorcycle wheel will make in travelling 2 km if the wheel's diameter is 85.1 cm.

17. A floodlight at a sports ground spread its illumination over an angle of 40° to a distance of 48 m. Determine (a) the angle in radians and (b) the maximum area that is floodlit.

18. Find the area swept out in 50 minutes by the minute hand of a large floral clock if the hand is 2 m long.

19. A helicopter landing pad is circular with a diameter of 5 m. Calculate its area.

20. A square plate of side 50 mm has 5 holes cut out of it. If the five holes are all circles of diameter 10 mm calculate the remaining area, correct to the nearest whole number.

21. A circular plastic plate has a radius of 60 mm. A sector subtends an angle of 1.6 radians at the centre. Calculate the area of the sector.

22. The end plate of a steel boiler is a flat circular plate of diameter 2.5 m. Holes are drilled in it as follows: two 500 mm diameter tube holes and 24 water tube holes each of diameter 40 mm. Calculate the remaining area of the end plate in square metres, correct to 3 significant figures.

23. A wheel of diameter 600 mm rolls without slipping along a horizontal surface. If the wheel rotates 4.5 revolutions calculate how far the wheel moves horizontally. Give the answer in metres correct to 3 significant figures.

24. A tight open belt passes directly over two pulleys, of diameters 1.2 m and 1.8 m respectively having their centres 2.4 m apart. Calculate the length of the belt, correct to 3 significant figures.

25. A rubber gasket for a cylinder is a circular annulus of area is 4084 mm² . If the outer diameter is 140 mm calculate the inner radius of the gasket, correct to the nearest whole number.

26. Determine (a) the shaded area in Fig. 27.13 and (b) the percentage of the whole sector that the shaded area represents.

Figure 27.13

27. Determine the length of steel strip required to make the clip shown in Fig. 27.14.

Figure 27.14

28. A 50° tapered hole is checked with a 40 mm diameter ball as shown in Fig. 27.15. Determine the length shown as x.

Figure 27.15

27.5 The equation of a circle

The simplest equation of a circle, centre at the origin and radius r, is given by

$$x^2 + y^2 = r^2$$

For example, Fig. 27.16 shows a circle $x^2 + y^2 = 9$

More generally, the equation of a circle, centre (a, b) and radius r, is given by

$$(x - a)^2 + (y - b)^2 = r^2 \qquad (1)$$

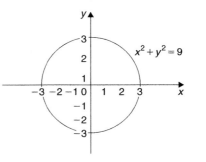

Figure 27.16

Fig. 27.17 shows a circle $(x - 2)^2 + (y - 3)^2 = 4$

The general equation of a circle is

$$x^2 + y^2 + 2ex + 2fy + c = 0 \qquad (2)$$

Multiplying out the bracketed terms in equation (1) gives

$$x^2 - 2ax + a^2 + y^2 - 2by + b^2 = r^2$$

Comparing this with equation (2) gives

$$2e = -2a, \text{ i.e. } \boldsymbol{a = -\frac{2e}{2}}$$

and

$$2f = -2b, \text{ i.e. } \boldsymbol{b = -\frac{2f}{2}}$$

and

$$c = a^2 + b^2 - r^2, \text{ i.e. } \boldsymbol{r = \sqrt{a^2 + b^2 - c}}$$

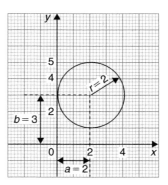

Figure 27.17

Thus, for example, the equation

$$x^2 + y^2 - 4x - 6y + 9 = 0$$

represents a circle with centre,

$$a = -\left(\frac{-4}{2}\right), b = -\left(\frac{-6}{2}\right) \text{ i.e. at (2, 3) and}$$

radius, $r = \sqrt{2^2 + 3^2 - 9} = 2$

Hence, $x^2 + y^2 - 4x - 6y + 9 = 0$ is the circle shown in Fig. 27.17 (which may be checked by multiplying out the brackets in the equation $(x-2)^2 + (y-3)^2 = 4$)

Problem 18. Determine (a) the radius and (b) the co-ordinates of the centre of the circle given by the equation $x^2 + y^2 + 8x - 2y + 8 = 0$

$x^2 + y^2 + 8x - 2y + 8 = 0$ is of the form shown in equation (2),

where $a = -\left(\frac{8}{2}\right) = -4, b = -\left(\frac{-2}{2}\right) = 1$

and $r = \sqrt{(-4)^2 + 1^2 - 8} = \sqrt{9} = 3$

Hence, $x^2 + y^2 + 8x - 2y + 8 = 0$ represents a circle **centre $(-4, 1)$** and **radius 3**, as shown in Fig. 27.18.

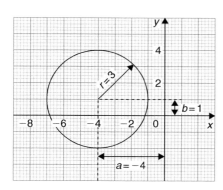

Figure 27.18

Alternatively, $x^2 + y^2 + 8x - 2y + 8 = 0$ may be re-arranged as

$$(x+4)^2 + (y-1)^2 - 9 = 0$$

i.e. $$(x+4)^2 + (y-1)^2 = 3^2$$

which represents a circle, **centre $(-4, 1)$** and **radius 3**, as stated above.

Problem 19. Sketch the circle given by the equation $x^2 + y^2 - 4x + 6y - 3 = 0$

The equation of a circle, centre (a, b), radius r is given by

$$(x - a)^2 + (y - b)^2 = r^2$$

The general equation of a circle is
$x^2 + y^2 + 2ex + 2fy + c = 0$

From above $a = -\dfrac{2e}{2}$, $b = -\dfrac{2f}{2}$ and $r = \sqrt{a^2 + b^2 - c}$

Hence, if $x^2 + y^2 - 4x + 6y - 3 = 0$

then $a = -\left(\dfrac{-4}{2}\right) = 2, b = -\left(\dfrac{6}{2}\right) = -3$ and

$$r = \sqrt{2^2 + (-3)^2 - (-3)} = \sqrt{16} = 4$$

Thus, the circle has **centre $(2, -3)$** and **radius 4**, as shown in Fig. 27.19.

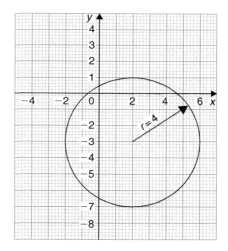

Figure 27.19

Alternatively, $x^2 + y^2 - 4x + 6y - 3 = 0$ may be re-arranged as

$$(x-2)^2 + (y+3)^2 - 3 - 13 = 0$$

i.e. $$(x-2)^2 + (y+3)^2 = 4^2$$

which represents a circle, **centre $(2, -3)$** and **radius 4**, as stated above.

Now try the following Practice Exercise

> **Practice Exercise 137 The equation of a circle (answers on page 456)**
>
> 1. Determine (a) the radius and (b) the co-ordinates of the centre of the circle given by the equation $x^2 + y^2 - 6x + 8y + 21 = 0$
>
> 2. Sketch the circle given by the equation $x^2 + y^2 - 6x + 4y - 3 = 0$
>
> 3. Sketch the curve $x^2 + (y-1)^2 - 25 = 0$
>
> 4. Sketch the curve $x = 6\sqrt{\left[1 - \left(\dfrac{y}{6}\right)^2\right]}$

> **Practice Exercise 138 Multiple-choice questions on the circle and its properties (answers on page 456)**
>
> **Each question has only one correct answer**
>
> 1. An arc of a circle of length 5.0 cm subtends an angle of 2 radians. The circumference of the circle is:
>   (a) 2.5 cm (b) 10.0 cm
>   (c) 5.0 cm (d) 15.7 cm
>
> 2. A pendulum of length 1.2 m swings through an angle of $12°$ in a single swing. The length of arc traced by the pendulum bob is:
>   (a) 14.40 cm (b) 25.13 cm
>   (c) 10.00 cm (d) 45.24 cm
>
> 3. A wheel on a car has a diameter of 800 mm. If the car travels 5 miles, the number of complete revolutions the wheel makes (given 1 km = $\dfrac{5}{8}$ mile) is:
>   (a) 1989 (b) 1591
>   (c) 3183 (d) 10000

> 4. The equation of a circle is
> $$x^2 + 2x + y^2 - 10y + 22 = 0.$$
>   The co-ordinates of its centre are:
>   (a) $(1, 5)$ (b) $(-2, 4)$ (c) $(-1, 5)$ (d) $(2, 4)$
>
> 5. A semicircle has a radius of 2 cm. Its perimeter is:
>   (a) $(8 + 2\pi)$ cm (b) 2π cm
>   (c) $(4 + 2\pi)$ cm (d) $(4\pi + 2)$ cm
>
> 6. The equation of a circle is
> $$x^2 + y^2 - 2x + 4y - 4 = 0.$$
>   Which of the following statements is correct?
>   (a) The circle has centre $(1, -2)$ and radius 4
>   (b) The circle has centre $(-1, 2)$ and radius 2
>   (c) The circle has centre $(-1, -2)$ and radius 4
>   (d) The circle has centre $(1, -2)$ and radius 3
>
> 7. The area of the sector of a circle of radius 9 cm is 27π cm^2. The angle of the sector, in degrees, is:
>   (a) $120°$ (b) $60°$ (c) $11°$ (d) $240°$
>
> 8. The length of the arc of a circle of radius 5 cm when the angle subtended at the centre is $32°$ is:
>   (a) 3.49 cm (b) 2.79 cm
>   (c) 0.44 cm (d) 6.98 cm
>
> 9. A sector of a circle subtends an angle of 1.2 radians and has an arc length of 6 cm. The circumference of the circle is:
>   (a) 10π cm (b) 5π cm
>   (c) 15π cm (d) 25π cm
>
> 10. A circle has its centre at co-ordinates $(3, -2)$ and has a radius of 4. The equation of the circle is:
>   (a) $(x+3)^2 + (y-2)^2 = 42$
>   (b) $(x-3)^2 + (y+2)^2 = 16$
>   (c) $x^2 + y^2 = 52$
>   (d) $(x-3)^2 + (y-2)^2 = 16$

For fully worked solutions to each of the problems in Practice Exercises 134 to 137 in this chapter, go to the website:
www.routledge.com/cw/bird

This assignment covers the material contained in Chapters 26 and 27. *The marks available are shown in brackets at the end of each question.*

1. A rectangular metal plate has an area of $9\,600\,\text{cm}^2$. If the length of the plate is 1.2 m, calculate the width, in centimetres. (3)

2. Calculate the cross-sectional area of the angle iron section shown in Fig. RT10.1, the dimensions being in millimetres. (4)

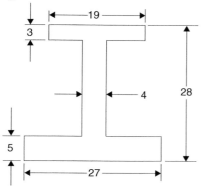

Figure RT10.1

3. Find the area of the trapezium *MNOP* shown in Fig. RT10.2 when $a = 6.3\,\text{cm}, b = 11.7\,\text{cm}$ and $h = 5.5\,\text{cm}$. (3)

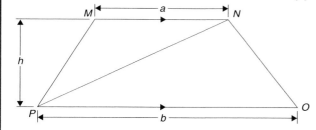

Figure RT10.2

4. Find the area of the triangle *DEF* in Fig. RT10.3, correct to 2 decimal places. (4)

5. A rectangular park measures 150 m by 70 m. A 2 m flower border is constructed round the two longer sides and one short side. A circular fish pond of diameter 15 m is in the centre of the park and the remainder of the park is grass. Calculate, correct to the nearest square metre, the area of (a) the fish pond, (b) the flower borders and (c) the grass. (6)

6. A swimming pool is 55 m long and 10 m wide. The perpendicular depth at the deep end is 5 m and at

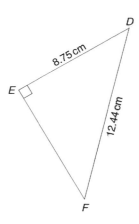

Figure RT10.3

the shallow end is 1.5 m, the slope from one end to the other being uniform. The inside of the pool needs two coats of a protective paint before it is filled with water. Determine how many litres of paint will be needed if 1 litre covers $10\,\text{m}^2$ (7)

7. Find the area of an equilateral triangle of side 20.0 cm. (4)

8. A steel template is of the shape shown in Fig. RT10.4, the circular area being removed. Determine the area of the template, in square centimetres, correct to 1 decimal place. (8)

Figure RT10.4

9. The area of a plot of land on a map is $400\,\text{mm}^2$. If the scale of the map is 1 to 50 000,

determine the true area of the land in hectares (1 hectare $= 10^4 \, \text{m}^2$) (4)

10. Determine the shaded area in Fig. RT10.5, correct to the nearest square centimetre. (3)

Figure RT10.5

11. Determine the diameter of a circle, correct to the nearest millimetre, whose circumference is $178.4 \, \text{cm}$. (2)

12. Calculate the area of a circle of radius $6.84 \, \text{cm}$, correct to 1 decimal place. (2)

13. The circumference of a circle is $250 \, \text{mm}$. Find its area, correct to the nearest square millimetre. (4)

14. Find the area of the sector of a circle having a radius of $50.0 \, \text{mm}$, with angle subtended at centre of $120°$. (3)

15. Determine the total area of the shape shown in Fig. RT10.6, correct to 1 decimal place. (7)

Figure RT10.6

16. The radius of a circular cricket ground is $75 \, \text{m}$. The boundary is painted with white paint and 1 tin of paint will paint a line $22.5 \, \text{m}$ long. How many tins of paint are needed? (3)

17. Find the area of a $1.5 \, \text{m}$ wide path surrounding a circular plot of land $100 \, \text{m}$ in diameter. (3)

18. A cyclometer shows 2530 revolutions in a distance of $3.7 \, \text{km}$. Find the diameter of the wheel in centimetres, correct to 2 decimal places. (4)

19. The minute hand of a wall clock is $10.5 \, \text{cm}$ long. How far does the tip travel in the course of 24 hours? (4)

20. Convert
   (a) $125°47'$ to radians.
   (b) 1.724 radians to degrees and minutes. (4)

21. Calculate the length of metal strip needed to make the clip shown in Fig. RT10.7. (7)

Figure RT10.7

22. A lorry has wheels of radius $50 \, \text{cm}$. Calculate the number of complete revolutions a wheel makes (correct to the nearest revolution) when travelling 3 miles (assume $1 \, \text{mile} = 1.6 \, \text{km}$). (4)

23. The equation of a circle is
   $x^2 + y^2 + 12x - 4y + 4 = 0$. Determine
   (a) the diameter of the circle.
   (b) the co-ordinates of the centre of the circle. (7)

Chapter 28

Volumes and surface areas of common solids

Why it is important to understand: Volumes and surface areas of common solids

There are many practical applications where volumes and surface areas of common solids are required. Examples include determining capacities of oil, water, petrol and fish tanks, ventilation shafts and cooling towers, determining volumes of blocks of metal, ball-bearings, boilers and buoys, and calculating the cubic metres of concrete needed for a path. Finding the surface areas of loudspeaker diaphragms and lampshades provide further practical examples. Understanding these calculations is essential for the many practical applications in engineering, construction, architecture and science.

At the end of this chapter you should be able to:

- state the SI unit of volume
- calculate the volumes and surface areas of cuboids, cylinders, prisms, pyramids, cones and spheres
- calculate volumes and surface areas of frusta of pyramids and cones
- appreciate that volumes of similar bodies are proportional to the cubes of the corresponding linear dimensions

28.1 Introduction

The **volume** of any solid is a measure of the space occupied by the solid. Volume is measured in **cubic units** such as mm^3, cm^3 and m^3.

This chapter deals with finding volumes of common solids; in engineering it is often important to be able to calculate volume or capacity to estimate, say, the amount of liquid, such as water, oil or petrol, in different shaped containers.

A **prism** is a solid with a constant cross-section and with two ends parallel. The shape of the end is used to describe the prism. For example, there are rectangular prisms (called cuboids), triangular prisms and circular prisms (called cylinders).

On completing this chapter you will be able to calculate the volumes and surface areas of rectangular and other prisms, cylinders, pyramids, cones and spheres, together with frusta of pyramids and cones. Volumes of similar shapes are also considered.

28.2 Volumes and surface areas of common shapes

Cuboids or rectangular prisms

A cuboid is a solid figure bounded by six rectangular faces; all angles are right angles and opposite faces are equal. A typical cuboid is shown in Fig. 28.1 with length l, breadth b and height h.

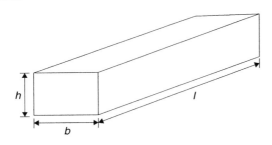

Figure 28.1

$$\text{Volume of cuboid} = l \times b \times h$$

and

$$\text{surface area} = 2bh + 2hl + 2lb = 2(bh + hl + lb)$$

A **cube** is a square prism. If all the sides of a cube are x then

$$\text{Volume} = x^3 \text{ and } \text{surface area} = 6x^2$$

Problem 1. A metal cuboid container has dimensions of 12 cm by 4 cm by 3 cm. Determine (a) its volume and (b) its total surface area

The cuboid is similar to that in Fig. 28.1, with $l = 12$ cm, $b = 4$ cm and $h = 3$ cm.

(a) **Volume of cuboid** $= l \times b \times h = 12 \times 4 \times 3$
$$= \mathbf{144 \, cm^3}$$

(b) **Surface area** $= 2(bh + hl + lb)$
$$= 2(4 \times 3 + 3 \times 12 + 12 \times 4)$$
$$= 2(12 + 36 + 48)$$
$$= 2 \times 96 = \mathbf{192 \, cm^2}$$

Problem 2. An oil tank is the shape of a cube, each edge being of length 1.5 m. Determine (a) the maximum capacity of the tank in m³ and litres and (b) its total surface area ignoring input and output orifices

(a) **Volume of oil tank** = volume of cube
$$= 1.5 \, \text{m} \times 1.5 \, \text{m} \times 1.5 \, \text{m}$$
$$= 1.5^3 \, \text{m}^3 = \mathbf{3.375 \, m^3}$$

$1 \, \text{m}^3 = 100 \, \text{cm} \times 100 \, \text{cm} \times 100 \, \text{cm} = 10^6 \, \text{cm}^3$.
Hence,
$$\text{volume of tank} = 3.375 \times 10^6 \, \text{cm}^3$$

$1 \, \text{litre} = 1000 \, \text{cm}^3$, hence **oil tank capacity**
$$= \frac{3.375 \times 10^6}{1000} \, \text{litres} = \mathbf{3375 \, litres}$$

(b) Surface area of one side $= 1.5 \, \text{m} \times 1.5 \, \text{m}$
$$= 2.25 \, \text{m}^2$$

A cube has six identical sides, hence

$$\textbf{total surface area of oil tank} = \mathbf{6} \times \mathbf{2.25}$$
$$= \mathbf{13.5 \, m^2}$$

Problem 3. A water tank is the shape of a rectangular prism having length 2 m, breadth 75 cm and height 500 mm. Determine the capacity of the tank in (a) m³ (b) cm³ (c) litres

Capacity means volume; when dealing with liquids, the word capacity is usually used.
The water tank is similar in shape to that in Fig. 28.1, with $l = 2$ m, $b = 75$ cm and $h = 500$ mm.

(a) Capacity of water tank $= l \times b \times h$. To use this formula, all dimensions **must** be in the same units. Thus, $l = 2$ m, $b = 0.75$ m and $h = 0.5$ m (since $1 \, \text{m} = 100 \, \text{cm} = 1000 \, \text{mm}$). Hence,

$$\textbf{capacity of tank} = 2 \times 0.75 \times 0.5 = \mathbf{0.75 \, m^3}$$

(b) $1 \, \text{m}^3 = 1 \, \text{m} \times 1 \, \text{m} \times 1 \, \text{m}$
$$= 100 \, \text{cm} \times 100 \, \text{cm} \times 100 \, \text{cm}$$
i.e. $\mathbf{1 \, m^3 = 1\,000\,000 = 10^6 \, cm^3}$. Hence,
$$\textbf{capacity} = 0.75 \, \text{m}^3 = 0.75 \times 10^6 \, \text{cm}^3$$
$$= \mathbf{750\,000 \, cm^3}$$

(c) $1 \, \text{litre} = 1000 \, \text{cm}^3$. Hence,

$$\mathbf{750\,000 \, cm^3} = \frac{750{,}000}{1000} = \mathbf{750 \, litres}$$

Cylinders

A cylinder is a circular prism. A cylinder of radius r and height h is shown in Fig. 28.2.

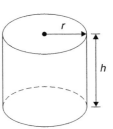

Figure 28.2

$$\textbf{Volume} = \pi r^2 h$$

$$\textbf{Curved surface area} = 2\pi r h$$

$$\textbf{Total surface area} = 2\pi r h + 2\pi r^2$$

Total surface area means the curved surface area plus the area of the two circular ends.

Problem 4. A solid metal cylinder has a base diameter of 12 cm and a perpendicular height of 20 cm. Calculate (a) the volume and (b) the total surface area

(a) **Volume** $= \pi r^2 h = \pi \times \left(\dfrac{12}{2}\right)^2 \times 20$

$$= 720\pi = \textbf{2262 cm}^3$$

(b) **Total surface area**

$$= 2\pi r h + 2\pi r^2$$

$$= (2 \times \pi \times 6 \times 20) + (2 \times \pi \times 6^2)$$

$$= 240\pi + 72\pi = 312\pi = \textbf{980 cm}^2$$

Problem 5. A copper pipe has the dimensions shown in Fig. 28.3. Calculate the volume of copper in the pipe, in cubic metres.

2.5 m

12 cm

25 cm

Figure 28.3

Outer diameter, $D = 25$ cm $= 0.25$ m and inner diameter, $d = 12$ cm $= 0.12$ m.

Area of cross-section of copper

$$= \frac{\pi D^2}{4} - \frac{\pi d^2}{4} = \frac{\pi(0.25)^2}{4} - \frac{\pi(0.12)^2}{4}$$

$$= 0.0491 - 0.0113 = 0.0378 \, \text{m}^2$$

Hence, **volume of copper**

$$= (\text{cross-sectional area}) \times \text{length of pipe}$$

$$= 0.0378 \times 2.5 = \textbf{0.0945 m}^3$$

More prisms

A right-angled **triangular prism** is shown in Fig. 28.4 with dimensions b, h and l.

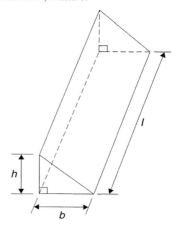

h

b

l

Figure 28.4

$$\textbf{Volume} = \frac{1}{2}bhl$$

and

$$\textbf{surface area} = \textbf{area of each end}$$

$$+ \textbf{area of three sides}$$

Notice that the volume is given by the area of the end (i.e. area of triangle $= \frac{1}{2}bh$) multiplied by the length l. In fact, the volume of any shaped prism is given by the area of an end multiplied by the length.

Problem 6. Determine the volume (in cm^3) of the wooden shape shown in Fig. 28.5

16 mm

12 mm

40 mm

Figure 28.5

The solid shown in Fig. 28.5 is a triangular prism. The volume V of any prism is given by $V = Ah$, where

A is the cross-sectional area and *h* is the perpendicular height. Hence,

$$\textbf{volume} = \frac{1}{2} \times 16 \times 12 \times 40 = 3840\,\text{mm}^3$$

$$= \textbf{3.840 cm}^3$$

(since $1\,\text{cm}^3 = 1000\,\text{mm}^3$)

Problem 7. Calculate the volume of the right-angled triangular prism component shown in Fig. 28.6. Also, determine its total surface area

Figure 28.6

Volume of right-angled triangular prism

$$= \frac{1}{2}bhl = \frac{1}{2} \times 8 \times 6 \times 40$$

i.e. **volume = 960 cm³**

Total surface area = area of each end + area of three sides.

In triangle *ABC*, $AC^2 = AB^2 + BC^2$

from which, $AC = \sqrt{AB^2 + BC^2} = \sqrt{6^2 + 8^2}$
$= 10\,\text{cm}$

Hence, total surface area

$$= 2\left(\frac{1}{2}bh\right) + (AC \times 40) + (BC \times 40) + (AB \times 40)$$

$$= (8 \times 6) + (10 \times 40) + (8 \times 40) + (6 \times 40)$$

$$= 48 + 400 + 320 + 240$$

i.e. **total surface area = 1008 cm²**

Problem 8. Calculate the volume and total surface area of the solid metal prism shown in Fig. 28.7

Figure 28.7

The solid shown in Fig. 28.7 is a **trapezoidal prism**.

Volume of prism = cross-sectional area × height

$$= \frac{1}{2}(11+5)4 \times 15 = 32 \times 15$$

$$= \textbf{480 cm}^3$$

Surface area of prism

= sum of two trapeziums + 4 rectangles

$= (2 \times 32) + (5 \times 15) + (11 \times 15) + 2(5 \times 15)$

$= 64 + 75 + 165 + 150 = \textbf{454 cm}^2$

Now try the following Practice Exercise

Practice Exercise 139 Volumes and surface areas of common shapes (answers on page 456)

1. Change a volume of $1\,200\,000\,\text{cm}^3$ to cubic metres.

2. Change a volume of $5000\,\text{mm}^3$ to cubic centimetres.

3. A metal cube has a surface area of $24\,\text{cm}^2$. Determine its volume.

4. A rectangular block of wood has dimensions of 40mm by 12mm by 8mm. Determine
   (a) its volume, in cubic millimetres
   (b) its total surface area in square millimetres.

5. Determine the capacity, in litres, of a fish tank measuring 90 cm by 60 cm by 1.8 m, given 1 litre $= 1000 \, cm^3$.

6. A rectangular block of metal has dimensions of 40 mm by 25 mm by 15 mm. Determine its volume in cm^3. Find also its mass if the metal has a density of $9 \, g/cm^3$.

7. Determine the maximum capacity, in litres, of a fish tank measuring 50 cm by 40 cm by 2.5 m (1 litre $= 1000 \, cm^3$).

8. Determine how many cubic metres of concrete are required for a 120 m long path, 150 mm wide and 80 mm deep.

9. A small cylindrical container has a diameter 30 mm and height 50 mm. Calculate
   (a) its volume in cubic centimetres, correct to 1 decimal place
   (b) the total surface area in square centimetres, correct to 1 decimal place.

10. Find (a) the volume and (b) the total surface area of a right-angled triangular prism workpiece of length 80 cm and whose triangular end has a base of 12 cm and perpendicular height 5 cm.

11. A steel ingot whose volume is $2 \, m^3$ is rolled out into a plate which is 30 mm thick and 1.80 m wide. Calculate the length of the plate in metres.

12. The volume of a cylindrical container is $75 \, cm^3$. If its height is 9.0 cm, find its radius.

13. Calculate the volume of a metal tube whose outside diameter is 8 cm and whose inside diameter is 6 cm, if the length of the tube is 4 m.

14. The volume of a cylindrical tin is $400 \, cm^3$. If its radius is 5.20 cm, find its height. Also determine its curved surface area.

15. A cylinder is cast from a rectangular piece of alloy 5 cm by 7 cm by 12 cm. If the length of the cylinder is to be 60 cm, find its diameter.

16. Find the volume and the total surface area of a regular hexagonal bar of metal of length 3 m if each side of the hexagon is 6 cm.

17. A block of lead 1.5 m by 90 cm by 750 mm is hammered out to make a square sheet 15 mm thick. Determine the dimensions of the square sheet, correct to the nearest centimetre.

18. How long will it take a tap dripping at a rate of $800 \, mm^3/s$ to fill a 3-litre can?

19. A cylinder is cast from a rectangular piece of alloy 5.20 cm by 6.50 cm by 19.33 cm. If the height of the cylinder is to be 52.0 cm, determine its diameter, correct to the nearest centimetre.

20. How much concrete is required for the construction of the path shown in Fig. 28.8, if the path is 12 cm thick?

Figure 28.8

21. A lorry has a closed cylindrical chemical storage tank of length 3.6 m and diameter 1.3 m. Calculate for the tank its (a) surface area, in m^2, and (b) storage capacity, in m^3 and in litres. (Note that 1 litre $= 1000 \, cm^3$).

22. How many cylindrical jars of radius 80 mm and height 250 mm can be filled completely from a rectangular vat 2.4 m long, 60 cm wide and 950 mm high?

Pyramids

Volume of any pyramid

$$= \frac{1}{3} \times \textbf{area of base} \times \textbf{perpendicular height}$$

A square-based pyramid is shown in Fig. 28.9 with base dimensions x by x and the perpendicular height of the pyramid h. For the square-base pyramid shown,

$$\text{volume} = \frac{1}{3}x^2h$$

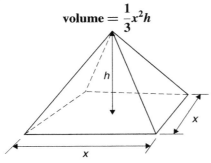

Figure 28.9

> **Problem 9.** A square pyramid has a perpendicular height of 16 cm. If a side of the base is 6 cm, determine the volume of the pyramid

Volume of pyramid

$$= \frac{1}{3} \times \text{area of base} \times \text{perpendicular height}$$

$$= \frac{1}{3} \times (6 \times 6) \times 16$$

$$= \mathbf{192\,cm^3}$$

> **Problem 10.** Determine the volume and the total surface area of the square pyramid paperweight shown in Fig. 28.10 if its perpendicular height is 12 cm.

Volume of pyramid

$$= \frac{1}{3}(\text{area of base}) \times \text{perpendicular height}$$

$$= \frac{1}{3}(5 \times 5) \times 12$$

$$= \mathbf{100\,cm^3}$$

Figure 28.10

The total surface area consists of a square base and 4 equal triangles.

Area of triangle ADE

$$= \frac{1}{2} \times \text{base} \times \text{perpendicular height}$$

$$= \frac{1}{2} \times 5 \times AC$$

The length AC may be calculated using Pythagoras' theorem on triangle ABC, where $AB = 12$ cm and $BC = \frac{1}{2} \times 5 = 2.5$ cm.

$$AC = \sqrt{AB^2 + BC^2} = \sqrt{12^2 + 2.5^2} = 12.26\,\text{cm}$$

Hence,

$$\text{area of triangle } ADE = \frac{1}{2} \times 5 \times 12.26 = 30.65\,\text{cm}^2$$

Total surface area of pyramid $= (5 \times 5) + 4(30.65)$

$$= \mathbf{147.6\,cm^2}$$

> **Problem 11.** A rectangular prism of metal having dimensions of 5 cm by 6 cm by 18 cm is melted down and recast into a pyramid having a rectangular base measuring 6 cm by 10 cm. Calculate the perpendicular height of the pyramid, assuming no waste of metal

Volume of rectangular prism $= 5 \times 6 \times 18 = 540\,\text{cm}^3$

Volume of pyramid

$$= \frac{1}{3} \times \text{area of base} \times \text{perpendicular height}$$

Hence, $\quad 540 = \frac{1}{3} \times (6 \times 10) \times h$

from which, $\quad h = \dfrac{3 \times 540}{6 \times 10} = 27\,\text{cm}$

i.e. **perpendicular height of pyramid $= 27$ cm**

Cones

A cone is a circular-based pyramid. A cone of base radius r and perpendicular height h is shown in Fig. 28.11.

$$\text{Volume} = \frac{1}{3} \times \text{area of base} \times \text{perpendicular height}$$

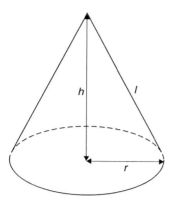

Figure 28.11

i.e.

$$\text{Volume} = \frac{1}{3}\pi r^2 h$$

$$\text{Curved surface area} = \pi rl$$

$$\text{Total surface area} = \pi rl + \pi r^2$$

> **Problem 12.** Calculate the volume, in cubic centimetres, of a solid wooden cone of radius 30 mm and perpendicular height 80 mm

Volume of cone $= \frac{1}{3}\pi r^2 h = \frac{1}{3} \times \pi \times 30^2 \times 80$

$$= 75398.2236 \ldots \text{ mm}^3$$

$1\,\text{cm} = 10\,\text{mm}$ and
$1\,\text{cm}^3 = 10\,\text{mm} \times 10\,\text{mm} \times 10\,\text{mm} = 10^3\,\text{mm}^3$, or

$$\mathbf{1\,mm^3 = 10^{-3}\,cm^3}$$

Hence, $75398.2236 \ldots \text{ mm}^3$

$$= 75398.2236 \ldots \times 10^{-3}\,\text{cm}^3$$

i.e.

$$\mathbf{volume = 75.40\,cm^3}$$

Alternatively, from the question, $r = 30\,\text{mm} = 3\,\text{cm}$ and $h = 80\,\text{mm} = 8\,\text{cm}$. Hence,

$$\mathbf{volume} = \frac{1}{3}\pi r^2 h = \frac{1}{3} \times \pi \times 3^2 \times 8 = \mathbf{75.40\,cm^3}$$

> **Problem 13.** Determine the volume and total surface area of a metal cone of radius 5 cm and perpendicular height 12 cm

The cone is shown in Fig. 28.12.

Figure 28.12

Volume of cone $= \frac{1}{3}\pi r^2 h = \frac{1}{3} \times \pi \times 5^2 \times 12$

$$= \mathbf{314.2\,cm^3}$$

Total surface area = curved surface area + area of base

$$= \pi rl + \pi r^2$$

From Fig. 28.12, slant height l may be calculated using Pythagoras' theorem:

$$l = \sqrt{12^2 + 5^2} = 13\,\text{cm}$$

Hence, **total surface area** $= (\pi \times 5 \times 13) + (\pi \times 5^2)$

$$= \mathbf{282.7\,cm^2}$$

Spheres

For the sphere shown in Fig. 28.13:

$$\mathbf{Volume} = \frac{4}{3}\pi r^3 \quad \text{and} \quad \mathbf{surface\ area} = 4\pi r^2$$

Figure 28.13

> **Problem 14.** Find the volume and surface area of a spherical shot put of diameter 10 cm

Since diameter $= 10\,\text{cm}$, radius, $r = 5\,\text{cm}$.

Volume of sphere $= \frac{4}{3}\pi r^3 = \frac{4}{3} \times \pi \times 5^3$

$$= \mathbf{523.6\,cm^3}$$

Surface area of sphere $= 4\pi r^2 = 4 \times \pi \times 5^2$

$$= \textbf{314.2 cm}^2$$

Problem 15. The surface area of a spherical cannon ball is $201.1\,\text{cm}^2$. Find the diameter of the cannon ball and hence its volume

Surface area of sphere $= 4\pi r^2$

Hence, $201.1\,\text{cm}^2 = 4 \times \pi \times r^2$

from which $\qquad r^2 = \dfrac{201.1}{4 \times \pi} = 16.0$

and $\qquad$ radius, $r = \sqrt{16.0} = 4.0\,\text{cm}$

from which, **diameter** $= 2 \times r = 2 \times 4.0 = \textbf{8.0 cm}$

Volume of cannon ball $= \dfrac{4}{3}\pi r^3 = \dfrac{4}{3} \times \pi \times (4.0)^3$

$$= \textbf{268.1 cm}^3$$

Now try the following Practice Exercise

Practice Exercise 140 Volumes and surface areas of common shapes (answers on page 456)

1. If a conical feature has a diameter of 80 mm and a perpendicular height of 120 mm, calculate its volume in cm^3 and its curved surface area.

2. A square metal pyramid has a perpendicular height of 4 cm. If a side of the base is 2.4 cm long, find the volume and total surface area of the pyramid.

3. A steel ball bearing has a diameter of 6 cm. Determine its volume and surface area.

4. If the volume of a wooden sphere is $566\,\text{cm}^3$, find its radius.

5. A metal pyramid feature having a square base has a perpendicular height of 25 cm and a volume of $75\,\text{cm}^3$. Determine, in centimetres, the length of each side of the base.

6. A conical paper weight has a base diameter of 16 mm and a perpendicular height of

40 mm. Find its volume correct to the nearest cubic millimetre.

7. Determine (a) the volume and (b) the surface area of a spherical steel ball of radius 40 mm.

8. The volume of a spherical bearing is $325\,\text{cm}^3$. Determine its diameter.

9. Given the radius of the earth is 6380 km, calculate, in engineering notation
   (a) its surface area in km^2
   (b) its volume in km^3

10. An ingot whose volume is $1.5\,\text{m}^3$ is to be made into ball bearings whose radii are 8.0 cm. How many bearings will be produced from the ingot, assuming 5% wastage?

11. A spherical chemical storage tank has an internal diameter of 5.6 m. Calculate the storage capacity of the tank, correct to the nearest cubic metre. If 1 litre $= 1000\,\text{cm}^3$, determine the tank capacity in litres.

12. A feed to a process line is in the form of a cone of radius 1.8 m and perpendicular height 2.5 m. The feed supplies flour to a bread mixing line. Calculate the maximum volume of flour that the feed can hold.

28.3 Summary of volumes and surface areas of common solids

A summary of volumes and surface areas of regular solids is shown in Table 28.1 on page 307.

28.4 More complex volumes and surface areas

Here are some worked problems involving more complex and composite solids.

Problem 16. A wooden section is shown in Fig. 28.14. Find (a) its volume in m^3 and (b) its total surface area

Table 28.1   **Volumes and surface areas of regular solids**

Rectangular prism (or cuboid)

Volume $= l \times b \times h$
Surface area $= 2(bh + hl + lb)$

Cylinder

Volume $= \pi r^2 h$
Total surface area $= 2\pi rh + 2\pi r^2$

Triangular prism

Volume $= \dfrac{1}{2}bhl$
Surface area $=$ area of each end $+$
             area of three sides

Pyramid

Volume $= \dfrac{1}{3} \times A \times h$
Total surface area $=$
sum of areas of triangles
forming sides $+$ area of base

Cone

Volume $= \dfrac{1}{3}\pi r^2 h$
Curved surface area $= \pi rl$
Total surface area $= \pi rl + \pi r^2$

Sphere

Volume $= \dfrac{4}{3}\pi r^3$
Curved surface area $= 4\pi r^2$

Figure 28.14

(a) The section of wood is a prism whose end comprises a rectangle and a semicircle. Since the radius of the semicircle is 8 cm, the diameter is 16 cm. Hence, the rectangle has dimensions 12 cm by 16 cm.

Area of end $= (12 \times 16) + \frac{1}{2}\pi 8^2 = 292.5\,\text{cm}^2$

Volume of wooden section

$= \text{area of end} \times \text{perpendicular height}$

$= 292.5 \times 300 = \mathbf{87\,750\,\text{cm}^3}$

$= \dfrac{87750}{10^6}\,\text{m}^3, \text{since } 1\,\text{m}^3 = 10^6\,\text{cm}^3$

$= \mathbf{0.08775\,\text{m}^3}$

(b) The total surface area comprises the two ends (each of area 292.5 cm^2), three rectangles and a curved surface (which is half a cylinder). Hence,

total surface area

$= (2 \times 292.5) + 2(12 \times 300)$

$\qquad + (16 \times 300) + \frac{1}{2}(2\pi \times 8 \times 300)$

$= 585 + 7200 + 4800 + 2400\pi$

$= \mathbf{20\,125\,\text{cm}^2} \quad \text{or} \quad \mathbf{2.0125\,\text{m}^2}$

> ⚑ **Problem 17.** A solid wooden pyramid has a rectangular base 3.60 cm by 5.40 cm. Determine the volume and total surface area of the pyramid if each of its sloping edges is 15.0 cm

The pyramid is shown in Fig. 28.15. To calculate the volume of the pyramid, the perpendicular height EF is required. Diagonal BD is calculated using Pythagoras' theorem,

i.e. $\quad BD = \sqrt{[3.60^2 + 5.40^2]} = 6.490\,\text{cm}$

Hence, $\quad EB = \frac{1}{2}BD = \dfrac{6.490}{2} = 3.245\,\text{cm}$

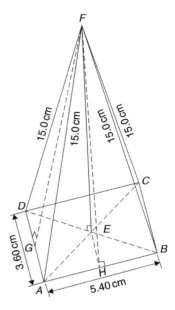

Figure 28.15

Using Pythagoras' theorem on triangle BEF gives

$$BF^2 = EB^2 + EF^2$$

from which $\quad EF = \sqrt{\left(BF^2 - EB^2\right)}$

$= \sqrt{15.0^2 - 3.245^2} = 14.64\,\text{cm}$

Volume of pyramid

$= \frac{1}{3}(\text{area of base})(\text{perpendicular height})$

$= \frac{1}{3}(3.60 \times 5.40)(14.64) = \mathbf{94.87\,\text{cm}^3}$

Area of triangle ADF (which equals triangle BCF) $= \frac{1}{2}(AD)(FG)$, where G is the mid-point of AD. Using Pythagoras' theorem on triangle FGA gives

$$FG = \sqrt{[15.0^2 - 1.80^2]} = 14.89\,\text{cm}$$

Hence, area of triangle $ADF = \frac{1}{2}(3.60)(14.89)$

$= 26.80\,\text{cm}^2$

Similarly, if H is the mid-point of AB,

$$FH = \sqrt{15.0^2 - 2.70^2} = 14.75\,\text{cm}$$

Hence, area of triangle ABF (which equals triangle CDF) $= \frac{1}{2}(5.40)(14.75) = 39.83\,\text{cm}^2$

Total surface area of pyramid

$$= 2(26.80) + 2(39.83) + (3.60)(5.40)$$

$$= 53.60 + 79.66 + 19.44$$

$$= \textbf{152.7 cm}^2$$

Problem 18. Calculate the volume and total surface area of a metal hemisphere of diameter 5.0 cm

Volume of hemisphere $= \dfrac{1}{2}$ (volume of sphere)

$$= \frac{2}{3}\pi r^3 = \frac{2}{3}\pi \left(\frac{5.0}{2}\right)^3$$

$$= \textbf{32.7 cm}^3$$

Total surface area

$$= \text{curved surface area} + \text{area of circle}$$

$$= \frac{1}{2}(\text{surface area of sphere}) + \pi r^2$$

$$= \frac{1}{2}(4\pi r^2) + \pi r^2$$

$$= 2\pi r^2 + \pi r^2 = 3\pi r^2 = 3\pi \left(\frac{5.0}{2}\right)^2$$

$$= \textbf{58.9 cm}^2$$

Problem 19. A rectangular piece of metal having dimensions 4 cm by 3 cm by 12 cm is melted down and recast into a pyramid having a rectangular base measuring 2.5 cm by 5 cm. Calculate the perpendicular height of the pyramid

Volume of rectangular prism of metal $= 4 \times 3 \times 12$

$$= 144 \text{ cm}^3$$

Volume of pyramid

$$= \frac{1}{3}(\text{area of base})(\text{perpendicular height})$$

Assuming no waste of metal,

$$144 = \frac{1}{3}(2.5 \times 5)(\text{height})$$

i.e. **perpendicular height of pyramid** $= \dfrac{144 \times 3}{2.5 \times 5}$

$$= \textbf{34.56 cm}$$

Problem 20. A rivet consists of a cylindrical head, of diameter 1 cm and depth 2 mm, and a shaft of diameter 2 mm and length 1.5 cm. Determine the volume of metal in 2000 such rivets

Radius of cylindrical head $= \dfrac{1}{2}$ cm $= 0.5$ cm and

height of cylindrical head $= 2$ mm $= 0.2$ cm.

Hence, volume of cylindrical head

$$= \pi r^2 h = \pi(0.5)^2(0.2) = 0.1571 \text{ cm}^3$$

Volume of cylindrical shaft

$$= \pi r^2 h = \pi \left(\frac{0.2}{2}\right)^2 (1.5) = 0.0471 \text{ cm}^3$$

Total volume of 1 rivet $= 0.1571 + 0.0471$

$$= 0.2042 \text{ cm}^3$$

Volume of metal in 2000 such rivets

$$= 2000 \times 0.2042 = \textbf{408.4 cm}^3$$

Problem 21. A solid metal cylinder of radius 6 cm and height 15 cm is melted down and recast into a plumb bob shape comprising a hemisphere surmounted by a cone. Assuming that 8% of the metal is wasted in the process, determine the height of the conical portion if its diameter is to be 12 cm

Volume of cylinder $= \pi r^2 h = \pi \times 6^2 \times 15$

$$= 540\pi \text{ cm}^3$$

If 8% of metal is lost then 92% of 540π gives the volume of the new shape, shown in Fig. 28.16.
Hence, the volume of (hemisphere + cone)

$$= 0.92 \times 540\pi \text{ cm}^3$$

i.e. $\dfrac{1}{2}\left(\dfrac{4}{3}\pi r^3\right) + \dfrac{1}{3}\pi r^2 h = 0.92 \times 540\pi$

Dividing throughout by π gives

$$\frac{2}{3}r^3 + \frac{1}{3}r^2 h = 0.92 \times 540$$

Figure 28.16

Since the diameter of the new shape is to be 12 cm, radius $r = 6$ cm,

then $\quad \dfrac{2}{3}(6)^3 + \dfrac{1}{3}(6)^2 h = 0.92 \times 540$

$$144 + 12h = 496.8$$

i.e. **height of conical portion**,

$$h = \dfrac{496.8 - 144}{12} = \mathbf{29.4\,cm}$$

Problem 22. A block of copper having a mass of 50 kg is drawn out to make 500 m of wire of uniform cross-section. Given that the density of copper is 8.91 g/cm^3, calculate (a) the volume of copper, (b) the cross-sectional area of the wire and (c) the diameter of the cross-section of the wire

(a) A density of 8.91 g/cm^3 means that 8.91 g of copper has a volume of 1 cm^3, or 1 g of copper has a volume of $(1 \div 8.91)$ cm^3

$$\text{Density} = \dfrac{\text{mass}}{\text{volume}}$$

from which $\quad \text{volume} = \dfrac{\text{mass}}{\text{density}}$

Hence, 50 kg, i.e. 50 000 g, has a

$$\textbf{volume} = \dfrac{\text{mass}}{\text{density}} = \dfrac{50000}{8.91}\,\text{cm}^3 = \mathbf{5612\,cm^3}$$

(b) Volume of wire = area of circular cross-section
$\times$ length of wire.

Hence, $5612\,\text{cm}^3 = \text{area} \times (500 \times 100\,\text{cm})$

from which, **area** $= \dfrac{5612}{500 \times 100}\,\text{cm}^2$

$$= \mathbf{0.1122\,cm^2}$$

(c) Area of circle $= \pi r^2$ or $\dfrac{\pi d^2}{4}$

hence, $0.1122 = \dfrac{\pi d^2}{4}$

from which, $d = \sqrt{\left(\dfrac{4 \times 0.1122}{\pi}\right)} = 0.3780\,\text{cm}$

i.e. **diameter of cross-section is 3.780 mm**.

Problem 23. A boiler consists of a cylindrical section of length 8 m and diameter 6 m, on one end of which is surmounted a hemispherical section of diameter 6 m and on the other end a conical section of height 4 m and base diameter 6 m. Calculate the volume of the boiler and the total surface area

The boiler is shown in Fig. 28.17.

Figure 28.17

Volume of hemisphere, $\quad P = \dfrac{2}{3}\pi r^3$

$$= \dfrac{2}{3} \times \pi \times 3^3 = 18\pi\,\text{m}^3$$

Volume of cylinder, $\quad Q = \pi r^2 h = \pi \times 3^2 \times 8$

$$= 72\pi\,\text{m}^3$$

Volume of cone, $\quad R = \dfrac{1}{3}\pi r^2 h = \dfrac{1}{3} \times \pi \times 3^2 \times 4$

$$= 12\pi\,\text{m}^3$$

Total volume of boiler $= 18\pi + 72\pi + 12\pi$

$$= 102\pi = \mathbf{320.4\,m^3}$$

Surface area of hemisphere, $P = \dfrac{1}{2}(4\pi r^2)$

$$= 2 \times \pi \times 3^2 = 18\pi\,\text{m}^2$$

Curved surface area of cylinder, $Q = 2\pi rh$

$$= 2 \times \pi \times 3 \times 8$$

$$= 48\pi\,\text{m}^2$$

The slant height of the cone, l, is obtained by Pythagoras' theorem on triangle ABC, i.e.

$$l = \sqrt{\left(4^2 + 3^2\right)} = 5$$

Curved surface area of cone,

$$R = \pi rl = \pi \times 3 \times 5 = 15\pi\,\text{m}^2$$

Total surface area of boiler $= 18\pi + 48\pi + 15\pi$

$$= 81\pi = \mathbf{254.5\ m^2}$$

Now try the following Practice Exercise

Practice Exercise 141 More complex volumes and surface areas (answers on page 456)

1. Find the total surface area of a wooden hemisphere of diameter 50 mm.

2. Find (a) the volume and (b) the total surface area of a metal hemisphere of diameter 6 cm.

3. Determine the mass of a hemispherical copper container whose external and internal radii are 12 cm and 10 cm, assuming that 1 cm^3 of copper weighs 8.9 g.

4. A metal plumb bob comprises a hemisphere surmounted by a cone. If the diameter of the hemisphere and cone are each 4 cm and the total length is 5 cm, find its total volume.

5. A marquee is in the form of a cylinder surmounted by a cone. The total height is 6 m and the cylindrical portion has a height of 3.5 m with a diameter of 15 m. Calculate the surface area of material needed to make the marquee assuming 12% of the material is wasted in the process.

6. Determine (a) the volume and (b) the total surface area of the following metal solids.

    (i) a cone of radius 8.0 cm and perpendicular height 10 cm.
    (ii) a sphere of diameter 7.0 cm.
    (iii) a hemisphere of radius 3.0 cm.
    (iv) a 2.5 cm by 2.5 cm square pyramid of perpendicular height 5.0 cm.
    (v) a 4.0 cm by 6.0 cm rectangular pyramid of perpendicular height 12.0 cm.
    (vi) a 4.2 cm by 4.2 cm square pyramid whose sloping edges are each 15.0 cm
    (vii) a pyramid having an octagonal base of side 5.0 cm and perpendicular height 20 cm.

7. A metal sphere weighing 24 kg is melted down and recast into a solid cone of base radius 8.0 cm. If the density of the metal is 8000 kg/m^3 determine

    (a) the diameter of the metal sphere
    (b) the perpendicular height of the cone, assuming that 15% of the metal is lost in the process.

8. Find the volume of a regular hexagonal pyramid feature if the perpendicular height is 16.0 cm and the side of the base is 3.0 cm.

9. A buoy consists of a hemisphere surmounted by a cone. The diameter of the cone and hemisphere is 2.5 m and the slant height of the cone is 4.0 m. Determine the volume and surface area of the buoy.

10. A petrol container is in the form of a central cylindrical portion 5.0 m long with a hemispherical section surmounted on each end. If the diameters of the hemisphere and cylinder are both 1.2 m, determine the capacity of the tank in litres (1 litre = 1000 cm^3).

11. Fig. 28.18 shows a metal rod section. Determine its volume and total surface area.

Figure 28.18

12. Find the volume (in cm³) of the die-casting shown in Fig. 28.19. The dimensions are in millimetres.

Figure 28.19

13. The cross-section of part of a circular ventilation shaft is shown in Fig. 28.20, ends *AB* and *CD* being open. Calculate

(a) the volume of the air, correct to the nearest litre, contained in the part of the system shown, neglecting the sheet metal thickness (given 1litre = 1000 cm³)

(b) the cross-sectional area of the sheet metal used to make the system, in square metres

(c) the cost of the sheet metal if the material costs £11.50 per square metre, assuming that 25% extra metal is required due to wastage.

14. In level ground, a drainage trench is excavated 0.5 m wide. The bottom is flat, and the

sides are vertical, but the depth varies uniformly over a length of 20 m from a depth of 0.5 m at one end to 1.0 m at the other. Calculate the material dug out of the trench.

Figure 28.20

15. A 3 km tunnel is to be dug under a hill. Initially the tunnel is cut by a boring machine which cuts a circular profile having a radius of 5 m. The floor of the tunnel is then laid to a maximum depth of 2 m. Calculate (a) the surface area of the tunnel wall, and (b) the volume of earth material to be removed.

16. Each edge of a hexagonal steel plate is 0.32 m in length. The plate is 15 mm thick and has a density of 7800 kg/m³. Calculate the mass of the steel plate.

28.5 Volumes and surface areas of frusta of pyramids and cones

The **frustum** of a pyramid or cone is the portion remaining when a part containing the vertex is cut off by a plane parallel to the base.

The **volume of a frustum of a pyramid or cone** is given by the volume of the whole pyramid or cone minus the volume of the small pyramid or cone cut off.

The **surface area of the sides of a frustum of a pyramid or cone** is given by the surface area of the whole pyramid or cone minus the surface area of the small pyramid or cone cut off. This gives the lateral surface area of the frustum. If the total surface area of the frustum is required then the surface area of the two parallel ends are added to the lateral surface area.

There is an alternative method for finding the volume and surface area of a **frustum of a cone**. With reference to Fig. 28.21,

Figure 28.21

$$\text{Volume} = \frac{1}{3}\pi h(R^2 + Rr + r^2)$$

$$\text{Curved surface area} = \pi l(R + r)$$

$$\text{Total surface area} = \pi l(R + r) + \pi r^2 + \pi R^2$$

> **Problem 24.** Determine the volume of a concrete frustum of a cone if the diameter of the ends are 6.0cm and 4.0cm and its perpendicular height is 3.6cm

(i) Method 1

A section through the vertex of a complete cone is shown in Fig. 28.22.

From Chapter 22, using similar triangles,

$$\frac{AP}{DP} = \frac{DR}{BR}$$

Hence, $$\frac{AP}{2.0} = \frac{3.6}{1.0}$$

from which $$AP = \frac{(2.0)(3.6)}{1.0}$$

$$= 7.2\,\text{cm}$$

The height of the large cone $= 3.6 + 7.2$

$$= 10.8\,\text{cm}$$

Volume of frustum of concrete cone

$= $ volume of large cone

$\qquad -$ volume of small cone cut off

$$= \frac{1}{3}\pi(3.0)^2(10.8) - \frac{1}{3}\pi(2.0)^2(7.2)$$

$$= 101.79 - 30.16 = \mathbf{71.6\,cm^3}$$

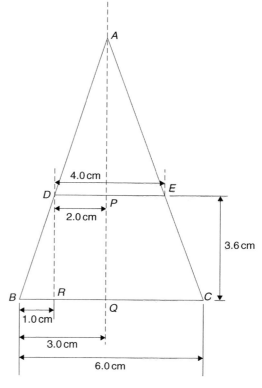

Figure 28.22

(ii) Method 2

From above, volume of the frustum of a cone

$$= \frac{1}{3}\pi h(R^2 + Rr + r^2)$$

where $R = 3.0\,\text{cm}$,

$$r = 2.0\,\text{cm} \quad \text{and} \quad h = 3.6\,\text{cm}$$

Hence, **volume of frustum**

$$= \frac{1}{3}\pi(3.6)\left[(3.0)^2 + (3.0)(2.0) + (2.0)^2\right]$$

$$= \frac{1}{3}\pi(3.6)(19.0) = \mathbf{71.6\,cm^3}$$

> **Problem 25.** Find the total surface area of the frustum of the concrete cone in Problem 24.

(i) Method 1

Curved surface area of frustum $=$ curved surface area of large cone $-$ curved surface area of small cone cut off.

From Fig. 28.22, using Pythagoras' theorem,

$$AB^2 = AQ^2 + BQ^2$$

from which $AB = \sqrt{[10.8^2 + 3.0^2]} = 11.21\,\text{cm}$

and $AD^2 = AP^2 + DP^2$

from which $AD = \sqrt{[7.2^2 + 2.0^2]} = 7.47\,\text{cm}$

Curved surface area of large cone $= \pi r l$

$$= \pi(BQ)(AB) = \pi(3.0)(11.21)$$

$$= 105.65\,\text{cm}^2$$

and curved surface area of small cone

$$= \pi(DP)(AD) = \pi(2.0)(7.47) = 46.94\,\text{cm}^2$$

Hence, curved surface area of frustum

$$= 105.65 - 46.94$$

$$= 58.71\,\text{cm}^2$$

Total surface area of frustum

$$= \text{curved surface area}$$
$$\quad + \text{area of two circular ends}$$

$$= 58.71 + \pi(2.0)^2 + \pi(3.0)^2$$

$$= 58.71 + 12.57 + 28.27 = \mathbf{99.6\,cm^2}$$

(ii) Method 2

From page 313, total surface area of frustum
$$= \pi l(R + r) + \pi r^2 + \pi R^2$$

where $l = BD = 11.21 - 7.47 = 3.74\,\text{cm}$, $R = 3.0\,\text{cm}$ and $r = 2.0\,\text{cm}$.

Hence, total surface area of frustum

$$= \pi(3.74)(3.0 + 2.0) + \pi(2.0)^2 + \pi(3.0)^2$$

$$= \mathbf{99.6\,cm^2}$$

> **Problem 26.** A storage hopper is in the shape of a frustum of a pyramid. Determine its volume if the ends of the frustum are squares of sides 8.0 m and 4.6 m, respectively, and the perpendicular height between its ends is 3.6 m

The frustum is shown shaded in Fig. 28.23(a) as part of a complete pyramid. A section perpendicular to the base through the vertex is shown in Fig. 28.23(b).

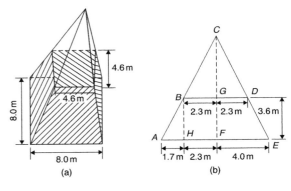

Figure 28.23

By similar triangles $\dfrac{CG}{BG} = \dfrac{BH}{AH}$

From which, height

$$CG = BG\left(\frac{BH}{AH}\right) = \frac{(2.3)(3.6)}{1.7} = 4.87\,\text{m}$$

Height of complete pyramid $= 3.6 + 4.87 = 8.47\,\text{m}$

Volume of large pyramid $= \dfrac{1}{3}(8.0)^2(8.47)$

$$= 180.69\,\text{m}^3$$

Volume of small pyramid cut off $= \dfrac{1}{3}(4.6)^2(4.87)$

$$= 34.35\,\text{m}^3$$

Hence, **volume of storage hopper** $= 180.69 - 34.35$

$$= \mathbf{146.3\,m^3}$$

> **Problem 27.** Determine the lateral surface area of the storage hopper in Problem 26

The lateral surface area of the storage hopper consists of four equal trapeziums. From Fig. 28.24,

$$\text{Area of trapezium } PRSU = \frac{1}{2}(PR + SU)(QT)$$

$OT = 1.7\,\text{m}$ (same as AH in Fig. 28.23(b) and $OQ = 3.6\,\text{m}$. By Pythagoras' theorem,

$$QT = \sqrt{(OQ^2 + OT^2)} = \sqrt{[3.6^2 + 1.7^2]} = 3.98\,\text{m}$$

Area of trapezium $PRSU$

$$= \frac{1}{2}(4.6 + 8.0)(3.98) = 25.07\,\text{m}^2$$

Lateral surface area of hopper $= 4(25.07)$

$$= \mathbf{100.3\,m^2}$$

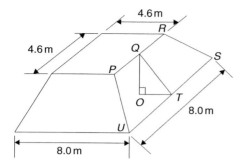

Figure 28.24

> **Problem 28.** A lampshade is in the shape of a frustum of a cone. The vertical height of the shade is 25.0 cm and the diameters of the ends are 20.0 cm and 10.0 cm, respectively. Determine the area of the material needed to form the lampshade, correct to 3 significant figures

The curved surface area of a frustum of a cone $= \pi l(R + r)$ from page 302. Since the diameters of the ends of the frustum are 20.0 cm and 10.0 cm, from Fig. 28.25,

Figure 28.25

$$r = 5.0 \text{ cm}, R = 10.0 \text{ cm}$$

and $\quad l = \sqrt{[25.0^2 + 5.0^2]} = 25.50 \text{ cm}$

from Pythagoras' theorem.
Hence, curved surface area
$$= \pi(25.50)(10.0 + 5.0) = 1201.7 \text{ cm}^2$$

i.e. the area of material needed to form the lampshade is **1200 cm²**, correct to 3 significant figures.

> **Problem 29.** A cooling tower is in the form of a cylinder surmounted by a frustum of a cone, as shown in Fig. 28.26. Determine the volume of air space in the tower if 40% of the space is used for pipes and other structures

Figure 28.26

Volume of cylindrical portion $= \pi r^2 h$

$$= \pi \left(\frac{25.0}{2}\right)^2 (12.0)$$

$$= 5890 \text{ m}^3$$

Volume of frustum of cone $= \frac{1}{3}\pi h(R^2 + Rr + r^2)$

where $\qquad h = 30.0 - 12.0 = 18.0 \text{ m}$,

$$R = 25.0 \div 2 = 12.5 \text{ m}$$

and $\qquad r = 12.0 \div 2 = 6.0 \text{ m}$.

Hence, volume of frustum of cone

$$= \frac{1}{3}\pi(18.0)\left[(12.5)^2 + (12.5)(6.0) + (6.0)^2\right]$$

$$= 5038 \text{ m}^3$$

Total volume of cooling tower $= 5890 + 5038$

$$= 10\,928 \text{ m}^3$$

If 40% of space is occupied then
volume of air space $= 0.6 \times 10928 = \textbf{6557 m}^3$

Now try the following Practice Exercise

Practice Exercise 142 Volumes and surface areas of frusta of pyramids and cones (answers on page 456)

1. The radii of the faces of a frustum of a conical paper weight are 2.0cm and 4.0cm and the thickness of the frustum is 5.0cm. Determine its volume and total surface area.

2. A frustum of a wooden pyramid has square ends, the squares having sides 9.0cm and 5.0cm, respectively. Calculate the volume and total surface area of the frustum if the perpendicular distance between its ends is 8.0cm.

3. A cooling tower is in the form of a frustum of a cone. The base has a diameter of 32.0m, the top has a diameter of 14.0m and the vertical height is 24.0m. Calculate the volume of the tower and the curved surface area.

4. A loudspeaker diaphragm is in the form of a frustum of a cone. If the end diameters are 28.0cm and 6.00cm and the vertical distance between the ends is 30.0cm, find the area of material needed to cover the curved surface of the speaker.

5. A rectangular prism of metal having dimensions 4.3cm by 7.2cm by 12.4cm is melted down and recast into a frustum of a square pyramid, 10% of the metal being lost in the process. If the ends of the frustum are squares of side 3cm and 8cm respectively, find the thickness of the frustum.

6. Determine the volume and total surface area of a bucket consisting of an inverted frustum of a cone, of slant height 36.0cm and end diameters 55.0cm and 35.0cm.

7. A cylindrical tank of diameter 2.0m and perpendicular height 3.0m is to be replaced by a tank of the same capacity but in the form of a frustum of a cone. If the diameters of the ends of the frustum are 1.0m and 2.0m, respectively, determine the vertical height required.

8. A concrete pillar is made up of two parts. The top section is a cylinder of radius 125 mm and height 750 mm. The base is a frustum of a cone with base radius 300 mm, top radius 125 mm and height 240 mm. Calculate the volume of the pillar in cubic centimetres and in cubic metres, each correct to 4 significant figures.

28.6 Volumes of similar shapes

Fig. 28.27 shows two cubes, one of which has sides three times as long as those of the other.

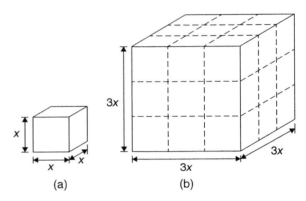

Figure 28.27

Volume of Fig. 28.27(a) $= (x)(x)(x) = x^3$

Volume of Fig. 28.27(b) $= (3x)(3x)(3x) = 27x^3$

Hence, Fig. 28.27(b) has a volume $(3)^3$, i.e. 27, times the volume of Fig. 28.27(a).

Summarising, **the volumes of similar bodies are proportional to the cubes of corresponding linear dimensions.**

Problem 30. A car has a mass of 1000kg. A model of the car is made to a scale of 1 to 50. Determine the mass of the model if the car and its model are made of the same material

$$\frac{\text{Volume of model}}{\text{Volume of car}} = \left(\frac{1}{50}\right)^3$$

since the volume of similar bodies are proportional to the cube of corresponding dimensions.

Mass = density × volume and, since both car and model are made of the same material,

$$\frac{\text{Mass of model}}{\text{Mass of car}} = \left(\frac{1}{50}\right)^3$$

Hence, mass of model

$$= (\text{mass of car})\left(\frac{1}{50}\right)^3 = \frac{1000}{50^3} = \textbf{0.008\,kg or 8\,g}$$

Now try the following Practice Exercise

Practice Exercise 143   Volumes of similar shapes (answers on page 457)

1. The diameter of two spherical bearings are in the ratio 2 : 5. What is the ratio of their volumes?

2. An engineering component has a mass of 400 g. If each of its dimensions are reduced by 30%, determine its new mass.

Practice Exercise 144   Multiple-choice questions on volumes and surface areas of common shapes (answers on page 457)

Each question has only one correct answer

1. A water tank is in the shape of a rectangular prism having length 1.5 m, breadth 60 cm and height 300 mm. If 1 litre = 1000 cm^3, the capacity of the tank is:

   (a) 27 litre   (b) 2.7 litre
   (c) 2700 litre   (d) 270 litre

2. A hollow shaft has an outside diameter of 6.0 cm and an inside diameter of 4.0 cm. The cross- sectional area of the shaft is:

   (a) 6283 mm^2 (b) 1257 mm^2
   (c) 1571 mm^2 (d) 628 mm^2

3. The surface area of a sphere of diameter 40 mm is:

   (a) 50.27 cm^2 (b) 33.51 cm^2
   (c) 268.08 cm^2 (d) 201.06 cm^2

4. The total surface area of a cylinder of length 20 cm and diameter 6 cm is:

   (a) 56.55 cm^2 (b) 433.54 cm^2
   (c) 980.18 cm^2 (d) 226.19 cm^2

5. The length of each edge of a cube is d. The surface area of the cube is given by:

   (a) 7d^2 (b) 5d^2 (c) 6d^2 (d) 4d^2

6. The total surface area of a solid hemisphere of diameter 6.0 cm is:

   (a) 84.82 cm^2 (b) 339.3 cm^2
   (c) 226.2 cm^2 (d) 56.55 cm^2

7. A cylindrical, copper pipe, 1.8 m long, has an outside diameter of 300 mm and an inside diameter of 180 mm. The volume of copper in the pipe, in cubic metres is:

   (a) 0.3257 m^2 (b) 0.0814 m^2
   (c) 8.143 m^2 (d) 814.3 m^2

8. A cylinder and a cone have the same base radius and the same height. The ratio of the volume of the cylinder to that of the cone is:
   (a) 3:2   (b) 2:1   (c) 3:1   (d) 2:3

9. A cube has a volume of 27 cm^3. The length of one edge is:

   (a) 2.25 cm   (b) 6.25 cm
   (c) 9 cm   (d) 3 cm

10. A vehicle has a mass of 2000 kg. A model of the vehicle is made to a scale of 1 to 100. If the vehicle and model are made of the same material, the mass of the model is:

    (a) 2 g   (b) 20 kg   (c) 200 g   (d) 20 g

For fully worked solutions to each of the problems in Practice Exercises 139 to 143 in this chapter, go to the website:

www.routledge.com/cw/bird

Chapter 29

Irregular areas and volumes and mean values

Why it is important to understand: Irregular areas and volumes and mean values of waveforms

Surveyors, farmers and landscapers often need to determine the area of irregularly shaped pieces of land to work with the land properly. There are many applications in business, economics and the sciences, including all aspects of engineering, where finding the areas of irregular shapes, the volumes of solids and the lengths of irregular shaped curves are important applications. Typical earthworks include roads, railway beds, causeways, dams and canals. Other common earthworks are land grading to reconfigure the topography of a site, or to stabilise slopes. Engineers need to concern themselves with issues of geotechnical engineering (such as soil density and strength) and with quantity estimation to ensure that soil volumes in the cuts match those of the fills, while minimising the distance of movement. Simpson's rule is a staple of scientific data analysis and engineering; it is widely used, for example, by naval architects to numerically determine hull offsets and cross-sectional areas to determine volumes and centroids of ships or lifeboats. There are therefore plenty of examples where irregular areas and volumes need to be determined by engineers.

At the end of this chapter you should be able to:

- use the trapezoidal rule to determine irregular areas
- use the mid-ordinate rule to determine irregular areas
- use Simpson's rule to determine irregular areas
- estimate the volume of irregular solids
- determine the mean values of waveforms

29.1 Areas of irregular figures

Areas of irregular plane surfaces may be approximately determined by using

(a) a planimeter,

(b) the trapezoidal rule,

(c) the mid-ordinate rule, or

(d) Simpson's rule.

Such methods may be used by, for example, engineers estimating areas of indicator diagrams of steam engines, surveyors estimating areas of plots of land or naval architects estimating areas of water planes or transverse sections of ships.

(a) **A planimeter** is an instrument for directly measuring small areas bounded by an irregular curve. There are many different kinds of planimeters but all operate in a similar way. A pointer on the planimeter is used to trace around the boundary of the shape. This induces a movement in another part of the instrument and a reading of this is used to establish the area of the shape.

(b) **Trapezoidal rule**

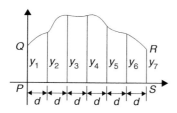

Figure 29.1

To determine the area $PQRS$ in Fig. 29.1,

(i) Divide base PS into any number of equal intervals, each of width d (the greater the number of intervals, the greater the accuracy).
(ii) Accurately measure ordinates y_1, y_2, y_3, etc.
(iii) Area $PQRS$
$$\approx d \left[\frac{y_1 + y_7}{2} + y_2 + y_3 + y_4 + y_5 + y_6 \right]$$

In general, the trapezoidal rule states

$$\textbf{Area} \approx \begin{pmatrix} \textbf{width of} \\ \textbf{interval} \end{pmatrix} \left[\frac{\textbf{1}}{\textbf{2}} \begin{pmatrix} \textbf{first} + \textbf{last} \\ \textbf{ordinate} \end{pmatrix} \right.$$
$$\left. + \begin{pmatrix} \textbf{sum of} \\ \textbf{remaining} \\ \textbf{ordinates} \end{pmatrix} \right]$$

(c) **Mid-ordinate rule**

Figure 29.2

To determine the area $ABCD$ of Fig. 29.2,
(i) Divide base AD into any number of equal intervals, each of width d (the greater the number of intervals, the greater the accuracy).
(ii) Erect ordinates in the middle of each interval (shown by broken lines in Fig. 29.2).

(iii) Accurately measure ordinates y_1, y_2, y_3, etc.
(iv) Area $ABCD$
$$\approx d(y_1 + y_2 + y_3 + y_4 + y_5 + y_6).$$

In general, the mid-ordinate rule states

Area ≈ (width of interval)(sum of mid-ordinates)

(d) **Simpson's rule***
To determine the area $PQRS$ of Fig. 29.1,

(i) Divide base PS into an **even** number of intervals, each of width d (the greater the number of intervals, the greater the accuracy).
(ii) Accurately measure ordinates y_1, y_2, y_3, etc.
(iii) Area $PQRS \approx$
$$\frac{d}{3} \left[(y_1 + y_7) + 4(y_2 + y_4 + y_6) \right.$$
$$\left. + 2(y_3 + y_5) \right]$$

In general, Simpson's rule states

$$\textbf{Area} \approx \frac{\textbf{1}}{\textbf{3}} \begin{pmatrix} \textbf{width of} \\ \textbf{interval} \end{pmatrix} \left[\begin{pmatrix} \textbf{first} + \textbf{last} \\ \textbf{ordinate} \end{pmatrix} \right.$$
$$\left. + 4 \begin{pmatrix} \textbf{sum of even} \\ \textbf{ordinates} \end{pmatrix} + 2 \begin{pmatrix} \textbf{sum of remaining} \\ \textbf{odd ordinates} \end{pmatrix} \right]$$

*Who was **Simpson**? – **Thomas Simpson** FRS (20 August 1710–14 May 1761) was the British mathematician who invented Simpson's rule to approximate definite integrals. To find out more go to **www.routledge.com/cw/bird**

Accuracy of approximate methods

For a function with an increasing gradient, the trapezoidal rule will tend to over-estimate and the mid-ordinate rule will tend to under-estimate (but by half as much). The appropriate combination of the two in Simpson's rule eliminates this error term, giving a rule which will perfectly model anything up to a cubic, and have a proportionately lower error for any function of greater complexity.

In general, for a given number of strips, Simpson's rule is considered the most accurate of the three numerical methods.

Problem 1. A car starts from rest and its speed is measured every second for 6 s.

Time t (s)	0	1	2	3	4	5	6

Speed v (m/s) 0 2.5 5.5 8.75 12.5 17.5 24.0

Determine the distance travelled in 6 seconds (i.e. the area under the v/t graph), using (a) the trapezoidal rule (b) the mid-ordinate rule (c) Simpson's rule

A graph of speed/time is shown in Fig. 29.3.

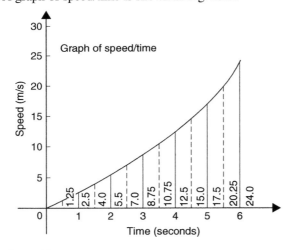

Figure 29.3

(a) **Trapezoidal rule** (see (b) above)
The time base is divided into 6 strips, each of width 1 s, and the length of the ordinates measured. Thus,

$$\text{area} \approx (1)\left[\left(\frac{0+24.0}{2}\right) + 2.5 + 5.5 \right.$$
$$\left. + 8.75 + 12.5 + 17.5\right] = \textbf{58.75 m}$$

(b) **Mid-ordinate rule** (see (c) above)
The time base is divided into 6 strips each of width 1 s. Mid-ordinates are erected as shown in Fig. 29.3 by the broken lines. The length of each mid-ordinate is measured. Thus,

$$\text{area} \approx (1)[1.25 + 4.0 + 7.0 + 10.75 + 15.0$$
$$+ 20.25] = \textbf{58.25 m}$$

(c) **Simpson's rule** (see (d) above)
The time base is divided into 6 strips each of width 1 s and the length of the ordinates measured. Thus,

$$\text{area} \approx \frac{1}{3}(1)[(0 + 24.0) + 4(2.5 + 8.75$$
$$+ 17.5) + 2(5.5 + 12.5)] = \textbf{58.33 m}$$

Problem 2. A river is 15 m wide. Soundings of the depth are made at equal intervals of 3 m across the river and are as shown below.

Depth (m)	0	2.2	3.3	4.5	4.2	2.4	0

Calculate the cross-sectional area of the flow of water at this point using Simpson's rule

From (d) above,

$$\text{Area} \approx \frac{1}{3}(3)[(0+0) + 4(2.2+4.5+2.4) + 2(3.3+4.2)]$$
$$= (1)[0 + 36.4 + 15] = \textbf{51.4 m}^2$$

Now try the following Practice Exercise

Practice Exercise 145   Areas of irregular figures (answers on page 457)

1. Plot a graph of $y = 3x - x^2$ by completing a table of values of y from $x = 0$ to $x = 3$. Determine the area enclosed by the curve, the x-axis and ordinates $x = 0$ and $x = 3$ by (a) the trapezoidal rule (b) the mid-ordinate rule (c) Simpson's rule.

2. Plot the graph of $y = 2x^2 + 3$ between $x = 0$ and $x = 4$. Estimate the area enclosed by the curve, the ordinates $x = 0$ and $x = 4$ and the x-axis by an approximate method.

3. The velocity of a car at 1 second intervals is given in the following table.

Time t (s)	0	1	2	3	4	5	6
Velocity v (m/s)	0	2.0	4.5	8.0	14.0	21.0	29.0

Determine the distance travelled in 6 seconds (i.e. the area under the v/t graph) using Simpson's rule.

4. The shape of a piece of land is shown in Fig. 29.4. To estimate the area of the land, a surveyor takes measurements at intervals of 50 m, perpendicular to the straight portion with the results shown (the dimensions being in metres). Estimate the area of the land in hectares ($1\,\text{ha} = 10^4\,\text{m}^2$).

140 | 160 | 200 | 190 | 180 | 130

50 50 50 50 50 50

Figure 29.4

5. The deck of a ship is 35 m long. At equal intervals of 5 m the width is given by the following table.

Width (m)	0	2.8	5.2	6.5	5.8	4.1	3.0	2.3

Estimate the area of the deck.

29.2 Volumes of irregular solids

If the cross-sectional areas $A_1, A_2, A_3, \ldots$ of an irregular solid bounded by two parallel planes are known at equal intervals of width d (as shown in Fig. 29.5), by Simpson's rule

Volume, $V \approx \dfrac{d}{3}[(A_1 + A_7) + 4(A_2 + A_4 + A_6)$
$$+ 2(A_3 + A_5)]$$

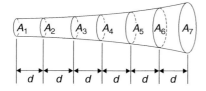

Figure 29.5

Problem 3. A tree trunk is 12 m in length and has a varying cross-section. The cross-sectional areas at intervals of 2 m measured from one end are 0.52, 0.55, 0.59, 0.63, 0.72, 0.84 and 0.97 m². Estimate the volume of the tree trunk

A sketch of the tree trunk is similar to that shown in Fig. 29.5 above, where $d = 2\,\text{m}, A_1 = 0.52\,\text{m}^2$, $A_2 = 0.55\,\text{m}^2$, and so on.
Using Simpson's rule for volumes gives

Volume $\approx \dfrac{2}{3}[(0.52 + 0.97) + 4(0.55 + 0.63 + 0.84)$
$$+ 2(0.59 + 0.72)]$$
$$= \dfrac{2}{3}[1.49 + 8.08 + 2.62] = \mathbf{8.13\,m^3}$$

Problem 4. The areas of seven horizontal cross-sections of a water reservoir at intervals of 10 m are 210, 250, 320, 350, 290, 230 and 170 m². Calculate the capacity of the reservoir in litres

Using Simpson's rule for volumes gives

Volume $\approx \dfrac{10}{3}[(210 + 170) + 4(250 + 350 + 230)$
$$+ 2(320 + 290)]$$
$$= \dfrac{10}{3}[380 + 3320 + 1220] = 16400\,\text{m}^3$$

$16400\,\text{m}^3 = 16400 \times 10^6\,\text{cm}^3$.

Since 1 litre $= 1000\,\text{cm}^3$,

capacity of reservoir $= \dfrac{16400 \times 10^6}{1000}$litres
$$= 16400000 = \mathbf{16.4 \times 10^6\,litres}$$

Now try the following Practice Exercise

1. The areas of equidistantly spaced sections of the underwater form of a small boat are as follows:
1.76, 2.78, 3.10, 3.12, 2.61, 1.24 and 0.85 m². Determine the underwater volume if the sections are 3 m apart.

2. To estimate the amount of earth to be removed when constructing a cutting, the

cross-sectional area at intervals of 8 m were estimated as follows:

0, 2.8, 3.7, 4.5, 4.1, 2.6 and 0 m²

Estimate the volume of earth to be excavated.

3. The circumference of a 12 m long log of timber of varying circular cross-section is measured at intervals of 2 m along its length and the results are as follows. Estimate the volume of the timber in cubic metres.

Distance from one end (m)	0	2	4	6
Circumference (m)	2.80	3.25	3.94	4.32

Distance from one end (m)	8	10	12
Circumference (m)	5.16	5.82	6.36

29.3 Mean or average values of waveforms

The mean or average value, y, of the waveform shown in Fig. 29.6 is given by

$$y = \frac{\text{area under curve}}{\text{length of base, } b}$$

If the mid-ordinate rule is used to find the area under the curve, then

$$y \approx \frac{\text{sum of mid-ordinates}}{\text{number of mid-ordinates}}$$

$$\left(= \frac{y_1 + y_2 + y_3 + y_4 + y_5 + y_6 + y_7}{7} \text{ for Fig. 29.6} \right)$$

Figure 29.6

For a **sine wave**, the mean or average value

(a) over one complete cycle is **zero** (see Fig. 29.7(a)),

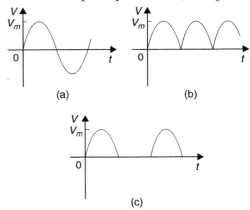

Figure 29.7

(b) over half a cycle is **0.637 × maximum value** or $\dfrac{2}{\pi}$ **× maximum value**,

(c) of a full-wave rectified waveform (see Fig. 29.7(b)) is **0.637 × maximum value**,

(d) of a half-wave rectified waveform (see Fig. 29.7(c)) is **0.318 × maximum value** or $\dfrac{1}{\pi}$ **× maximum value**.

Problem 5. Determine the average values over half a cycle of the periodic waveforms shown in Fig. 29.8

Figure 29.8

(a) Area under triangular waveform (a) for a half cycle is given by

$$\text{Area} = \frac{1}{2}(\text{base})(\text{perpendicular height})$$

$$= \frac{1}{2}(2 \times 10^{-3})(20) = 20 \times 10^{-3} \text{ Vs}$$

Average value of waveform

$$= \frac{\text{area under curve}}{\text{length of base}}$$

$$= \frac{20 \times 10^{-3}\,\text{Vs}}{2 \times 10^{-3}\,\text{s}} = \textbf{10 V}$$

(b) Area under waveform (b) for a half cycle
$$= (1 \times 1) + (3 \times 2) = 7\,\text{As}$$

Average value of waveform $= \dfrac{\text{area under curve}}{\text{length of base}}$

$$= \frac{7\,\text{As}}{3\,\text{s}} = \textbf{2.33 A}$$

(c) A half cycle of the voltage waveform (c) is completed in 4 ms.

Area under curve $= \dfrac{1}{2}\{(3-1)10^{-3}\}(10)$

$$= 10 \times 10^{-3}\,\text{Vs}$$

Average value of waveform $= \dfrac{\text{area under curve}}{\text{length of base}}$

$$= \frac{10 \times 10^{-3}\,\text{Vs}}{4 \times 10^{-3}\,\text{s}} = \textbf{2.5 V}$$

Problem 6. Determine the mean value of current over one complete cycle of the periodic waveforms shown in Fig. 29.9

Figure 29.9

(a) One cycle of the trapezoidal waveform (a) is completed in 10 ms (i.e. the periodic time is 10 ms).

Area under curve $=$ area of trapezium

$$= \frac{1}{2}(\text{sum of parallel sides})(\text{perpendicular}$$
$$\text{distance between parallel sides})$$

$$= \frac{1}{2}\{(4+8) \times 10^{-3}\}(5 \times 10^{-3})$$

$$= 30 \times 10^{-6}\,\text{As}$$

Mean value over one cycle $= \dfrac{\text{area under curve}}{\text{length of base}}$

$$= \frac{30 \times 10^{-6}\,\text{As}}{10 \times 10^{-3}\,\text{s}}$$

$$= \textbf{3 mA}$$

(b) One cycle of the sawtooth waveform (b) is completed in 5 ms.

Area under curve $= \dfrac{1}{2}(3 \times 10^{-3})(2)$

$$= 3 \times 10^{-3}\,\text{As}$$

Mean value over one cycle $= \dfrac{\text{area under curve}}{\text{length of base}}$

$$= \frac{3 \times 10^{-3}\,\text{As}}{5 \times 10^{-3}\,\text{s}}$$

$$= \textbf{0.6 A}$$

Problem 7. The power used in a manufacturing process during a 6 hour period is recorded at intervals of 1 hour as shown below.

Time (h)	0	1	2	3	4	5	6
Power (kW)	0	14	29	51	45	23	0

Plot a graph of power against time and, by using the mid-ordinate rule, determine (a) the area under the curve and (b) the average value of the power

The graph of power/time is shown in Fig. 29.10.

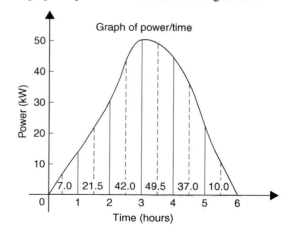

Figure 29.10

(a) The time base is divided into 6 equal intervals, each of width 1 hour. Mid-ordinates are erected (shown by broken lines in Fig. 29.10) and measured. The values are shown in Fig. 29.10.

Area under curve

$\approx$ (width of interval)(sum of mid-ordinates)

$= (1)[7.0 + 21.5 + 42.0 + 49.5 + 37.0 + 10.0]$

$= \mathbf{167\,kWh}$ (i.e. a measure of electrical energy)

(b) **Average value of waveform** $= \dfrac{\text{area under curve}}{\text{length of base}}$

$$= \frac{167\,\text{kWh}}{6\,\text{h}}$$

$$= \mathbf{27.83\,kW}$$

Alternatively, average value

$$= \frac{\text{sum of mid-ordinates}}{\text{number of mid-ordinates}}$$

> **Problem 8.** Fig. 29.11 shows a sinusoidal output voltage of a full-wave rectifier. Determine, using the mid-ordinate rule with 6 intervals, the mean output voltage

Figure 29.11

One cycle of the output voltage is completed in π radians or $180°$. The base is divided into 6 intervals, each of width $30°$. The mid-ordinate of each interval will lie at $15°, 45°, 75°$, etc.

At $15°$ the height of the mid-ordinate is

$$10\sin 15° = 2.588\,\text{V},$$

At $45°$ the height of the mid-ordinate is

$$10\sin 45° = 7.071\,\text{V, and so on.}$$

The results are tabulated below.

Mid-ordinate	Height of mid-ordinate
$15°$	$10\sin 15° = 2.588$ V
$45°$	$10\sin 45° = 7.071$ V
$75°$	$10\sin 75° = 9.659$ V
$105°$	$10\sin 105° = 9.659$ V
$135°$	$10\sin 135° = 7.071$ V
$165°$	$10\sin 165° = 2.588$ V
Sum of mid-ordinates = 38.636 V	

Mean or average value of output voltage

$$\approx \frac{\text{sum of mid-ordinates}}{\text{number of mid-ordinates}} = \frac{38.636}{6} = \mathbf{6.439\,V}$$

(With a larger number of intervals a more accurate answer may be obtained.)

For a sine wave the actual mean value is $0.637 \times$ maximum value, which in this problem gives 6.37 V.

> **Problem 9.** An indicator diagram for a steam engine is shown in Fig. 29.12. The base line has been divided into 6 equally spaced intervals and the lengths of the 7 ordinates measured with the results shown in centimetres. Determine
>
> (a) the area of the indicator diagram using Simpson's rule
>
> (b) the mean pressure in the cylinder given that 1 cm represents 100 kPa

Figure 29.12

(a) The width of each interval is $\dfrac{12.0}{6}$ cm. Using Simpson's rule,

$$\mathbf{area} \approx \frac{1}{3}(2.0)[(3.6 + 1.6) + 4(4.0 + 2.9 + 1.7)$$
$$+ 2(3.5 + 2.2)]$$

$$= \frac{2}{3}[5.2 + 34.4 + 11.4] = \mathbf{34\,cm^2}$$

(b) Mean height of ordinates $= \dfrac{\text{area of diagram}}{\text{length of base}}$

$$= \frac{34}{12} = 2.83\,\text{cm}$$

Since 1 cm represents 100 kPa,

mean pressure in the cylinder
$$= 2.83\,\text{cm} \times 100\,\text{kPa/cm} = \mathbf{283\,kPa}$$

Now try the following Practice Exercise

Practice Exercise 147   Mean or average values of waveforms (answers on page 457)

1. Determine the mean value of the periodic waveforms shown in Fig. 29.13 over a half cycle.

(a)

(b)

(c)

Figure 29.13

2. Find the average value of the periodic waveforms shown in Fig. 29.14 over one complete cycle.

(a)

(b)

Figure 29.14

3. An alternating current has the following values at equal intervals of 5 ms:

Time (ms)	0	5	10	15	20	25	30
Current (A)	0	0.9	2.6	4.9	5.8	3.5	0

Plot a graph of current against time and estimate the area under the curve over the 30 ms period, using the mid-ordinate rule, and determine its mean value.

4. Determine, using an approximate method, the average value of a sine wave of maximum value 50 V for (a) a half cycle (b) a complete cycle.

5. An indicator diagram of a steam engine is 12 cm long. Seven evenly spaced ordinates, including the end ordinates, are measured as follows:

5.90, 5.52, 4.22, 3.63, 3.32, 3.24 and 3.16 cm.

Determine the area of the diagram and the mean pressure in the cylinder if 1 cm represents 90 kPa.

Practice Exercise 148   Multiple-choice questions on irregular areas and volumes and mean values (Answers on page 457)

Each question has only one correct answer

1. The speed of a car at 1 second intervals is given in the following table:

Time t (s)	0	1	2	3	4	5	6
Speed v(m/s)	0	2.5	5.0	9.0	15.0	22.0	30.0

The distance travelled in 6 s (i.e. the area under the v/t graph) using the trapezoidal rule is:
(a) 83.5 m   (b) 68 m
(c) 68.5 m   (d) 204 m

2. The mean value of a sine wave over half a cycle is:
(a) $0.318 \times$ maximum value
(b) $0.707 \times$ maximum value
(c) the peak value
(d) $0.637 \times$ maximum value

3. An indicator diagram for a steam engine is as shown in Figure 29.15. The base has been divided into 6 equally spaced intervals and the lengths of the 7 ordinates measured, with the results shown in centimetres. Using Simpson's rule, the area of the indicator diagram is:
(a) 17.9 cm^2 (b) 32 cm^2
(c) 16 cm^2 (d) 96 cm^2

3.1 3.9 3.5 2.8 2.0 1.5 1.2

12.0 cm

Figure 29.15

4. The mean value over half a cycle of a sine wave is 7.5 volts. The peak value of the sine wave is:
   (a) 15 V   (b) 23.58 V
   (c) 11.77 V   (d) 10.61 V

5. The areas of seven horizontal cross-sections of a water reservoir at intervals of 9 m are 70, 160, 210, 280, 250, 220 and 170 m^2. Given that 1 litre = 1000 cm^3, the capacity of the reservoir is:
   (a) 11.4×10^6 litres   (b) 34.2×10^6 litres
   (c) 11.4×10^3 litres   (d) 32.4×10^3 litres

For fully worked solutions to each of the problems in Practice Exercises 145 to 147 in this chapter, go to the website:
www.routledge.com/cw/bird

This assignment covers the material contained in Chapters 28 and 29. *The marks available are shown in brackets at the end of each question.*

1. A rectangular block of alloy has dimensions of 60 mm by 30 mm by 12 mm. Calculate the volume of the alloy in cubic centimetres. (3)

2. Determine how many cubic metres of concrete are required for a 120 m long path, 400 mm wide and 10 cm deep. (3)

3. Find the volume of a cylinder of radius 5.6 cm and height 15.5 cm. Give the answer correct to the nearest cubic centimetre. (3)

4. A garden roller is 0.35 m wide and has a diameter of 0.20 m. What area will it roll in making 40 revolutions? (4)

5. Find the volume of a cone of height 12.5 cm and base diameter 6.0 cm, correct to 1 decimal place. (3)

6. Find (a) the volume and (b) the total surface area of the right-angled triangular prism shown in Fig. RT11.1. (9)

Figure RT11.1

9.70 cm
4.80 cm
11.6 cm

7. A pyramid having a square base has a volume of 86.4 cm^3. If the perpendicular height is 20 cm, determine the length of each side of the base. (4)

8. A copper pipe is 80 m long. It has a bore of 80 mm and an outside diameter of 100 mm. Calculate, in cubic metres, the volume of copper in the pipe. (4)

9. Find (a) the volume and (b) the surface area of a sphere of diameter 25 mm. (4)

10. A piece of alloy with dimensions 25 mm by 60 mm by 1.60 m is melted down and recast into a cylinder whose diameter is 150 mm. Assuming no wastage, calculate the height of the cylinder in centimetres, correct to 1 decimal place. (4)

11. Determine the volume (in cubic metres) and the total surface area (in square metres) of a solid metal cone of base radius 0.5 m and perpendicular height 1.20 m. Give answers correct to 2 decimal places. (6)

12. A rectangular storage container has dimensions 3.2 m by 90 cm by 60 cm. Determine its volume in (a) m^3 (b) cm^3 (4)

13. Calculate (a) the volume and (b) the total surface area of a 10 cm by 15 cm rectangular pyramid of height 20 cm. (8)

14. A water container is of the form of a central cylindrical part 3.0 m long and diameter 1.0 m, with a hemispherical section surmounted at each end as shown in Fig. RT11.2. Determine the maximum capacity of the container, correct to the nearest litre. (1 litre = 1000 cm^3) (5)

3.0 m
1.0 m

Figure RT11.2

15. Find the total surface area of a bucket consisting of an inverted frustum of a cone of slant height 35.0 cm and end diameters 60.0 cm and 40.0 cm. (4)

16. A boat has a mass of 20 000 kg. A model of the boat is made to a scale of 1 to 80. If the model is made of the same material as the boat, determine the mass of the model (in grams). (3)

17. Plot a graph of $y = 3x^2 + 5$ from $x = 1$ to $x = 4$. Estimate, correct to 2 decimal places, using 6 intervals, the area enclosed by the curve, the ordinates $x = 1$ and $x = 4$, and the x-axis by
    (a) the trapezoidal rule
    (b) the mid-ordinate rule
    (c) Simpson's rule. (16)

18. A circular cooling tower is 20 m high. The inside diameter of the tower at different heights is given in the following table.

Height (m)	0	5.0	10.0	15.0	20.0
Diameter (m)	16.0	13.3	10.7	8.6	8.0

Determine the area corresponding to each diameter and hence estimate the capacity of the tower in cubic metres. (7)

19. A vehicle starts from rest and its velocity is measured every second for 6 seconds, with the following results.

Time t (s)	0	1	2	3	4	5	6
Velocity v (m/s)	0	1.2	2.4	3.7	5.2	6.0	9.2

Using Simpson's rule, calculate
(a) the distance travelled in 6 s (i.e. the area under the v/t graph),
(b) the average speed over this period. (6)

Chapter 30

Vectors

Why it is important to understand: Vectors

Vectors are an important part of the language of science, mathematics and engineering. They are used to discuss multivariable calculus, electrical circuits with oscillating currents, stress and strain in structures and materials, flows of atmospheres and fluids, and in many other applications. Resolving a vector into components is a precursor to computing things with or about a vector quantity. Because position, velocity, acceleration, force, momentum and angular momentum are all vector quantities, resolving vectors into components is a most important skill required in any engineering studies.

At the end of this chapter you should be able to:

- distinguish between scalars and vectors
- recognise how vectors are represented
- add vectors using the nose-to-tail method
- add vectors using the parallelogram method
- resolve vectors into their horizontal and vertical components
- add vectors by calculation using horizontal and vertical components
- perform vector subtraction
- understand relative velocity
- understand i, j, k notation

30.1 Introduction

This chapter initially explains the difference between scalar and vector quantities and shows how a vector is drawn and represented.

Any object that is acted upon by an external force will respond to that force by moving in the line of the force. However, if two or more forces act simultaneously, the result is more difficult to predict; the ability to add two or more vectors then becomes important.

This chapter shows how vectors are added and subtracted, both by drawing and by calculation, and how finding the resultant of two or more vectors has many uses in engineering. (Resultant means the single vector which would have the same effect as the individual vectors.) Relative velocities and vector i, j, k notation are also briefly explained.

30.2 Scalars and vectors

The time taken to fill a water tank may be measured as, say, 50 s. Similarly, the temperature in a room may be measured as, say, 16°C or the mass of a bearing may be measured as, say, 3 kg. Quantities such as time, temperature and mass are entirely defined by a numerical value and are called **scalars** or **scalar quantities**.

Not all quantities are like this. Some are defined by more than just size; some also have direction. For example, the velocity of a car may be 90 km/h due west, a force of 20 N may act vertically downwards, or an acceleration of 10 m/s^2 may act at 50° to the horizontal.

Quantities such as velocity, force and acceleration, which **have both a magnitude and a direction**, are called **vectors**.

Now try the following Practice Exercise

Practice Exercise 149   Scalar and vector quantities (answers on page 457)

1. State the difference between scalar and vector quantities.

In Problems 2 to 9, state whether the quantities given are scalar or vector.

2. A temperature of 70°C

3. 5 m^3 volume

4. A downward force of 20 N

5. 500 J of work

6. 30 cm^2 area

7. A south-westerly wind of 10 knots

8. 50 m distance

9. An acceleration of 15 m/s^2 at 60° to the horizontal

30.3 Drawing a vector

A vector quantity can be represented graphically by a line, drawn so that

(a) the **length** of the line denotes the magnitude of the quantity, and

(b) the **direction** of the line denotes the direction in which the vector quantity acts.

An arrow is used to denote the sense, or direction, of the vector.

The arrow end of a vector is called the 'nose' and the other end the 'tail'. For example, a force of 9 N acting at 45° to the horizontal is shown in Fig. 30.1. Note that an angle of +**45°** is drawn from the horizontal and moves **anticlockwise**.

Figure 30.1

A velocity of 20 m/s at −60° is shown in Fig. 30.2. Note that an angle of −**60°** is drawn from the horizontal and moves **clockwise**.

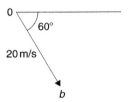

Figure 30.2

Representing a vector

There are a number of ways of representing vector quantities. These include

(a) Using bold print.

(b) $\overrightarrow{AB}$ where an arrow above two capital letters denotes the sense of direction, where A is the starting point and B the end point of the vector.

(c) $\overline{AB}$ or $\overline{a}$; i.e. a line over the top of letter.

(d) $\underline{a}$; i.e. underlined letter.

The force of 9 N at 45° shown in Fig. 30.1 may be represented as

$$\textbf{0a} \quad \text{or} \quad \overrightarrow{0a} \quad \text{or} \quad \overline{0a}$$

The magnitude of the force is $0a$

Similarly, the velocity of 20 m/s at −60° shown in Fig. 30.2 may be represented as

$$\textbf{0b} \quad \text{or} \quad \overrightarrow{0b} \quad \text{or} \quad \overline{0b}$$

The magnitude of the velocity is $0b$

In this chapter a vector quantity is denoted by **bold print**.

30.4 Addition of vectors by drawing

Adding two or more vectors by drawing assumes that a ruler, pencil and protractor are available. Results obtained by drawing are naturally not as accurate as those obtained by calculation.

(a) **Nose-to-tail method**

Two force vectors, F_1 and F_2, are shown in Fig. 30.3. When an object is subjected to more than one force, the resultant of the forces is found by the addition of vectors.

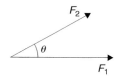

Figure 30.3

To add forces F_1 and F_2,
(i) Force F_1 is drawn to scale horizontally, shown as **0a** in Fig. 30.4.
(ii) From the nose of F_1, force F_2 is drawn at angle θ to the horizontal, shown as **ab**
(iii) The resultant force is given by length **0b**, which may be measured.
This procedure is called the **'nose-to-tail'** or **'triangle' method**.

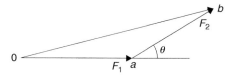

Figure 30.4

(b) **Parallelogram method**

To add the two force vectors, F_1 and F_2 of Fig. 30.3,
(i) A line cb is constructed which is parallel to and equal in length to **0a** (see Fig. 30.5).
(ii) A line ab is constructed which is parallel to and equal in length to **0c**
(iii) The resultant force is given by the diagonal of the parallelogram; i.e. length **0b**
This procedure is called the **'parallelogram' method**.

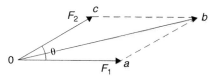

Figure 30.5

> **Problem 1.** A force of 5 N is inclined at an angle of 45° to a second force of 8 N, both forces acting at a point. Find the magnitude of the resultant of these two forces and the direction of the resultant with respect to the 8 N force by (a) the nose-to-tail method and (b) the parallelogram method

The two forces are shown in Fig. 30.6. (Although the 8 N force is shown horizontal, it could have been drawn in any direction.)

Figure 30.6

(a) **Nose-to-tail method**
(i) The 8 N force is drawn horizontally 8 units long, shown as **0a** in Fig. 30.7.
(ii) From the nose of the 8 N force, the 5 N force is drawn 5 units long at an angle of 45° to the horizontal, shown as **ab**
(iii) The resultant force is given by length **0b** and is measured as **12 N** and angle θ is measured as **17°**

Figure 30.7

(b) **Parallelogram method**
(i) In Fig. 30.8, a line is constructed which is parallel to and equal in length to the 8 N force.
(ii) A line is constructed which is parallel to and equal in length to the 5 N force.

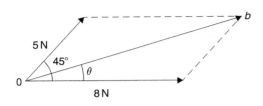

Figure 30.8

(iii) The resultant force is given by the diagonal of the parallelogram, i.e. length 0*b*, and is measured as **12 N** and angle θ is measured as **17°**

Thus, **the resultant of the two force vectors in Fig. 30.6 is 12 N at 17° to the 8 N force**.

Problem 2. Forces of 15 N and 10 N are at an angle of 90° to each other as shown in Fig. 30.9. Find, by drawing, the magnitude of the resultant of these two forces and the direction of the resultant with respect to the 15 N force

Figure 30.9

Using the nose-to-tail method,

(i) The 15 N force is drawn horizontally 15 units long, as shown in Fig. 30.10.

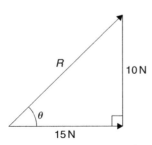

Figure 30.10

(ii) From the nose of the 15 N force, the 10 N force is drawn 10 units long at an angle of 90° to the horizontal as shown.

(iii) The resultant force is shown as *R* and is measured as **18 N** and angle θ is measured as **34°**

Thus, **the resultant of the two force vectors is 18 N at 34° to the 15 N force**.

Problem 3. Velocities of 10 m/s, 20 m/s and 15 m/s act as shown in Fig. 30.11. Determine, by drawing, the magnitude of the resultant velocity and its direction relative to the horizontal

Figure 30.11

When more than 2 vectors are being added the nose-to-tail method is used. The order in which the vectors are added does not matter. In this case the order taken is ν_1, then ν_2, then ν_3. However, if a different order is taken the same result will occur.

(i) ν_1 is drawn 10 units long at an angle of 30° to the horizontal, shown as **0*a*** in Fig. 30.12.

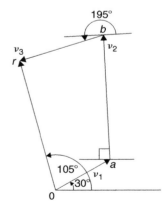

Figure 30.12

(ii) From the nose of ν_1, ν_2 is drawn 20 units long at an angle of 90° to the horizontal, shown as ***ab***

(iii) From the nose of v_2, v_3 is drawn 15 units long at an angle of 195° to the horizontal, shown as **br**

(iv) The resultant velocity is given by length **0r** and is measured as **22 m/s** and the angle measured to the horizontal is **105°**

Thus, **the resultant of the three velocities is 22 m/s at 105° to the horizontal.**

Worked examples 1 to 3 have demonstrated how vectors are added to determine their resultant and their direction. However, drawing to scale is time-consuming and not highly accurate. The following sections demonstrate how to determine resultant vectors by calculation using horizontal and vertical components and, where possible, by Pythagoras' theorem.

30.5 Resolving vectors into horizontal and vertical components

A force vector F is shown in Fig. 30.13 at angle θ to the horizontal. Such a vector can be resolved into two components such that the vector addition of the components is equal to the original vector.

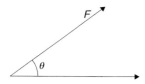

Figure 30.13

The two components usually taken are a **horizontal component** and a **vertical component**. If a right-angled triangle is constructed as shown in Fig. 30.14, $0a$ is called the horizontal component of F and ab is called the vertical component of F.

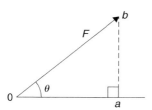

Figure 30.14

From trigonometry (see Chapter 22 and remember SOH CAH TOA),

$$\cos\theta = \frac{0a}{0b}, \text{ from which } 0a = 0b\cos\theta = F\cos\theta$$

i.e. **the horizontal component of $F = F\cos\theta$**, and

$$\sin\theta = \frac{ab}{0b} \text{ from which, } ab = 0b\sin\theta = F\sin\theta$$

i.e. **the vertical component of $F = F\sin\theta$.**

Problem 4. Resolve the force vector of 50 N at an angle of 35° to the horizontal into its horizontal and vertical components

The **horizontal component** of the 50 N force, $0a = 50\cos35° = $ **40.96 N**
The **vertical component** of the 50 N force, $ab = 50\sin35° = $ **28.68 N**
The horizontal and vertical components are shown in Fig. 30.15.

Figure 30.15

(To check: by Pythagoras,

$$0b = \sqrt{40.96^2 + 28.68^2} = 50\,\text{N}$$

and $$\theta = \tan^{-1}\left(\frac{28.68}{40.96}\right) = 35°$$

Thus, the vector addition of components 40.96 N and 28.68 N is 50 N at 35°)

Problem 5. Resolve the velocity vector of 20 m/s at an angle of −30° to the horizontal into horizontal and vertical components

The **horizontal component** of the 20 m/s velocity, $0a = 20\cos(-30°) = $ **17.32 m/s**
The **vertical component** of the 20 m/s velocity, $ab = 20\sin(-30°) = $ **−10 m/s**
The horizontal and vertical components are shown in Fig. 30.16.

Problem 6. Resolve the displacement vector of 40 m at an angle of 120° into horizontal and vertical components

Figure 30.16

The **horizontal component** of the 40 m displacement,
$$0a = 40\cos 120° = \mathbf{-20.0\,m}$$
The **vertical component** of the 40 m displacement,
$$ab = 40\sin 120° = \mathbf{34.64\,m}$$
The horizontal and vertical components are shown in Fig. 30.17.

Figure 30.17

30.6 Addition of vectors by calculation

Two force vectors, F_1 and F_2, are shown in Fig. 30.18, F_1 being at an angle of θ_1 and F_2 at an angle of θ_2

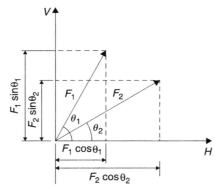

Figure 30.18

A method of adding two vectors together is to use horizontal and vertical components.
The horizontal component of force F_1 is $F_1\cos\theta_1$ and the horizontal component of force F_2 is $F_2\cos\theta_2$. The

total horizontal component of the two forces,
$$H = F_1\cos\theta_1 + F_2\cos\theta_2$$

The vertical component of force F_1 is $F_1\sin\theta_1$ and the vertical component of force F_2 is $F_2\sin\theta_2$. The total vertical component of the two forces,
$$V = F_1\sin\theta_1 + F_2\sin\theta_2$$

Since we have H and V, the resultant of F_1 and F_2 is obtained by using the theorem of Pythagoras. From Fig. 30.19,
$$0b^2 = H^2 + V^2$$
i.e. **resultant** $= \sqrt{H^2 + V^2}$ at an angle given by $\theta = \tan^{-1}\left(\dfrac{V}{H}\right)$

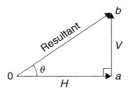

Figure 30.19

Problem 7. A force of 5 N is inclined at an angle of 45° to a second force of 8 N, both forces acting at a point. Calculate the magnitude of the resultant of these two forces and the direction of the resultant with respect to the 8 N force

The two forces are shown in Fig. 30.20.

Figure 30.20

The horizontal component of the 8 N force is $8\cos 0°$ and the horizontal component of the 5 N force is $5\cos 45°$. The total horizontal component of the two forces,
$$H = 8\cos 0° + 5\cos 45° = 8 + 3.5355 = \mathbf{11.5355}$$

The vertical component of the 8 N force is $8\sin 0°$ and the vertical component of the 5 N force is $5\sin 45°$. The total vertical component of the two forces,

$$V = 8\sin 0° + 5\sin 45° = 0 + 3.5355 = \mathbf{3.5355}$$

Figure 30.21

From Fig. 30.21, magnitude of resultant vector

$$= \sqrt{H^2 + V^2}$$
$$= \sqrt{11.5355^2 + 3.5355^2} = \mathbf{12.07\,N}$$

The direction of the resultant vector,

$$\theta = \tan^{-1}\left(\frac{V}{H}\right) = \tan^{-1}\left(\frac{3.5355}{11.5355}\right)$$
$$= \tan^{-1} 0.30648866\ldots = \mathbf{17.04°}$$

Thus, **the resultant of the two forces is a single vector of 12.07 N at 17.04° to the 8 N vector.**

Problem 8. Forces of 15 N and 10 N are at an angle of 90° to each other as shown in Fig. 30.22. Calculate the magnitude of the resultant of these two forces and its direction with respect to the 15 N force

Figure 30.22

The horizontal component of the 15 N force is $15\cos 0°$ and the horizontal component of the 10 N force is $10\cos 90°$. The total horizontal component of the two forces,

$$H = 15\cos 0° + 10\cos 90° = 15 + 0 = \mathbf{15}$$

The vertical component of the 15 N force is $15\sin 0°$ and the vertical component of the 10 N force is $10\sin 90°$. The total vertical component of the two forces,

$$V = 15\sin 0° + 10\sin 90° = 0 + 10 = \mathbf{10}$$

Magnitude of resultant vector
$$= \sqrt{H^2 + V^2} = \sqrt{15^2 + 10^2} = \mathbf{18.03\,N}$$
The direction of the resultant vector,
$$\theta = \tan^{-1}\left(\frac{V}{H}\right) = \tan^{-1}\left(\frac{10}{15}\right) = \mathbf{33.69°}$$

Thus, **the resultant of the two forces is a single vector of 18.03 N at 33.69° to the 15 N vector.**

There is an alternative method of calculating the resultant vector in this case. If we used the triangle method, the diagram would be as shown in Fig. 30.23.

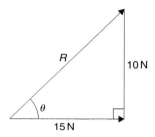

Figure 30.23

Since a right-angled triangle results, we could use Pythagoras' theorem without needing to go through the procedure for horizontal and vertical components. In fact, the horizontal and vertical components are 15 N and 10 N respectively.

This is, of course, a special case. **Pythagoras can only be used when there is an angle of 90° between vectors.** This is demonstrated in worked Problem 9.

Problem 9. Calculate the magnitude and direction of the resultant of the two acceleration vectors shown in Fig. 30.24.

Figure 30.24

The 15 m/s^2 acceleration is drawn horizontally, shown as **0a** in Fig. 30.25.

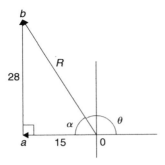

Figure 30.25

From the nose of the 15 m/s^2 acceleration, the 28 m/s^2 acceleration is drawn at an angle of 90° to the horizontal, shown as **ab**
The resultant acceleration, R, is given by length **0b**
Since a right-angled triangle results, the theorem of Pythagoras may be used.

$$0b = \sqrt{15^2 + 28^2} = \mathbf{31.76\,m/s^2}$$

and

$$\alpha = \tan^{-1}\left(\frac{28}{15}\right) = \mathbf{61.82°}$$

Measuring from the horizontal,
$\theta = 180° - 61.82° = \mathbf{118.18°}$
Thus, **the resultant of the two accelerations is a single vector of 31.76 m/s^2 at 118.18° to the horizontal**.

Problem 10. Velocities of 10 m/s, 20 m/s and 15 m/s act as shown in Fig. 30.26. Calculate the magnitude of the resultant velocity and its direction relative to the horizontal

Figure 30.26

The horizontal component of the 10 m/s velocity
$= 10\cos 30° = 8.660$ m/s,

the horizontal component of the 20 m/s velocity
is $20\cos 90° = 0$ m/s
and the horizontal component of the 15 m/s velocity is
$15\cos 195° = -14.489$ m/s
The total horizontal component of the three velocities,

$$H = 8.660 + 0 - 14.489 = \mathbf{-5.829\,m/s}$$

The vertical component of the 10 m/s velocity
$= 10\sin 30° = 5$ m/s,
the vertical component of the 20 m/s velocity is
$20\sin 90° = 20$ m/s
and the vertical component of the 15 m/s velocity is
$15\sin 195° = -3.882$ m/s
The total vertical component of the three velocities,
$$V = 5 + 20 - 3.882 = \mathbf{21.118\,m/s}$$

Figure 30.27

From Fig. 30.27, magnitude of resultant vector,

$$R = \sqrt{H^2 + V^2} = \sqrt{5.829^2 + 21.118^2} = \mathbf{21.91\,m/s}$$

The direction of the resultant vector,

$$\alpha = \tan^{-1}\left(\frac{V}{H}\right) = \tan^{-1}\left(\frac{21.118}{5.829}\right) = \mathbf{74.57°}$$

Measuring from the horizontal,
$$\theta = 180° - 74.57° = \mathbf{105.43°}$$
Thus, **the resultant of the three velocities is a single vector of 21.91 m/s at 105.43° to the horizontal**.

Now try the following Practice Exercise

Practice Exercise 150   Addition of vectors by calculation (answers on page 457)

1. A force of 7 N is inclined at an angle of 50° to a second force of 12 N, both forces acting at a point. Calculate the magnitude of the resultant of the two forces and the direction of the resultant with respect to the 12 N force.

2. Velocities of 5 m/s and 12 m/s act at a point at 90° to each other. Calculate the resultant velocity and its direction relative to the 12 m/s velocity.

3. Calculate the magnitude and direction of the resultant of the two force vectors shown in Fig. 30.28.

Figure 30.28

4. Calculate the magnitude and direction of the resultant of the two force vectors shown in Fig. 30.29.

Figure 30.29

5. A displacement vector s_1 is 30 m at 0°. A second displacement vector s_2 is 12 m at 90°. Calculate the magnitude and direction of the resultant vector $s_1 + s_2$

6. Three forces of 5 N, 8 N and 13 N act as shown in Fig. 30.30. Calculate the magnitude and direction of the resultant force.

Figure 30.30

7. If velocity $v_1 = 25$ m/s at 60° and $v_2 = 15$ m/s at −30°, calculate the magnitude and direction of $v_1 + v_2$

8. Calculate the magnitude and direction of the resultant vector of the force system shown in Fig. 30.31.

Figure 30.31

9. Calculate the magnitude and direction of the resultant vector of the system shown in Fig. 30.32.

Figure 30.32

10. An object is acted upon by two forces of magnitude 10 N and 8 N at an angle of 60° to each other. Determine the resultant force on the object.

11. A ship heads in a direction of E 20°S at a speed of 20 knots while the current is 4 knots in a direction of N 30°E. Determine the speed and actual direction of the ship.

30.7 Vector subtraction

In Fig. 30.33, a force vector F is represented by oa. The vector $(-oa)$ can be obtained by drawing a vector from o in the opposite sense to oa but having the same magnitude, shown as ob in Fig. 30.33; i.e. $ob = (-oa)$

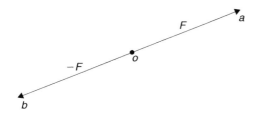

Figure 30.33

For two vectors acting at a point, as shown in Fig. 30.34(a), the resultant of vector addition is

$$os = oa + ob$$

Fig. 30.34(b) shows vectors $ob + (-oa)$ that is, $ob - oa$ and the vector equation is $ob - oa = od$. Comparing od in Fig. 30.34(b) with the broken line ab in Fig. 30.34(a) shows that the second diagonal of the parallelogram method of vector addition gives the magnitude and direction of vector subtraction of oa from ob

(a)   (b)

Figure 30.34

⚑ **Problem 11.** Accelerations of $a_1 = 1.5\,\text{m/s}^2$ at $90°$ and $a_2 = 2.6\,\text{m/s}^2$ at $145°$ act at a point. Find $a_1 + a_2$ and $a_1 - a_2$ by (a) drawing a scale vector diagram and (b) calculation

---

(a) The scale vector diagram is shown in Fig. 30.35. By measurement,

$$a_1 + a_2 = \textbf{3.7 m/s}^2 \textbf{ at } \textbf{126}°$$

$$a_1 - a_2 = \textbf{2.1 m/s}^2 \textbf{ at } \textbf{0}°$$

(b) Resolving horizontally and vertically gives Horizontal component of $a_1 + a_2$,
$$H = 1.5\cos 90° + 2.6\cos 145° = -\textbf{2.13}$$

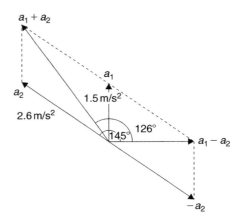

Figure 30.35

Vertical component of $a_1 + a_2$,
$$V = 1.5\sin 90° + 2.6\sin 145° = \textbf{2.99}$$
From Fig. 30.36, the magnitude of $a_1 + a_2$,
$$R = \sqrt{(-2.13)^2 + 2.99^2} = \textbf{3.67 m/s}^2$$
In Fig. 30.36, $\alpha = \tan^{-1}\left(\dfrac{2.99}{2.13}\right) = 54.53°$ and
$$\theta = 180° - 54.53° = 125.47°$$
Thus, $a_1 + a_2 = \textbf{3.67 m/s}^2 \textbf{ at } \textbf{125.47}°$

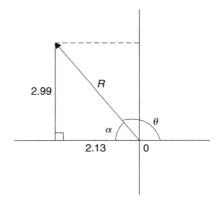

Figure 30.36

Horizontal component of $a_1 - a_2$
$$= 1.5\cos 90° - 2.6\cos 145° = \textbf{2.13}$$
Vertical component of $a_1 - a_2$
$$= 1.5\sin 90° - 2.6\sin 145° = 0$$
Magnitude of $a_1 - a_2 = \sqrt{2.13^2 + 0^2}$
$$= \textbf{2.13 m/s}^2$$
Direction of $a_1 - a_2 = \tan^{-1}\left(\dfrac{0}{2.13}\right) = \textbf{0}°$
Thus, $a_1 - a_2 = \textbf{2.13 m/s}^2 \textbf{ at } \textbf{0}°$

Problem 12.   Calculate the resultant of (a) $v_1 - v_2 + v_3$ and (b) $v_2 - v_1 - v_3$ when $v_1 = 22$ units at $140°$, $v_2 = 40$ units at $190°$ and $v_3 = 15$ units at $290°$

(a)   The vectors are shown in Fig. 30.37.

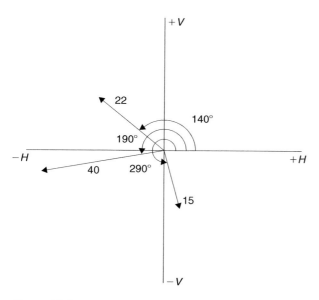

Figure 30.37

The horizontal component of $v_1 - v_2 + v_3$
$= (22 \cos 140°) - (40 \cos 190°) + (15 \cos 290°)$
$= (-16.85) - (-39.39) + (5.13) = \mathbf{27.67\ units}$
The vertical component of $v_1 - v_2 + v_3$
$= (22 \sin 140°) - (40 \sin 190°) + (15 \sin 290°)$
$= (14.14) - (-6.95) + (-14.10) = \mathbf{6.99\ units}$
The magnitude of the resultant,
$R = \sqrt{27.67^2 + 6.99^2} = \mathbf{28.54\ units}$
The direction of the resultant R
$$= \tan^{-1}\left(\frac{6.99}{27.67}\right) = 14.18°$$
Thus, $v_1 - v_2 + v_3 = \mathbf{28.54\ units\ at\ 14.18°}$

(b)   The horizontal component of $v_2 - v_1 - v_3$
$= (40 \cos 190°) - (22 \cos 140°) - (15 \cos 290°)$
$= (-39.39) - (-16.85) - (5.13) = \mathbf{-27.67\ units}$
The vertical component of $v_2 - v_1 - v_3$
$= (40 \sin 190°) - (22 \sin 140°) - (15 \sin 290°)$
$= (-6.95) - (14.14) - (-14.10) = \mathbf{-6.99\ units}$
From Fig. 30.38, the magnitude of the resultant, $R = \sqrt{(-27.67)^2 + (-6.99)^2} = \mathbf{28.54\ units}$
and $\alpha = \tan^{-1}\left(\frac{6.99}{27.67}\right) = 14.18°$, from which,
$\theta = 180° + 14.18° = 194.18°$

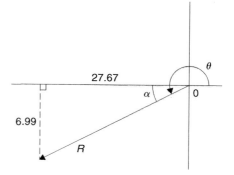

Figure 30.38

Thus, $v_2 - v_1 - v_3 = \mathbf{28.54\ units\ at\ 194.18°}$
This result is as expected, since $v_2 - v_1 - v_3 = -(v_1 - v_2 + v_3)$ and the vector 28.54 units at 194.18° is minus times (i.e. is 180° out of phase with) the vector 28.54 units at 14.18°

Now try the following Practice Exercise

⚑ 1.   Forces of $F_1 = 40$ N at $45°$ and $F_2 = 30$ N at $125°$ act at a point. Determine by drawing and by calculation (a) $F_1 + F_2$ (b) $F_1 - F_2$

⚑ 2.   Calculate the resultant of (a) $v_1 + v_2 - v_3$ (b) $v_3 - v_2 + v_1$ when $v_1 = 15$ m/s at $85°$, $v_2 = 25$ m/s at $175°$ and $v_3 = 12$ m/s at $235°$

30.8 Relative velocity

For relative velocity problems, some fixed datum point needs to be selected. This is often a fixed point on the earth's surface. In any vector equation, only the start and finish points affect the resultant vector of a system. Two different systems are shown in Fig. 30.39, but, in each of the systems, the resultant vector is $\mathbf{ad}$.

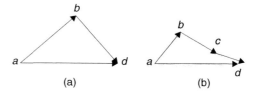

Figure 30.39

The vector equation of the system shown in Fig. 30.39(a) is

$$ad = ab + bd$$

and that for the system shown in Fig. 30.39(b) is

$$ad = ab + bc + cd$$

Thus, in vector equations of this form, only the first and last letters, a and d, respectively, fix the magnitude and direction of the resultant vector. This principle is used in relative velocity problems.

> **Problem 13.** Two cars, P and Q, are travelling towards the junction of two roads which are at right angles to one another. Car P has a velocity of 45 km/h due east and car Q a velocity of 55 km/h due south. Calculate (a) the velocity of car P relative to car Q and (b) the velocity of car Q relative to car P

(a) The directions of the cars are shown in Fig. 30.40(a), which is called a **space diagram**. The velocity diagram is shown in Fig. 30.40(b), in which pe is taken as the velocity of car P relative to point e on the earth's surface. The velocity of P relative to Q is vector pq and the vector equation is $pq = pe + eq$. Hence, the vector directions are as shown, eq being in the opposite direction to qe

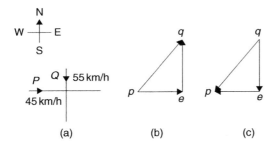

Figure 30.40

From the geometry of the vector triangle, the magnitude of $pq = \sqrt{45^2 + 55^2} = 71.06$ km/h and the direction of $pq = \tan^{-1}\left(\dfrac{55}{45}\right) = 50.71°$

That is, **the velocity of car P relative to car Q is 71.06 km/h at 50.71°**

(b) The velocity of car Q relative to car P is given by the vector equation $qp = qe + ep$ and the vector diagram is as shown in Fig. 30.40(c), having ep opposite in direction to pe

From the geometry of this vector triangle, the magnitude of $qp = \sqrt{45^2 + 55^2} = 71.06$ km/h and the direction of $qp = \tan^{-1}\left(\dfrac{55}{45}\right) = 50.71°$ but must lie in the third quadrant; i.e. the required angle is $180° + 50.71° = 230.71°$

That is, **the velocity of car Q relative to car P is 71.06 km/h at 230.71°**

Now try the following Practice Exercise

> **Practice Exercise 152   Relative velocity (answers on page 457)**
>
> 1. A car is moving along a straight horizontal road at 79.2 km/h and rain is falling vertically downwards at 26.4 km/h. Find the velocity of the rain relative to the driver of the car.
>
> 2. Calculate the time needed to swim across a river 142 m wide when the swimmer can swim at 2 km/h in still water and the river is flowing at 1 km/h. At what angle to the bank should the swimmer swim?
>
> 3. A ship is heading in a direction N 60°E at a speed which in still water would be 20 km/h. It is carried off course by a current of 8 km/h in a direction of E 50°S. Calculate the ship's actual speed and direction.

30.9 i, j and k notation

A method of completely specifying the direction of a vector in space relative to some reference point is to use three unit vectors, i, j and k, mutually at right angles to each other, as shown in Fig. 30.41.

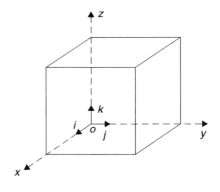

Figure 30.41

Calculations involving vectors given in i, j, k notation are carried out in exactly the same way as standard algebraic calculations, as shown in the worked examples below.

Problem 14. Determine
$(3i + 2j + 2k) - (4i - 3j + 2k)$

$(3i + 2j + 2k) - (4i - 3j + 2k)$
$= 3i + 2j + 2k - 4i + 3j - 2k$
$= -i + 5j$

Problem 15. Given $p = 3i + 2k$,
$q = 4i - 2j + 3k$ and $r = -3i + 5j - 4k$, determine
(a) $-r$ (b) $3p$ (c) $2p + 3q$ (d) $-p + 2r$
(e) $0.2p + 0.6q - 3.2r$

(a) $\quad -r = -(-3i + 5j - 4k) = +3i - 5j + 4k$

(b) $\quad 3p = 3(3i + 2k) = 9i + 6k$

(c) $\quad 2p + 3q = 2(3i + 2k) + 3(4i - 2j + 3k)$
$\qquad = 6i + 4k + 12i - 6j + 9k$
$\qquad = 18i - 6j + 13k$

(d) $\quad -p + 2r = -(3i + 2k) + 2(-3i + 5j - 4k)$
$\qquad = -3i - 2k + (-6i + 10j - 8k)$
$\qquad = -3i - 2k - 6i + 10j - 8k$
$\qquad = -9i + 10j - 10k$

(e) $\quad 0.2p + 0.6q - 3.2r$
$\qquad = 0.2(3i + 2k) + 0.6(4i - 2j + 3k)$
$\qquad \qquad \qquad - 3.2(-3i + 5j - 4k)$
$\qquad = 0.6i + 0.4k + 2.4i - 1.2j + 1.8k + 9.6i$
$\qquad \qquad \qquad - 16j + 12.8k$
$\qquad = 12.6i - 17.2j + 15k$

Now try the following Practice Exercise

Practice Exercise 153 i, j, k notation
(answers on page 457)

Given that $p = 2i + 0.5j - 3k, q = -i + j + 4k$ and $r = 6j - 5k$, evaluate and simplify the following vectors in i, j, k form.

1. $-q$ 2. $2p$
3. $q + r$ 4. $-q + 2p$
5. $3q + 4r$ 6. $q - 2p$
7. $p + q + r$ 8. $p + 2q + 3r$
9. $2p + 0.4q + 0.5r$ 10. $7r - 2q$

Practice Exercise 154   Multiple-choice questions on vectors (answers on page 457)

Each question has only one correct answer

1. Which of the following quantities is considered a vector?

   (a) time   (b) temperature
   (c) force   (d) mass

2. A force vector of 40 N is at an angle of 72° to the horizontal. Its vertical component is:

   (a) 40 N   (b) 12.36 N
   (c) 38.04 N   (d) 123.11 N

3. A displacement vector of 30 m is at an angle of −120° to the horizontal. Its horizontal component is:
   (a) −15 m   (b) −25.98 m
   (c) 15 m   (d) 25.98 m

4. Two voltage phasors are shown in Figure 30.42. If $V_1 = 40$ volts and $V_2 = 100$ volts, the resultant (i.e. length OA) is:

   (a) 131.4 volts at 32.55° to V_1
   (b) 105.0 volts at 32.55° to V_1
   (c) 131.4 volts at 68.30° to V_1
   (d) 105.0 volts at 42.31° to V_1

Figure 30.42

5. A force of 4 N is inclined at an angle of 45° to a second force of 7 N, both forces acting at a point, as shown in Figure 30.43. The magnitude of the resultant of these two forces and the direction of the resultant with respect to the 7 N force is:

   (a) 3 N at 45°   (b) 5 N at 146°
   (c) 11 N at 135°   (d) 10.2 N at 16°

Figure 30.43

Questions 6 and 7 relate to the following information.

Two voltage phasors V_1 and V_2 are shown in Figure 30.44.

Figure 30.44

6. The resultant $V_1 + V_2$ is given by:

   (a) 14.16 V at 62° to V_1
   (b) 38.72 V at − 19° to V_1
   (c) 38.72 V at 161° to V_1
   (d) 14.16 V at 118° to V_1

7. The resultant $V_1 - V_2$ is given by:

   (a) 38.72 V at - 19° to V_1
   (b) 14.16 V at 62° to V_1
   (c) 38.72 V at 161° to V_1
   (d) 14.16 V at 118° to V_1

8. The magnitude of the resultant of velocities of 3 m/s at 20° and 7 m/s at 120° when acting simultaneously at a point is:
   (a) 7.12 m/s   (b) 10 m/s
   (c) 21 m/s   (d) 4 m/s

9. Three forces of 2 N, 3 N and 4 N act as shown in Figure 30.45. The magnitude of the resultant force is:
   (a) 8.08 N   (b) 7.17 N
   (c) 9 N   (d) 1 N

Figure 30.45

10. The following three forces are acting upon an object in space:

$$p = 4i - 3j + 2k \quad q = -5k$$
$$r = -2i + 7j + 6k$$

The resultant of the forces $p + q - r$ is:

   (a) $-3i + 4j + 4k$ (b) $2i + 4j + 3k$
   (c) $6i - 10j - 9k$ (d) $2i + 4j - 3k$

Chapter 31

Methods of adding alternating waveforms

Why it is important to understand: Methods of adding alternating waveforms

In electrical engineering, a phasor is a rotating vector representing a quantity such as an alternating current or voltage that varies sinusoidally. Sometimes it is necessary when studying sinusoidal quantities to add together two alternating waveforms, for example in an a.c. series circuit, that are not in phase with each other. Electrical engineers, electronics engineers, electronic engineering technicians and aircraft engineers all use phasor diagrams to visualise complex constants and variables. So, given oscillations to add and subtract, the required rotating vectors are constructed, called a phasor diagram, and graphically the resulting sum and/or difference oscillations are added or calculated. Phasors may be used to analyse the behaviour of electrical and mechanical systems that have reached a kind of equilibrium called sinusoidal steady state. Hence, discovering different methods of combining sinusoidal waveforms is of some importance in certain areas of engineering.

At the end of this chapter you should be able to:

- determine the resultant of two phasors by graph plotting
- determine the resultant of two or more phasors by drawing
- determine the resultant of two phasors by the sine and cosine rules
- determine the resultant of two or more phasors by horizontal and vertical components

31.1 Combining two periodic functions

There are a number of instances in engineering and science where waveforms have to be combined and where it is required to determine the single phasor (called the resultant) that could replace two or more separate phasors. Uses are found in electrical alternating current theory, in mechanical vibrations, in the addition of forces and with sound waves.

There are a number of methods of determining the resultant waveform. These include

(a) drawing the waveforms and adding graphically

(b) drawing the phasors and measuring the resultant

(c) using the cosine and sine rules

(d) using horizontal and vertical components.

31.2 Plotting periodic functions

This may be achieved by sketching the separate functions on the same axes and then adding (or subtracting) ordinates at regular intervals. This is demonstrated in the following worked problems.

> **Problem 1.** Plot the graph of $y_1 = 3\sin A$ from $A = 0°$ to $A = 360°$. On the same axes plot $y_2 = 2\cos A$. By adding ordinates, plot $y_R = 3\sin A + 2\cos A$ and obtain a sinusoidal expression for this resultant waveform

$y_1 = 3\sin A$ and $y_2 = 2\cos A$ are shown plotted in Fig. 31.1. Ordinates may be added at, say, $15°$ intervals. For example,

at $0°, y_1 + y_2 = 0 + 2 = 2$

at $15°, y_1 + y_2 = 0.78 + 1.93 = 2.71$

at $120°, y_1 + y_2 = 2.60 + -1 = 1.6$

at $210°, y_1 + y_2 = -1.50 - 1.73 = -3.23,$ and so on.

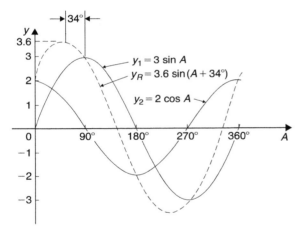

Figure 31.1

The resultant waveform, shown by the broken line, has the same period, i.e. $360°$, and thus the same frequency as the single phasors. The maximum value, or amplitude, of the resultant is 3.6. The resultant waveform **leads** $y_1 = 3\sin A$ by $34°$ or $34 \times \dfrac{\pi}{180}$ rad $= 0.593$ rad. The sinusoidal expression for the resultant waveform is

$$y_R = 3.6\sin(A + 34°) \text{ or } y_R = 3.6\sin(A + 0.593)$$

> **Problem 2.** Plot the graphs of $y_1 = 4\sin\omega t$ and $y_2 = 3\sin(\omega t - \pi/3)$ on the same axes, over one cycle. By adding ordinates at intervals plot $y_R = y_1 + y_2$ and obtain a sinusoidal expression for the resultant waveform

$y_1 = 4\sin\omega t$ and $y_2 = 3\sin(\omega t - \pi/3)$ are shown plotted in Fig. 31.2.

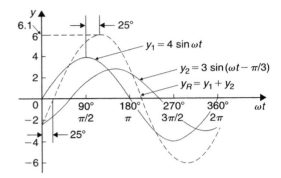

Figure 31.2

Ordinates are added at $15°$ intervals and the resultant is shown by the broken line. The amplitude of the resultant is 6.1 and it **lags** y_1 by $25°$ or 0.436 rad.
Hence, the sinusoidal expression for the resultant waveform is

$$y_R = 6.1\sin(\omega t - 0.436)$$

Figure 31.3

> **Problem 3.** Determine a sinusoidal expression for $y_1 - y_2$ when $y_1 = 4\sin\omega t$ and $y_2 = 3\sin(\omega t - \pi/3)$

y_1 and y_2 are shown plotted in Fig. 31.3. At $15°$ intervals y_2 is subtracted from y_1. For example, at $0°, y_1 - y_2 = 0 - (-2.6) = +2.6$

at $30°, y_1 - y_2 = 2 - (-1.5) = +3.5$

at $150°, y_1 - y_2 = 2 - 3 = -1$, and so on.

The amplitude, or peak, value of the resultant (shown by the broken line) is 3.6 and it leads y_1 by 45° or 0.79 rad. Hence,

$$y_1 - y_2 = 3.6\sin(\omega t + 0.79)$$

Problem 4. Two alternating currents are given by $i_1 = 20\sin\omega t$ amperes and $i_2 = 10\sin\left(\omega t + \dfrac{\pi}{3}\right)$ amperes. By drawing the waveforms on the same axes and adding, determine the sinusoidal expression for the resultant $i_1 + i_2$

i_1 and i_2 are shown plotted in Fig. 31.4. The resultant waveform for $i_1 + i_2$ is shown by the broken line. It has the same period, and hence frequency, as i_1 and i_2

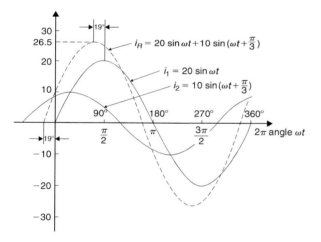

Figure 31.4

The amplitude or peak value is 26.5 A. The resultant waveform leads the waveform of $i_1 = 20\sin\omega t$ by 19° or 0.33 rad.

Hence, the sinusoidal expression for the resultant $i_1 + i_2$ is given by

$$i_R = i_1 + i_2 = 26.5\sin(\omega t + 0.33)\,\text{A}$$

Now try the following Practice Exercise

31.3 Determining resultant phasors by drawing

The resultant of two periodic functions may be found from their relative positions when the time is zero. For example, if $y_1 = 4\sin\omega t$ and $y_2 = 3\sin(\omega t - \pi/3)$ then each may be represented as **phasors** as shown in Fig. 31.5, y_1 being 4 units long and drawn horizontally and y_2 being 3 units long, lagging y_1 by $\pi/3$ radians or 60°. To determine the resultant of $y_1 + y_2, y_1$ is drawn horizontally as shown in Fig. 31.6 and y_2 is joined to the end of y_1 at 60° to the horizontal. The resultant is given by y_R. This is the same as the diagonal of a parallelogram that is shown completed in Fig. 31.7.

Figure 31.5

Resultant y_R, in Figs 31.6 and 31.7, may be determined by drawing the phasors and their directions to scale by the nose-to-tail or parallelogram methods shown in chapter 30, and measuring using a ruler and protractor. In this example, y_R is measured as 6 units long and angle ϕ is measured as 25°

$$25° = 25 \times \frac{\pi}{180}\text{radians} = 0.44 \text{ rad}$$

Figure 31.6

Figure 31.7

Hence, summarising, by drawing,
$$y_R = y_1 + y_2 = 4\sin\omega t + 3\sin(\omega t - \pi/3)$$
$$= 6\sin(\omega t - 0.44)$$
If the resultant phasor, $y_R = y_1 - y_2$ is required then y_2 is still 3 units long but is drawn in the opposite direction, as shown in Fig. 31.8.

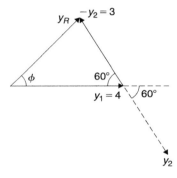

Figure 31.8

It may be shown that $= 3.6\sin(\omega t + 0.80)$

Problem 5. Two alternating currents are given by $i_1 = 20\sin\omega t$ amperes and $i_2 = 10\sin\left(\omega t + \dfrac{\pi}{3}\right)$ amperes. Determine $i_1 + i_2$ by drawing phasors

The relative positions of i_1 and i_2 at time $t = 0$ are shown as phasors in Fig. 31.9, where $\dfrac{\pi}{3}$rad $= 60°$. The phasor diagram in Fig. 31.10 is drawn to scale with a ruler and protractor.
The resultant i_R is shown and is measured as 26 A and angle ϕ as 19° or 0.33 rad leading i_1. Hence, by drawing and measuring,

$$i_R = i_1 + i_2 = 26\sin(\omega t + 0.33)\,\text{A}$$

Figure 31.9

Figure 31.10

Problem 6. For the currents in Problem 5, determine $i_1 - i_2$ by drawing phasors

At time $t = 0$, current i_1 is drawn 20 units long horizontally as shown by $0a$ in Fig. 31.11. Current i_2 is shown, drawn 10 units long and leading by 60°. The current $-i_2$ is drawn in the opposite direction as a broken line, shown as ab in Fig. 31.11. The resultant i_R is given by $0b$ lagging by angle ϕ

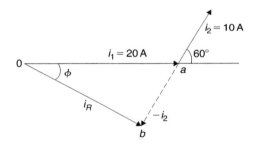

Figure 31.11

By measurement, $i_R = 17$ A and $\phi = 30°$ or 0.52 rad. Hence, by drawing phasors,

$$i_R = i_1 - i_2 = 17\sin(\omega t - 0.52)\text{A}$$

Now try the following Practice Exercise

Practice Exercise 156   Determining resultant phasors by drawing (answers on page 457)

1. Determine a sinusoidal expression for $2\sin\theta + 4\cos\theta$ by drawing phasors.

2. If $v_1 = 10\sin\omega t$ volts and $v_2 = 14\sin(\omega t + \pi/3)$ volts, determine by drawing phasors sinusoidal expressions for (a) $v_1 + v_2$ (b) $v_1 - v_2$

3. Express $12\sin\omega t + 5\cos\omega t$ in the form $A\sin(\omega t \pm \alpha)$ by drawing phasors.

31.4 Determining resultant phasors by the sine and cosine rules

As stated earlier, the resultant of two periodic functions may be found from their relative positions when the time is zero. For example, if $y_1 = 5\sin\omega t$ and $y_2 = 4\sin(\omega t - \pi/6)$ then each may be represented by phasors as shown in Fig. 31.12, y_1 being 5 units long and drawn horizontally and y_2 being 4 units long, lagging y_1 by $\pi/6$ radians or $30°$. To determine the resultant of $y_1 + y_2$, y_1 is drawn horizontally as shown in Fig. 31.13 and y_2 is joined to the end of y_1 at $\pi/6$ radians; i.e. $30°$ to the horizontal. The resultant is given by y_R

Using the cosine rule (from Chapter 24) on triangle $0ab$ of Fig. 31.13 gives

$$y_R^2 = 5^2 + 4^2 - [2(5)(4)\cos 150°]$$
$$= 25 + 16 - (-34.641) = 75.641$$

from which $\quad\quad y_R = \sqrt{75.641} = 8.697$

Using the sine rule, $\quad \dfrac{8.697}{\sin 150°} = \dfrac{4}{\sin\phi}$

from which $\quad \sin\phi = \dfrac{4\sin 150°}{8.697} = 0.22996$

and $\quad\quad \phi = \sin^{-1} 0.22996$
$$= 13.29° \text{ or } 0.232\,\text{rad}$$

Hence, $\boldsymbol{y_R = y_1 + y_2 = 5\sin\omega t + 4\sin(\omega t - \pi/6)}$
$$= \boldsymbol{8.697\sin(\omega t - 0.232)}$$

Figure 31.12

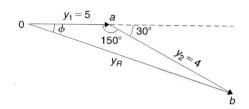

Figure 31.13

Given $y_1 = 2\sin\omega t$ and $y_2 = 3\sin(\omega t + \pi/4)$, obtain an expression, by calculation, for the resultant, $y_R = y_1 + y_2$

When time $t = 0$, the position of phasors y_1 and y_2 are as shown in Fig. 31.14(a). To obtain the resultant, y_1 is drawn horizontally, 2 units long, and y_2 is drawn 3 units long at an angle of $\pi/4$ rad or $45°$ and joined to the end of y_1 as shown in Fig. 31.14(b).

From Fig. 31.14(b), and using the cosine rule,

$$y_R^2 = 2^2 + 3^2 - [2(2)(3)\cos 135°]$$
$$= 4 + 9 - [-8.485] = 21.485$$

(a)

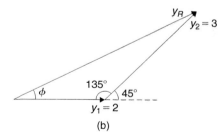

(b)

Figure 31.14

Hence, $\quad\quad y_R = \sqrt{21.485} = 4.6352$

Using the sine rule $\quad \dfrac{3}{\sin\phi} = \dfrac{4.6352}{\sin 135°}$

from which $\qquad \sin\phi = \dfrac{3\sin 135°}{4.6352} = 0.45765$

Hence, $\qquad\qquad \phi = \sin^{-1} 0.45765$

$$= 27.24° \text{ or } 0.475\,\text{rad}$$

Thus, by calculation, $\quad y_R = 4.635\sin(\omega t + 0.475)$

Problem 8. Determine
$$20\sin\omega t + 10\sin\left(\omega t + \dfrac{\pi}{3}\right)A$$
using the cosine and sine rules

From the phasor diagram of Fig. 31.15 and using the cosine rule,

$$i_R^2 = 20^2 + 10^2 - [2(20)(10)\cos 120°] = 700$$

Figure 31.15

Hence, $\qquad\qquad i_R = \sqrt{700} = 26.46\,\text{A}$

Using the sine rule gives $\quad \dfrac{10}{\sin\phi} = \dfrac{26.46}{\sin 120°}$

from which $\qquad \sin\phi = \dfrac{10\sin 120°}{26.46}$

$$= 0.327296$$

and $\qquad \phi = \sin^{-1} 0.327296 = 19.10°$

$$= 19.10 \times \dfrac{\pi}{180} = 0.333\,\text{rad}$$

Hence, by cosine and sine rules,
$$i_R = i_1 + i_2 = 26.46\sin(\omega t + 0.333)\,\text{A}$$

Now try the following Practice Exercise

Practice Exercise 157 Resultant phasors by the sine and cosine rules (answers on page 457)

1. Determine, using the cosine and sine rules, a sinusoidal expression for
$$y = 2\sin A + 4\cos A$$

2. Given $v_1 = 10\sin\omega t$ volts and $v_2 = 14\sin(\omega t + \pi/3)$ volts, use the cosine

and sine rules to determine sinusoidal expressions for (a) $v_1 + v_2$ (b) $v_1 - v_2$

In Problems 3 to 5, express the given expressions in the form $A\sin(\omega t \pm \alpha)$ by using the cosine and sine rules.

3. $12\sin\omega t + 5\cos\omega t$

4. $7\sin\omega t + 5\sin\left(\omega t + \dfrac{\pi}{4}\right)$

5. $6\sin\omega t + 3\sin\left(\omega t - \dfrac{\pi}{6}\right)$

6. The sinusoidal currents in two parallel branches of an electrical network are $400\sin\omega t$ and $750\sin(\omega t - \pi/3)$, both measured in milliamperes. Determine the total current flowing into the parallel arrangement. Give the answer in sinusoidal form and in amperes.

7. When a mass is attached to a spring and oscillates in a vertical plane, the displacement, s, at time t is given by: $s = (75\sin\omega t + 40\cos\omega t)$ mm. Express this in the form $s = A\sin(\omega t + \phi)$

31.5 Determining resultant phasors by horizontal and vertical components

If a right-angled triangle is constructed as shown in Fig. 31.16, $0a$ is called the horizontal component of F and ab is called the vertical component of F

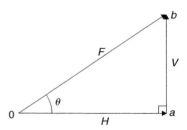

Figure 31.16

From trigonometry (see Chapter 22 and remember SOH CAH TOA),

$$\cos\theta = \dfrac{0a}{0b}, \text{ from which } 0a = 0b\cos\theta = F\cos\theta$$

i.e. **the horizontal component of $F, H = F\cos\theta$**, and

$$\sin\theta = \dfrac{ab}{0b}, \text{ from which } ab = 0b\sin\theta = F\sin\theta$$

i.e. **the vertical component of** F, $V = F \sin\theta$.
Determining resultant phasors by horizontal and vertical components is demonstrated in the following worked problems.

📌 **Problem 9.** Two alternating voltages are given by $v_1 = 15 \sin\omega t$ volts and $v_2 = 25 \sin(\omega t - \pi/6)$ volts. Determine a sinusoidal expression for the resultant $v_R = v_1 + v_2$ by finding horizontal and vertical components

The relative positions of v_1 and v_2 at time $t = 0$ are shown in Fig. 31.17(a) and the phasor diagram is shown in Fig. 31.17(b).

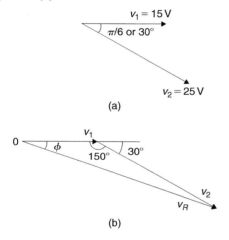

(a)

(b)

Figure 31.17

The horizontal component of v_R,
$H = 15 \cos 0° + 25 \cos(-30°) = \mathbf{36.65\,V}$

The vertical component of v_R,
$V = 15 \sin 0° + 25 \sin(-30°) = \mathbf{-12.50\,V}$

Hence, $v_R = \sqrt{36.65^2 + (-12.50)^2}$

by Pythagoras' theorem

$= \mathbf{38.72\ volts}$

$$\tan\phi = \frac{V}{H} = \frac{-12.50}{36.65} = -0.3411$$

from which $\phi = \tan^{-1}(-0.3411)$

$= \mathbf{-18.83°}$ or $\mathbf{-0.329\ radians}.$

Hence, $v_R = v_1 + v_2 = \mathbf{38.72 \sin(\omega t - 0.329)\,V}$

📌 **Problem 10.** For the voltages in Problem 9, determine the resultant $v_R = v_1 - v_2$ using horizontal and vertical components

The horizontal component of v_R,
$H = 15 \cos 0° - 25 \cos(-30°) = \mathbf{-6.65\,V}$
The vertical component of v_R,
$V = 15 \sin 0° - 25 \sin(-30°) = \mathbf{12.50\,V}$

Hence, $v_R = \sqrt{(-6.65)^2 + (12.50)^2}$

by Pythagoras' theorem

$= \mathbf{14.16\ volts}$

$$\tan\phi = \frac{V}{H} = \frac{12.50}{-6.65} = -1.8797$$

from which $\phi = \tan^{-1}(-1.8797)$

$= \mathbf{118.01°}$ or $\mathbf{2.06\ radians}$

(i.e. in second quadrant)

Hence, $v_R = v_1 - v_2 = \mathbf{14.16 \sin(\omega t + 2.06)\,V}$

The phasor diagram is shown in Fig. 31.18.

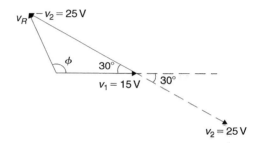

Figure 31.18

📌 **Problem 11.** Determine

$$20 \sin\omega t + 10 \sin\left(\omega t + \frac{\pi}{3}\right) A$$

using horizontal and vertical components

Figure 31.19

From the phasors shown in Fig. 31.19,
Total horizontal component,
$$H = 20 \cos 0° + 10 \cos 60° = \mathbf{25.0}$$
Total vertical component,
$$V = 20 \sin 0° + 10 \sin 60° = \mathbf{8.66}$$
By Pythagoras, the resultant,
$$i_R = \sqrt{[25.0^2 + 8.66^2]} = \mathbf{26.46\,A}$$

Phase angle, $\phi = \tan^{-1}\left(\dfrac{8.66}{25.0}\right)$

$\qquad = \mathbf{19.11°}$ or $\mathbf{0.333\,rad}$

Hence, by using horizontal and vertical components,

$$20\sin\omega t + 10\sin\left(\omega t + \dfrac{\pi}{3}\right) = \mathbf{26.46\sin(\omega t + 0.333)A}$$

Now try the following Practice Exercise

Practice Exercise 158   Resultant phasors by horizontal and vertical components (answers on page 457)

In Problems 1 to 5, express the combination of periodic functions in the form $A\sin(\omega t \pm \alpha)$ by horizontal and vertical components.

1. $7\sin\omega t + 5\sin\left(\omega t + \dfrac{\pi}{4}\right)A$

2. $6\sin\omega t + 3\sin\left(\omega t - \dfrac{\pi}{6}\right)V$

3. $i = 25\sin\omega t - 15\sin\left(\omega t + \dfrac{\pi}{3}\right)A$

4. $v = 8\sin\omega t - 5\sin\left(\omega t - \dfrac{\pi}{4}\right)V$

5. $x = 9\sin\left(\omega t + \dfrac{\pi}{3}\right) - 7\sin\left(\omega t - \dfrac{3\pi}{8}\right)m$

6. The voltage drops across two components when connected in series across an a.c. supply are $v_1 = 200\sin 314.2t$ and $v_2 = 120\sin(314.2t - \pi/5)$ volts respectively. Determine
    (a) the voltage of the supply (given by $v_1 + v_2$) in the form $A\sin(\omega t \pm \alpha)$
    (b) the frequency of the supply

7. If the supply to a circuit is $v = 20\sin 628.3t$ volts and the voltage drop across one of the components is $v_1 = 15\sin(628.3t - 0.52)$ volts, calculate
    (a) the voltage drop across the remainder of the circuit, given by $v - v_1$, in the form $A\sin(\omega t \pm \alpha)$
    (b) the supply frequency
    (c) the periodic time of the supply.

8. The voltages across three components in a series circuit when connected across an a.c. supply are $v_1 = 25\sin\left(300\pi t + \dfrac{\pi}{6}\right)$ volts,

$v_2 = 40\sin\left(300\pi t - \dfrac{\pi}{4}\right)$ volts and

$v_3 = 50\sin\left(300\pi t + \dfrac{\pi}{3}\right)$ volts. Calculate

(a) the supply voltage, in sinusoidal form, in the form $A\sin(\omega t \pm \alpha)$

(b) the frequency of the supply

(c) the periodic time.

9. In an electrical circuit, two components are connected in series. The voltage across the first component is given by $80\sin(\omega t + \pi/3)$ volts, and the voltage across the second component is given by $150\sin(\omega t - \pi/4)$ volts. Determine the total supply voltage to the two components. Give the answer in sinusoidal form.

Practice Exercise 159   Multiple-choice questions on methods of adding alternating waveforms (answers on page 458)

Each question has only one correct answer

Questions 1 and 2 relate to the following information.
Two alternating voltages are given by:
$v_1 = 2\sin\omega t$ and $v_2 = 3\sin\left(\omega t + \dfrac{\pi}{4}\right)$ volts.

1. Which of the phasor diagrams shown in Figure 31.20 represents $v_R = v_1 + v_2$?
    (a) (i)   (b) (ii)   (c) (iii)   (d) (iv)

Figure 31.20

2. Which of the phasor diagrams shown in Figure 31.20 represents $v_R = v_1 - v_2$?
    (a) (i)   (b) (ii)   (c) (iii)   (d) (iv)

3. Two alternating currents are given by $i_1 = 4\sin\omega t$ amperes and $i_2 = 3\sin\left(\omega t + \frac{\pi}{6}\right)$ amperes. The sinusoidal expression for the resultant $i_1 + i_2$ is:

   (a) $5\sin\left(\omega t + 30\right)A$
   (b) $5\sin\left(\omega t + \frac{\pi}{6}\right)A$
   (c) $6.77\sin\left(\omega t + 0.22\right)A$
   (d) $6.77\sin\left(\omega t + 12.81\right)A$

4. Two alternating voltages are given by
   $v_1 = 5\sin\omega t$ volts and
   $v_2 = 12\sin\left(\omega t - \frac{\pi}{3}\right)$ volts.
   The sinusoidal expression for the resultant $v_1 + v_2$ is:

   (a) $13\sin\left(\omega t - 60\right)V$
   (b) $15.13\sin\left(\omega t - 0.76\right)V$
   (c) $13\sin\left(\omega t - 1.05\right)V$
   (d) $15.13\sin\left(\omega t - 43.37\right)V$

5. Two alternating currents are given by $i_1 = 3\sin\omega t$ amperes and
   $i_2 = 4\sin\left(\omega t + \frac{\pi}{4}\right)$ amperes. The sinusoidal expression for the resultant $i_1 - i_2$ is:

   (a) $1.00\sin\left(\omega t - 3.09\right)A$
   (b) $6.48\sin\left(\omega t + 0.45\right)A$
   (c) $1.00\sin(\omega t - 176.96)A$
   (d) $6.48\sin\left(\omega t - 25.89\right)A$

For fully worked solutions to each of the problems in Practice Exercises 155 to 158 in this chapter, go to the website:
www.routledge.com/cw/bird

This assignment covers the material contained in Chapters 30 and 31. *The marks available are shown in brackets at the end of each question.*

1. State the difference between scalar and vector quantities. (2)

2. State whether the following are scalar or vector quantities.
   (a) A temperature of 50°C
   (b) 2 m³ volume
   (c) A downward force of 10 N
   (d) 400 J of work
   (e) 20 cm² area
   (f) A south-easterly wind of 20 knots
   (g) 40 m distance
   (h) An acceleration of 25 m/s² at 30° to the horizontal. (8)

3. A velocity vector of 16 m/s acts at an angle of −40° to the horizontal. Calculate its horizontal and vertical components, correct to 3 significant figures. (4)

4. Calculate the resultant and direction of the displacement vectors shown in Fig. RT12.1, correct to 2 decimal places. (6)

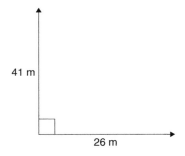

Figure RT12.1

5. Calculate the resultant and direction of the force vectors shown in Fig. RT12.2, correct to 2 decimal places. (6)

6. If acceleration $a_1 = 11$ m/s² at 70° and $a_2 = 19$ m/s² at −50°, calculate the magnitude and direction of $a_1 + a_2$, correct to 2 decimal places. (8)

7. If velocity $v_1 = 36$ m/s at 52° and $v_2 = 17$ m/s at −15°, calculate the magnitude and direction of $v_1 - v_2$, correct to 2 decimal places. (8)

Figure RT12.2

8. Forces of 10 N, 16 N and 20 N act as shown in Fig. RT12.3. Determine the magnitude of the resultant force and its direction relative to the 16 N force
   (a) by scaled drawing
   (b) by calculation. (13)

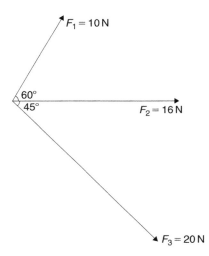

Figure RT12.3

9. For the three forces shown in Fig. RT12.3, calculate the resultant of $F_1 - F_2 - F_3$ and its direction relative to force F_2 (9)

10. Two cars, A and B, are travelling towards crossroads. A has a velocity of 60 km/h due south and B a velocity of 75 km/h due west. Calculate the velocity of A relative to B. (6)

11. Given $a = -3i + 3j + 5k$, $b = 2i - 5j + 7k$ and $c = 3i + 6j - 4k$, determine the following: (i) $-4b$ (ii) $a + b - c$ (iii) $5b - 3a$ (6)

12. Calculate the magnitude and direction of the resultant vector of the displacement system shown in Fig. RT12.4.   (9)

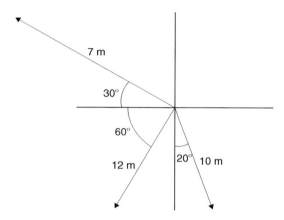

Fig. RT12.4

13. The instantaneous values of two alternating voltages are given by

$$v_1 = 150 \sin\left(\omega t + \frac{\pi}{3}\right) \text{ volts}$$

and

$$v_2 = 90 \sin\left(\omega t - \frac{\pi}{6}\right) \text{ volts}$$

Plot the two voltages on the same axes to scales of 1 cm = 50 volts and 1 cm = π/6. Obtain a sinusoidal expression for the resultant of v_1 and v_2 in the form $R \sin\left(\omega t + \alpha\right)$ (a) by adding ordinates at intervals and (b) by calculation.   (15)

Chapter 32

Presentation of statistical data

Why it is important to understand: Presentation of statistical data

Statistics is the study of the collection, organisation, analysis and interpretation of data. It deals with all aspects of this, including the planning of data collection in terms of the design of surveys and experiments. Statistics is applicable to a wide variety of academic disciplines, including natural and social sciences, engineering, government and business. Statistical methods can be used for summarising or describing a collection of data. Engineering statistics combines engineering and statistics. Design of experiments is a methodology for formulating scientific and engineering problems using statistical models. Quality control and process control use statistics as a tool to manage conformance to specifications of manufacturing processes and their products. Time and methods engineering use statistics to study repetitive operations in manufacturing in order to set standards and find optimum manufacturing procedures. Reliability engineering measures the ability of a system to perform for its intended function (and time) and has tools for improving performance. Probabilistic design involves the use of probability in product and system design. System identification uses statistical methods to build mathematical models of dynamical systems from measured data. System identification also includes the optimal design of experiments for efficiently generating informative data for fitting such models. This chapter introduces the presentation of statistical data.

At the end of this chapter you should be able to:

- distinguish between discrete and continuous data
- present data diagrammatically – pictograms, horizontal and vertical bar charts, percentage component bar charts, pie diagrams
- produce a tally diagram for a set of data
- form a frequency distribution from a tally diagram
- construct a histogram from a frequency distribution
- construct a frequency polygon from a frequency distribution
- produce a cumulative frequency distribution from a set of grouped data
- construct an ogive from a cumulative frequency distribution

32.1 Some statistical terminology

Discrete and continuous data

Data are obtained largely by two methods:

(a) By counting – for example, the number of stamps sold by a post office in equal periods of time.

(b) By measurement – for example, the heights of a group of people.

When data are obtained by counting and only whole numbers are possible, the data are called **discrete**. Measured data can have any value within certain limits and are called **continuous**.

> **Problem 1.**   Data are obtained on the topics given below. State whether they are discrete or continuous data.
>
> (a) The number of days on which rain falls in a month for each month of the year.
>
> (b) The mileage travelled by each of a number of salesmen.
>
> (c) The time that each of a batch of similar batteries lasts.
>
> (d) The amount of money spent by each of several families on food.

(a) The number of days on which rain falls in a given month must be an integer value and is obtained by **counting** the number of days. Hence, these data are **discrete**.

(b) A salesman can travel any number of miles (and parts of a mile) between certain limits and these data are **measured**. Hence, the data are **continuous**.

(c) The time that a battery lasts is **measured** and can have any value between certain limits. Hence, these data are **continuous**.

(d) The amount of money spent on food can only be expressed correct to the nearest pence, the amount being **counted**. Hence, these data are **discrete**.

Now try the following Practice Exercise

> **Practice Exercise 160   Discrete and continuous data (answers on page 458)**
>
> In the following problems, state whether data relating to the topics given are discrete or continuous.
>
> 1. (a) The amount of petrol produced daily, for each of 31 days, by a refinery.
>   (b) The amount of coal produced daily by each of 15 miners.
>   (c) The number of bottles of milk delivered daily by each of 20 milkmen.
>   (d) The size of 10 samples of rivets produced by a machine.
>
> 2. (a) The number of people visiting an exhibition on each of 5 days.
>   (b) The time taken by each of 12 athletes to run 100 metres.
>   (c) The value of stamps sold in a day by each of 20 post offices.
>   (d) The number of defective items produced in each of 10 one-hour periods by a machine.

Further statistical terminology

A **set** is a group of data and an individual value within the set is called a **member** of the set. Thus, if the masses of five people are measured correct to the nearest 0.1 kilogram and are found to be 53.1 kg, 59.4 kg, 62.1 kg, 77.8 kg and 64.4 kg then the set of masses in kilograms for these five people is

$$\{53.1, 59.4, 62.1, 77.8, 64.4\}$$

and one of the members of the set is 59.4

A set containing all the members is called a **population**. A **sample** is a set of data collected and/or selected from a statistical population by a defined procedure. Typically, the population is very large, making a census or a complete enumeration of all the values in the population impractical or impossible. The sample usually represents a subset of manageable size. Samples are collected and statistics are calculated from the samples so that inferences or extrapolations can be made from the sample to the population. Thus, all car registration numbers form a population but the registration numbers of, say, 20 cars taken at random throughout the country are a sample drawn from that population.

The number of times that the value of a member occurs in a set is called the **frequency** of that member. Thus, in the set $\{2, 3, 4, 5, 4, 2, 4, 7, 9\}$, member 4 has a frequency of three, member 2 has a frequency of 2 and the other members have a frequency of one.

The **relative frequency** with which any member of a set occurs is given by the ratio

$$\frac{\text{frequency of member}}{\text{total frequency of all members}}$$

For the set $\{2, 3, 5, 4, 7, 5, 6, 2, 8\}$, the relative frequency of member 5 is $\frac{2}{9}$. Often, relative frequency is expressed as a percentage and the **percentage relative frequency** is

$$(\text{relative frequency} \times 100)\,\%$$

32.2   Presentation of ungrouped data

Ungrouped data can be presented diagrammatically in several ways and these include

(a)   **pictograms**, in which pictorial symbols are used to represent quantities (see Problem 2),

(b)   **horizontal bar charts**, having data represented by equally spaced horizontal rectangles (see Problem 3), and

(c)   **vertical bar charts**, in which data are represented by equally spaced vertical rectangles (see Problem 4).

Trends in ungrouped data over equal periods of time can be presented diagrammatically by a **percentage component bar chart**. In such a chart, equally spaced rectangles of any width, but whose height corresponds to 100%, are constructed. The rectangles are then subdivided into values corresponding to the percentage relative frequencies of the members (see Problem 5).

A **pie diagram** is used to show diagrammatically the parts making up the whole. In a pie diagram, the area of a circle represents the whole and the areas of the sectors of the circle are made proportional to the parts which make up the whole (see Problem 6).

> **Problem 2.**   The number of television sets repaired in a workshop by a technician in 6 one-month periods is as shown below. Present these data as a pictogram

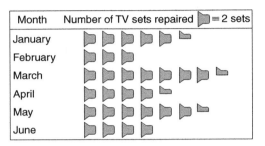

Month	January	February	March
Number repaired	11	6	15

Month	April	May	June
Number repaired	9	13	8

Each symbol shown in Fig. 32.1 represents two television sets repaired. Thus, in January, $5\frac{1}{2}$ symbols are used to represent the 11 sets repaired; in February, 3 symbols are used to represent the 6 sets repaired, and so on.

Figure 32.1

> **Problem 3.**   The distance in miles travelled by four salesmen in a week are as shown below.

Salesman	P	Q	R	S
Distance travelled (miles)	413	264	597	143

Use a horizontal bar chart to represent these data diagrammatically

Equally spaced horizontal rectangles of any width, but whose length is proportional to the distance travelled, are used. Thus, the length of the rectangle for salesman P is proportional to 413 miles, and so on. The horizontal bar chart depicting these data is shown in Fig. 32.2.

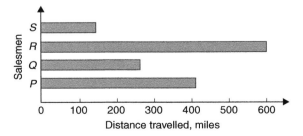

Figure 32.2

Problem 4. The number of issues of tools or materials from a store in a factory is observed for 7 one-hour periods in a day and the results of the survey are as follows.

Period	1	2	3	4	5	6	7
Number of issues	34	17	9	5	27	13	6

Present these data on a vertical bar chart

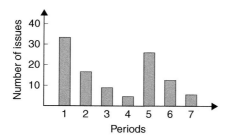

Figure 32.3

In a vertical bar chart, equally spaced vertical rectangles of any width, but whose height is proportional to the quantity being represented, are used. Thus, the height of the rectangle for period 1 is proportional to 34 units, and so on. The vertical bar chart depicting these data is shown in Fig. 32.3.

Problem 5. The numbers of various types of dwellings sold by a company annually over a three-year period are as shown below. Draw percentage component bar charts to present these data

	Year 1	Year 2	Year 3
4-roomed bungalows	24	17	7
5-roomed bungalows	38	71	118
4-roomed houses	44	50	53
5-roomed houses	64	82	147
6-roomed houses	30	30	25

A table of percentage relative frequency values, correct to the nearest 1%, is the first requirement. Since

percentage relative frequency

$$= \frac{\text{frequency of member} \times 100}{\text{total frequency}}$$

then for 4-roomed bungalows in year 1

percentage relative frequency

$$= \frac{24 \times 100}{24 + 38 + 44 + 64 + 30} = 12\%$$

The percentage relative frequencies of the other types of dwellings for each of the three years are similarly calculated and the results are as shown in the table below.

	Year 1	Year 2	Year 3
4-roomed bungalows	12%	7%	2%
5-roomed bungalows	19%	28%	34%
4-roomed houses	22%	20%	15%
5-roomed houses	32%	33%	42%
6-roomed houses	15%	12%	7%

The percentage component bar chart is produced by constructing three equally spaced rectangles of any width, corresponding to the three years. The heights of the rectangles correspond to 100% relative frequency and are subdivided into the values in the table of percentages shown above. A key is used (different types of shading or different colour schemes) to indicate corresponding percentage values in the rows of the table of percentages. The percentage component bar chart is shown in Fig. 32.4.

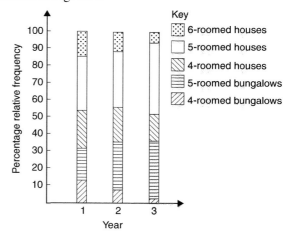

Figure 32.4

Problem 6. The retail price of a product costing £2 is made up as follows: materials 10 p, labour 20 p, research and development 40 p, overheads 70 p, profit 60 p. Present these data on a pie diagram

A circle of any radius is drawn. The area of the circle represents the whole, which in this case is £2. The circle is subdivided into sectors so that the areas of the sectors are proportional to the parts; i.e. the parts which make up the total retail price. For the area of a sector to be proportional to a part, the angle at the centre of the circle must be proportional to that part. The whole, £2 or 200 p, corresponds to 360°. Therefore,

$$10 \text{p corresponds to } 360 \times \frac{10}{200} \text{ degrees, i.e. } 18°$$

$$20 \text{p corresponds to } 360 \times \frac{20}{200} \text{ degrees, i.e. } 36°$$

and so on, giving the angles at the centre of the circle for the parts of the retail price as $18°, 36°, 72°, 126°$ and $108°$, respectively.

The pie diagram is shown in Fig. 32.5.

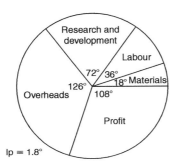

Figure 32.5

Problem 7.

(a) Using the data given in Fig. 32.2 only, calculate the amount of money paid to each salesman for travelling expenses if they are paid an allowance of 37 p per mile.

(b) Using the data presented in Fig. 32.4, comment on the housing trends over the three-year period.

(c) Determine the profit made by selling 700 units of the product shown in Fig. 32.5

(a) By measuring the length of rectangle P, the mileage covered by salesman P is equivalent to

413 miles. Hence **salesman P** receives a travelling allowance of

$$\frac{£413 \times 37}{100} \text{ i.e. } \textbf{£152.81}$$

Similarly, for **salesman Q**, the miles travelled are 264 and his allowance is

$$\frac{£264 \times 37}{100} \text{ i.e. } \textbf{£97.68}$$

Salesman R travels 597 miles and he receives

$$\frac{£597 \times 37}{100} \text{ i.e. } \textbf{£220.89}$$

Finally, **salesman S** receives

$$\frac{£143 \times 37}{100} \text{ i.e. } \textbf{£52.91}$$

(b) An analysis of Fig. 32.4 shows that 5-roomed bungalows and 5-roomed houses are becoming more popular, the greatest change in the three years being a 15% increase in the sales of 5-roomed bungalows.

(c) Since 1.8° corresponds to 1 p and the profit occupies 108° of the pie diagram, the profit per unit is

$$\frac{108 \times 1}{1.8} \text{ i.e. } 60 \text{p}$$

The profit when selling 700 units of the product is

$$£\frac{700 \times 60}{100} \text{ i.e. } \textbf{£420}$$

Now try the following Practice Exercise

Practice Exercise 161   Presentation of ungrouped data (answers on page 458)

1. The number of vehicles passing a stationary observer on a road in 6 ten-minute intervals is as shown. Draw a pictogram to represent these data.

Period of time	1	2	3	4	5	6
Number of vehicles	35	44	62	68	49	41

2. The number of components produced by a factory in a week is as shown below.

Day	Mon	Tues	Wed	Thur	Fri
Number of components	1580	2190	1840	2385	1280

Show these data on a pictogram.

3. For the data given in Problem 1 above, draw a horizontal bar chart.

4. Present the data given in Problem 2 above on a horizontal bar chart.

5. For the data given in Problem 1 above, construct a vertical bar chart.

6. Depict the data given in Problem 2 above on a vertical bar chart.

7. A factory produces three different types of components. The percentages of each of these components produced for 3 one-month periods are as shown below. Show this information on percentage component bar charts and comment on the changing trend in the percentages of the types of component produced.

Month	1	2	3
Component P	20	35	40
Component Q	45	40	35
Component R	35	25	25

8. A company has five distribution centres and the mass of goods in tonnes sent to each centre during 4 one-week periods is as shown.

Week	1	2	3	4
Centre A	147	160	174	158
Centre B	54	63	77	69
Centre C	283	251	237	211
Centre D	97	104	117	144
Centre E	224	218	203	194

Use a percentage component bar chart to present these data and comment on any trends.

9. The employees in a company can be split into the following categories: managerial 3, supervisory 9, craftsmen 21, semi-skilled 67, others 44. Show these data on a pie diagram.

10. The way in which an apprentice spent his time over a one-month period is as follows: drawing office 44 hours, production 64 hours, training 12 hours, at college 28 hours. Use a pie diagram to depict this information.

11. (a) With reference to Fig. 32.5, determine the amount spent on labour and materials to produce 1650 units of the product.
   (b) If in year 2 of Fig. 32.4 1% corresponds to 2.5 dwellings, how many bungalows are sold in that year?

12. (a) If the company sells 23 500 units per annum of the product depicted in Fig. 32.5, determine the cost of their overheads per annum.
   (b) If 1% of the dwellings represented in year 1 of Fig. 32.4 corresponds to 2 dwellings, find the total number of houses sold in that year.

32.3 Presentation of grouped data

When the number of members in a set is small, say ten or less, the data can be represented diagrammatically without further analysis, by means of pictograms, bar charts, percentage components bar charts or pie diagrams (as shown in Section 32.2).

For sets having more than ten members, those members having similar values are grouped together in **classes** to form a **frequency distribution**. To assist in accurately counting members in the various classes, a **tally diagram** is used (see Problems 8 and 12).

A **frequency distribution** is merely a table showing classes and their corresponding frequencies (see Problems 8 and 12). The new set of values obtained by forming a frequency distribution is called **grouped data**. The terms used in connection with grouped data are shown in Fig. 32.6(a). The size or range of a class is given by the **upper class boundary value** minus the **lower class boundary value** and in Fig. 32.6(b) is $7.65 - 7.35$;

i.e. 0.30. The **class interval** for the class shown in Fig. 32.6(b) is 7.4 to 7.6 and the class mid-point value is given by

$$\frac{(\text{upper class boundary value}) + (\text{lower class boundary value})}{2}$$

and in Fig. 32.6(b) is

$$\frac{7.65 + 7.35}{2} \quad \text{i.e. } 7.5$$

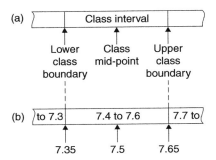

Figure 32.6

One of the principal ways of presenting grouped data diagrammatically is to use a **histogram**, in which the **areas** of vertical, adjacent rectangles are made proportional to frequencies of the classes (see Problem 9). When class intervals are equal, the heights of the rectangles of a histogram are equal to the frequencies of the classes. For histograms having unequal class intervals, the area must be proportional to the frequency. Hence, if the class interval of class A is twice the class interval of class B, then for equal frequencies the height of the rectangle representing A is half that of B (see Problem 11).

Another method of presenting grouped data diagrammatically is to use a **frequency polygon**, which is the graph produced by plotting frequency against class mid-point values and joining the co-ordinates with straight lines (see Problem 12).

A **cumulative frequency distribution** is a table showing the cumulative frequency for each value of upper class boundary. The cumulative frequency for a particular value of upper class boundary is obtained by adding the frequency of the class to the sum of the previous frequencies. A cumulative frequency distribution is formed in Problem 13.

The curve obtained by joining the co-ordinates of cumulative frequency (vertically) against upper class boundary (horizontally) is called an **ogive** or a **cumulative frequency distribution curve** (see Problem 13).

Problem 8. The data given below refer to the gain of each of a batch of 40 transistors, expressed correct to the nearest whole number. Form a frequency distribution for these data having seven classes

81	83	87	74	76	89	82	84
86	76	77	71	86	85	87	88
84	81	80	81	73	89	82	79
81	79	78	80	85	77	84	78
83	79	80	83	82	79	80	77

The **range** of the data is the value obtained by taking the value of the smallest member from that of the largest member. Inspection of the set of data shows that range $= 89 - 71 = 18$. The size of each class is given approximately by the range divided by the number of classes. Since seven classes are required, the size of each class is $18 \div 7$; that is, approximately 3. To achieve seven equal classes spanning a range of values from 71 to 89, the class intervals are selected as 70–72, 73–75, and so on. To assist with accurately determining the number in each class, a **tally diagram** is produced, as shown in Table 32.1(a). This is obtained by listing the classes in the left-hand column and then inspecting each of the 40 members of the set in turn and allocating them to the appropriate classes by putting '1's in the appropriate rows. Every fifth '1' allocated to a particular row is shown as an oblique line crossing the four previous '1's, to help with final counting.

Table 32.1(a)

Class	Tally
70–72	1
73–75	11
76–78	⊬⊬⊤ 11
79–81	⊬⊬⊤ ⊬⊬⊤ 11
82–84	⊬⊬⊤ 1111
85–87	⊬⊬⊤ 1
88–90	111

A **frequency distribution** for the data is shown in Table 32.1(b) and lists classes and their corresponding frequencies, obtained from the tally diagram. (Class mid-point values are also shown in the table, since they are used for constructing the histogram for these data (see Problem 9).)

Table 32.1(b)

Class	Class mid-point	Frequency
70–72	71	1
73–75	74	2
76–78	77	7
79–81	80	12
82–84	83	9
85–87	86	6
88–90	89	3

Problem 9. Construct a histogram for the data given in Table 32.1(b)

The histogram is shown in Fig. 32.7. The width of the rectangles corresponds to the upper class boundary values minus the lower class boundary values and the heights of the rectangles correspond to the class frequencies. The easiest way to draw a histogram is to mark the class mid-point values on the horizontal scale and draw the rectangles symmetrically about the appropriate class mid-point values and touching one another.

Figure 32.7

Problem 10. The amount of money earned weekly by 40 people working part-time in a factory, correct to the nearest £10, is shown below. Form a frequency distribution having 6 classes for these data

80	90	70	110	90	160	110	80
140	30	90	50	100	110	60	100
80	90	110	80	100	90	120	70
130	170	80	120	100	110	40	110
50	100	110	90	100	70	110	80

Inspection of the set given shows that the majority of the members of the set lie between £80 and £110 and that there is a much smaller number of extreme values ranging from £30 to £170. If equal class intervals are selected, the frequency distribution obtained does not give as much information as one with unequal class intervals. Since the majority of the members lie between £80 and £100, the class intervals in this range are selected to be smaller than those outside of this range. There is no unique solution and one possible solution is shown in Table 32.2.

Table 32.2

Class	Frequency
20–40	2
50–70	6
80–90	12
100–110	14
120–140	4
150–170	2

Problem 11. Draw a histogram for the data given in Table 32.2

When dealing with unequal class intervals, the histogram must be drawn so that the areas (and not the heights) of the rectangles are proportional to the frequencies of the classes. The data given are shown in columns 1 and 2 of Table 32.3. Columns 3 and 4 give the upper and lower class boundaries, respectively. In column 5, the class ranges (i.e. upper class boundary minus lower class boundary values) are listed. The heights of the rectangles are proportional to the ratio $\dfrac{\text{frequency}}{\text{class range}}$, as shown in column 6. The histogram is shown in Fig. 32.8.

Table 32.3

1 Class	2 Frequency	3 Upper class boundary	4 Lower class boundary	5 Class range	6 Height of rectangle
20–40	2	45	15	30	$\frac{2}{30} = \frac{1}{15}$
50–70	6	75	45	30	$\frac{6}{30} = \frac{3}{15}$
80–90	12	95	75	20	$\frac{12}{20} = \frac{9}{15}$
100–110	14	115	95	20	$\frac{14}{20} = \frac{10\frac{1}{2}}{15}$
120–140	4	145	115	30	$\frac{4}{30} = \frac{2}{15}$
150–170	2	175	145	30	$\frac{2}{30} = \frac{1}{15}$

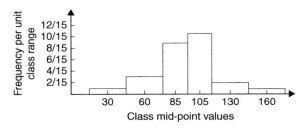

Figure 32.8

> **Problem 12.** The masses of 50 ingots in kilograms are measured correct to the nearest 0.1 kg and the results are as shown below. Produce a frequency distribution having about seven classes for these data and then present the grouped data as a frequency polygon and a histogram

8.0	8.6	8.2	7.5	8.0	9.1	8.5	7.6	8.2	7.8
8.3	7.1	8.1	8.3	8.7	7.8	8.7	8.5	8.2	8.5
7.7	8.4	7.9	8.8	7.2	8.1	7.8	8.2	7.7	7.5
8.1	7.4	8.8	8.0	8.4	8.5	8.1	7.3	9.0	8.6
7.4	8.2	8.4	7.7	8.3	8.2	7.9	8.5	7.9	8.0

The **range** of the data is the member having the largest value minus the member having the smallest value. Inspection of the set of data shows that
range $= 9.1 - 7.1 = $ **2.0**

The size of each class is given approximately by

$$\frac{\text{range}}{\text{number of classes}}$$

Since about seven classes are required, the size of each class is $2.0 \div 7$, i.e. approximately 0.3, and thus the **class limits** are selected as 7.1 to 7.3, 7.4 to 7.6, 7.7 to 7.9, and so on. The **class mid-point** for the 7.1 to 7.3 class is

$$\frac{7.35 + 7.05}{2} \text{ i.e. } 7.2$$

the class mid-point for the 7.4 to 7.6 class is

$$\frac{7.65 + 7.35}{2} \text{ i.e. } 7.5$$

and so on.

To assist with accurately determining the number in each class, a **tally diagram** is produced as shown in Table 32.4. This is obtained by listing the classes in the left-hand column and then inspecting each of the 50 members of the set of data in turn and allocating it to the appropriate class by putting a '1' in the appropriate row. Each fifth '1' allocated to a particular row is marked as an oblique line to help with final counting.

A **frequency distribution** for the data is shown in Table 32.5 and lists classes and their corresponding frequencies. Class mid-points are also shown in this table since they are used when constructing the frequency polygon and histogram.

Table 32.4

Class	Tally
7.1 to 7.3	111
7.4 to 7.6	̶J̶H̶T̶
7.7 to 7.9	̶J̶H̶T̶ 1111
8.0 to 8.2	̶J̶H̶T̶ ̶J̶H̶T̶ 1111
8.3 to 8.5	̶J̶H̶T̶ ̶J̶H̶T̶ 1
8.6 to 8.8	̶J̶H̶T̶ 1
8.9 to 9.1	11

Table 32.5

Class	Class mid-point	Frequency
7.1 to 7.3	7.2	3
7.4 to 7.6	7.5	5
7.7 to 7.9	7.8	9
8.0 to 8.2	8.1	14
8.3 to 8.5	8.4	11
8.6 to 8.8	8.7	6
8.9 to 9.1	9.0	2

A **frequency polygon** is shown in Fig. 32.9, the co-ordinates corresponding to the class mid-point/frequency values given in Table 32.5. The co-ordinates are joined by straight lines and the polygon is 'anchored-down' at each end by joining to the next class mid-point value and zero frequency.

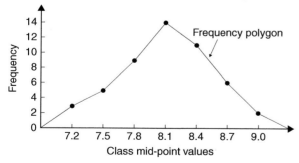

Figure 32.9

A **histogram** is shown in Fig. 32.10, the width of a rectangle corresponding to (upper class boundary value – lower class boundary value) and height corresponding to the class frequency. The easiest way to draw a histogram is to mark class mid-point values on the hori-

zontal scale and to draw the rectangles symmetrically about the appropriate class mid-point values and touching one another. A histogram for the data given in Table 32.5 is shown in Fig. 32.10.

Figure 32.10

> ⚑ **Problem 13.** The frequency distribution for the masses in kilograms of 50 ingots is
>
> | | |
> |---|---|
> | 7.1 to 7.3 | 3 |
> | 7.4 to 7.6 | 5 |
> | 7.7 to 7.9 | 9 |
> | 8.0 to 8.2 | 14 |
> | 8.3 to 8.5 | 11 |
> | 8.6 to 8.8 | 6 |
> | 8.9 to 9.1 | 2 |
>
> Form a cumulative frequency distribution for these data and draw the corresponding ogive

A **cumulative frequency distribution** is a table giving values of cumulative frequency for the values of upper class boundaries and is shown in Table 32.6. Columns 1 and 2 show the classes and their frequencies. Column 3 lists the upper class boundary values for the classes given in column 1. Column 4 gives the cumulative frequency values for all frequencies less than the upper class boundary values given in column 3. Thus, for example, for the 7.7 to 7.9 class shown in row 3, the cumulative frequency value is the sum of all frequencies having values of less than 7.95, i.e. $3 + 5 + 9 = 17$, and so on.

Table 32.6

1	2	3	4
Class	Frequency	Upper class boundary less than	Cumulative frequency
7.1–7.3	3	7.35	3
7.4–7.6	5	7.65	8
7.7–7.9	9	7.95	17
8.0–8.2	14	8.25	31
8.3–8.5	11	8.55	42
8.6–8.8	6	8.85	48
8.9–9.1	2	9.15	50

The **ogive** for the cumulative frequency distribution given in Table 32.6 is shown in Fig. 32.11. The co-ordinates corresponding to each upper class boundary/cumulative frequency value are plotted and the co-ordinates are joined by straight lines (not the best curve drawn through the co-ordinates as in experimental work). The ogive is 'anchored' at its start by adding the co-ordinate (7.05, 0)

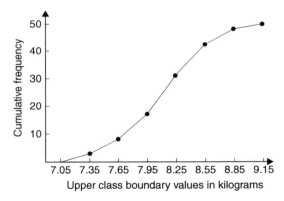

Figure 32.11

Now try the following Practice Exercise

Practice Exercise 162   Presentation of grouped data (answers on page 458)

1. The mass in kilograms, correct to the nearest one-tenth of a kilogram, of 60 bars of metal are as shown. Form a frequency distribution of about eight classes for these data.

39.8	40.1	40.3	40.0	40.6	39.7	40.0	40.4	39.6	39.3
39.6	40.7	40.2	39.9	40.3	40.2	40.4	39.9	39.8	40.0
40.2	40.1	40.3	39.7	39.9	40.5	39.9	40.5	40.0	39.9
40.1	40.8	40.0	40.0	40.1	40.2	40.1	40.0	40.2	39.9
39.7	39.8	40.4	39.7	39.9	39.5	40.1	40.1	39.9	40.2
39.5	40.6	40.0	40.1	39.8	39.7	39.5	40.2	39.9	40.3

2. Draw a histogram for the frequency distribution given in the solution of Problem 1.

3. The information given below refers to the value of resistance in ohms of a batch of 48 resistors of similar value. Form a frequency distribution for the data, having about six classes, and draw a frequency polygon and histogram to represent these data diagrammatically.

21.0	22.4	22.8	21.5	22.6	21.1	21.6	22.3
22.9	20.5	21.8	22.2	21.0	21.7	22.5	20.7
23.2	22.9	21.7	21.4	22.1	22.2	22.3	21.3
22.1	21.8	22.0	22.7	21.7	21.9	21.1	22.6
21.4	22.4	22.3	20.9	22.8	21.2	22.7	21.6
22.2	21.6	21.3	22.1	21.5	22.0	23.4	21.2

4. The time taken in hours to the failure of 50 specimens of a metal subjected to fatigue failure tests are as shown. Form a frequency distribution, having about eight classes and unequal class intervals, for these data.

28	22	23	20	12	24	37	28	21	25
21	14	30	23	27	13	23	7	26	19
24	22	26	3	21	24	28	40	27	24
20	25	23	26	47	21	29	26	22	33
27	9	13	35	20	16	20	25	18	22

5. Form a cumulative frequency distribution and hence draw the ogive for the frequency distribution given in the solution to Problem 3.

6. Draw a histogram for the frequency distribution given in the solution to Problem 4.

7. The frequency distribution for a batch of 50 capacitors of similar value, measured in microfarads, is

10.5–10.9	2
11.0–11.4	7
11.5–11.9	10
12.0–12.4	12
12.5–12.9	11
13.0–13.4	8

Form a cumulative frequency distribution for these data.

8. Draw an ogive for the data given in the solution of Problem 7.

9. The diameter in millimetres of a reel of wire is measured in 48 places and the results are as shown.

2.10	2.29	2.32	2.21	2.14	2.22
2.28	2.18	2.17	2.20	2.23	2.13
2.26	2.10	2.21	2.17	2.28	2.15
2.16	2.25	2.23	2.11	2.27	2.34
2.24	2.05	2.29	2.18	2.24	2.16
2.15	2.22	2.14	2.27	2.09	2.21
2.11	2.17	2.22	2.19	2.12	2.20
2.23	2.07	2.13	2.26	2.16	2.12

(a) Form a frequency distribution of diameters having about 6 classes.

(b) Draw a histogram depicting the data.

(c) Form a cumulative frequency distribution.

(d) Draw an ogive for the data.

Practice Exercise 163   Multiple-choice questions on presentation of statistical data (answers on page 459)

Each question has only one correct answer

1. A pie diagram is shown in Figure 32.12 where P, Q, R and S represent the salaries of four employees of a construction company. P earns £24000 p.a. Employee S earns:
    (a) £20000   (b) £36000
    (c) £40000   (d) £24000

Figure 32.12

2. The curve obtained by joining the co-ordinates of cumulative frequency against upper class boundary values is called;
    (a) a histogram
    (b) a frequency polygon
    (c) a tally diagram
    (d) an ogive

3. Which of the following is a discrete variable?
    (a) The diameter of a workpiece
    (b) The temperature of a furnace
    (c) The number of people working in a factory
    (d) The volume of air in an office

4. A histogram is drawn for a frequency distribution with unequal class intervals. The frequency of the class is represented by the:
    (a) height of the rectangles (b) perimeter of the rectangles (c) area of the rectangles (d) width of the rectangles

5. An ogive is a:
    (a) measure of dispersion
    (b) cumulative frequency diagram
    (c) measure of location
    (d) type of histogram

For fully worked solutions to each of the problems in Practice Exercises 160 to 162 in this chapter, go to the website:

Chapter 33

Mean, median, mode and standard deviation

Why it is important to understand: Mean, median, mode and standard deviation

Statistics is a field of mathematics that pertains to data analysis. In many real-life situations, it is helpful to describe data by a single number that is most representative of the entire collection of numbers. Such a number is called a measure of central tendency; the most commonly used measures are mean, median, mode and standard deviation, the latter being the average distance between the actual data and the mean. Statistics is important in the field of engineering since it provides tools to analyse collected data. For example, a chemical engineer may wish to analyse temperature measurements from a mixing tank. Statistical methods can be used to determine how reliable and reproducible the temperature measurements are, how much the temperature varies within the data set, what future temperatures of the tank may be and how confident the engineer can be in the temperature measurements made. When performing statistical analysis on a set of data, the mean, median, mode and standard deviation are all helpful values to calculate; this chapter explains how to determine these measures of central tendency.

At the end of this chapter you should be able to:

- determine the mean, median and mode for a set of ungrouped data
- determine the mean, median and mode for a set of grouped data
- draw a histogram from a set of grouped data
- determine the mean, median and mode from a histogram
- calculate the standard deviation from a set of ungrouped data
- calculate the standard deviation from a set of grouped data
- determine the quartile values from an ogive
- determine quartile, decile and percentile values from a set of data

---

33.1 Measures of central tendency

A single value, which is representative of a set of values, may be used to give an indication of the general size of the members in a set, the word '**average**' often being used to indicate the single value. The statistical term used for 'average' is the 'arithmetic mean' or just the '**mean**'.

Other measures of central tendency may be used and these include the **median** and the **modal** values.

33.2 Mean, median and mode for discrete data

Mean

The **arithmetic mean value** is found by adding together the values of the members of a set and dividing by the number of members in the set. Thus, the mean of the set of numbers $\{4, 5, 6, 9\}$ is

$$\frac{4 + 5 + 6 + 9}{4} \text{ i.e. } 6$$

In general, the mean of the set $\{x_1, x_2, x_3, \ldots x_n\}$ is

$$\bar{x} = \frac{x_1 + x_2 + x_3 + \cdots + x_n}{n} \text{ written as } \frac{\sum x}{n}$$

where $\sum$ is the Greek letter 'sigma' and means 'the sum of' and $\bar{x}$ (called x-bar) is used to signify a mean value.

Median

The **median value** often gives a better indication of the general size of a set containing extreme values. The set $\{7, 5, 74, 10\}$ has a mean value of 24, which is not really representative of any of the values of the members of the set. The median value is obtained by

(a) **ranking** the set in ascending order of magnitude, and

(b) selecting the value of the **middle member** for sets containing an odd number of members or finding the value of the mean of the two middle members for sets containing an even number of members.

For example, the set $\{7, 5, 74, 10\}$ is ranked as $\{5, 7, 10, 74\}$ and, since it contains an even number of members (four in this case), the mean of 7 and 10 is taken, giving a median value of 8.5. Similarly, the set $\{3, 81, 15, 7, 14\}$ is ranked as $\{3, 7, 14, 15, 81\}$ and the median value is the value of the middle member, i.e. 14

Mode

The **modal value**, or **mode**, is the most commonly occurring value in a set. If two values occur with the same frequency, the set is '**bi-modal**'. The set $\{5, 6, 8, 2, 5, 4, 6, 5, 3\}$ has a modal value of 5, since the member having a value of 5 occurs the most, i.e. three times.

Problem 1. Determine the mean, median and mode for the set $\{2, 3, 7, 5, 5, 13, 1, 7, 4, 8, 3, 4, 3\}$

The mean value is obtained by adding together the values of the members of the set and dividing by the number of members in the set. Thus,

mean value, $\bar{x}$

$$= \frac{2 + 3 + 7 + 5 + 5 + 13 + 1 + 7 + 4 + 8 + 3 + 4 + 3}{13}$$

$$= \frac{65}{13} = 5$$

To obtain the median value the set is ranked, that is, placed in ascending order of magnitude, and since the set contains an odd number of members the value of the middle member is the median value. Ranking the set gives $\{1, 2, 3, 3, 3, 4, 4, 5, 5, 7, 7, 8, 13\}$. The middle term is the seventh member; i.e. 4. Thus, the **median value is 4**

The **modal value** is the value of the most commonly occurring member and is **3**, which occurs three times, all other members only occurring once or twice.

Problem 2. The following set of data refers to the amount of money in £s taken by a news vendor for 6 days. Determine the mean, median and modal values of the set

$$\{27.90, 34.70, 54.40, 18.92, 47.60, 39.68\}$$

Mean value

$$= \frac{27.90 + 34.70 + 54.40 + 18.92 + 47.60 + 39.68}{6}$$

$$= \pounds 37.20$$

The ranked set is

$$\{18.92, 27.90, 34.70, 39.68, 47.60, 54.40\}$$

Since the set has an even number of members, the mean of the middle two members is taken to give the median value; i.e.

$$\text{median value} = \frac{34.70 + 39.68}{2} = \pounds 37.19$$

Since no two members have the same value, this set has **no mode**.

Now try the following Practice Exercise

> **Practice Exercise 164 Mean, median and mode for discrete data (answers on page 459)**
>
> In Problems 1 to 4, determine the mean, median and modal values for the sets given.
> 1. {3, 8, 10, 7, 5, 14, 2, 9, 8}
> 2. {26, 31, 21, 29, 32, 26, 25, 28}
> 3. {4.72, 4.71, 4.74, 4.73, 4.72, 4.71, 4.73, 4.72}
> 4. {73.8, 126.4, 40.7, 141.7, 28.5, 237.4, 157.9}

33.3 Mean, median and mode for grouped data

The mean value for a set of grouped data is found by determining the sum of the (frequency × class mid-point values) and dividing by the sum of the frequencies; i.e.

$$\text{mean value}, \bar{x} = \frac{f_1 x_1 + f_2 x_2 + \cdots f_n x_n}{f_1 + f_2 + \cdots + f_n} = \frac{\sum (fx)}{\sum f}$$

where f is the frequency of the class having a mid-point value of x, and so on.

> ⚑ **Problem 3.** The frequency distribution for the value of resistance in ohms of 48 resistors is as shown. Determine the mean value of resistance
>
> | 20.5–20.9 | 3 |
> | 21.0–21.4 | 10 |
> | 21.5–21.9 | 11 |
> | 22.0–22.4 | 13 |
> | 22.5–22.9 | 9 |
> | 23.0–23.4 | 2 |

The class mid-point/frequency values are 20.7 3, 21.2 10, 21.7 11, 22.2 13, 22.7 9 and 23.2 2
For grouped data, the mean value is given by

$$\bar{x} = \frac{\sum (fx)}{\sum f}$$

where f is the class frequency and x is the class mid-point value. Hence mean value,

$$\bar{x} = \frac{\begin{array}{c}(3 \times 20.7) + (10 \times 21.2) + (11 \times 21.7) \\ + (13 \times 22.2) + (9 \times 22.7) + (2 \times 23.2)\end{array}}{48}$$
$$= \frac{1052.1}{48} = 21.919\ldots$$

i.e. **the mean value is 21.9 ohms**, correct to 3 significant figures.

Histograms

The mean, median and modal values for grouped data may be determined from a **histogram**. In a histogram, frequency values are represented vertically and variable values horizontally. The mean value is given by the value of the variable corresponding to a vertical line drawn through the centroid of the histogram. The median value is obtained by selecting a variable value such that the area of the histogram to the left of a vertical line drawn through the selected variable value is equal to the area of the histogram on the right of the line. The modal value is the variable value obtained by dividing the width of the highest rectangle in the histogram in proportion to the heights of the adjacent rectangles. The method of determining the mean, median and modal values from a histogram is shown in Problem 4.

> ⚑ **Problem 4.** The time taken in minutes to assemble a device is measured 50 times and the results are as shown. Draw a histogram depicting the data and hence determine the mean, median and modal values of the distribution
>
> | 14.5–15.5 | 5 |
> | 16.5–17.5 | 8 |
> | 18.5–19.5 | 16 |
> | 20.5–21.5 | 12 |
> | 22.5–23.5 | 6 |
> | 24.5–25.5 | 3 |

The histogram is shown in Fig. 33.1. The mean value lies at the centroid of the histogram. With reference to

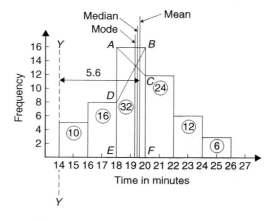

Figure 33.1

any arbitrary axis, say *YY* shown at a time of 14 minutes, the position of the horizontal value of the centroid can be obtained from the relationship $AM = \sum (am)$, where *A* is the area of the histogram, *M* is the horizontal distance of the centroid from the axis *YY*, *a* is the area of a rectangle of the histogram and *m* is the distance of the centroid of the rectangle from *YY*. The areas of the individual rectangles are shown circled on the histogram giving a total area of 100 square units. The positions, *m*, of the centroids of the individual rectangles are 1, 3, 5, ... units from *YY*. Thus

$$100M = (10 \times 1) + (16 \times 3) + (32 \times 5) + (24 \times 7)$$
$$+ (12 \times 9) + (6 \times 11)$$

i.e. $M = \dfrac{560}{100} = 5.6$ units from *YY*

Thus, the position of the **mean** with reference to the time scale is $14 + 5.6$, i.e. **19.6 minutes**.

The median is the value of time corresponding to a vertical line dividing the total area of the histogram into two equal parts. The total area is 100 square units, hence the vertical line must be drawn to give 50 units of area on each side. To achieve this with reference to Fig. 33.1, rectangle *ABFE* must be split so that $50 - (10 + 16)$ units of area lie on one side and $50 - (24 + 12 + 6)$ units of area lie on the other. This shows that the area of *ABFE* is split so that 24 units of area lie to the left of the line and 8 units of area lie to the right; i.e. the vertical line must pass through 19.5 minutes. Thus, the **median value** of the distribution is **19.5 minutes**.

The mode is obtained by dividing the line *AB*, which is the height of the highest rectangle, proportionally to the heights of the adjacent rectangles. With reference to Fig. 33.1, this is achieved by joining *AC* and *BD* and drawing a vertical line through the point of intersection of these two lines. This gives the **mode** of the distribution, which is **19.3 minutes**.

Now try the following Practice Exercise

Practice Exercise 165   Mean, median and mode for grouped data (answers on page 459)

1. 21 bricks have a mean mass of 24.2 kg and 29 similar bricks have a mass of 23.6 kg. Determine the mean mass of the 50 bricks.

2. The frequency distribution given below refers to the heights in centimetres of 100 people.

Determine the mean value of the distribution, correct to the nearest millimetre.

150–156	5
157–163	18
164–170	20
171–177	27
178–184	22
185–191	8

3. The gain of 90 similar transistors is measured and the results are as shown. By drawing a histogram of this frequency distribution, determine the mean, median and modal values of the distribution.

83.5–85.5	6
86.5–88.5	39
89.5–91.5	27
92.5–94.5	15
95.5–97.5	3

4. The diameters, in centimetres, of 60 holes bored in engine castings are measured and the results are as shown. Draw a histogram depicting these results and hence determine the mean, median and modal values of the distribution.

2.011–2.014	7
2.016–2.019	16
2.021–2.024	23
2.026–2.029	9
2.031–2.034	5

33.4 Standard deviation

Discrete data

The standard deviation of a set of data gives an indication of the amount of dispersion, or the scatter, of members of the set from the measure of central tendency. Its value is the root-mean-square value of the members of the set and for discrete data is obtained as follows.

(i) Determine the measure of central tendency, usually the mean value, (occasionally the median or modal values are specified).

(ii) Calculate the deviation of each member of the set from the mean, giving

$$(x_1 - \bar{x}), (x_2 - \bar{x}), (x_3 - \bar{x}), \ldots$$

(iii) Determine the squares of these deviations; i.e. $(x_1 - \bar{x})^2, (x_2 - \bar{x})^2, (x_3 - \bar{x})^2, \ldots$

(iv) Find the sum of the squares of the deviations, i.e. $(x_1 - \bar{x})^2 + (x_2 - \bar{x})^2 + (x_3 - \bar{x})^2, \ldots$

(v) Divide by the number of members in the set, n, giving
$$\frac{\left(x_1 - \bar{x}\right)^2 + \left(x_2 - \bar{x}\right)^2 + \left(x^3 - \bar{x}\right)^2 + \cdots}{n}$$

(vi) Determine the square root of (v)

The standard deviation is indicated by σ (the Greek letter small 'sigma') and is written mathematically as

$$\textbf{standard deviation}, \sigma = \sqrt{\left\{ \frac{\sum (\mathbf{x} - \bar{\mathbf{x}})^2}{\mathbf{n}} \right\}}$$

where x is a member of the set, $\bar{x}$ is the mean value of the set and n is the number of members in the set. The value of standard deviation gives an indication of the distance of the members of a set from the mean value. The set {1, 4, 7, 10, 13} has a mean value of 7 and a standard deviation of about 4.2. The set {5, 6, 7, 8, 9} also has a mean value of 7 but the standard deviation is about 1.4. This shows that the members of the second set are mainly much closer to the mean value than the members of the first set. The method of determining the standard deviation for a set of discrete data is shown in Problem 5.

Problem 5. Determine the standard deviation from the mean of the set of numbers {5, 6, 8, 4, 10, 3}, correct to 4 significant figures

The arithmetic mean, $\bar{x} = \dfrac{\sum x}{n}$

$$= \frac{5 + 6 + 8 + 4 + 10 + 3}{6} = 6$$

Standard deviation, $\sigma = \sqrt{\left\{ \dfrac{\sum \left(x - \bar{x}\right)^2}{n} \right\}}$

The $(x - \bar{x})^2$ values are $(5-6)^2, (6-6)^2, (8-6)^2,$ $(4-6)^2, (10-6)^2$ and $(3-6)^2$

The sum of the $(x - \bar{x})^2$ values,

i.e. $\sum \left(x - \bar{x}\right)^2$, is $1 + 0 + 4 + 4 + 16 + 9 = 34$

and $\dfrac{\sum \left(x - \bar{x}\right)^2}{n} = \dfrac{34}{6} = 5.\dot{6}$

since there are 6 members in the set.
Hence, **standard deviation,**

$$\sigma = \sqrt{\left\{ \frac{\sum (x - \bar{x})^2}{n} \right\}} = \sqrt{5.\dot{6}} = \textbf{2.380},$$

correct to 4 significant figures.

Grouped data

For grouped data,

$$\textbf{standard deviation}, \sigma = \sqrt{\left\{ \frac{\sum \{ f(x - \bar{x})^2 \}}{\sum f} \right\}}$$

where f is the class frequency value, x is the class mid-point value and $\bar{x}$ is the mean value of the grouped data. The method of determining the standard deviation for a set of grouped data is shown in Problem 6.

Problem 6. The frequency distribution for the values of resistance in ohms of 48 resistors is as shown. Calculate the standard deviation from the mean of the resistors, correct to 3 significant figures

20.5–20.9	3
21.0–21.4	10
21.5–21.9	11
22.0–22.4	13
22.5–22.9	9
23.0–23.4	2

The standard deviation for grouped data is given by

$$\sigma = \sqrt{\left\{ \frac{\sum \{ f(x - \bar{x})^2 \}}{\sum f} \right\}}$$

From Problem 3, the distribution mean value is $\bar{x} = 21.92$, correct to 2 significant figures.

The 'x-values' are the class mid-point values, i.e. $20.7, 21.2, 21.7, \ldots$

Thus, the $(x - \bar{x})^2$ values are $(20.7 - 21.92)^2,$ $(21.2 - 21.92)^2, (21.7 - 21.92)^2, \ldots$

and the $f(x - \bar{x})^2$ values are $3(20.7 - 21.92)^2,$ $10(21.2 - 21.92)^2, 11(21.7 - 21.92)^2, \ldots$

The $\sum f\left(x - \bar{x}\right)^2$ values are

$$4.4652 + 5.1840 + 0.5324 + 1.0192$$
$$+5.4756 + 3.2768 = 19.9532$$

$$\frac{\sum \left\{ f \left(x - \bar{x} \right)^2 \right\}}{\sum f} = \frac{19.9532}{48} = 0.41569$$

and **standard deviation**,

$$\sigma = \sqrt{\left\{ \frac{\sum \left\{ f \left(x - \bar{x} \right)^2 \right\}}{\sum f} \right\}} = \sqrt{0.41569}$$

$$= \mathbf{0.645}, \text{correct to 3 significant figures.}$$

Now try the following Practice Exercise

Practice Exercise 166   Standard deviation (answers on page 459)

1. Determine the standard deviation from the mean of the set of numbers

   $$\{35, 22, 25, 23, 28, 33, 30\}$$

   correct to 3 significant figures.

2. The values of capacitances, in microfarads, of ten capacitors selected at random from a large batch of similar capacitors are

   34.3, 25.0, 30.4, 34.6, 29.6, 28.7,

   33.4, 32.7, 29.0 and 31.3

   Determine the standard deviation from the mean for these capacitors, correct to 3 significant figures.

3. The tensile strength in megapascals for 15 samples of tin were determined and found to be

   34.61, 34.57, 34.40, 34.63, 34.63, 34.51,

   34.49, 34.61, 34.52, 34.55, 34.58, 34.53,

   34.44, 34.48 and 34.40

   Calculate the mean and standard deviation from the mean for these 15 values, correct to 4 significant figures.

4. Calculate the standard deviation from the mean for the mass of the 50 bricks given in Problem 1 of Practice Exercise 165, page 370, correct to 3 significant figures.

5. Determine the standard deviation from the mean, correct to 4 significant figures, for the heights of the 100 people given in Problem 2 of Practice Exercise 165, page 370.

6. Calculate the standard deviation from the mean for the data given in Problem 4 of Practice Exercise 165, page 370, correct to 3 decimal places.

33.5 Quartiles, deciles and percentiles

Other measures of dispersion which are sometimes used are the quartile, decile and percentile values. The **quartile values** of a set of discrete data are obtained by selecting the values of members which divide the set into four equal parts. Thus, for the set $\{2, 3, 4, 5, 5, 7, 9, 11, 13, 14, 17\}$ there are 11 members and the values of the members dividing the set into four equal parts are 4, 7 and 13. These values are signified by Q_1, Q_2 and Q_3 and called the first, second and third quartile values, respectively. It can be seen that the second quartile value, Q_2, is the value of the middle member and hence is the median value of the set.

For grouped data the ogive may be used to determine the quartile values. In this case, points are selected on the vertical cumulative frequency values of the ogive, such that they divide the total value of cumulative frequency into four equal parts. Horizontal lines are drawn from these values to cut the ogive. The values of the variable corresponding to these cutting points on the ogive give the quartile values (see Problem 7).

When a set contains a large number of members, the set can be split into ten parts, each containing an equal number of members. These ten parts are then called **deciles**. For sets containing a very large number of members, the set may be split into one hundred parts, each containing an equal number of members. One of these parts is called a **percentile**.

Problem 7. The frequency distribution given below refers to the overtime worked by a group of craftsmen during each of 48 working weeks in a year. Draw an ogive for these data and hence determine the quartile values.

25–29	5
30–34	4
35–39	7
40–44	11
45–49	12
50–54	8
55–59	1

The cumulative frequency distribution (i.e. upper class boundary/cumulative frequency values) is

29.5 5, 34.5 9, 39.5 16, 44.5 27,
49.5 39, 54.5 47, 59.5 48

The ogive is formed by plotting these values on a graph, as shown in Fig. 33.2. The total frequency is divided into four equal parts, each having a range of $48 \div 4$, i.e. 12. This gives cumulative frequency values of 0 to

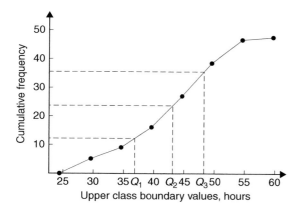

Figure 33.2

12 corresponding to the first quartile, 12 to 24 corresponding to the second quartile, 24 to 36 corresponding to the third quartile and 36 to 48 corresponding to the fourth quartile of the distribution; i.e. the distribution is divided into four equal parts. The quartile values are those of the variable corresponding to cumulative frequency values of 12, 24 and 36, marked Q_1, Q_2 and Q_3 in Fig. 33.2. These values, correct to the nearest hour, are **37 hours, 43 hours and 48 hours**, respectively. The Q_2 value is also equal to the median value of the distribution. One measure of the dispersion of a distribution is called the **semi-interquartile range** and is given by $(Q_3 - Q_1) \div 2$ and is $(48 - 37) \div 2$ in this case; i.e. $\mathbf{5\frac{1}{2}}$ **hours**.

> **Problem 8.** Determine the numbers contained in the (a) 41st to 50th percentile group and (b) 8th decile group of the following set of numbers.
>
> 14 22 17 21 30 28 37 7 23 32
> 24 17 20 22 27 19 26 21 15 29

The set is ranked, giving

 7 14 15 17 17 19 20 21 21 22
22 23 24 26 27 28 29 30 32 37

(a) There are 20 numbers in the set, hence the first 10% will be the two numbers 7 and 14, the second 10% will be 15 and 17, and so on. Thus, the 41st to 50th percentile group will be the numbers **21 and 22**

(b) The first decile group is obtained by splitting the ranked set into 10 equal groups and selecting the first group; i.e. the numbers 7 and 14. The second decile group is the numbers 15 and 17, and so on. Thus, the 8th decile group contains the numbers **27 and 28**

Now try the following Practice Exercise

> **Practice Exercise 167   Quartiles, deciles and percentiles (answers on page 459)**
>
> 1. The number of working days lost due to accidents for each of 12 one-month periods are as shown. Determine the median and first and third quartile values for this data.
>
>   27 37 40 28 23 30 35 24 30 32 31 28
>
> 2. The number of faults occurring on a production line in a nine-week period are as shown below. Determine the median and quartile values for the data.
>
>   30 27 25 24 27 37 31 27 35
>
> 3. Determine the quartile values and semi-interquartile range for the frequency distribution given in Problem 2 of Practice Exercise 133, page 358.
>
> 4. Determine the numbers contained in the 5th decile group and in the 61st to 70th percentile groups for the following set of numbers.
>
>   40 46 28 32 37 42 50 31 48 45
>   32 38 27 33 40 35 25 42 38 41
>
> 5. Determine the numbers in the 6th decile group and in the 81st to 90th percentile group for the following set of numbers.
>
>   43 47 30 25 15 51 17 21 37 33 44 56 40 49 22
>   36 44 33 17 35 58 51 35 44 40 31 41 55 50 16

> **Practice Exercise 168   Multiple-choice questions on mean, median, mode and standard deviation (answers on page 459)**
>
> **Each question has only one correct answer**
>
> Questions 1 and 2 relate to the following information.

A set of measurements (in mm) is as follows: {4, 5, 2, 11, 7, 6, 5, 1, 5, 8, 12, 6}

1. The median is:

    (a) 6 mm (b) 5 mm (c) 72 mm (d) 5.5 mm

2. The mean is:

    (a) 6 mm (b) 5 mm (c) 72 mm (d) 5.5 mm

3. The mean of the numbers 5, 13, 9, x and 10 is 11. The value of x is:

    (a) 20 (b) 18 (c) 17 (d) 16

Questions 4 to 7 relate to the following information:
The capacitance (in pF) of six capacitors is as follows: {5, 6, 8, 5, 10, 2}

4. The median value is:

    (a) 36 pF (b) 6 pF (c) 5.5 pF (d) 5 pF

5. The modal value is:

    (a) 36 pF (b) 6 pF (c) 5.5 pF (d) 5 pF

6. The mean value is:

    (a) 36 pF (b) 6 pF (c) 5.5 pF (d) 5 pF

7. The standard deviation is:

    (a) 2.66 pF (b) 2.52 pF
    (c) 2.45 pF (d) 6.33 pF

8. The number of faults occurring on a production line in a 11-week period are as shown:

    32 29 27 26 29 39 33 29 37 28 35

    The third quartile value is:

    (a) 29 (b) 31 (c) 35 (d) 33

Questions 9 and 10 relate to the following information:
The frequency distribution for the values of resistance in ohms of 40 transistors is as follows:
15.5 − 15.9 3   16.0 − 16.4 10 16.5 − 16.9 13   17.0 − 17.4 8   17.5 − 17.9 6

9. The mean value of the resistance is:

    (a) 16.75 Ω (b) 1.0 Ω
    (c) 15.85 Ω (d) 16.95 Ω

10. The standard deviation is:

    (a)] 0.335 Ω (b) 0.251 Ω
    (c) 0.682 Ω (d) 0.579 Ω

Chapter 34

Probability

Why it is important to understand: Probability

Engineers deal with uncertainty in their work, often with precision and analysis, and probability theory is widely used to model systems in engineering and scientific applications. There are a number of examples of where probability is used in engineering. For example, with electronic circuits, scaling down the power and energy of such circuits reduces the reliability and predictability of many individual elements, but the circuits must nevertheless be engineered so that the overall circuit is reliable. Centres for disease control need to decide whether to institute massive vaccination or other preventative measures in the face of globally threatening, possibly mutating diseases in humans and animals. System designers must weigh the costs and benefits of measures for reliability and security, such as levels of backups and firewalls, in the face of uncertainty about threats from equipment failures or malicious attackers. Models incorporating probability theory have been developed and are continuously being improved for understanding the brain, gene pools within populations, weather and climate forecasts, microelectronic devices and imaging systems such as computer aided tomography (CAT) scan and radar. The electric power grid, including power generating stations, transmission lines and consumers, is a complex system with many redundancies; however, breakdowns occur, and guidance for investment comes from modelling the most likely sequences of events that could cause outage. Similar planning and analysis is done for communication networks, transportation networks, water and other infrastructure. Probabilities, permutations and combinations are used daily in many different fields that range from gambling and games, to mechanical or structural failure rates, to rates of detection in medical screening. Uncertainty is clearly all around us, in our daily lives and in many professions. Standard deviation is widely used when results of opinion polls are described. The language of probability theory lets people break down complex problems, and argue about pieces of them with each other, and then aggregate information about subsystems to analyse a whole system. This chapter briefly introduces the important subject of probability.

At the end of this chapter you should be able to:

- define probability
- define expectation, dependent event, independent event and conditional probability
- state the addition and multiplication laws of probability
- use the laws of probability in simple calculations
- use the laws of probability in practical situations

34.1 Introduction to probability

Probability

Probability and statistics are quite extensively linked. For instance, when a scientist performs a measurement, the outcome of that measurement has a certain amount of 'chance' associated with it: factors such as electronic noise in the equipment, minor fluctuations in environmental conditions, and even human error have a random effect on the measurement. Often, the variations caused by these factors are minor, but they do have a significant effect in many cases. As a result, the scientist cannot expect to get the exact same measurement result in every case, and this variation requires that he describe his measurements *statistically* (for instance, using a mean and standard deviation). Likewise, when an anthropologist considers a small group of people from a larger population, the results of the study (assuming they involve numerical data) will involve some random variations that must be taken into account using statistics.

The **probability** of something happening is the likelihood or chance of it happening. Values of probability lie between 0 and 1, where 0 represents an absolute impossibility and 1 represents an absolute certainty. The probability of an event happening usually lies somewhere between these two extreme values and is expressed as either a proper or decimal fraction. Examples of probability are

that a length of copper wire has zero resistance at 100°C	0
that a fair, six-sided dice will stop with a 3 upwards	$\frac{1}{6}$ or 0.1667
that a fair coin will land with a head upwards	$\frac{1}{2}$ or 0.5
that a length of copper wire has some resistance at 100°C	1

If p is the probability of an event happening and q is the probability of the same event not happening, then the total probability is $p + q$ and is equal to unity, since it is an absolute certainty that the event either will or will not occur; i.e. $p + q = 1$

> **Problem 1.** Determine the probabilities of selecting at random (a) a man and (b) a woman from a crowd containing 20 men and 33 women

(a) The probability of selecting at random a man, p, is given by the ratio

$$\frac{\text{number of men}}{\text{number in crowd}}$$

i.e. $p = \dfrac{20}{20 + 33} = \dfrac{\mathbf{20}}{\mathbf{53}}$ or **0.3774**

(b) The probability of selecting at random a woman, q, is given by the ratio

$$\frac{\text{number of women}}{\text{number in crowd}}$$

i.e. $q = \dfrac{33}{20 + 33} = \dfrac{\mathbf{33}}{\mathbf{53}}$ or **0.6226**

(Check: the total probability should be equal to 1:

$p = \dfrac{20}{53}$ and $q = \dfrac{33}{53}$, thus the total probability,

$$p + q = \frac{20}{53} + \frac{33}{53} = 1$$

hence no obvious error has been made.)

> **Problem 2.** A fair coin is tossed three times. What is the probability of obtaining two heads and one tail?

There are 8 possible ways the coin can land: (H, H, H), (H, H, T), (H, T, H), (H, T, T), (T, H, H), (T, H, T), (T, T, H) and (T, T, T) where H represents a head and T represents a tail. Of these, 3 have two heads and one tail: (H, H, T), (H, T, H) and (T, H, H)
Thus, the number of ways it can happen is 3 and the total number of outcomes is 8
The probability of an event happening

$$= \frac{\text{number of ways it can happen}}{\text{total number of outcomes}}$$

Therefore, the probability of obtaining one head and two tails $= \dfrac{3}{8}$

> **Problem 3.** A committee of three is chosen from five candidates −A, B, C, D and E. What is the probability that B will be on the committee?

There are 10 possible committees: (A, B, C), (A, B, D), (A, B, E), (A, C, D), (A, C, E), (A, D, E), (B, C, D), (B, C, E), (B, D, E) and (C, D, E).

Of these, B is included in 6: (A, B, C), (A, B, E), (A, B, E), (B, C, D), (B, C, E) and (B, D, E)

Thus, the number of ways it can happen is 6 and the total number of outcomes is 10

The probability of an event happening

$$= \frac{\text{number of ways it can happen}}{\text{total number of outcomes}}$$

Therefore, the probability of B being on the committee

$$= \frac{6}{10} = \frac{3}{5}$$

Expectation

The **expectation**, E, of an event happening is defined in general terms as the product of the probability p of an event happening and the number of attempts made, n; i.e. $E = pn$

Thus, since the probability of obtaining a 3 upwards when rolling a fair dice is 1/6, the expectation of getting a 3 upwards on four throws of the dice is

$$\frac{1}{6} \times 4, \text{i.e.} \frac{2}{3}$$

Thus **expectation is the average occurrence of an event**.

> **Problem 4.** Find the expectation of obtaining a 4 upwards with 3 throws of a fair dice

Expectation is the average occurrence of an event and is defined as the probability times the number of attempts. The probability, p, of obtaining a 4 upwards for one throw of the dice is 1/6

If 3 attempts are made, $n = 3$ and the expectation, E, is pn, i.e.

$$E = \frac{1}{6} \times 3 = \frac{1}{2} \quad \text{or} \quad \textbf{0.50}$$

Dependent events

A **dependent event** is one in which the probability of an event happening affects the probability of another event happening. Let 5 transistors be taken at random from a batch of 100 transistors for test purposes and the probability of there being a defective transistor, p_1, be determined. At some later time, let another 5 transistors be taken at random from the 95 remaining transistors in the batch and the probability of there being a defective transistor, p_2, be determined. The value of p_2 is different from p_1 since the batch size has effectively altered from 100 to 95; i.e. probability p_2 is dependent on probability p_1. Since 5 transistors are drawn and then another 5 transistors are drawn without replacing the first 5, the second random selection is said to be **without replacement**.

Independent events

An independent event is one in which the probability of an event happening does not affect the probability of another event happening. If 5 transistors are taken at random from a batch of transistors and the probability of a defective transistor, p_1, is determined and the process is repeated after the original 5 have been replaced in the batch to give p_2, then p_1 is equal to p_2. Since the 5 transistors are replaced between draws, the second selection is said to be **with replacement**.

34.2 Laws of probability

The addition law of probability

The addition law of probability is recognised by the word 'or' joining the probabilities.

If p_A is the probability of event A happening and p_B is the probability of event B happening, the probability of **event A or event B** happening is given by $p_A + p_B$ Similarly, the probability of events **A or B or C or ...N** happening is given by

$$p_A + p_B + p_C + \cdots + p_N$$

Suppose that we draw a card from a well-shuffled standard deck of cards. We want to determine the probability that the card drawn is a two or a face card. The event 'a face card is drawn' is mutually exclusive with the event 'a two is drawn,' so we will simply need to add the probabilities of these two events together.

There is a total of 12 face cards (king, queen and jack for four suites), and so the probability of drawing a face card is 12/52. There are four twos in the deck, and so the probability of drawing a two is 4/52. This means that the probability of drawing a two **or** a face card is $12/52 + 4/52 = 16/52$ or 8/26 or 4/13.

The multiplication law of probability

The multiplication law of probability is recognised by the word '**and**' joining the probabilities.

If p_A is the probability of event A happening and p_B is the probability of event B happening, the probability of **event A and event B** happening is given by $p_A \times p_B$

Similarly, the probability of events *A* **and** *B* **and** *C* **and** ...*N* happening is given by

$$p_A \times p_B \times p_C \times \cdots \times p_N$$

Here are some worked problems to demonstrate probability.

Problem 5. There are 30 buttons in total in a bag; 13 are blue and 17 are red. What is the probability of taking two red buttons in a row out of the bag without looking?

The probability of taking a red button from the bag the first time is 17/30. Note that the numerator is 17, the total number of red buttons and the denominator is 30, the total number of buttons.

If this button is not put back into the bag, then the probability of taking a red button from the bag now is 16/29. The numerator is 16, the total number of red buttons remaining in the bag and the denominator is 29, the total number of buttons remaining.

The probability of taking two red buttons in a row without looking is given by the multiplication rule of probability, i.e.

$$\textbf{probability} = \frac{17}{30} \times \frac{16}{29} = \frac{272}{870} = \frac{136}{435} = 0.313 = 31.3\%$$

Problem 6. Calculate the probabilities of selecting at random
(a) the winning horse in a race in which 10 horses are running and
(b) the winning horses in both the first and second races if there are 10 horses in each race

(a) Since only one of the ten horses can win, the probability of selecting at random the winning horse is

$$\frac{\text{number of winners}}{\text{number of horses}} \text{ i.e. } \frac{\textbf{1}}{\textbf{10}} \text{ or } \textbf{0.10}$$

(b) The probability of selecting the winning horse in the first race is $\frac{1}{10}$

The probability of selecting the winning horse in the second race is $\frac{1}{10}$

The probability of selecting the winning horses in the first **and** second race is given by the multiplication law of probability; i.e.

$$\textbf{probability} = \frac{1}{10} \times \frac{1}{10} = \frac{\textbf{1}}{\textbf{100}} \text{ or } \textbf{0.01}$$

Problem 7. The probability of a component failing in one year due to excessive temperature is 1/20, that due to excessive vibration is 1/25 and that due to excessive humidity is 1/50. Determine the probabilities that during a one-year period a component

(a) fails due to excessive temperature and excessive vibration,
(b) fails due to excessive vibration or excessive humidity,
(c) will not fail because of both excessive temperature and excessive humidity

Let p_A be the probability of failure due to excessive temperature, then

$$p_A = \frac{1}{20} \quad \text{and} \quad \overline{p_A} = \frac{19}{20}$$

(where $\overline{p_A}$ is the probability of not failing)

Let p_B be the probability of failure due to excessive vibration, then

$$p_B = \frac{1}{25} \quad \text{and} \quad \overline{p_B} = \frac{24}{25}$$

Let p_C be the probability of failure due to excessive humidity, then

$$p_C = \frac{1}{50} \quad \text{and} \quad \overline{p_C} = \frac{49}{50}$$

(a) The probability of a component failing due to excessive temperature **and** excessive vibration is given by

$$p_A \times p_B = \frac{1}{20} \times \frac{1}{25} = \frac{\textbf{1}}{\textbf{500}} \text{ or } \textbf{0.002}$$

(b) The probability of a component failing due to excessive vibration **or** excessive humidity is

$$p_B + p_C = \frac{1}{25} + \frac{1}{50} = \frac{\textbf{3}}{\textbf{50}} \text{ or } \textbf{0.06}$$

(c) The probability that a component will not fail due excessive temperature **and** will not fail due to excess humidity is

$$\overline{p_A} \times \overline{p_C} = \frac{19}{20} \times \frac{49}{50} = \frac{\textbf{931}}{\textbf{1000}} \text{ or } \textbf{0.931}$$

Problem 8. A batch of 100 capacitors contains 73 which are within the required tolerance values and 17 which are below the required tolerance values, the remainder being above the required tolerance values. Determine the probabilities that, when randomly selecting a capacitor and then a second capacitor,

(a) both are within the required tolerance values when selecting with replacement,

(b) the first one drawn is below and the second one drawn is above the required tolerance value, when selection is without replacement

(a) The probability of selecting a capacitor within the required tolerance values is 73/100. The first capacitor drawn is now replaced and a second one is drawn from the batch of 100. The probability of this capacitor being within the required tolerance values is also 73/100. Thus, the probability of selecting a capacitor within the required tolerance values for both the first **and** the second draw is

$$\frac{73}{100} \times \frac{73}{100} = \frac{5329}{10000} \quad \text{or} \quad \textbf{0.5329}$$

(b) The probability of obtaining a capacitor below the required tolerance values on the first draw is 17/100. There are now only 99 capacitors left in the batch, since the first capacitor is not replaced. The probability of drawing a capacitor above the required tolerance values on the second draw is 10/99, since there are $(100 - 73 - 17)$, i.e. 10, capacitors above the required tolerance value. Thus, the probability of randomly selecting a capacitor below the required tolerance values and subsequently randomly selecting a capacitor above the tolerance values is

$$\frac{17}{100} \times \frac{10}{99} = \frac{170}{9900} = \frac{17}{990} \quad \text{or} \quad \textbf{0.0172}$$

Now try the following Practice Exercise

Practice Exercise 169   Laws of probability (answers on page 459)

1. In a batch of 45 lamps 10 are faulty. If one lamp is drawn at random, find the probability of it being (a) faulty (b) satisfactory.

2. A box of fuses are all of the same shape and size and comprises 23 2 A fuses, 47 5 A fuses and 69 13 A fuses. Determine the probability

of selecting at random (a) a 2 A fuse (b) a 5 A fuse (c) a 13 A fuse.

3. (a) Find the probability of having a 2 upwards when throwing a fair 6-sided dice.

   (b) Find the probability of having a 5 upwards when throwing a fair 6-sided dice.

   (c) Determine the probability of having a 2 and then a 5 on two successive throws of a fair 6-sided dice.

4. There are 10 counters in a bag, 3 are red, 2 are blue and 5 are green. The contents of the bag are shaken before one is randomly chosen from the bag. What is the probability that a red counter is not chosen?

5. Determine the probability that the total score is 8 when two like dice are thrown.

6. The probability of event A happening is $\frac{3}{5}$ and the probability of event B happening is $\frac{2}{3}$. Calculate the probabilities of

   (a) both A and B happening

   (b) only event A happening, i.e. event A happening and event B not happening

   (c) only event B happening

   (d) either A, or B, or A and B happening

7. When testing 1000 soldered joints, 4 failed during a vibration test and 5 failed due to having a high resistance. Determine the probability of a joint failing due to

   (a) vibration

   (b) high resistance

   (c) vibration or high resistance

   (d) vibration and high resistance

Here are some further worked problems on probability.

Problem 9. A batch of 40 components contains 5 which are defective. A component is drawn at random from the batch and tested and then a second component is drawn. Determine the probability that neither of the components is defective when drawn (a) with replacement and (b) without replacement

(a) **With replacement**

The probability that the component selected on the first draw is satisfactory is 35/40 i.e. 7/8. The component is now replaced and a second draw is made. The probability that this component is also satisfactory is 7/8. Hence, the probability that both the first component drawn and the second component drawn are satisfactory is

$$\frac{7}{8} \times \frac{7}{8} = \frac{49}{64} \quad \text{or} \quad \textbf{0.7656}$$

(b) **Without replacement**

The probability that the first component drawn is satisfactory is 7/8. There are now only 34 satisfactory components left in the batch and the batch number is 39. Hence, the probability of drawing a satisfactory component on the second draw is 34/39. Thus, the probability that the first component drawn and the second component drawn are satisfactory i.e. neither is defective is

$$\frac{7}{8} \times \frac{34}{39} = \frac{238}{312} \quad \text{or} \quad \textbf{0.7628}$$

Problem 10. A batch of 40 components contains 5 which are defective. If a component is drawn at random from the batch and tested and then a second component is drawn at random, calculate the probability of having one defective component, both (a) with replacement and (b) without replacement

The probability of having one defective component can be achieved in two ways. If p is the probability of drawing a defective component and q is the probability of drawing a satisfactory component, then the probability of having one defective component is given by drawing a satisfactory component and then a defective component **or** by drawing a defective component and then a satisfactory one; i.e. by $q \times p + p \times q$

(a) **With replacement**

$$p = \frac{5}{40} = \frac{1}{8} \quad \text{and} \quad q = \frac{35}{40} = \frac{7}{8}$$

Hence, the probability of having one defective component is

$$\frac{1}{8} \times \frac{7}{8} + \frac{7}{8} \times \frac{1}{8}$$

i.e. $$\frac{7}{64} + \frac{7}{64} = \frac{7}{32} \quad \text{or} \quad \textbf{0.2188}$$

(b) **Without replacement**

$p_1 = \frac{1}{8}$ and $q_1 = \frac{7}{8}$ on the first of the two draws

The batch number is now 39 for the second draw, thus,

$$p_2 = \frac{5}{39} \quad \text{and} \quad q_2 = \frac{35}{39}$$

$$p_1 q_2 + q_1 p_2 = \frac{1}{8} \times \frac{35}{39} + \frac{7}{8} \times \frac{5}{39}$$

$$= \frac{35 + 35}{312}$$

$$= \frac{70}{312} \quad \text{or} \quad \textbf{0.2244}$$

Problem 11. A box contains 74 brass washers, 86 steel washers and 40 aluminium washers. Three washers are drawn at random from the box without replacement. Determine the probability that all three are steel washers

Assume, for clarity of explanation, that a washer is drawn at random, then a second, then a third (although this assumption does not affect the results obtained). The total number of washers is

$$74 + 86 + 40, \quad \text{i.e.} \quad 200$$

The probability of randomly selecting a steel washer on the first draw is 86/200. There are now 85 steel washers in a batch of 199. The probability of randomly selecting a steel washer on the second draw is 85/199. There are now 84 steel washers in a batch of 198. The probability of randomly selecting a steel washer on the third draw is 84/198. Hence, the probability of selecting a steel washer on the first draw **and** the second draw **and** the third draw is

$$\frac{86}{200} \times \frac{85}{199} \times \frac{84}{198} = \frac{614040}{7880400} = \textbf{0.0779}$$

Problem 12. For the box of washers given in Problem 11 above, determine the probability that there are no aluminium washers drawn when three washers are drawn at random from the box without replacement

The probability of not drawing an aluminium washer on the first draw is $1 - \left(\dfrac{40}{200}\right)$ i.e. 160/200. There are now 199 washers in the batch of which 159 are not made of aluminium. Hence, the probability of not drawing an aluminium washer on the second draw is 159/199. Similarly, the probability of not drawing an aluminium washer on the third draw is 158/198. Hence the probability of not drawing an aluminium washer on the first **and** second **and** third draws is

$$\frac{160}{200} \times \frac{159}{199} \times \frac{158}{198} = \frac{4019520}{7880400} = \mathbf{0.5101}$$

> **Problem 13.** For the box of washers in Problem 11 above, find the probability that there are two brass washers and either a steel or an aluminium washer when three are drawn at random, without replacement

Two brass washers (A) and one steel washer (B) can be obtained in any of the following ways.

1st draw	2nd draw	3rd draw
A	A	B
A	B	A
B	A	A

Two brass washers and one aluminium washer (C) can also be obtained in any of the following ways.

1st draw	2nd draw	3rd draw
A	A	C
A	C	A
C	A	A

Thus, there are six possible ways of achieving the combinations specified. If A represents a brass washer, B a steel washer and C an aluminium washer, the combinations and their probabilities are as shown.

First	Second	Third	Probability		
A	A	B	$\dfrac{74}{200}$	$\times \dfrac{73}{199} \times \dfrac{86}{198}$	$= 0.0590$
A	B	A	$\dfrac{74}{200}$	$\times \dfrac{86}{199} \times \dfrac{73}{198}$	$= 0.0590$
B	A	A	$\dfrac{86}{200}$	$\times \dfrac{74}{199} \times \dfrac{73}{198}$	$= 0.0590$
A	A	C	$\dfrac{74}{200}$	$\times \dfrac{73}{199} \times \dfrac{40}{198}$	$= 0.0274$
A	C	A	$\dfrac{74}{200}$	$\times \dfrac{40}{199} \times \dfrac{73}{198}$	$= 0.0274$
C	A	A	$\dfrac{40}{200}$	$\times \dfrac{74}{199} \times \dfrac{73}{198}$	$= 0.0274$

The probability of having the first combination **or** the second **or** the third, and so on, is given by the sum of the probabilities; i.e. by $3 \times 0.0590 + 3 \times 0.0274$, i.e. **0.2592**

Now try the following Practice Exercise

Practice Exercise 170 Laws of probability (answers on page 459)

> 1. The probability that component A will operate satisfactorily for 5 years is 0.8 and that B will operate satisfactorily over that same period of time is 0.75. Find the probabilities that in a 5 year period
>   (a) both components will operate satisfactorily
>   (b) only component A will operate satisfactorily
>   (c) only component B will operate satisfactorily

> 2. In a particular street, 80% of the houses have landline telephones. If two houses selected at random are visited, calculate the probabilities that

(a) they both have a telephone

(b) one has a telephone but the other does not

3. Veroboard pins are packed in packets of 20 by a machine. In a thousand packets, 40 have less than 20 pins. Find the probability that if 2 packets are chosen at random, one will contain less than 20 pins and the other will contain 20 pins or more.

4. A batch of 1 kW fire elements contains 16 which are within a power tolerance and 4 which are not. If 3 elements are selected at random from the batch, calculate the probabilities that

(a) all three are within the power tolerance

(b) two are within but one is not within the power tolerance

5. An amplifier is made up of three transistors, A, B and C. The probabilities of A, B or C being defective are 1/20, 1/25 and 1/50, respectively. Calculate the percentage of amplifiers produced

(a) which work satisfactorily

(b) which have just one defective transistor

6. A box contains 14 40 W lamps, 28 60 W lamps and 58 25 W lamps, all the lamps being of the same shape and size. Three lamps are drawn at random from the box, first one, then a second, then a third. Determine the probabilities of

(a) getting one 25 W, one 40 W and one 60 W lamp with replacement

(b) getting one 25 W, one 40 W and one 60 W lamp without replacement

(c) getting either one 25 W and two 40 W or one 60 W and two 40 W lamps with replacement

Practice Exercise 171 Multiple-choice questions on probability (answers on page 459)

Each question has only one correct answer

1. In a box of 50 nails, 4 are faulty. One nail is taken from the box at random. The probability that the nail is faulty is:

(a) $\dfrac{1}{25}$ (b) $\dfrac{2}{27}$ (c) $\dfrac{2}{25}$ (d) $\dfrac{23}{25}$

2. Using a standard deck of 52 cards, the percentage probability of drawing a three or a king or a queen is:

(a) 5.77% (b) 1.18%
(c) 17.31% (d) 23.08%

3. The percentage probability of selecting at random the winning horses in all of three consecutive races if there are 8 horses in each race is:

(a) 12.50% (b) 1.563%
(c) 0.195% (d) 37.5%

4. In a bag are 12 red balls and 8 blue balls. If one ball is drawn from the bag, then a second ball drawn without returning the first ball to the bag, the percentage probability of drawing two blue balls is:

(a) 76.84% (b) 14.74%
(c) 80% (d) 16%

Questions 5 to 7 relate to the following information.
The probability of a component failing in one year due to excessive temperature is $\dfrac{1}{16}$, due to excessive vibration is $\dfrac{1}{20}$ and due to excessive humidity is $\dfrac{1}{40}$

5. The probability that a component fails due to excessive temperature and excessive vibration is:

(a) $\dfrac{285}{320}$ (b) $\dfrac{1}{320}$ (c) $\dfrac{9}{80}$ (d) $\dfrac{1}{800}$

6. The probability that a component fails due to excessive vibration or excessive humidity is:

(a) 0.00125 (b) 0.00257
(c) 0.1125 (d) 0.0750

7. The probability that a component will not fail because of both excessive temperature and excessive humidity is:

(a) 0.914 (b) 1.913
(c) 0.00156 (d) 0.0875

Questions 8 to 10 relate to the following information:

A box contains 35 brass washers, 40 steel washers and 25 aluminium washers.

3 washers are drawn at random from the box without replacement.

8. The probability that all three are steel washers is:

   (a) 0.0611   (b) 1.200
   (c) 0.0640   (d) 1.182

9. The probability that there are no aluminium washers is:

   (a) 2.250   (b) 0.418   (c) 0.014   (d) 0.422

10. The probability that there are two brass washers and either a steel or an aluminium washer is:

    (a) 0.071   (b) 0.687   (c) 0.239   (d) 0.343

For fully worked solutions to each of the problems in Practice Exercises 169 and 170 in this chapter, go to the website:
www.routledge.com/cw/bird

This assignment covers the material contained in Chapters 32 to 34. *The marks available are shown in brackets at the end of each question.*

1. A company produces five products in the following proportions:

Product A	24
Product B	6
Product C	15
Product D	9
Product E	18

Draw (a) a horizontal bar chart and (b) a pie diagram to represent these data visually. (9)

2. State whether the data obtained on the following topics are likely to be discrete or continuous.
   (a) the number of books in a library
   (b) the speed of a car
   (c) the time to failure of a light bulb (3)

3. Draw a histogram, frequency polygon and ogive for the data given below which refer to the diameter of 50 components produced by a machine.

Class intervals	Frequency
1.30–1.32 mm	4
1.33–1.35 mm	7
1.36–1.38 mm	10
1.39–1.41 mm	12
1.42–1.44 mm	8
1.45–1.47 mm	5
1.48–1.50 mm	4

(16)

4. Determine the mean, median and modal values for the following lengths given in metres:
   28, 20, 44, 30, 32, 30, 28, 34, 26, 28 (6)

5. The length in millimetres of 100 bolts is as shown below.

50–56	6
57–63	16
64–70	22
71–77	30
78–84	19
85–91	7

Determine for the sample
(a) the mean value
(b) the standard deviation, correct to 4 significant figures (10)

6. The number of faulty components in a factory in a 12 week period is

   14 12 16 15 10 13 15 11 16 19 17 19

   Determine the median and the first and third quartile values. (7)

7. Determine the probability of winning a prize in a lottery by buying 10 tickets when there are 10 prizes and a total of 5000 tickets sold. (4)

8. A sample of 50 resistors contains 44 which are within the required tolerance value, 4 which are below and the remainder which are above. Determine the probability of selecting from the sample a resistor which is

   (a) below the required tolerance
   (b) above the required tolerance

   Now two resistors are selected at random from the sample. Determine the probability, correct to 3 decimal places, that neither resistor is defective when drawn

   (c) with replacement
   (d) without replacement
   (e) If a resistor is drawn at random from the batch and tested and then a second resistor is drawn from those left, calculate the probability of having one defective component when selection is without replacement. (15)

Chapter 35

Introduction to differentiation

Why it is important to understand: **Introduction to differentiation**

There are many practical situations engineers have to analyse which involve quantities that are varying. Typical examples include the stress in a loaded beam, the temperature of an industrial chemical, the rate at which the speed of a vehicle is increasing or decreasing, the current in an electrical circuit or the torque on a turbine blade. Differential calculus, or differentiation, is a mathematical technique for analysing the way in which functions change. A good knowledge of algebra, in particular, laws of indices, is essential. This chapter explains how to differentiate the five most common functions, providing an important base for future studies.

At the end of this chapter you should be able to:

- state that calculus comprises two parts − differential and integral calculus
- understand functional notation
- describe the gradient of a curve and limiting value
- differentiate $y = ax^n$ by the general rule
- differentiate sine and cosine functions
- differentiate exponential and logarithmic functions

35.1 Introduction to calculus

Calculus is a branch of mathematics involving or leading to calculations dealing with continuously varying functions such as velocity and acceleration, rates of change and maximum and minimum values of curves.

Calculus has widespread applications in science and engineering and is used to solve complicated problems for which algebra alone is insufficient.

Calculus is a subject that falls into two parts:

(a) **differential calculus** (or **differentiation**),

(b) **integral calculus** (or **integration**).

This chapter provides an introduction to differentiation and applies differentiation to rates of change. Chapter 36 introduces integration and applies it to determine areas under curves.

Further applications of differentiation and integration are explored in *Bird's Engineering Mathematics* 9th Edition (2021).

35.2 Functional notation

In an equation such as $y = 3x^2 + 2x - 5, y$ is said to be a function of x and may be written as $y = f(x)$

An equation written in the form $f(x) = 3x^2 + 2x - 5$ is termed **functional notation**. The value of $f(x)$ when $x = 0$ is denoted by $f(0)$, and the value of $f(x)$ when $x = 2$ is denoted by $f(2)$, and so on. Thus, when $f(x) = 3x^2 + 2x - 5$,

$$f(0) = 3(0)^2 + 2(0) - 5 = -5$$

and $f(2) = 3(2)^2 + 2(2) - 5 = 11$, and so on.

Problem 1. If $f(x) = 4x^2 - 3x + 2$, find $f(0), f(3), f(-1)$ and $f(3) - f(-1)$

$$f(x) = 4x^2 - 3x + 2$$

$$f(0) = 4(0)^2 - 3(0) + 2 = \mathbf{2}$$

$$f(3) = 4(3)^2 - 3(3) + 2 = 36 - 9 + 2 = \mathbf{29}$$

$$f(-1) = 4(-1)^2 - 3(-1) + 2 = 4 + 3 + 2 = \mathbf{9}$$

$$f(3) - f(-1) = 29 - 9 = \mathbf{20}$$

Problem 2. Given that $f(x) = 5x^2 + x - 7$, determine (a) $f(-2)$ (b) $f(2) \div f(1)$

(a) $f(-2) = 5(-2)^2 + (-2) - 7 = 20 - 2 - 7 = \mathbf{11}$

(b) $f(2) = 5(2)^2 + 2 - 7 = 15$

$\qquad f(1) = 5(1)^2 + 1 - 7 = -1$

$$f(2) \div f(1) = \frac{15}{-1} = \mathbf{-15}$$

Now try the following Practice Exercise

Practice Exercise 172   Functional notation (answers on page 459)

1. If $f(x) = 6x^2 - 2x + 1$, find $f(0), f(1), f(2), f(-1)$ and $f(-3)$

2. If $f(x) = 2x^2 + 5x - 7$, find $f(1), f(2), f(-1), f(2) - f(-1)$

3. Given $f(x) = 3x^3 + 2x^2 - 3x + 2$, prove that $f(1) = \frac{1}{7}f(2)$

35.3 The gradient of a curve

If a tangent is drawn at a point P on a curve, the gradient of this tangent is said to be the **gradient of the curve** at P. In Fig. 35.1, the gradient of the curve at P is equal to the gradient of the tangent PQ.

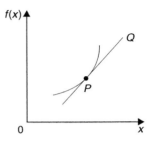

Figure 35.1

For the curve shown in Fig. 35.2, let the points A and B have co-ordinates (x_1, y_1) and (x_2, y_2), respectively. In functional notation, $y_1 = f(x_1)$ and $y_2 = f(x_2)$, as shown.

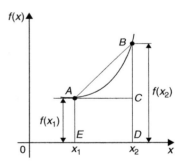

Figure 35.2

The gradient of the chord AB

$$= \frac{BC}{AC} = \frac{BD - CD}{ED} = \frac{f(x_2) - f(x_1)}{(x_2 - x_1)}$$

For the curve $f(x) = x^2$ shown in Fig. 35.3,

(a) the gradient of chord AB

$$= \frac{f(3) - f(1)}{3 - 1} = \frac{9 - 1}{2} = 4$$

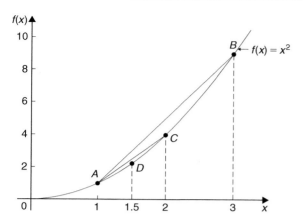

Figure 35.3

(b)   the gradient of chord AC

$$= \frac{f(2) - f(1)}{2 - 1} = \frac{4 - 1}{1} = 3$$

(c)   the gradient of chord AD

$$= \frac{f(1.5) - f(1)}{1.5 - 1} = \frac{2.25 - 1}{0.5} = 2.5$$

(d)   if E is the point on the curve $(1.1, f(1.1))$ then the gradient of chord AE

$$= \frac{f(1.1) - f(1)}{1.1 - 1} = \frac{1.21 - 1}{0.1} = 2.1$$

(e)   if F is the point on the curve $(1.01, f(1.01))$ then the gradient of chord AF

$$= \frac{f(1.01) - f(1)}{1.01 - 1} = \frac{1.0201 - 1}{0.01} = 2.01$$

Thus, as point B moves closer and closer to point A, the gradient of the chord approaches nearer and nearer to the value 2. This is called the **limiting value** of the gradient of the chord AB and when B coincides with A the chord becomes the tangent to the curve.

Now try the following Practice Exercise

1.   Plot the curve $f(x) = 4x^2 - 1$ for values of x from $x = -1$ to $x = +4$. Label the co-ordinates $(3, f(3))$ and $(1, f(1))$ as J and K,

respectively. Join points J and K to form the chord JK. Determine the gradient of chord JK. By moving J nearer and nearer to K, determine the gradient of the tangent of the curve at K.

35.4   Differentiation from first principles

In Fig. 35.4, A and B are two points very close together on a curve, δx (delta x) and δy (delta y) representing small increments in the x and y directions, respectively.

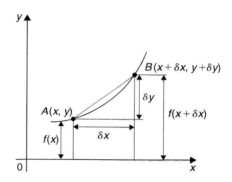

Figure 35.4

Gradient of chord $AB = \dfrac{\delta y}{\delta x}$

however, $\delta y = f(x + \delta x) - f(x)$

Hence, $\dfrac{\delta y}{\delta x} = \dfrac{f(x + \delta x) - f(x)}{\delta x}$

As δx approaches zero, $\dfrac{\delta y}{\delta x}$ approaches a limiting value and the gradient of the chord approaches the gradient of the tangent at A.

When determining the gradient of a tangent to a curve there **are two notations** used. The gradient of the curve at A in Fig. 35.4 can either be written as

$$\underset{\delta x \to 0}{\text{limit}} \frac{\delta y}{\delta x} \quad \text{or} \quad \underset{\delta x \to 0}{\text{limit}} \left\{ \frac{f(x + \delta x) - f(x)}{\delta x} \right\}$$

In **Leibniz*** notation, $\dfrac{dy}{dx} = \underset{\delta x \to 0}{\text{limit}} \dfrac{\delta y}{\delta x}$

In **functional notation,**

$$f'(x) = \underset{\delta x \to 0}{\text{limit}} \left\{ \frac{f(x + \delta x) - f(x)}{\delta x} \right\}$$

$\dfrac{dy}{dx}$ is the same as $f'(x)$ and is called the **differential coefficient** or the **derivative**. The process of finding the differential coefficient is called **differentiation**. Summarising, the differential coefficient,

$$\frac{dy}{dx} = f'(x) = \underset{\delta x \to 0}{\text{limit}} \frac{\delta y}{\delta x} = \underset{\delta x \to 0}{\text{limit}} \left\{ \frac{f(x + \delta x) - f(x)}{\delta x} \right\}$$

> **Problem 3.**   Differentiate from first principles $f(x) = x^2$

To 'differentiate from first principles' means 'to find $f'(x)$' using the expression

$$f'(x) = \underset{\delta x \to 0}{\text{limit}} \left\{ \frac{f(x + \delta x) - f(x)}{\delta x} \right\}$$

*Who was Leibniz? – **Gottfried Wilhelm Leibniz** (sometimes von Leibniz) (1 July 1646 – 14 November 1716) was a German mathematician and philosopher. Leibniz developed infinitesimal calculus and invented the Leibniz wheel. To find out more go to **www.routledge.com/cw/bird**

$f(x) = x^2$ and substituting $(x + \delta x)$ for x gives $f(x + \delta x) = (x + \delta x)^2 = x^2 + 2x\delta x + \delta x^2$, hence,

$$f'(x) = \underset{\delta x \to 0}{\text{limit}} \left\{ \frac{(x^2 + 2x\delta x + \delta x^2) - (x^2)}{\delta x} \right\}$$

$$= \underset{\delta x \to 0}{\text{limit}} \left\{ \frac{2x\delta x + \delta x^2}{\delta x} \right\} = \underset{\delta x \to 0}{\text{limit}} \{2x + \delta x\}$$

As $\delta x \to 0, \{2x + \delta x\} \to \{2x + 0\}$.

Thus, $f'(x) = 2x$ i.e. **the differential coefficient of x^2 is $2x$**

This means that the general expression for the gradient of the curve $f(x) = x^2$ is $2x$. If the gradient is required at, say, $x = 3$, then gradient $= 2(3) = $ **6**

Differentiation from first principles can be a lengthy process and we do not want to have to go through this procedure every time we want to differentiate a function. In reality we do not have to because from the above procedure has evolved a set general rule, which we consider in the following section.

35.5   Differentiation of $y = ax^n$ by the general rule

From differentiation by first principles, a general rule for differentiating ax^n emerges where a and n are any constants. This rule is

$$\text{if } y = ax^n \text{ then } \frac{dy}{dx} = anx^{n-1}$$

or   if $f(x) = ax^n$ then $f'(x) = anx^{n-1}$

When differentiating, results can be expressed in a number of ways. For example,

(a)   if $y = 3x^2$ then $\dfrac{dy}{dx} = 6x$

(b)   if $f(x) = 3x^2$ then $f'(x) = 6x$

(c)   the differential coefficient of $3x^2$ is $6x$

(d)   the derivative of $3x^2$ is $6x$

(e) $\dfrac{d}{dx}(3x^2) = 6x$

Revision of some laws of indices

$\dfrac{1}{x^a} = x^{-a}$ For example, $\dfrac{1}{x^2} = x^{-2}$ and $x^{-5} = \dfrac{1}{x^5}$

$\sqrt{x} = x^{\frac{1}{2}}$ For example, $\sqrt{5} = 5^{\frac{1}{2}}$ and
$16^{\frac{1}{2}} = \sqrt{16} = \pm 4$ and $\dfrac{1}{\sqrt{x}} = \dfrac{1}{x^{\frac{1}{2}}} = x^{-\frac{1}{2}}$

$\sqrt[a]{x^b} = x^{\frac{b}{a}}$ For example, $\sqrt[3]{x^5} = x^{\frac{5}{3}}$ and $x^{\frac{4}{3}} = \sqrt[3]{x^4}$

and $\dfrac{1}{\sqrt[3]{x^7}} = \dfrac{1}{x^{\frac{7}{3}}} = x^{-\frac{7}{3}}$

$x^0 = 1$ For example, $7^0 = 1$ and $43.5^0 = 1$

Here are some worked problems to demonstrate the general rule for differentiating $y = ax^n$

> **Problem 4.** Differentiate the following with respect to x: $y = 4x^7$

Comparing $y = 4x^7$ with $y = ax^n$ shows that $a = 4$ and $n = 7$. Using the general rule,

$$\frac{dy}{dx} = anx^{n-1} = (4)(7)x^{7-1} = \mathbf{28x^6}$$

> **Problem 5.** If $y = 5x^4 - 3x^3 - 2x^2 + 7$ write down the expression for $\dfrac{dy}{dx}$

If $y = ax^n$ then $\dfrac{dy}{dx} = anx^{n-1}$ and when the values of a and n are small we can get used to writing down the expression for $\dfrac{dy}{dx}$ straightaway without showing working.

So, if $y = 5x^4 - 3x^3 - 2x^2 + 7$

then $\dfrac{dy}{dx} = \mathbf{20x^3 - 9x^2 - 4x}$

> **Problem 6.** Differentiate the following with respect to x: $y = \dfrac{3}{x^2}$

$y = \dfrac{3}{x^2} = 3x^{-2}$, hence $a = 3$ and $n = -2$ in the general rule.

$$\frac{dy}{dx} = anx^{n-1} = (3)(-2)x^{-2-1} = -6x^{-3} = -\frac{\mathbf{6}}{\mathbf{x^3}}$$

> **Problem 7.** Differentiate the following with respect to x: $y = 5\sqrt{x}$

$y = 5\sqrt{x} = 5x^{\frac{1}{2}}$, hence $a = 5$ and $n = \dfrac{1}{2}$ in the general rule.

$$\frac{dy}{dx} = anx^{n-1} = (5)\left(\frac{1}{2}\right)x^{\frac{1}{2}-1}$$

$$= \frac{5}{2}x^{-\frac{1}{2}} = \frac{5}{2x^{\frac{1}{2}}} = \frac{\mathbf{5}}{\mathbf{2\sqrt{x}}}$$

> **Problem 8.** Differentiate $y = 4$

$y = 4$ may be written as $y = 4x^0$; i.e. in the general rule $a = 4$ and $n = 0$. Hence,

$$\frac{dy}{dx} = (4)(0)x^{0-1} = \mathbf{0}$$

The equation $y = 4$ represents a **straight horizontal line** and the gradient of a horizontal line is zero, hence the result could have been determined on inspection. In general, **the differential coefficient of a constant is always zero.**

> **Problem 9.** Differentiate $y = 7x$

Since $y = 7x$, i.e. $y = 7x^1$, in the general rule $a = 7$ and $n = 1$. Hence,

$$\frac{dy}{dx} = (7)(1)x^{1-1} = 7x^0 = \mathbf{7}\qquad \text{since } x^0 = 1$$

The gradient of the line $y = 7x$ is 7 (from $y = mx + c$), hence the result could have been obtained by inspection. In general, **the differential coefficient of kx, where k is a constant, is always k.**

> **Problem 10.** Find the differential coefficient of $y = \dfrac{2}{3}x^4 - \dfrac{4}{x^3} + 9$

$$y = \frac{2}{3}x^4 - \frac{4}{x^3} + 9$$

i.e. $$y = \frac{2}{3}x^4 - 4x^{-3} + 9$$

$$\frac{dy}{dx} = \left(\frac{2}{3}\right)(4)x^{4-1} - (4)(-3)x^{-3-1} + 0$$

$$= \frac{8}{3}x^3 + 12x^{-4}$$

i.e. $$\frac{dy}{dx} = \frac{\mathbf{8}}{\mathbf{3}}\mathbf{x^3} + \frac{\mathbf{12}}{\mathbf{x^4}}$$

> **Problem 11.** If $f(t) = 4t + \dfrac{1}{\sqrt{t^3}}$ find $f'(t)$

$$f(t) = 4t + \frac{1}{\sqrt{t^3}} = 4t + \frac{1}{t^{\frac{3}{2}}} = 4t^1 + t^{-\frac{3}{2}}$$

Hence, $$f'(t) = (4)(1)t^{1-1} + \left(-\frac{3}{2}\right)t^{-\frac{3}{2}-1}$$

$$= 4t^0 - \frac{3}{2}t^{-\frac{5}{2}}$$

i.e. $$f'(t) = 4 - \frac{3}{2t^{\frac{5}{2}}} = 4 - \frac{3}{2\sqrt{t^5}}$$

Problem 12. Determine $\frac{dy}{dx}$ given $y = \frac{3x^2 - 5x}{2x}$

$$y = \frac{3x^2 - 5x}{2x} = \frac{3x^2}{2x} - \frac{5x}{2x} = \frac{3}{2}x - \frac{5}{2}$$

Hence, $\frac{dy}{dx} = \frac{3}{2}$ or **1.5**

Problem 13. Differentiate $y = \frac{(x+2)^2}{x}$ with respect to x

$$y = \frac{(x+2)^2}{x} = \frac{x^2 + 4x + 4}{x} = \frac{x^2}{x} + \frac{4x}{x} + \frac{4}{x}$$

i.e. $y = x^1 + 4 + 4x^{-1}$

Hence, $\frac{dy}{dx} = 1x^{1-1} + 0 + (4)(-1)x^{-1-1}$

$$= x^0 - 4x^{-2} = 1 - \frac{4}{x^2} \qquad \text{(since } x^0 = 1\text{)}$$

Problem 14. Find the gradient of the curve $y = 2x^2 - \frac{3}{x}$ at $x = 2$

$$y = 2x^2 - \frac{3}{x} = 2x^3 - 3x^{-1}$$

$$\text{Gradient} = \frac{dy}{dx} = (2)(2)x^{2-1} - (3)(-1)x^{-1-1}$$

$$= 4x + 3x^{-2}$$

$$= 4x + \frac{3}{x^2}$$

When $x = 2$, gradient $= 4x + \frac{3}{x^2} = 4(2) + \frac{3}{(2)^2}$

$$= 8 + \frac{3}{4} = \textbf{8.75}$$

Problem 15. Find the gradient of the curve $y = 3x^4 - 2x^2 + 5x - 2$ at the points $(0, -2)$ and $(1, 4)$

The gradient of a curve at a given point is given by the corresponding value of the derivative.

Thus, since $y = 3x^4 - 2x^2 + 5x - 2$,

the **gradient** $= \frac{dy}{dx} = 12x^3 - 4x + 5$

At the point $(0, -2)$, $x = 0$, thus

the **gradient** $= 12(0)^3 - 4(0) + 5 = \textbf{5}$

At the point $(1, 4)$, $x = 1$, thus

the **gradient** $= 12(1)^3 - 4(1) + 5 = \textbf{13}$

Now try the following Practice Exercise

Practice Exercise 174   Differentiation of $y = ax^n$ by the general rule (answers on page 459)

In Problems 1 to 20, determine the differential coefficients with respect to the variable.

1. $y = 7x^4$ 2. $y = 2x + 1$

3. $y = x^2 - x$ 4. $y = 2x^3 - 5x + 6$

5. $y = \frac{1}{x}$ 6. $y = 12$

7. $y = x - \frac{1}{x^2}$

8. $y = 3x^5 - 2x^4 + 5x^3 + x^2 - 1$

9. $y = \frac{2}{x^3}$ 10. $y = 4x(1 - x)$

11. $y = \sqrt{x}$ 12. $y = \sqrt{t^3}$

13. $y = 6 + \frac{1}{x^3}$ 14. $y = 3x - \frac{1}{\sqrt{x}} + \frac{1}{x}$

15. $y = (x + 1)^2$ 16. $y = x + 3\sqrt{x}$

17. $y = (1 - x)^2$ 18. $y = \frac{5}{x^2} - \frac{1}{\sqrt{x^7}} + 2$

19. $y = 3(t - 2)^2$ 20. $y = \frac{(x+2)^2}{x}$

21. Find the gradient of the following curves at the given points.

   (a) $y = 3x^2$ at $x = 1$

   (b) $y = \sqrt{x}$ at $x = 9$

(c) $y = x^3 + 3x - 7$ at $x = 0$

(d) $y = \dfrac{1}{\sqrt{x}}$ at $x = 4$

(e) $y = \dfrac{1}{x}$ at $x = 2$

(f) $y = (2x + 3)(x - 1)$ at $x = -2$

22.   Differentiate $f(x) = 6x^2 - 3x + 5$ and find the gradient of the curve at

     (a) $x = -1$ (b) $x = 2$

23.   Find the differential coefficient of $y = 2x^3 + 3x^2 - 4x - 1$ and determine the gradient of the curve at $x = 2$

24.   Determine the derivative of $y = -2x^3 + 4x + 7$ and determine the gradient of the curve at $x = -1.5$

35.6   Differentiation of sine and cosine functions

Fig. 35.5(a) shows a graph of $y = \sin x$. The gradient is continually changing as the curve moves from 0 to A to B to C to D. The gradient, given by $\dfrac{dy}{dx}$, may be plotted in a corresponding position below $y = \sin x$, as shown in Fig. 35.5(b).

At 0, the gradient is positive and is at its steepest. Hence, $0'$ is a maximum positive value. Between 0 and A the gradient is positive but is decreasing in value until at A the gradient is zero, shown as A'. Between A and B the gradient is negative but is increasing in value until at B the gradient is at its steepest. Hence B' is a maximum negative value.

If the gradient of $y = \sin x$ is further investigated between B and C and C and D then the resulting graph of $\dfrac{dy}{dx}$ is seen to be a **cosine wave**.

Hence the rate of change of $\sin x$ is $\cos x$, i.e.

$$\textbf{if } y = \sin x \textbf{ then } \frac{dy}{dx} = \cos x$$

It may also be shown that

$$\textbf{if } y = \sin ax, \; \frac{dy}{dx} = a \cos ax \qquad (1)$$

(where a is a constant)

$$\text{and } \textbf{if } y = \sin(ax + \alpha), \; \frac{dy}{dx} = a \cos(ax + \alpha) \qquad (2)$$

(where a and α are constants).

If a similar exercise is followed for $y = \cos x$ then the graphs of Fig. 35.6 result, showing $\dfrac{dy}{dx}$ to be a graph of $\sin x$ but displaced by π radians.

If each point on the curve $y = \sin x$ (as shown in Fig. 35.5(a)) were to be made negative (i.e. $+\dfrac{\pi}{2}$ made $-\dfrac{\pi}{2}$, $-\dfrac{3\pi}{2}$ made $+\dfrac{3\pi}{2}$, and so on) then the graph shown in Fig. 35.6(b) would result. This latter graph therefore represents the curve of $-\sin x$

Figure 35.5

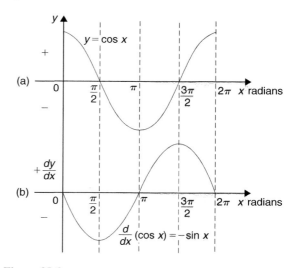

Figure 35.6

Thus,

$$\textbf{if } y = \cos x, \frac{dy}{dx} = -\sin x$$

It may also be shown that

$$\textbf{if } y = \cos ax, \frac{dy}{dx} = -a\sin ax \qquad (3)$$

(where a is a constant)

and $\textbf{if } y = \cos(ax + \alpha), \frac{dy}{dx} = -a\sin(ax + \alpha)$ (4)

(where a and α are constants).

Problem 16. Find the differential coefficient of $y = 7\sin 2x - 3\cos 4x$

$$\frac{dy}{dx} = (7)(2\cos 2x) - (3)(-4\sin 4x)$$

from equations (1) and (3)

$$= \textbf{14}\cos 2x + \textbf{12}\sin 4x$$

Problem 17. Differentiate the following with respect to the variable (a) $y = 2\sin 5\theta$
(b) $f(t) = 3\cos 2t$

(a) $y = 2\sin 5\theta$

$$\frac{dy}{d\theta} = (2)(5\cos 5\theta) = \textbf{10}\cos 5\theta$$

from equation (1)

(b) $f(t) = 3\cos 2t$

$$f'(t) = (3)(-2\sin 2t) = -\textbf{6}\sin 2t$$

from equation (3)

Problem 18. Differentiate the following with respect to the variable
(a) $f(\theta) = 5\sin(100\pi\theta - 0.40)$
(b) $f(t) = 2\cos(5t + 0.20)$

(a) If $f(\theta) = 5\sin(100\pi\theta - 0.40)$

$$f'(\theta) = 5[100\pi\cos(100\pi\theta - 0.40)]$$

from equation (2), where $a = 100\pi$

$$= \textbf{500}\pi\cos(100\pi\theta - 0.40)$$

(b) If $f(t) = 2\cos(5t + 0.20)$

$$f'(t) = 2[-5\sin(5t + 0.20)]$$

from equation (4), where $a = 5$

$$= -\textbf{10}\sin(5t + 0.20)$$

Problem 19. An alternating voltage is given by $v = 100\sin 200t$ volts, where t is the time in seconds. Calculate the rate of change of voltage when (a) $t = 0.005$ s and (b) $t = 0.01$ s

$v = 100\sin 200t$ volts. The rate of change of v is given by $\dfrac{dv}{dt}$

$$\frac{dv}{dt} = (100)(200\cos 200t) = 20000\cos 200t$$

(a) When $t = 0.005$ s,
$$\frac{dv}{dt} = 20000\cos(200)(0.005) = 20000\cos 1$$
$\cos 1$ means 'the cosine of 1 radian' (make sure your calculator is on radians, not degrees). Hence,

$$\frac{dv}{dt} = \textbf{10806 volts per second}$$

(b) When $t = 0.01$ s,
$$\frac{dv}{dt} = 20000\cos(200)(0.01) = 20000\cos 2$$
Hence,

$$\frac{dv}{dt} = -\textbf{8323 volts per second}$$

Now try the following Practice Exercise

Practice Exercise 175 Differentiation of sine and cosine functions (answers on page 460)

1. Differentiate with respect to x: (a) $y = 4\sin 3x$ (b) $y = 2\cos 6x$

2. Given $f(\theta) = 2\sin 3\theta - 5\cos 2\theta$, find $f'(\theta)$

3. Find the gradient of the curve $y = 2\cos\dfrac{1}{2}x$ at $x = \dfrac{\pi}{2}$

4. Determine the gradient of the curve $y = 3\sin 2x$ at $x = \dfrac{\pi}{3}$

5. An alternating current is given by $i = 5\sin 100t$ amperes, where t is the time in seconds. Determine the rate of change of current $\left(\text{i.e. } \dfrac{di}{dt}\right)$ when $t = 0.01$ seconds.

6. $v = 50\sin 40t$ volts represents an alternating voltage, v, where t is the time in seconds. At

a time of 20×10^{-3} seconds, find the rate of change of voltage $\left(\text{i.e. } \dfrac{dv}{dt}\right)$

7. If $f(t) = 3\sin(4t + 0.12) - 2\cos(3t - 0.72)$, determine $f'(t)$

35.7   Differentiation of e^{ax} and $\ln ax$

A graph of $y = e^x$ is shown in Fig. 35.7(a). The gradient of the curve at any point is given by $\dfrac{dy}{dx}$ and is continually changing. By drawing tangents to the curve at many points on the curve and measuring the gradient of the tangents, values of $\dfrac{dy}{dx}$ for corresponding values of x may be obtained. These values are shown graphically in Fig. 35.7(b).

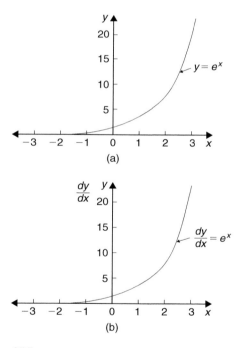

Figure 35.7

The graph of $\dfrac{dy}{dx}$ against x is identical to the original graph of $y = e^x$. It follows that

$$\text{if } \boldsymbol{y = e^x, \text{ then }} \dfrac{dy}{dx} = e^x$$

It may also be shown that

$$\text{if } \boldsymbol{y = e^{ax}, \text{ then }} \dfrac{dy}{dx} = ae^{ax}$$

Therefore,

$$\text{if } y = 2e^{6x}, \text{then } \dfrac{dy}{dx} = (2)(6e^{6x}) = \boldsymbol{12e^{6x}}$$

A graph of $y = \ln x$ is shown in Fig. 35.8(a). The gradient of the curve at any point is given by $\dfrac{dy}{dx}$ and is continually changing. By drawing tangents to the curve at many points on the curve and measuring the gradient of the tangents, values of $\dfrac{dy}{dx}$ for corresponding values of x may be obtained. These values are shown graphically in Fig. 35.8(b).

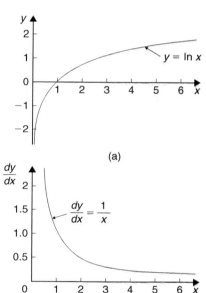

Figure 35.8

The graph of $\dfrac{dy}{dx}$ against x is the graph of $\dfrac{dy}{dx} = \dfrac{1}{x}$. It follows that

$$\text{if } \boldsymbol{y = \ln x, \text{ then }} \dfrac{dy}{dx} = \dfrac{1}{x}$$

It may also be shown that

$$\text{if } \boldsymbol{y = \ln ax, \text{ then }} \dfrac{dy}{dx} = \dfrac{1}{x}$$

(Note that, in the latter expression, the constant a does not appear in the $\dfrac{dy}{dx}$ term.) Thus,

$$\text{if } y = \ln 4x, \text{ then } \dfrac{dy}{dx} = \dfrac{1}{x}$$

Problem 20. Differentiate the following with respect to the variable (a) $y = 3e^{2x}$ (b) $f(t) = \dfrac{4}{3e^{5t}}$

(a) If $y = 3e^{2x}$ then $\dfrac{dy}{dx} = (3)(2e^{2x}) = \mathbf{6e^{2x}}$

(b) If $f(t) = \dfrac{4}{3e^{5t}} = \dfrac{4}{3}e^{-5t}$, then

$$f'(t) = \frac{4}{3}(-5e^{-5t}) = -\frac{20}{3}e^{-5t} = -\frac{\mathbf{20}}{\mathbf{3e^{5t}}}$$

Problem 21. Differentiate $y = 5\ln 3x$

If $y = 5\ln 3x$, then $\dfrac{dy}{dx} = (5)\left(\dfrac{1}{x}\right) = \dfrac{5}{x}$

Now try the following Practice Exercise

Practice Exercise 176   Differentiation of e^{ax} and $\ln ax$ (answers on page 460)

1. Differentiate with respect to x: (a) $y = 5e^{3x}$ (b) $y = \dfrac{2}{7e^{2x}}$

2. Given $f(\theta) = 5\ln 2\theta - 4\ln 3\theta$, determine $f'(\theta)$

3. If $f(t) = 4\ln t + 2$, evaluate $f'(t)$ when $t = 0.25$

4. Find the gradient of the curve $y = 2e^x - \dfrac{1}{4}\ln 2x$ at $x = \dfrac{1}{2}$ correct to 2 decimal places.

5. Evaluate $\dfrac{dy}{dx}$ when $x = 1$, given $y = 3e^{4x} - \dfrac{5}{2e^{3x}} + 8\ln 5x$. Give the answer correct to 3 significant figures.

35.8   Summary of standard derivatives

The standard derivatives used in this chapter are summarised in Table 35.1 and are true for all real values of x.

Problem 22. Find the gradient of the curve $y = 3x^2 - 7x + 2$ at the point $(1, -2)$

Table 35.1

y or $f(x)$	$\dfrac{dy}{dx}$ or $f'(x)$
ax^n	anx^{n-1}
$\sin ax$	$a\cos ax$
$\cos ax$	$-a\sin ax$
e^{ax}	ae^{ax}
$\ln ax$	$\dfrac{1}{x}$

If $y = 3x^2 - 7x + 2$, then gradient $= \dfrac{dy}{dx} = 6x - 7$
At the point $(1, -2), x = 1$,
hence **gradient** $= 6(1) - 7 = \mathbf{-1}$

Problem 23. If $y = \dfrac{3}{x^2} - 2\sin 4x + \dfrac{2}{e^x} + \ln 5x$, determine $\dfrac{dy}{dx}$

$$y = \frac{3}{x^2} - 2\sin 4x + \frac{2}{e^x} + \ln 5x$$

$$= 3x^{-2} - 2\sin 4x + 2e^{-x} + \ln 5x$$

$$\frac{dy}{dx} = 3(-2x^{-3}) - 2(4\cos 4x) + 2(-e^{-x}) + \frac{1}{x}$$

$$= -\frac{6}{x^3} - 8\cos 4x - \frac{2}{e^x} + \frac{1}{x}$$

Now try the following Practice Exercise

Practice Exercise 177   Standard derivatives (answers on page 460)

1. Find the gradient of the curve $y = 2x^4 + 3x^3 - x + 4$ at the points (a) $(0, 4)$ (b) $(1, 8)$

2. Differentiate with respect to x: $y = \dfrac{2}{x^2} + 2\ln 2x - 2(\cos 5x + 3\sin 2x) - \dfrac{2}{e^{3x}}$

35.9 Successive differentiation

When a function $y = f(x)$ is differentiated with respect to x, the differential coefficient is written as $\dfrac{dy}{dx}$ or $f'(x)$. If the expression is differentiated again, the second differential coefficient is obtained and is written as $\dfrac{d^2y}{dx^2}$ (pronounced dee two y by dee x squared) or $f''(x)$ (pronounced f double-dash x). By successive differentiation further higher derivatives such as $\dfrac{d^3y}{dx^3}$ and $\dfrac{d^4y}{dx^4}$ may be obtained. Thus,

if $y = 5x^4$, $\dfrac{dy}{dx} = 20x^3$, $\dfrac{d^2y}{dx^2} = 60x^2$, $\dfrac{d^3y}{dx^3} = 120x$,

$\dfrac{d^4y}{dx^4} = 120$ and $\dfrac{d^5y}{dx^5} = 0$

Problem 24. If $f(x) = 4x^5 - 2x^3 + x - 3$, find $f''(x)$

$$f(x) = 4x^5 - 2x^3 + x - 3$$
$$f'(x) = 20x^4 - 6x^2 + 1$$
$$\mathbf{f''(x) = 80x^3 - 12x} \quad \text{or} \quad \mathbf{4x(20x^2 - 3)}$$

Problem 25. Given $y = \dfrac{2}{3}x^3 - \dfrac{4}{x^2} + \dfrac{1}{2x} - \sqrt{x}$, determine $\dfrac{d^2y}{dx^2}$

$$y = \frac{2}{3}x^3 - \frac{4}{x^2} + \frac{1}{2x} - \sqrt{x}$$
$$= \frac{2}{3}x^3 - 4x^{-2} + \frac{1}{2}x^{-1} - x^{\frac{1}{2}}$$
$$\frac{dy}{dx} = \left(\frac{2}{3}\right)(3x^2) - 4(-2x^{-3})$$
$$+ \left(\frac{1}{2}\right)(-1x^{-2}) - \frac{1}{2}x^{-\frac{1}{2}}$$

i.e. $\dfrac{dy}{dx} = 2x^2 + 8x^{-3} - \dfrac{1}{2}x^{-2} - \dfrac{1}{2}x^{-\frac{1}{2}}$

$$\frac{d^2y}{dx^2} = 4x + (8)(-3x^{-4}) - \left(\frac{1}{2}\right)(-2x^{-3})$$
$$- \left(\frac{1}{2}\right)\left(-\frac{1}{2}x^{-\frac{3}{2}}\right)$$

$$= 4x - 24x^{-4} + 1x^{-3} + \frac{1}{4}x^{-\frac{3}{2}}$$

i.e. $\dfrac{d^2y}{dx^2} = 4x - \dfrac{24}{x^4} + \dfrac{1}{x^3} + \dfrac{1}{4\sqrt{x^3}}$

Now try the following Practice Exercise

Practice Exercise 178   Successive differentiation (answers on page 460)

1. If $y = 3x^4 + 2x^3 - 3x + 2$, find (a) $\dfrac{d^2y}{dx^2}$ (b) $\dfrac{d^3y}{dx^3}$

2. If $y = 4x^2 + \dfrac{1}{x}$ find $\dfrac{d^2y}{dx^2}$

3. (a) Given $f(t) = \dfrac{2}{5}t^2 - \dfrac{1}{t^3} + \dfrac{3}{t} - \sqrt{t} + 1$, determine $f''(t)$

    (b) Evaluate $f''(t)$ in part (a) when $t = 1$

4. If $y = 3\sin 2t + \cos t$, find $\dfrac{d^2y}{dx^2}$

5. If $f(\theta) = 2\ln 4\theta$, show that $f''(\theta) = -\dfrac{2}{\theta^2}$

35.10 Rates of change

If a quantity y depends on and varies with a quantity x then the rate of change of y with respect to x is $\dfrac{dy}{dx}$. Thus, for example, the rate of change of pressure p with height h is $\dfrac{dp}{dh}$

A rate of change with respect to time is usually just called 'the rate of change', the 'with respect to time' being assumed. Thus, for example, a rate of change of current, i, is $\dfrac{di}{dt}$ and a rate of change of temperature, θ, is $\dfrac{d\theta}{dt}$, and so on.

Here are some worked problems to demonstrate practical examples of rates of change.

Problem 26. The length L metres of a certain metal rod at temperature $t°C$ is given by $L = 1 + 0.00003t + 0.0000004t^2$. Determine the rate of change of length, in mm/°C, when the temperature is (a) $100°C$ (b) $250°C$

The rate of change of length means $\dfrac{dL}{dt}$

Since length $L = 1 + 0.00003t + 0.0000004t^2$, then

$$\frac{dL}{dt} = 0.00003 + 0.0000008t.$$

(a) **When $t = 100°C$,**

$$\frac{dL}{dt} = 0.00003 + (0.0000008)(100)$$

$$= 0.00011\,\mathrm{m/°C} = \mathbf{0.11\,mm/°C}.$$

(b) **When $t = 250°C$,**

$$\frac{dL}{dt} = 0.00003 + (0.0000008)(250)$$

$$= 0.00023\,\mathrm{m/°C} = \mathbf{0.23\,mm/°C}.$$

Problem 27. The luminous intensity I candelas of a lamp at varying voltage V is given by $I = 5 \times 10^{-4} V^2$. Determine the voltage at which the light is increasing at a rate of 0.4 candelas per volt

---

The rate of change of light with respect to voltage is given by $\dfrac{dI}{dV}$

Since $I = 5 \times 10^{-4} V^2$

$$\frac{dI}{dV} = (5 \times 10^{-4})(2V) = 10 \times 10^{-4} V = 10^{-3} V$$

When the light is increasing at 0.4 candelas per volt then

$$+0.4 = 10^{-3} V, \text{ from which}$$

$$\textbf{voltage}, V = \frac{0.4}{10^{-3}} = 0.4 \times 10^{+3} = \mathbf{400\,volts}$$

Problem 28. Newton's law of cooling is given by $\theta = \theta_0 e^{-kt}$, where the excess of temperature at zero time is $\theta_0\,°C$ and at time t seconds is $\theta\,°C$. Determine the rate of change of temperature after 50 s, given that $\theta_0 = 15°C$ and $k = -0.02$

The rate of change of temperature is $\dfrac{d\theta}{dt}$

Since $\theta = \theta_0 e^{-kt}$, then $\dfrac{d\theta}{dt} = (\theta_0)(-ke^{-kt})$

$$= -k\theta_0 e^{-kt}$$

When $\theta_0 = 15, k = -0.02$ and $t = 50$, then

$$\frac{d\theta}{dt} = -(-0.02)(15)e^{-(-0.02)(50)}$$

$$= 0.30\,e^1 = \mathbf{0.815°C/s}$$

Problem 29. The pressure p of the atmosphere at height h above ground level is given by $p = p_0 e^{-h/c}$, where p_0 is the pressure at ground level and c is a constant. Determine the rate of change of pressure with height when $p_0 = 10^5$ pascals and $c = 6.2 \times 10^4$ at 1550 metres

---

The rate of change of pressure with height is $\dfrac{dp}{dh}$

Since $p = p_0 e^{-h/c}$, then

$$\frac{dp}{dh} = (p_0)\left(-\frac{1}{c}e^{-h/c}\right) = -\frac{p_0}{c}e^{-h/c}$$

When $p_0 = 10^5, c = 6.2 \times 10^4$ and $h = 1550$, then

rate of change of pressure,

$$\frac{dp}{dh} = -\frac{10^5}{6.2 \times 10^4}e^{-(1550/6.2 \times 10^4)}$$

$$= -\frac{10}{6.2}e^{-0.025} = \mathbf{-1.573\,Pa/m}$$

Now try the following Practice Exercise

Practice Exercise 179   Rates of change (answers on page 460)

1. An alternating current, i amperes, is given by $i = 10\sin 2\pi ft$, where f is the frequency in hertz and t is the time in seconds. Determine the rate of change of current when $t = 12$ ms, given that $f = 50$ Hz.

2. The luminous intensity, I candelas, of a lamp is given by $I = 8 \times 10^{-4} V^2$, where V is the voltage. Find

   (a) the rate of change of luminous intensity with voltage when $V = 100$ volts

   (b) the voltage at which the light is increasing at a rate of 0.5 candelas per volt.

3. The voltage across the plates of a capacitor at any time t seconds is given by $v = Ve^{-t/CR}$, where V, C and R are constants. Given $V = 200\,\text{V}, C = 0.10\,\mu\text{F}$ and $R = 2\,\text{M}\Omega$, find

   (a) the initial rate of change of voltage
   (b) the rate of change of voltage after 0.2 s

4. The pressure p of the atmosphere at height h above ground level is given by $p = p_0 e^{-h/c}$, where p_0 is the pressure at ground level and c is a constant. Determine the rate of change of pressure with height when $p_0 = 1.013 \times 10^5$ pascals and $c = 6.05 \times 10^4$ at 1450 metres.

5. The height, h metres, of a liquid in a chemical storage tank is given by: $h = 100(1 - e^{-\frac{t}{40}})$ where t seconds is the time from which the tank starts to be filled. Calculate the speed in m/s at which the liquid level rises in the tank after 10 s.

35.11 Differentiation of a product

When y = uv, and u and v are both functions of x,

then
$$\frac{dy}{dx} = u\frac{dv}{dx} + v\frac{du}{dx}$$

This is known as the **product rule**.
When differentiating a product of two terms, the product rule formula must be used.

Problem 30. Find the differential coefficient of: $y = 2x\sin 3x$

$2x\sin 3x$ is a product of two terms $2x$ and $\sin 3x$
Let $u = 2x$ and $v = \sin 3x$
Using the product rule: $\dfrac{dy}{dx} = u\dfrac{dv}{dx} + v\dfrac{du}{dx}$

gives: $\dfrac{dy}{dx} = (2x)(3\cos 3x) + (\sin 3x)(2)$

i.e. $\dfrac{dy}{dx} = \textbf{6x cos 3x} + \textbf{2 sin 3x}$

$$= 2(3x\cos 3x + \sin 3x)$$

Note that the differential coefficient of a product is **not** obtained by merely differentiating each term and multiplying the two answers together. The product rule formula **must** be used when differentiating products.

Problem 31. Find the rate of change of y with respect to x given: $y = 3x\ln 2x$

The rate of change of y with respect to x is given by $\dfrac{dy}{dx}$.
$y = 3x\ln 2x$ is a product.
Let $u = 3x$ and $v = \ln 2x$

Then $\dfrac{dy}{dx} = u\dfrac{dv}{dx} + v\dfrac{du}{dx}$

$$= (3x)\left(\frac{1}{x}\right) + (\ln 2x)[3]$$

i.e. $\dfrac{dy}{dx} = \textbf{3} + \textbf{3 ln 2x}$

$$= \textbf{3(1 + ln 2x)}$$

Problem 32. Determine the rate of change of voltage, given $v = 4t\sin 2t$ volts when $t = 0.2$s

Rate of change of voltage

$$= \frac{dv}{dt} = (4t)(2\cos 2t) + (\sin 2t)(4)$$

$$= 8t\cos 2t + 4\sin 2t$$

When $t = 0.2$, $\dfrac{dv}{dt} = 8(0.2)\cos[2(0.2)] + 4\sin[2(0.2)]$

$$= 1.6\cos 0.4 + 4\sin 0.4$$

(where $\cos 0.4$ means the cosine of 0.4 radians $= 0.92106$)

Hence, $\dfrac{dv}{dt} = 1.6(0.92106) + 4(0.38942)$

$$= 1.473696 + 1.55768 = 3.0314$$

i.e. **the rate of change of voltage when t = 0.2 s is 3.03 volts/s, correct to 3 significant figures.**

Now try the following exercise

Practice Exercise 180   Further problems on differentiating products (answers on page 460)

In Problems 1 to 5, differentiate the given products with respect to the variable

1. $y = x\sin x$ 2. $y = x^2\cos 3x$

3. $y = 4x\ln 3x$ 4. $y = e^{3t}\sin 4t$

5. $y = e^{4\theta}\ln 3\theta$

6. Evaluate $\dfrac{di}{dt}$, correct to 4 significant figures, when $t = 0.1$, and $i = 10\,t\sin 2t$

7. Evaluate $\dfrac{dz}{dt}$, correct to 4 significant figures, when $t = 0.5$, given that $z = 3e^{2t}\sin 4t$

35.12 Differentiation of a quotient

When $y = \dfrac{u}{v}$, and u and v are both functions of x

then

$$\frac{dy}{dx} = \frac{v\dfrac{du}{dx} - u\dfrac{dv}{dx}}{v^2}$$

This is known as the **quotient rule**.

Problem 33. Find the differential coefficient of:

$$y = \frac{4\sin 2x}{x^3}$$

$\dfrac{4\sin 2x}{x^3}$ is a quotient. Let $u = 4\sin 2x$ and $v = x^3$ (Note that v is **always** the denominator and u the numerator)

$$\frac{dy}{dx} = \frac{v\dfrac{du}{dx} - u\dfrac{dv}{dx}}{v^2}$$

where $\dfrac{du}{dx} = (4)(2)\cos 2x = 8\cos 2x$

and $\dfrac{dv}{dx} = 3x^2$

Hence, $\dfrac{dy}{dx} = \dfrac{\left(x^3\right)\left(8\cos 2x\right) - \left(4\sin 2x\right)\left(3x^2\right)}{\left(x^3\right)^2}$

$$= \frac{8x^3\cos 2x - 12x^2\sin 2x}{x^6}$$

$$= \frac{4x^2\left[2x\cos 2x - 3\sin 2x\right]}{x^6}$$

i.e. $\dfrac{dy}{dx} = \dfrac{4}{x^4}(2x\cos 2x - 3\sin 2x)$

Note that the differential coefficient is **not** obtained by merely differentiating each term in turn and then dividing the numerator by the denominator. The quotient formula **must** be used when differentiating quotients.

Problem 34. Find the differential coefficient of:

$$y = \frac{2e^{3x}}{5x^2}$$

$\dfrac{2e^{3x}}{5x^2}$ is a quotient. Let $u = 2e^{3x}$ and $v = 5x^2$

$$\frac{dy}{dx} = \frac{v\dfrac{du}{dx} - u\dfrac{dv}{dx}}{v^2}$$

where $\dfrac{du}{dx} = (2)(3e^{3x}) = 6e^{3x}$

and $\dfrac{dv}{dx} = 10x$

Hence, $\dfrac{dy}{dx} = \dfrac{(5x^2)(6e^{3x}) - (2e^{3x})(10x)}{(5x^2)^2}$

$$= \frac{30x^2e^{3x} - 20xe^{3x}}{25x^4} = \frac{10xe^{3x}[3x-2]}{25x^4}$$

i.e. $\dfrac{dy}{dx} = \dfrac{2e^{3x}}{5x^3}(3x - 2)$

Problem 35. Differentiate: $y = \dfrac{xe^{2x}}{\cos x}$

The function $\dfrac{xe^{2x}}{\cos x}$ is a quotient, whose numerator is a product.

Let $u = xe^{2x}$ and $v = \cos x$

then $\dfrac{du}{dx} = (x)(2e^{2x}) + (e^{2x})(1)$

and $\dfrac{dv}{dt} = -\sin x$

Hence, $\dfrac{dy}{dx} = \dfrac{v\dfrac{du}{dx} - u\dfrac{dv}{dx}}{v^2}$

$$= \frac{(\cos x)\left[2xe^{2x} + e^{2x}\right] - \left(xe^{2x}\right)\left(-\sin x\right)}{\left(\cos x\right)^2}$$

$$= \frac{2xe^{2x}\cos x + e^{2x}\cos x + xe^{2x}\sin x}{\cos^2 x}$$

$$= \frac{e^{2x}\left[2x\cos x + \cos x + x\sin x\right]}{\cos^2 x}$$

i.e. $\dfrac{dy}{dx} = \dfrac{e^{2x}}{\cos^2 t}(2x\cos x + \cos x + x\sin x)$

Problem 36. Determine the gradient of the curve $y = \dfrac{5x}{2x+1}$ at the point $(2, 2)$

Let $u = 5x$ and $v = 2x + 1$

$$\frac{dy}{dx} = \frac{v\dfrac{du}{dx} - u\dfrac{dv}{dx}}{v^2}$$

$$= \frac{(2x+1)(5) - (5x)(2)}{(2x+1)^2}$$

$$= \frac{10x + 5 - 10x}{(2x+1)^2} = \frac{5}{(2x+1)^2}$$

At the point $(2, 2)$, $x = 2$,

hence,   the **gradient** $= \dfrac{dy}{dx} = \dfrac{5}{(2(2)+1)^2}$

$$= \frac{5}{5^2} = \frac{5}{25} = \frac{1}{5} \text{ or } \mathbf{0.2}$$

Now try the following exercise

Practice Exercise 181   Further problems on differentiating quotients (answers on page 460)

In Problems 1 to 5, differentiate the quotients with respect to the variable.

1. $y = \dfrac{\sin x}{x}$ 2. $y = \dfrac{2\cos 3x}{x^3}$

3. $y = \dfrac{3x}{x^2 + 1}$ 4. $y = \dfrac{\ln 2x}{2x}$

5. $y = \dfrac{x e^x}{\cos x}$

6. Find the gradient of the curve $y = \dfrac{2x}{x^2 - 5}$ at the point $(2, -4)$

7. Evaluate $\dfrac{dy}{dx}$ at $x = 2$, correct to 3 significant figures, given $y = \dfrac{2x^2 + 5}{\ln x}$

35.13 Function of a function

It is often easier to make a substitution before differentiating.

If y is a function of x then $\dfrac{dy}{dx} = \dfrac{dy}{du} \times \dfrac{du}{dx}$

This is known as the **'function of a function'** rule (or sometimes the **chain rule**).

For example, if $y = (3x - 1)^9$ then, by making the substitution $u = (3x - 1)$, $y = u^9$, which is of the 'standard' form.

Hence, $\dfrac{dy}{du} = 9u^8$ and $\dfrac{du}{dx} = 3$

Then $\dfrac{dy}{dx} = \dfrac{dy}{du} \times \dfrac{du}{dx} = (9u^8)(3) = 27u^8$

Rewriting u as $(3x - 1)$ gives: $\dfrac{dy}{dx} = \mathbf{27(3x-1)^8}$

Since y is a function of u, and u is a function of x, then y is a function of a function of x.

Problem 37.   Differentiate: $y = 3\cos(5x^2 + 2)$

Let $u = 5x^2 + 2$ then $y = 3\cos u$

Hence, $\dfrac{du}{dx} = 10x$ and $\dfrac{dy}{du} = -3\sin u$

Using the function of a function rule,

$$\frac{dy}{dx} = \frac{dy}{du} \times \frac{du}{dx} = (-3\sin u)(10x) = -30x\sin u$$

Rewriting u as $5x^2 + 2$ gives:

$$\frac{dy}{dx} = \mathbf{-30\, x \sin\, (5x^2 + 2)}$$

Problem 38.   Find the derivative of:

$$y = (4t^3 - 3t)^6$$

Let $u = 4t^3 - 3t$, then $y = u^6$

Hence, $\dfrac{du}{dt} = 12t^2 - 3$ and $\dfrac{dy}{dt} = 6u^5$

Using the function of a function rule,

$$\frac{dy}{dx} = \frac{dy}{du} \times \frac{du}{dx} = (6u^5)(12t^2 - 3)$$

Rewriting u as $(4t^3 - 3t)$ gives:

$$\frac{dy}{dx} = 6(4t^3 - 3t)^5(12t^2 - 3)$$

$$= 18(4t^2 - 1)(4t^3 - 3t)^5$$

Problem 39.   Determine the differential coefficient of: $y = \sqrt{3x^2 + 4x - 1}$

$y = \sqrt{3x^2 + 4x - 1} = (3x^2 + 4x - 1)^{1/2}$

Let $u = 3x^2 + 4x - 1$ then $y = u^{1/2}$

Hence $\dfrac{du}{dx} = 6x + 4$ and $\dfrac{dy}{du} = \dfrac{1}{2}u^{-1/2} = \dfrac{1}{2\sqrt{u}}$

Using the function of a function rule,

$$\frac{dy}{dx} = \frac{dy}{du} \times \frac{du}{dx} = \left(\frac{1}{2\sqrt{u}}\right)(6x+4) = \frac{3x+2}{\sqrt{u}}$$

i.e. $\qquad \dfrac{dy}{dx} = \dfrac{3x+2}{\sqrt{(3x^2+4x-1)}}$

Now try the following exercise

Practice Exercise 182 Further problems on the function of a function (answers on page 460)

In Problems 1 to 7, find the differential coefficients with respect to the variable.

1. $y = (2x - 1)^4$ 2. $y = (2x^3 - 5x)^5$

3. $y = 5\cos(4x + 3)$ 4. $y = 2\sin(3\theta - 2)$

5. $y = 2\cos^5 x$
   [Note that $\cos^5 x$ means $(\cos x)^5$]

6. $y = \dfrac{1}{(x^3 - 2x + 1)^5}$ 7. $y = 4e^{3t+1}$

8. Differentiate $y = x \sin\left(x - \dfrac{\pi}{3}\right)$ with respect to x, and evaluate, correct to 3 significant figures, when $x = \dfrac{\pi}{2}$
   (Note that y is a product)

Practice Exercise 183 Multiple-choice questions on introduction to differentiation (answers on page 460)

Each question has only one correct answer.

1. Differentiating $y = 4x^5$ gives:
   (a) $\dfrac{dy}{dx} = \dfrac{2}{3}x^6$ (b) $\dfrac{dy}{dx} = 20x^4$
   (c) $\dfrac{dy}{dx} = 4x^6$ (d) $\dfrac{dy}{dx} = 5x^4$

2. Differentiating $i = 3 \sin 2t - 2 \cos 3t$ with respect to t gives:
   (a) $3\cos 2t + 2\sin 3t$ (b) $6(\sin 2t - \cos 3t)$
   (c) $\dfrac{3}{2}\cos 2t + \dfrac{2}{3}\sin 3t$ (d) $6(\cos 2t + \sin 3t)$

3. Given that $y = 3e^x + 2\ln 3x$, $\dfrac{dy}{dx}$ is equal to:
   (a) $6e^x + \dfrac{2}{3x}$ (b) $3e^x + \dfrac{2}{x}$
   (c) $6e^x + \dfrac{2}{x}$ (d) $3e^x + \dfrac{2}{3}$

4. Given $f(t) = 3t^4 - 2$, $f'(t)$ is equal to:
   (a) $12t^3 - 2$ (b) $\dfrac{3}{4}t^5 - 2t + c$
   (c) $12t^3$ (d) $3t^5 - 2$

5. The gradient of the curve $y = 4x^2 - 7x + 3$ at the point $(1, 0)$ is
   (a) 3 (b) 0 (c) 1 (d) -7

6. If $y = 5\sqrt{x^3} - 2$, $\dfrac{dy}{dx}$ is equal to:
   (a) $\dfrac{15}{2}\sqrt{x}$ (b) $y = 2\sqrt{x^5} - 2x + c$
   (c) $\dfrac{5}{2}\sqrt{x} - 2$ (d) $5\sqrt{x} - 2x$

7. An alternating current is given by $i = 4 \sin 150t$ amperes, where t is the time in seconds. The rate of change of current at $t = 0.025$ s is:
   (a) 3.99 A/s (b) -492.3 A/s
   (c) -3.28 A/s (d) 598.7 A/s

8. The gradient of the curve $y = -2x^3 + 3x + 5$ at $x = 2$ is:
   (a) -21 (b) 27 (c) -16 (d) -5

9. If $y = 3x^2 - \ln 5x$ then $\dfrac{d^2y}{dx^2}$ is equal to:
   (a) $6 + \dfrac{1}{5x^2}$ (b) $6x - \dfrac{1}{x}$
   (c) $6 - \dfrac{1}{5x}$ (d) $6 + \dfrac{1}{x^2}$

10. An alternating voltage is given by $v = 10 \sin 300t$ volts, where t is the time in seconds. The rate of change of voltage when $t = 0.01 s$ is:
    (a) -2996 V/s (b) 157 V/s
    (c) -2970 V/s (d) 0.523 V/s

11. If $f(t) = 5t - \dfrac{1}{\sqrt{t}}$, then $f'(t)$ is equal to:
    (a) $5 - \dfrac{1}{2\sqrt{t^3}}$ (b) $5 - 2\sqrt{t}$
    (c) $\dfrac{5t^2}{2} - 2\sqrt{t} + c$ (d) $5 + \dfrac{1}{\sqrt{t^3}}$

12. The length ℓ metres of a certain metal rod at temperature $t°C$ is given by:

$$\ell = 1 + 4 \times 10^{-5}t + 4 \times 10^{-7}t^2$$

The rate of change of length, in mm/°C, when the temperature is 400°C, is:

(a) 3.6×10^{-4} (b) 1.00036
(c) 0.36   (d) 3.2×10^{-4}

13. Given $y = 4x^3 - 3\cos x - e^{-x} + \ln x - 30$
then $\dfrac{dy}{dx}$ is equal to:

(a) $12x^2 + 3\sin x - e^{-x} + \dfrac{1}{x}$

(b) $12x^2 - 3\sin x - e^{-x} + \dfrac{1}{x}$

(c) $x^4 - 3\sin x - e^{-x} + x(1 + \ln x) + 30x + c$

(d) $12x^2 + 3\sin x + e^{-x} + \dfrac{1}{x}$

14. If $s = 2t\ln t$ then $\dfrac{ds}{dt}$ is equal to:

(a) $2 + 2\ln t$

(b) $2t + \dfrac{2}{t}$

(c) $2 - 2\ln t$
(d) $2 + 2e^t$

15. If $f(t) = e^{2t}\ln 2t$, $f'(t)$ is equal to:

(a) $\dfrac{2e^{2t}}{t}$

(b) $e^{2t}\left(\dfrac{1}{t} + 2\ln 2t\right)$

(c) $\dfrac{e^{2t}}{2t}$

(d) $\dfrac{e^{2t}}{2t} + 2e^{2t}\ln 2t$

For fully worked solutions to each of the problems in Practice Exercises 172 to 182 in this chapter, go to the website:
www.routledge.com/cw/bird

Chapter 36

Standard integration

Why it is important to understand: Standard integration

Engineering is all about problem solving and many problems in engineering can be solved using calculus. Physicists, chemists, engineers and many other scientific and technical specialists use calculus in their everyday work; it is a technique of fundamental importance. Both integration and differentiation have numerous applications in engineering and science and some typical examples include determining areas, mean and rms values, volumes of solids of revolution, centroids, second moments of area, differential equations and Fourier series. Standard integrals are introduced in this chapter, together with one application – finding the area under a curve. For any further studies in engineering, differential and integral calculus are unavoidable.

At the end of this chapter you should be able to:

- understand that integration is the reverse process of differentiation
- determine integrals of the form ax^n where n is fractional, zero, or a positive or negative integer
- integrate standard functions – $\cos ax$, $\sin ax$, e^{ax}, $\dfrac{1}{x}$
- evaluate definite integrals
- determine the area under a curve

36.1 The process of integration

The process of integration reverses the process of differentiation. In differentiation, if $f(x) = 2x^2$ then $f'(x) = 4x$. Thus, the integral of $4x$ is $2x^2$; i.e. integration is the process of moving from $f'(x)$ to $f(x)$. By similar reasoning, the integral of $2t$ is t^2

Integration is a process of summation or adding parts together and an elongated S, shown as $\int$, is used to replace the words 'the integral of'. Hence, from above, $\int 4x = 2x^2$ and $\int 2t$ is t^2

In differentiation, the differential coefficient $\dfrac{dy}{dx}$ indicates that a function of x is being differentiated with respect to x, the dx indicating that it is 'with respect to x'

In integration the variable of integration is shown by adding d (the variable) after the function to be integrated. Thus,

$$\int 4x\,dx \text{ means 'the integral of } 4x \text{ with respect to } x\text{',}$$

and $\displaystyle\int 2t\,dt$ means 'the integral of $2t$ with respect to t'

As stated above, the differential coefficient of $2x^2$ is $4x$, hence; $\int 4x\,dx = 2x^2$. However, the differential coefficient of $2x^2 + 7$ is also $4x$. Hence, $\int 4x\,dx$ could also be equal to $2x^2 + 7$. To allow for the possible presence of a constant, whenever the process of integration is performed a constant c is added to the result. Thus,

$$\int 4x\,dx = 2x^2 + c \quad \text{and} \quad \int 2t\,dt = t^2 + c$$

c is called the **arbitrary constant of integration**.

36.2 The general solution of integrals of the form ax^n

The general solution of integrals of the form $\int ax^n\,dx$, where a and n are constants and $n \neq -1$ is given by

$$\int ax^n\,dx = \frac{ax^{n+1}}{n+1} + c$$

Using this rule gives

(i) $\int 3x^4\,dx = \frac{3x^{4+1}}{4+1} + c = \frac{3}{5}x^5 + c$

(ii) $\int \frac{4}{9}t^3\,dt\,dx = \frac{4}{9}\left(\frac{t^{3+1}}{3+1}\right) + c = \frac{4}{9}\left(\frac{t^4}{4}\right) + c$

$$= \frac{1}{9}t^4 + c$$

(iii) $\int \frac{2}{x^2}\,dx = \int 2x^{-2}\,dx = \frac{2x^{-2+1}}{-2+1} + c$

$$= \frac{2x^{-1}}{-1} + c = -\frac{2}{x} + c$$

(iv) $\int \sqrt{x}\,dx = \int x^{\frac{1}{2}}\,dx = \frac{x^{\frac{1}{2}+1}}{\frac{1}{2}+1} + c = \frac{x^{\frac{3}{2}}}{\frac{3}{2}} + c$

$$= \frac{2}{3}\sqrt{x^3} + c$$

Each of these results may be checked by differentiation.

(a) The integral of a constant k is $kx + c$. For example,

$$\int 8\,dx = 8x + c \quad \text{and} \quad \int 5\,dt = 5t + c$$

(b) When a sum of several terms is integrated the result is the sum of the integrals of the separate terms. For example,

$$\int (3x + 2x^2 - 5)\,dx$$

$$= \int 3x\,dx + \int 2x^2\,dx - \int 5\,dx$$

$$= \frac{3x^2}{2} + \frac{2x^3}{3} - 5x + c$$

36.3 Standard integrals

From Chapter 35, $\frac{d}{dx}(\sin ax) = a\cos ax$. Since integration is the reverse process of differentiation, it follows that

$$\int a\cos ax\,dx = \sin ax + c$$

or $$\int \cos ax\,dx = \frac{1}{a}\sin ax + c$$

By similar reasoning

$$\int \sin ax\,dx = -\frac{1}{a}\cos ax + c$$

$$\int e^{ax}\,dx = \frac{1}{a}e^{ax} + c$$

and $$\int \frac{1}{x}\,dx = \ln x + c$$

From above, $\int ax^n\,dx = \frac{ax^{n+1}}{n+1} + c$ **except when**

$$n = -1$$

When $n = -1$, $\int x^{-1}\,dx = \int \frac{1}{x}\,dx = \ln x + c$

A list of **standard integrals** is summarised in Table 36.1.

Table 36.1 Standard integrals

y	$\int y\,dx$
1. $\int ax^n$	$\frac{ax^{n+1}}{n+1} + c$ (except when $n = -1$)
2. $\int \cos ax\,dx$	$\frac{1}{a}\sin ax + c$
3. $\int \sin ax\,dx$	$-\frac{1}{a}\cos ax + c$
4. $\int e^{ax}\,dx$	$\frac{1}{a}e^{ax} + c$
5. $\int \frac{1}{x}\,dx$	$\ln x + c$

Problem 1. Determine $\int 7x^2\,dx$

The standard integral, $\int ax^n\,dx = \frac{ax^{n+1}}{n+1} + c$

When $a = 7$ and $n = 2$,

$$\int 7x^2\,dx = \frac{7x^{2+1}}{2+1} + c = \frac{7x^3}{3} + c \quad \text{or} \quad \frac{7}{3}x^3 + c$$

Problem 2. Determine $\int 2t^3\,dt$

When $a = 2$ and $n = 3$,

$$\int 2t^3\,dt = \frac{2t^{3+1}}{3+1} + c = \frac{2t^4}{4} + c = \frac{1}{2}t^4 + c$$

Note that each of the results in worked examples 1 and 2 may be checked by differentiating them.

Problem 3. Determine $\int 8\,dx$

$\int 8\,dx$ is the same as $\int 8x^0\,dx$ and, using the general rule when $a = 8$ and $n = 0$, gives

$$\int 8x^0\,dx = \frac{8x^{0+1}}{0+1} + c = \mathbf{8x + c}$$

In general, if k is a constant then $\int k\,dx = kx + c$

Problem 4. Determine $\int 2x\,dx$

When $a = 2$ and $n = 1$,

$$\int 2x\,dx = \int 2x^1\,dx = \frac{2x^{1+1}}{1+1} + c = \frac{2x^2}{2} + c$$
$$= x^2 + c$$

Problem 5. Determine $\int \left(3 + \frac{2}{5}x - 6x^2\right)dx$

$\int \left(3 + \frac{2}{5}x - 6x^2\right)dx$ may be written as

$\int 3\,dx + \int \frac{2}{5}x\,dx - \int 6x^2\,dx$

i.e. each term is integrated separately. (This splitting up of terms only applies, however, for addition and subtraction.) Hence,

$$\int \left(3 + \frac{2}{5}x - 6x^2\right)dx$$

$$= 3x + \left(\frac{2}{5}\right)\frac{x^{1+1}}{1+1} - (6)\frac{x^{2+1}}{2+1} + c$$

$$= 3x + \left(\frac{2}{5}\right)\frac{x^2}{2} - (6)\frac{x^3}{3} + c = \mathbf{3x + \frac{1}{5}x^2 - 2x^3 + c}$$

Note that when an integral contains more than one term there is no need to have an arbitrary constant for each; just a single constant c at the end is sufficient.

Problem 6. Determine $\int \left(\frac{2x^3 - 3x}{4x}\right)dx$

Rearranging into standard integral form gives

$$\int \left(\frac{2x^3 - 3x}{4x}\right)dx = \int \left(\frac{2x^3}{4x} - \frac{3x}{4x}\right)dx$$

$$= \int \left(\frac{1}{2}x^2 - \frac{3}{4}\right)dx = \left(\frac{1}{2}\right)\frac{x^{2+1}}{2+1} - \frac{3}{4}x + c$$

$$= \left(\frac{1}{2}\right)\frac{x^3}{3} - \frac{3}{4}x + c = \mathbf{\frac{1}{6}x^3 - \frac{3}{4}x + c}$$

Problem 7. Determine $\int (1-t)^2\,dt$

Rearranging $\int (1-t)^2\,dt$ gives

$$\int (1 - 2t + t^2)dt = t - \frac{2t^{1+1}}{1+1} + \frac{t^{2+1}}{2+1} + c$$

$$= t - \frac{2t^2}{2} + \frac{t^3}{3} + c$$

$$= \mathbf{t - t^2 + \frac{1}{3}t^3 + c}$$

This example shows that functions often have to be rearranged into the standard form of $\int ax^n\,dx$ before it is possible to integrate them.

Problem 8. Determine $\int \frac{5}{x^2}\,dx$

$$\int \frac{5}{x^2}\,dx = \int 5x^{-2}\,dx$$

Using the standard integral, $\int ax^n\,dx$, when $a = 5$ and $n = -2$, gives

$$\int 5x^{-2}\,dx = \frac{5x^{-2+1}}{-2+1} + c = \frac{5x^{-1}}{-1} + c$$

$$= -5x^{-1} + c = \mathbf{-\frac{5}{x} + c}$$

Problem 9. Determine $\int 3\sqrt{x}\,dx$

For fractional powers it is necessary to appreciate $\sqrt[n]{a^m} = a^{\frac{m}{n}}$

$$\int 3\sqrt{x}\,dx = \int 3x^{\frac{1}{2}}\,dx = \frac{3x^{\frac{1}{2}+1}}{\frac{1}{2}+1} + c = \frac{3x^{\frac{3}{2}}}{\frac{3}{2}} + c$$

$$= 2x^{\frac{3}{2}} + c = \mathbf{2\sqrt{x^3} + c}$$

Problem 10. Determine $\int \dfrac{-5}{9\sqrt[4]{t^3}}\,dt$

$$\int \frac{-5}{9\sqrt[4]{t^3}}\,dt = \int \frac{-5}{9t^{\frac{3}{4}}}\,dt = \int \left(-\frac{5}{9}\right) t^{-\frac{3}{4}}\,dt$$

$$= \left(-\frac{5}{9}\right) \frac{t^{-\frac{3}{4}+1}}{-\frac{3}{4}+1} + c$$

$$= \left(-\frac{5}{9}\right) \frac{t^{\frac{1}{4}}}{\frac{1}{4}} + c = \left(-\frac{5}{9}\right)\left(\frac{4}{1}\right) t^{\frac{1}{4}} + c$$

$$= -\frac{20}{9}\sqrt[4]{t} + c$$

Problem 11. Determine $\int 4\cos 3x\,dx$

From 2 of Table 36.1,

$$\int 4\cos 3x\,dx = (4)\left(\frac{1}{3}\right)\sin 3x + c$$

$$= \frac{4}{3}\sin 3x + c$$

Problem 12. Determine $\int 5\sin 2\theta\,d\theta$

From 3 of Table 36.1,

$$\int 5\sin 2\theta\,d\theta = (5)\left(-\frac{1}{2}\right)\cos 2\theta + c$$

$$= -\frac{5}{2}\cos 2\theta + c$$

Problem 13. Determine $\int 5e^{3x}\,dx$

From 4 of Table 36.1,

$$\int 5e^{3x}\,dx = (5)\left(\frac{1}{3}\right)e^{3x} + c$$

$$= \frac{5}{3}e^{3x} + c$$

Problem 14. Determine $\int \dfrac{2}{3e^{4t}}\,dt$

$$\int \frac{2}{3e^{4t}}\,dt = \int \frac{2}{3}e^{-4t}\,dt$$

$$= \left(\frac{2}{3}\right)\left(-\frac{1}{4}\right)e^{-4t} + c$$

$$= -\frac{1}{6}e^{-4t} + c = -\frac{1}{6e^{4t}} + c$$

Problem 15. Determine $\int \dfrac{3}{5x}\,dx$

From 5 of Table 36.1,

$$\int \frac{3}{5x}\,dx = \int \left(\frac{3}{5}\right)\left(\frac{1}{x}\right)\,dx$$

$$= \frac{3}{5}\ln x + c$$

Problem 16. Determine $\int \left(\dfrac{2x^2+1}{x}\right)\,dx$

$$\int \left(\frac{2x^2+1}{x}\right)\,dx = \int \left(\frac{2x^2}{x} + \frac{1}{x}\right)\,dx$$

$$= \int \left(2x + \frac{1}{x}\right)\,dx = \frac{2x^2}{2} + \ln x + c$$

$$= x^2 + \ln x + c$$

Now try the following Practice Exercise

Practice Exercise 184 Standard integrals (answers on page 460)

Determine the following integrals.

1. (a) $\int 4\,dx$ (b) $\int 7x\,dx$

2. (a) $\int 5x^3\,dx$ (b) $\int 3t^7\,dt$

3. (a) $\int \frac{2}{5}x^2\,dx$ (b) $\int \frac{5}{6}x^3\,dx$

4. (a) $\int (2x^4 - 3x)\,dx$ (b) $\int (2 - 3t^3)\,dt$

5. (a) $\int \left(\frac{3x^2 - 5x}{x}\right)\,dx$ (b) $\int (2+\theta)^2\,d\theta]$

6. (a) $\int (2+\theta)(3\theta - 1)\,d\theta$

    (b) $\int (3x-2)(x^2+1)\,dx$

7. (a) $\int \dfrac{4}{3x^2}\,dx$ (b) $\int \dfrac{3}{4x^4}\,dx$

8. (a) $2\int \sqrt{x^3}\,dx$ (b) $\int \dfrac{1}{4}\sqrt[4]{x^5}\,dx$

9. (a) $\int \dfrac{-5}{\sqrt{t^3}}\,dt$ (b) $\int \dfrac{3}{7\sqrt[5]{x^4}}\,dx$

10. (a) $\int 3\cos 2x\,dx$ (b) $\int 7\sin 3\theta\,d\theta$

11. (a) $\int 3\sin \dfrac{1}{2}x\,dx$ (b) $\int 6\cos \dfrac{1}{3}x\,dx$

12. (a) $\int \dfrac{3}{4}e^{2x}\,dx$ (b) $\dfrac{2}{3}\int \dfrac{dx}{e^{5x}}$

13. (a) $\int \dfrac{2}{3x}\,dx$ (b) $\int \left(\dfrac{u^2-1}{u}\right)du$

14. (a) $\int \dfrac{(2+3x)^2}{\sqrt{x}}\,dx$ (b) $\int \left(\dfrac{1}{t}+2t\right)^2 dt$

36.4 Definite integrals

Integrals containing an arbitrary constant c in their results are called **indefinite integrals** since their precise value cannot be determined without further information. **Definite integrals** are those in which limits are applied.

If an expression is written as $[x]_a^b$, b is called the **upper limit** and a the **lower limit**. The operation of applying the limits is defined as $[x]_a^b = (b) - (a)$

For example, the increase in the value of the integral x^2 as x increases from 1 to 3 is written as $\int_1^3 x^2 dx$. Applying the limits gives

$$\int_1^3 x^2 dx = \left[\dfrac{x^3}{3}+c\right]_1^3 = \left(\dfrac{3^3}{3}+c\right) - \left(\dfrac{1^3}{3}+c\right)$$

$$= (9+c) - \left(\dfrac{1}{3}+c\right) = 8\dfrac{2}{3}$$

Note that the c term always cancels out when limits are applied and it need not be shown with definite integrals.

Problem 17. Evaluate $\displaystyle\int_1^2 3x\,dx$

$$\int_1^2 3x\,dx = \left[\dfrac{3x^2}{2}\right]_1^2 = \left\{\dfrac{3}{2}(2)^2\right\} - \left\{\dfrac{3}{2}(1)^2\right\}$$

$$= 6 - 1\dfrac{1}{2} = \mathbf{4\dfrac{1}{2}}$$

Problem 18. Evaluate $\displaystyle\int_{-2}^{3} (4-x^2)\,dx$

$$\int_{-2}^{3}(4-x^2)dx = \left[4x - \dfrac{x^3}{3}\right]_{-2}^{3}$$

$$= \left\{4(3) - \dfrac{(3)^3}{3}\right\} - \left\{4(-2) - \dfrac{(-2)^3}{3}\right\}$$

$$= \{12 - 9\} - \left\{-8 - \dfrac{-8}{3}\right\}$$

$$= \{3\} - \left\{-5\dfrac{1}{3}\right\} = \mathbf{8\dfrac{1}{3}}$$

Problem 19. Evaluate $\displaystyle\int_0^2 x(3+2x)\,dx$

$$\int_0^2 x(3+2x)dx = \int_0^2 (3x+2x^2)dx = \left[\dfrac{3x^2}{2}+\dfrac{2x^3}{3}\right]_0^2$$

$$= \left\{\dfrac{3(2)^2}{2}+\dfrac{2(2)^3}{3}\right\} - \{0+0\}$$

$$= 6 + \dfrac{16}{3} = 11\dfrac{1}{3} \text{ or } \mathbf{11.33}$$

Problem 20. Evaluate $\displaystyle\int_{-1}^{1}\left(\dfrac{x^4-5x^2+x}{x}\right)dx$

$$\int_{-1}^{1}\left(\dfrac{x^4-5x^2+x}{x}\right)dx$$

$$= \int_{-1}^{1}\left(\dfrac{x^4}{x}-\dfrac{5x^2}{x}+\dfrac{x}{x}\right)dx$$

$$= \int_{-1}^{1}(x^3-5x+1)\,dx = \left[\dfrac{x^4}{4}-\dfrac{5x^2}{2}+x\right]_{-1}^{1}$$

$$= \left\{\frac{1}{4} - \frac{5}{2} + 1\right\} - \left\{\frac{(-1)^4}{4} - \frac{5(-1)^2}{2} + (-1)\right\}$$

$$= \left\{\frac{1}{4} - \frac{5}{2} + 1\right\} - \left\{\frac{1}{4} - \frac{5}{2} - 1\right\} = \mathbf{2}$$

Problem 21. Evaluate $\int_1^2 \left(\frac{1}{x^2} + \frac{2}{x}\right) dx$ correct to 3 decimal places

$$\int_1^2 \left(\frac{1}{x^2} + \frac{2}{x}\right) dx$$

$$= \int_1^2 \left\{x^{-2} + 2\left(\frac{1}{x}\right)\right\} dx = \left[\frac{x^{-2+1}}{-2+1} + 2\ln x\right]_1^2$$

$$= \left[\frac{x^{-1}}{-1} + 2\ln x\right]_1^2 = \left[-\frac{1}{x} + 2\ln x\right]_1^2$$

$$= \left(-\frac{1}{2} + 2\ln 2\right) - \left(-\frac{1}{1} + 2\ln 1\right) = \mathbf{1.886}$$

Problem 22. Evaluate $\int_0^{\pi/2} 3\sin 2x\,dx$

$$\int_0^{\pi/2} 3\sin 2x\,dx$$

$$= \left[(3)\left(-\frac{1}{2}\right)\cos 2x\right]_0^{\pi/2} = \left[-\frac{3}{2}\cos 2x\right]_0^{\pi/2}$$

$$= \left\{-\frac{3}{2}\cos 2\left(\frac{\pi}{2}\right)\right\} - \left\{-\frac{3}{2}\cos 2(0)\right\}$$

$$= \left\{-\frac{3}{2}\cos \pi\right\} - \left\{-\frac{3}{2}\cos 0\right\}$$

$$= \left\{-\frac{3}{2}(-1)\right\} - \left\{-\frac{3}{2}(1)\right\}$$

$$= \frac{3}{2} + \frac{3}{2} = \mathbf{3}$$

Problem 23. Evaluate $\int_1^2 4\cos 3t\,dt$

$$\int_1^2 4\cos 3t\,dt = \left[(4)\left(\frac{1}{3}\right)\sin 3t\right]_1^2 = \left[\frac{4}{3}\sin 3t\right]_1^2$$

$$= \left\{\frac{4}{3}\sin 6\right\} - \left\{\frac{4}{3}\sin 3\right\}$$

Note that limits of trigonometric functions are always expressed in **radians** – thus, for example, sin 6 means the sine of 6 radians $= -0.279415\dots$ Hence,

$$\int_1^2 4\cos 3t\,dt = \left\{\frac{4}{3}(-0.279415\dots)\right\} - \left\{\frac{4}{3}(0.141120\dots)\right\}$$

$$= (-0.37255) - (0.18816)$$

$$= \mathbf{-0.5607}$$

Problem 24. Evaluate $\int_1^2 4e^{2x}\,dx$ correct to 4 significant figures

$$\int_1^2 4e^{2x}\,dx = \left[\frac{4}{2}e^{2x}\right]_1^2 = 2\left[e^{2x}\right]_1^2 = 2[e^4 - e^2]$$

$$= 2[54.5982 - 7.3891]$$

$$= \mathbf{94.42}$$

Problem 25. Evaluate $\int_1^4 \frac{3}{4u}\,du$ correct to 4 significant figures

$$\int_1^4 \frac{3}{4u}\,du = \left[\frac{3}{4}\ln u\right]_1^4 = \frac{3}{4}[\ln 4 - \ln 1]$$

$$= \frac{3}{4}[1.3863 - 0] = \mathbf{1.040}$$

Now try the following Practice Exercise

Practice Exercise 185   Definite integrals (answers on page 461)

In Problems 1 to 10, evaluate the definite integrals (where necessary, correct to 4 significant figures).

1. (a) $\int_1^2 x\,dx$ (b) $\int_1^2 (x-1)\,dx$

2. (a) $\int_1^4 5x^2\,dx$ (b) $\int_{-1}^1 -\frac{3}{4}t^2\,dt$

3. (a) $\int_{-1}^2 (3-x^2)\,dx$ (b) $\int_1^3 (x^2 - 4x + 3)\,dx$

4. (a) $\int_1^2 (x^3 - 3x)\,dx$ (b) $\int_1^2 (x^2 - 3x + 3)\,dx$

5. (a) $\int_0^4 2\sqrt{x}\,dx$ (b) $\int_2^3 \frac{1}{x^2}\,dx$

6. (a) $\displaystyle\int_0^{\pi} \frac{3}{2}\cos\theta\, d\theta$ (b) $\displaystyle\int_0^{\pi/2} 4\cos\theta\, d\theta$

7. (a) $\displaystyle\int_{\pi/6}^{\pi/3} 2\sin 2\theta\, d\theta$ (b) $\displaystyle\int_0^2 3\sin t\, dt$

8. (a) $\displaystyle\int_0^1 5\cos 3x\, dx$

(b) $\displaystyle\int_{\pi/4}^{\pi/2} (3\sin 2x - 2\cos 3x)\, dx$

9. (a) $\displaystyle\int_0^1 3e^{3t}\, dt$ (b) $\displaystyle\int_{-1}^2 \frac{2}{3e^{2x}}\, dx$

10. (a) $\displaystyle\int_2^3 \frac{2}{3x}\, dx$ (b) $\displaystyle\int_1^3 \frac{2x^2+1}{x}\, dx$

11. The volume of liquid in a tank is given by: $v = \int_{t_1}^{t_2} q\, dt$.
Determine the volume of a chemical, given $q = (5 - 0.05t + 0.003t^2)\,\text{m}^3/\text{s}$, $t_1 = 0$ and $t_2 = 16\,\text{s}$

36.5 The area under a curve

The area shown shaded in Fig. 36.1 may be determined using approximate methods such as the trapezoidal rule, the mid-ordinate rule or Simpson's rule (see Chapter 29) or, more precisely, by using integration.

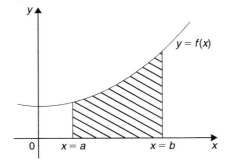

Figure 36.1

The shaded area in Fig. 36.1 is given by

$$\textbf{shaded area} = \int_a^b y\, dx = \int_a^b f(x)\, dx$$

Thus, determining the area under a curve by integration merely involves evaluating a definite integral, as shown in Section 36.4.

There are several instances in engineering and science where the area beneath a curve needs to be accurately determined. For example, the areas between the limits of a

velocity/time graph gives distance travelled,

force/distance graph gives work done,

voltage/current graph gives power, and so on.

Should a curve drop below the x-axis then $y(= f(x))$ becomes negative and $\int f(x)\, dx$ is negative. When determining such areas by integration, a negative sign is placed before the integral. For the curve shown in Fig. 36.2, the total shaded area is given by (area $E +$ area $F +$ area G).

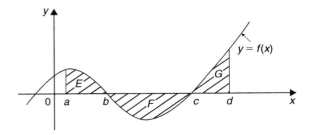

Figure 36.2

By integration,

$$\textbf{total shaded area} = \int_a^b f(x)\, dx - \int_b^c f(x)\, dx$$
$$+ \int_c^d f(x)\, dx$$

(Note that this is **not** the same as $\displaystyle\int_a^d f(x)\, dx$)

It is usually necessary to sketch a curve in order to check whether it crosses the x-axis.

Problem 26. Determine the area enclosed by $y = 2x + 3$, the x-axis and ordinates $x = 1$ and $x = 4$

$y = 2x + 3$ is a straight line graph as shown in Fig. 36.3, in which the required area is shown shaded.
By integration,

$$\text{shaded area} = \int_1^4 y\, dx = \int_1^4 (2x+3)\, dx = \left[\frac{2x^2}{2} + 3x\right]_1^4$$
$$= [(16+12) - (1+3)] = \textbf{24 square units}$$

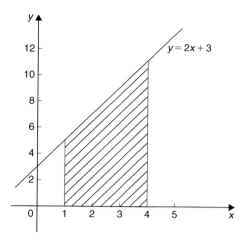

Figure 36.3

(This answer may be checked since the shaded area is a trapezium: area of trapezium $= \dfrac{1}{2}$ (sum of parallel sides)(perpendicular distance between parallel sides) $= \dfrac{1}{2}(5+11)(3) =$ **24 square units**.)

⚑ **Problem 27.**   The velocity v of a body t seconds after a certain instant is given by $v = (2t^2 + 5)$ m/s. Find by integration how far it moves in the interval from $t = 0$ to $t = 4$ s

Since $2t^2 + 5$ is a quadratic expression, the curve $v = 2t^2 + 5$ is a parabola cutting the v-axis at $v = 5$, as shown in Fig. 36.4.

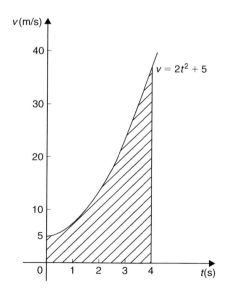

Figure 36.4

The distance travelled is given by the area under the v/t curve (shown shaded in Fig. 36.4). By integration,

$$\text{shaded area} = \int_0^4 v\, dt = \int_0^4 (2t^2 + 5)\, dt = \left[\frac{2t^3}{3} + 5t\right]_0^4$$

$$= \left(\frac{2(4^3)}{3} + 5(4)\right) - (0)$$

i.e. **distance travelled $= 62.67$ m**

Problem 28.   Sketch the graph $y = x^3 + 2x^2 - 5x - 6$ between $x = -3$ and $x = 2$ and determine the area enclosed by the curve and the x-axis

A table of values is produced and the graph sketched as shown in Fig. 36.5, in which the area enclosed by the curve and the x-axis is shown shaded.

x	-3	-2	-1	0	1	2
y	0	4	0	-6	-8	0

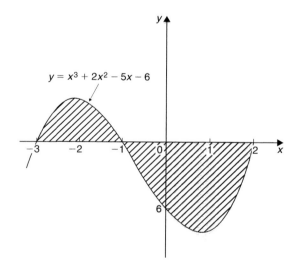

Figure 36.5

Shaded area $= \displaystyle\int_{-3}^{-1} y\, dx - \int_{-1}^{2} y\, dx$, the minus sign before the second integral being necessary since the enclosed area is below the x-axis. Hence,

$$\text{shaded area} = \int_{-3}^{-1} (x^3 + 2x^2 - 5x - 6)\,dx$$

$$- \int_{-1}^{2} (x^3 + 2x^2 - 5x - 6)\,dx$$

$$= \left[\frac{x^4}{4} + \frac{2x^3}{3} - \frac{5x^2}{2} - 6x \right]_{-3}^{-1}$$

$$- \left[\frac{x^4}{4} + \frac{2x^3}{3} - \frac{5x^2}{2} - 6x \right]_{-1}^{2}$$

$$= \left[\left\{ \frac{1}{4} - \frac{2}{3} - \frac{5}{2} + 6 \right\} - \left\{ \frac{81}{4} - 18 - \frac{45}{2} + 18 \right\} \right]$$

$$- \left[\left\{ 4 + \frac{16}{3} - 10 - 12 \right\} - \left\{ \frac{1}{4} - \frac{2}{3} - \frac{5}{2} + 6 \right\} \right]$$

$$= \left[\left\{ 3\frac{1}{12} \right\} - \left\{ -2\frac{1}{4} \right\} \right] - \left[\left\{ -12\frac{2}{3} \right\} - \left\{ 3\frac{1}{12} \right\} \right]$$

$$= \left[5\frac{1}{3} \right] - \left[-15\frac{3}{4} \right]$$

$$= 21\frac{1}{12} \quad \text{or} \quad \textbf{21.08 square units}$$

Problem 29. Determine the area enclosed by the curve $y = 3x^2 + 4$, the x-axis and ordinates $x = 1$ and $x = 4$ by (a) the trapezoidal rule, (b) the mid-ordinate rule, (c) Simpson's rule and (d) integration.

The curve $y = 3x^2 + 4$ is shown plotted in Fig. 36.6. The trapezoidal rule, the mid-ordinate rule and Simpson's rule are discussed in Chapter 29, page 319.

(a) **By the trapezoidal rule**

$$\text{area} = \left(\begin{array}{c} \text{width of} \\ \text{interval} \end{array} \right) \left[\frac{1}{2} \left(\begin{array}{c} \text{first} + \text{last} \\ \text{ordinate} \end{array} \right) \right.$$

$$\left. + \begin{array}{c} \text{sum of} \\ \text{remaining} \\ \text{ordinates} \end{array} \right]$$

Selecting six intervals each of width 0.5 gives

$$\text{area} \approx (0.5) \left[\frac{1}{2}(7 + 52) + 10.75 + 16 \right.$$

$$\left. + 22.75 + 31 + 40.75 \right]$$

$$= \textbf{75.375 square units}$$

x	0	1.0	1.5	2.0	2.5	3.0	3.5	4.0
y	4	7	10.75	16	22.75	31	40.75	52

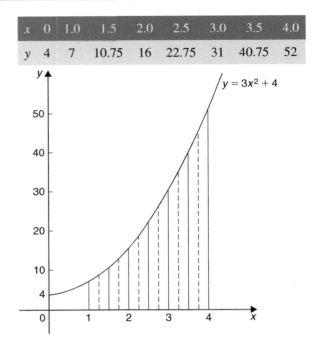

Figure 36.6

(b) **By the mid-ordinate rule**

area = (width of interval)(sum of mid-ordinates) Selecting six intervals, each of width 0.5, gives the mid-ordinates as shown by the broken lines in Fig. 36.6. Thus,

$$\text{area} \approx (0.5)(8.7 + 13.2 + 19.2 + 26.7$$

$$+ 35.7 + 46.2)$$

$$= \textbf{74.85 square units}$$

(c) **By Simpson's rule**[*]

$$\text{area} \approx \frac{1}{3} \left(\begin{array}{c} \text{width of} \\ \text{interval} \end{array} \right) \left[\left(\begin{array}{c} \text{first} + \text{last} \\ \text{ordinates} \end{array} \right) \right.$$

$$+ 4 \left(\begin{array}{c} \text{sum of even} \\ \text{ordinates} \end{array} \right)$$

$$\left. + 2 \left(\begin{array}{c} \text{sum of remaining} \\ \text{odd ordinates} \end{array} \right) \right]$$

---

[*]Who was Simpson? See page 319. To find out more go to www.routledge.com/cw/bird

Selecting six intervals, each of width 0.5, gives

$$\text{area} \approx \frac{1}{3}(0.5)[(7+52)+4(10.75+22.75$$

$$+40.75)+2(16+31)]$$

$$= \textbf{75 square units}$$

(d) **By integration**

$$\text{shaded area} = \int_1^4 y\,dx$$

$$= \int_1^4 (3x^2+4)\,dx = \left[x^3+4x\right]_1^4$$

$$= (64+16)-(1+4)$$

$$= \textbf{75 square units}$$

Integration gives the precise value for the area under a curve. In this case, Simpson's rule is seen to be the most accurate of the three approximate methods.

Problem 30. Find the area enclosed by the curve $y = \sin 2x$, the x-axis and the ordinates $x = 0$ and $x = \dfrac{\pi}{3}$

A sketch of $y = \sin 2x$ is shown in Fig. 36.7. (Note that $y = \sin 2x$ has a period of $\dfrac{2\pi}{2}$ i.e. π radians.)

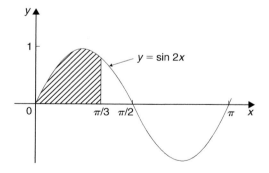

Figure 36.7

$$\text{Shaded area} = \int_0^{\pi/3} y\,dx$$

$$= \int_0^{\pi/3} \sin 2x\,dx = \left[-\frac{1}{2}\cos 2x\right]_0^{\pi/3}$$

$$= \left\{-\frac{1}{2}\cos\frac{2\pi}{3}\right\} - \left\{-\frac{1}{2}\cos 0\right\}$$

$$= \left\{-\frac{1}{2}\left(-\frac{1}{2}\right)\right\} - \left\{-\frac{1}{2}(1)\right\}$$

$$= \frac{1}{4}+\frac{1}{2} = \frac{3}{4} \text{ or } \textbf{0.75 square units}$$

Now try the following Practice Exercise

Practice Exercise 186   Area under curves (answers on page 461)

Unless otherwise stated all answers are in square units.

1. Show by integration that the area of a rectangle formed by the line $y = 4$, the ordinates $x = 1$ and $x = 6$ and the x-axis is 20 square units.

2. Show by integration that the area of the triangle formed by the line $y = 2x$, the ordinates $x = 0$ and $x = 4$ and the x-axis is 16 square units.

3. Sketch the curve $y = 3x^2+1$ between $x = -2$ and $x = 4$. Determine by integration the area enclosed by the curve, the x-axis and ordinates $x = -1$ and $x = 3$. Use an approximate method to find the area and compare your result with that obtained by integration.

4. The force F newtons acting on a body at a distance x metres from a fixed point is given by $F = 3x + 2x^2$. If work done $= \int_{x_1}^{x_2} F\,dx$, determine the work done when the body

moves from the position where $x_1 = 1$ m to that when $x_2 = 3$m.

In Problems 5 to 9, sketch graphs of the given equations and then find the area enclosed between the curves, the horizontal axis and the given ordinates.

5. $y = 5x; \quad x = 1, x = 4$

6. $y = 2x^2 - x + 1; \quad x = -1, x = 2$

7. $y = 2\sin 2x; \quad x = 0, x = \dfrac{\pi}{4}$

8. $y = 5\cos 3t; \quad t = 0, t = \dfrac{\pi}{6}$

9. $y = (x - 1)(x - 3); \quad x = 0, x = 3$

10. The velocity v of a vehicle t seconds after a certain instant is given by $v = (3t^2 + 4)$ m/s. Determine how far it moves in the interval from $t = 1$ s to $t = 5$ s.

Practice Exercise 187   Multiple-choice questions on standard integration (answers on page 461)

Each question has only one correct answer

1. $\int (5 - 3t^2)dt$ is equal to:

    (a) $5 - t^3 + c$ (b) $-3t^3 + c$
    (c) $-6t + c$ (d) $5t - t^3 + c$

2. $\int \left(\dfrac{5x - 1}{x} \right) dx$ is equal to:

    (a) $5x - \ln x + c$ (b) $\dfrac{5x^2 - x}{\dfrac{x^2}{2}}$

    (c) $\dfrac{5x^2}{2} + \dfrac{1}{x^2} + c$ (d) $5x + \dfrac{1}{x^2} + c$

3. The value of $\int_0^1 (3\sin 2\theta - 4\cos\theta)d\theta$, correct to 4 significant figures, is:

    (a) -0.06890 (b) -1.242
    (c) -2.742 (d) -1.569

4. A vehicle has a velocity $v = (2 + 3t)$ m/s after t seconds. The distance travelled is equal to the area under the v/t graph. In the first 3 seconds the vehicle has travelled:

    (a) 19.5 m   (b) 11 m   (c) 13.5 m   (d) 33 m

5. The area, in square units, enclosed by the curve $y = 2x + 3$, the x-axis and ordinates $x = 1$ and $x = 4$ is:

    (a) 28   (b) 2   (c) 24   (d) 39

6. $\int \dfrac{2}{9}t^3\, dt$ is equal to:

    (a) $\dfrac{t^4}{18} + c$ (b) $\dfrac{2}{3}t^2 + c$

    (c) $\dfrac{2}{9}t^4 + c$ (d) $\dfrac{2}{9}t^3 + c$

7. Evaluating $\int_0^{\pi/3} 3\sin 3x\, dx$ gives:

    (a) 1.503   (b) 2   (c) -18 (d) 6

8. $\int (5\sin 3t - 3\cos 5t)dt$ is equal to:

    (a) $-5\cos 3t + 3\sin 5t + c$
    (b) $15(\cos 3t + \sin 3t) + c$
    (c) $\dfrac{3}{5}\cos 3t - \dfrac{5}{3}\sin 5t + c$
    (d) $-\dfrac{5}{3}\cos 3t - \dfrac{3}{5}\sin 5t + c$

9. $\int (\sqrt{x} - 3)dx$ is equal to:

    (a) $\dfrac{3}{2}\sqrt{x^3} - 3x + c$ (b) $\dfrac{2}{3}\sqrt{x^3} + c$

    (c) $\dfrac{1}{2\sqrt{x}} + c$ (d) $\dfrac{2}{3}\sqrt{x^3} - 3x + c$

10. The area enclosed by the curve $y = 3\cos 2\theta$, the ordinates $\theta = 0$ and $\theta = \dfrac{\pi}{4}$ and the θ axis is:

    (a) -3 (b) 6   (c) 1.5   (d) 3

11. $\int \left(1 + \dfrac{4}{e^{2x}}\right) dx$ is equal to:

    (a) $\dfrac{8}{e^{2x}} + c$ (b) $x - \dfrac{2}{e^{2x}} + c$

    (c) $x + \dfrac{4}{e^{2x}} + c$ (d) $x - \dfrac{8}{e^{2x}} + c$

12. Evaluating $\int_{1}^{2} 2e^{3t} dt$, correct to 4 significant figures, gives:

    (a) 2300   (b) 255.6   (c) 766.7   (d) 282.3

13. $\int_{\frac{\pi}{2}}^{\frac{3\pi}{2}} (2\sin\theta - 3\cos\theta)$ is equal to:

    (a) 0   (b) 1   (c) 2   (d) 6

14. The area enclosed between the curve $y = x^2$, the x-axis, and the lines $x = 1$ and $x = 4$ is:

    (a) 63 square units
    (b) 27 square units
    (c) 21 square units
    (d) 9 square units

15. Evaluating $\int_{-1}^{2} (3x^2 + 4x^3) dx$ gives:

    (a) 54   (b) 22   (c) 24   (d) 26

For fully worked solutions to each of the problems in Practice Exercises 184 to 186 in this chapter, go to the website:

www.routledge.com/cw/bird

This assignment covers the material contained in Chapters 35 and 36. *The marks available are shown in brackets at the end of each question.*

1. Differentiate the following functions with respect to x.
   (a) $y = 5x^2 - 4x + 9$ (b) $y = x^4 - 3x^2 - 2$ (4)

2. Given $y = 2(x-1)^2$, find $\dfrac{dy}{dx}$ (3)

3. If $y = \dfrac{3}{x}$ determine $\dfrac{dy}{dx}$ (2)

4. Given $f(t) = \sqrt{t^5}$, find $f'(t)$ (2)

5. Determine the derivative of $y = 5 - 3x + \dfrac{4}{x^2}$ (3)

6. Calculate the gradient of the curve $y = 3\cos\dfrac{x}{3}$ at $x = \dfrac{\pi}{4}$, correct to 3 decimal places. (4)

7. Find the gradient of the curve $f(x) = 7x^2 - 4x + 2$ at the point $(1, 5)$ (3)

8. If $y = 5\sin 3x - 2\cos 4x$ find $\dfrac{dy}{dx}$ (2)

9. Determine the value of the differential coefficient of $y = 5\ln 2x - \dfrac{3}{e^{2x}}$ when $x = 0.8$, correct to 3 significant figures. (4)

10. If $y = 5x^4 - 3x^3 + 2x^2 - 6x + 5$, find (a) $\dfrac{dy}{dx}$ (b) $\dfrac{d^2y}{dx^2}$ (4)

11. Differentiate the following with respect to x:
    (a) $y = 3x\cos 2x$ (b) $y = \dfrac{2\sin 3x}{x}$
    (c) $y = (5x^2 - 2x)^4$ (10)

12. Newton's law of cooling is given by $\theta = \theta_0 e^{-kt}$, where the excess of temperature at zero time is $\theta_0\,°\mathrm{C}$ and at time t seconds is $\theta\,°\mathrm{C}$. Determine the rate of change of temperature after $40\,\mathrm{s}$, correct to 3 decimal places, given that $\theta_0 = 16°\mathrm{C}$ and $k = -0.01$ (4)

In Problems 13 to 16, determine the indefinite integrals.

13. (a) $\displaystyle\int (x^2 + 4)dx$ (b) $\displaystyle\int \dfrac{1}{x^3}dx$ (4)

14. (a) $\displaystyle\int \left(\dfrac{2}{\sqrt{x}} + 3\sqrt{x}\right) dx$ (b) $\displaystyle\int 3\sqrt{t^5}dt$ (4)

15. (a) $\displaystyle\int \dfrac{2}{\sqrt[3]{x^2}}dx$ (b) $\displaystyle\int \left(e^{0.5x} + \dfrac{1}{3x} - 2\right) dx$ (6)

16. (a) $\displaystyle\int (2+\theta)^2 d\theta$

    (b) $\displaystyle\int \left(\cos\dfrac{1}{2}x + \dfrac{3}{x} - e^{2x}\right) dx$ (6)

Evaluate the integrals in Problems 17 to 20, each, where necessary, correct to 4 significant figures.

17. (a) $\displaystyle\int_1^3 (t^2 - 2t)dt$ (b) $\displaystyle\int_{-1}^2 \left(2x^3 - 3x^2 + 2\right) dx$ (6)

18. (a) $\displaystyle\int_0^{\pi/3} 3\sin 2t\, dt$ (b) $\displaystyle\int_{\pi/4}^{3\pi/4} \cos\dfrac{1}{3}x\, dx$ (7)

19. (a) $\displaystyle\int_1^2 \left(\dfrac{2}{x^2} + \dfrac{1}{x} + \dfrac{3}{4}\right) dx$ (b) $\displaystyle\int_1^2 \left(\dfrac{3}{x} - \dfrac{1}{x^3}\right) dx$ (8)

20. (a) $\displaystyle\int_0^1 \left(\sqrt{x} + 2e^x\right) dx$ (b) $\displaystyle\int_1^2 \left(r^3 - \dfrac{1}{r}\right) dr$ (6)

In Problems 21 to 23, find the area bounded by the curve, the x-axis and the given ordinates. Assume answers are in square units. Give answers correct to 2 decimal places where necessary.

21. $y = x^2$; $x = 0, x = 2$ (3)

22. $y = 3x - x^2$; $x = 0, x = 3$ (3)

23. $y = (x-2)^2$; $x = 1, x = 2$ (4)

24. Find the area enclosed between the curve $y = \sqrt{x} + \dfrac{1}{\sqrt{x}}$, the horizontal axis and the ordinates $x = 1$ and $x = 4$. Give the answer correct to 2 decimal places. (5)

25. The force F newtons acting on a body at a distance x metres from a fixed point is given by $F = 2x + 3x^2$. If work done $= \displaystyle\int_{x_1}^{x_2} F\,dx$, determine the work done when the body moves from the position when $x = 1\,\text{m}$ to that when $x = 4\,\text{m}$. (3)

Chapter 37

Number sequences

Why it is important to understand: **Number sequences**

Number sequences are widely used in engineering applications including computer data structure and sorting algorithms, financial engineering, audio compression and architectural engineering. Thanks to engineers, robots have migrated from factory shop floors, as industrial manipulators, to outer space, as interplanetary explorers, in hospitals – as minimally invasive surgical assistants, in homes – as vacuum cleaners and lawn mowers, and in battlefields – as unmanned air, underwater and ground vehicles. Arithmetic progressions are used in simulation engineering and in the reproductive cycle of bacteria. Further uses of APs in daily life include uniform increase in the speed at regular intervals, completing patterns of objects, calculating simple interest, speed of an aircraft, increase or decrease in the costs of goods, sales and production, and so on. Geometric progressions are used in compound interest and in calculating the range of speeds on a drilling machine. GPs are used throughout mathematics, and they have many important applications in physics, engineering, biology, economics, computer science, queuing theory and finance.

At the end of this chapter you should be able to:

- define and determine simple number sequences
- calculate the nth term of a series
- calculate the nth term of an AP
- calculate the sum of n terms of an AP
- calculate the nth term of a GP
- calculate the sum of n terms of a GP
- calculate the sum to infinity of a GP

37.1 Simple sequences

A set of numbers which are connected by a definite law is called a **series** or a **sequence of numbers**. Each of the numbers in the series is called a **term** of the series.

For example, $1, 3, 5, 7, \ldots$ is a series obtained by adding 2 to the previous term, and $2, 8, 32, 128, \ldots$ is a sequence obtained by multiplying the previous term by 4

Problem 1. Determine the next two terms in the series: $3, 6, 9, 12, \ldots$

We notice that the sequence $3, 6, 9, 12, \ldots$ progressively increases by 3, thus the next two terms will be **15 and 18**

Problem 2. Find the next two terms in the series: $9, 5, 1, \ldots$

We notice that each term in the series $9, 5, 1, \ldots$ progressively decreases by 4, thus the next two terms will be $1 - 4$, i.e. $\mathbf{-3}$ and $-3 - 4$, i.e. $\mathbf{-7}$

Problem 3. Determine the next two terms in the series: $2, 6, 18, 54, \ldots$

We notice that the second term, 6, is three times the first term, the third term, 18, is three times the second term, and that the fourth term, 54, is three times the third term. Hence the fifth term will be $3 \times 54 = \mathbf{162}$, and the sixth term will be $3 \times 162 = \mathbf{486}$

Now try the following Practice Exercise

Practice Exercise 188   Simple sequences (answers on page 461)

Determine the next two terms in each of the following series:

1. $5, 9, 13, 17, \ldots$ 2. $3, 6, 12, 24, \ldots$

3. $112, 56, 28, \ldots$ 4. $12, 7, 2, \ldots$

5. $2, 5, 10, 17, 26, 37, \ldots$ 6. $1, 0.1, 0.01, \ldots$

7. $4, 9, 19, 34, \ldots$

37.2 The nth term of a series

If a series is represented by a general expression, say, $2n + 1$, where n is an integer (i.e. a whole number), then by substituting $n = 1, 2, 3, \ldots$ the terms of the series can be determined; in this example, the first three terms will be:

$2(1) + 1, \quad 2(2) + 1, \quad 2(3) + 1, \ldots, \quad$ i.e. $\quad 3, 5, 7, \ldots$

What is the nth term of the sequence $1, 3, 5, 7, \ldots$? First, we notice that the gap between each term is 2, hence the law relating the numbers is: '$2n +$ something'.
The second term, $3 = 2n +$ something, hence when $n = 2$ (i.e. the second term of the series), then $3 = 4 +$ something, and the 'something' must be -1
Thus, **the nth term of $1, 3, 5, 7, \ldots$ is $2n - 1$**

Hence the fifth term is given by $2(5) - 1 = 9$, and the twentieth term is $2(20) - 1 = 39$, and so on.

Problem 4. The nth term of a sequence is given by $3n + 1$. Write down the first four terms.

The first four terms of the series $3n + 1$ will be:

$3(1) + 1, \quad 3(2) + 1, \quad 3(3) + 1 \quad$ and $\quad 3(4) + 1$

i.e. **4, 7, 10 and 13**

Problem 5. The nth term of a series is given by $4n - 1$. Write down the first four terms.

The first four terms of the series $4n - 1$ will be:

$4(1) - 1, \quad 4(2) - 1, \quad 4(3) - 1 \quad$ and $\quad 4(4) - 1$

i.e. **3, 7, 11 and 15**

Problem 6. Find the nth term of the series: $1, 4, 7, \ldots$

We notice that the gap between each of the given three terms is 3, hence the law relating the numbers is: '$3n +$ something'.
The second term, $\quad 4 = 3n +$ something,
so when $n = 2$, then $\quad 4 = 6 +$ something,
so the 'something' must be -2 (from simple equations)
Thus, **the nth term of the series $1, 4, 7, \ldots$ is: $3n - 2$**

Problem 7. Find the nth term of the sequence: $3, 9, 15, 21, \ldots$ Hence determine the 15th term of the series.

We notice that the gap between each of the given four terms is 6, hence the law relating the numbers is: '$6n +$ something'.
The second term, $\quad 9 = 6n +$ something,
so when $n = 2$, then $\quad 9 = 12 +$ something,
so the 'something' must be -3

Thus, **the nth term of the series $3, 9, 15, 21, \ldots$ is: $6n - 3$**
The 15th term of the series is given by $6n - 3$ when $n = 15$
Hence, **the 15th term of the series $3, 9, 15, 21, \ldots$ is:** $6(15) - 3 = \mathbf{87}$

Problem 8. Find the nth term of the series: $1, 4, 9, 16, 25, \ldots$

This is a special series and does not follow the pattern of the previous examples. Each of the terms in the given series are square numbers,

i.e. $1, 4, 9, 16, 25, \ldots \equiv 1^2, 2^2, 3^2, 4^2, 5^2, \ldots$

Hence the nth term is: n^2

Now try the following Practice Exercise

Practice Exercise 189 The nth term of a series (answers on page 461)

1. The nth term of a sequence is given by $2n - 1$. Write down the first four terms.

2. The nth term of a sequence is given by $3n + 4$. Write down the first five terms.

3. Write down the first four terms of the sequence given by $5n + 1$.

In Problems 4 to 8, find the nth term in the series:

4. $5, 10, 15, 20, \ldots$ 5. $4, 10, 16, 22, \ldots$

6. $3, 5, 7, 9, \ldots$ 7. $2, 6, 10, 14, \ldots$

8. $9, 12, 15, 18, \ldots$

9. Write down the next two terms of the series: $1, 8, 27, 64, 125, \ldots$

37.3 Arithmetic progressions

When a sequence has a constant difference between successive terms it is called an **arithmetic progression** (often abbreviated to AP).

Examples include:

  (i) $1, 4, 7, 10, 13, \ldots$ where the **common difference** is 3

and (ii) $a, a + d, a + 2d, a + 3d, \ldots$ where the common difference is d.

General expression for the nth term of an AP

If the first term of an AP is 'a' and the common difference is 'd' then:

the nth term is: $a + (n - 1)d$

In example (i) above, the 7th term is given by $1 + (7 - 1)3 = \mathbf{19}$, which may be readily checked.

Sum of n terms of an AP

The sum S of an AP can be obtained by multiplying the average of all the terms by the number of terms.

The average of all the terms $= \dfrac{a + l}{2}$, where 'a' is the first term and 'l' is the last term, i.e. $l = a + (n - 1)d$, for n terms.

Hence, the sum of n terms, $S_n = n \left(\dfrac{a + l}{2} \right)$

$$= \frac{n}{2} \{ a + [a + (n - 1)d] \}$$

i.e. $$S_n = \frac{n}{2}[2a + (n - 1)d]$$

For example, the sum of the first 7 terms of the series $1, 4, 7, 10, 13, \ldots$ is given by:

$$S_7 = \frac{7}{2}[2(1) + (7 - 1)3] \quad \text{since } a = 1 \quad \text{and} \quad d = 3$$

$$= \frac{7}{2}[2 + 18] = \frac{7}{2}[20] = \mathbf{70}$$

Here are some worked problems to help understanding of arithmetic progressions.

Problem 9. Determine (a) the ninth, and (b) the sixteenth term of the series $2, 7, 12, 17, \ldots$

$2, 7, 12, 17, \ldots$ is an arithmetic progression with a common difference, d, of 5

(a) The nth term of an AP is given by $a + (n - 1)d$
Since the first term $a = 2, d = 5$ and $n = 9$ then the 9th term is: $2 + (9 - 1)5 = 2 + (8)(5)$
$$= 2 + 40 = \mathbf{42}$$

(b) The 16th term is: $2 + (16 - 1)5 = 2 + (15)(5)$
$$= 2 + 75 = \mathbf{77}$$

Problem 10. The 6th term of an AP is 17 and the 13th term is 38. Determine the 19th term

The nth term of an AP is: $a + (n - 1)d$

The 6th term is: $a + 5d = 17$ (1)

The 13th term is: $a + 12d = 38$ (2)

Equation (2) − equation (1) gives: $7d = 21$ from which,
$$d = \frac{21}{7} = 3$$

Substituting in equation (1) gives: $a + 15 = 17$ from which, $a = 2$

Hence, the 19th term is: $a + (n-1)d = 2 + (19-1)3$

$$= 2 + (18)(3)$$

$$= 2 + 54 = \mathbf{56}$$

Problem 11. Determine the number of the term whose value is 22 in the series $2.5, 4, 5.5, 7, \ldots$

$2.5, 4, 5.5, 7, \ldots$ is an AP where $a = 2.5$ and $d = 1.5$

Hence, if the nth term is 22 then:

$$a + (n-1)d = 22$$

i.e. $2.5 + (n-1)(1.5) = 22$

i.e. $(n-1)(1.5) = 22 - 2.5 = 19.5$

and $n - 1 = \dfrac{19.5}{1.5} = 13$

from which, $\boldsymbol{n = 13 + 1 = 14}$

i.e.   **the 14th term of the AP is 22**

Problem 12. Find the sum of the first 12 terms of the series $5, 9, 13, 17, \ldots$

$5, 9, 13, 17, \ldots$ is an AP where $a = 5$ and $d = 4$
The sum of n terms of an AP, $S_n = \dfrac{n}{2}[2a + (n-1)d]$

Hence, the sum of the first 12 terms,

$$S_{12} = \frac{12}{2}[2(5) + (12-1)4]$$

$$= 6[10 + 44] = 6(54) = \mathbf{324}$$

Problem 13. Find the sum of the first 21 terms of the series $3.5, 4.1, 4.7, 5.3, \ldots$

$3.5, 4.1, 4.7, 5.3, \ldots$ is an AP where $a = 3.5$ and $d = 0.6$

The sum of the first 21 terms,

$$S_{21} = \frac{21}{2}[2a + (n-1)d]$$

$$= \frac{21}{2}[2(3.5) + (21-1)0.6]$$

$$= \frac{21}{2}[7 + 12]$$

$$= \frac{21}{2}(19) = \frac{399}{2} = \mathbf{199.5}$$

Now try the following Practice Exercise

Practice Exercise 190   Arithmetic progressions (answers on page 461)

1. Find the 11th term of the series $8, 14, 20, 26, \ldots$

2. Find the 17th term of the series $11, 10.7, 10.4, 10.1, \ldots$

3. The 7th term of a series is 29 and the 11th term is 54. Determine the 16th term.

4. Find the 15th term of an arithmetic progression of which the first term is 2.5 and the 10th term is 16.

5. Determine the number of the term which is 29 in the series $7, 9.2, 11.4, 13.6, \ldots$

6. Find the sum of the first 11 terms of the series $4, 7, 10, 13, \ldots$

7. Determine the sum of the series $6.5, 8.0, 9.5, 11.0, \ldots, 32$

Here are some further worked problems on arithmetic progressions.

Problem 14. The sum of 7 terms of an AP is 35 and the common difference is 1.2. Determine the first term of the series.

$n = 7, d = 1.2$ and $S_7 = 35$. Since the sum of n terms of an AP is given by:

$$S_n = \frac{n}{2}[2a + (n-1)d]$$

Then $35 = \dfrac{7}{2}[2a + (7-1)1.2] = \dfrac{7}{2}[2a + 7.2]$

Hence, $\dfrac{35 \times 2}{7} = 2a + 7.2$

i.e. $10 = 2a + 7.2$

Thus, $2a = 10 - 7.2 = 2.8$

from which, $a = \dfrac{2.8}{2} = 1.4$

i.e.   **the first term, $a = 1.4$**

Problem 15. Three numbers are in arithmetic progression. Their sum is 15 and their product is 80. Determine the three numbers.

Let the three numbers be $(a - d), a$ and $(a + d)$

Then,	$(a - d) + a + (a + d) = 15$
i.e.	$3a = 15$
from which,	$a = 5$
Also,	$a(a - d)(a + d) = 80$
i.e.	$a(a^2 - d^2) = 80$
Since $a = 5$,	$5(5^2 - d^2) = 80$
i.e.	$125 - 5d^2 = 80$
and	$125 - 80 = 5d^2$
i.e.	$45 = 5d^2$
from which,	$d^2 = \dfrac{45}{5} = 9$
Hence,	$d = \sqrt{9} = \pm 3$

The three numbers are thus: $(5 - 3), 5$ and $(5 + 3)$, i.e. **2, 5 and 8**

Problem 16. Find the sum of all the numbers between 0 and 207 which are exactly divisible by 3.

The series $3, 6, 9, 12, \ldots 207$ is an AP whose first term, $a = 3$ and common difference, $d = 3$

The last term is:	$a + (n - 1)d = 207$
i.e.	$3 + (n - 1)3 = 207$
from which,	$(n - 1) = \dfrac{207 - 3}{3} = 68$
Hence,	$n = 68 + 1 = 69$

The sum of all 69 terms is given by:

$$S_{69} = \frac{n}{2}[2a + (n - 1)d]$$
$$= \frac{69}{2}[2(3) + (69 - 1)3]$$
$$= \frac{69}{2}[6 + 204] = \frac{69}{2}(210) = \mathbf{7245}$$

Problem 17. The first, 12th and last term of an arithmetic progression are: $4, 31.5$, and 376.5, respectively. Determine (a) the number of terms in the series, (b) the sum of all the terms and (c) the 80th term.

(a) Let the AP be $a, a + d, a + 2d, \ldots, a + (n - 1)d,$ where $a = 4$

The 12th term is: $a + (12 - 1)d = 31.5$

i.e.	$4 + 11d = 31.5$
from which,	$11d = 31.5 - 4 = 27.5$
Hence,	$d = \dfrac{27.5}{11} = 2.5$

The last term is: $a + (n - 1)d$

i.e. $4 + (n - 1)(2.5) = 376.5$

i.e. $(n - 1) = \dfrac{376.5 - 4}{2.5} = \dfrac{372.5}{2.5} = 149$

Hence, the number of terms in the series, $n = 149 + 1 = 150$

(b) **Sum of all the terms,**

$$S_{150} = \frac{n}{2}[2a + (n - 1)d]$$
$$= \frac{150}{2}[2(4) + (150 - 1)(2.5)]$$
$$= 75[8 + (149)(2.5)] = 75[8 + 372.5]$$
$$= 75(380.5) = \mathbf{28537.5}$$

(c) **The 80th term is:** $a + (n - 1)d = 4 + (80 - 1)(2.5)$
$$= 4 + (79)(2.5)$$
$$= 4 + 197.5$$
$$= \mathbf{201.5}$$

Now try the following Practice Exercise

Practice Exercise 191   Arithmetic progressions (answers on page 461)

1. The sum of 15 terms of an arithmetic progression is 202.5 and the common difference is 2. Find the first term of the series.

2. Three numbers are in arithmetic progression. Their sum is 9 and their product is 20.25. Determine the three numbers.

3. Find the sum of all the numbers between 5 and 250 which are exactly divisible by 4.

4. Find the number of terms of the series $5, 8, 11, \ldots$ of which the sum is 1025.

5. Insert four terms between 5 and 22.5 to form an arithmetic progression.

6. The first, tenth and last terms of an arithmetic progression are $9, 40.5$ and 425.5, respectively. Find (a) the number of terms, (b) the sum of all the terms and (c) the 70th term.

7. On commencing employment a man is paid a salary of £16 000 per annum and receives annual increments of £480. Determine his salary in the 9th year and calculate the total he will have received in the first 12 years.

8. An oil company bores a hole 80 m deep. Estimate the cost of boring if the cost is £30 for drilling the first metre with an increase in cost of £2 per metre for each succeeding metre.

37.4   Geometric progressions

When a sequence has a constant ratio between successive terms it is called a **geometric progression** (often abbreviated to GP). The constant is called the **common ratio, r**.
Examples include

(i) $1, 2, 4, 8, \ldots$ where the common ratio is 2

and (ii) $a, ar, ar^2, ar^3, \ldots$ where the common ratio is r

General expression for the nth term of a GP

If the first term of a GP is 'a' and the common ratio is 'r', then

the nth term is: ar^{n-1}

which can be readily checked from the above examples. For example, the 8th term of the GP $1, 2, 4, 8, \ldots$ is $(1)(2)^7 = \textbf{128}$, since 'a' $= 1$ and 'r' $= 2$

Sum of n terms of a GP

Let a GP be $a, ar, ar^2, ar^3, \ldots, ar^{n-1}$

then the sum of n terms,

$$S_n = a + ar + ar^2 + ar^3 + \cdots + ar^{n-1} \cdots \quad (1)$$

Multiplying throughout by r gives:

$$rS_n = ar + ar^2 + ar^3 + ar^4 + \cdots ar^{n-1} + ar^n \cdots \quad (2)$$

Subtracting equation (2) from equation (1) gives:

$$S_n - rS_n = a - ar^n$$

i.e. $\qquad S_n(1 - r) = a(1 - r^n)$

Thus, **the sum of n terms, $S_n = \dfrac{a(1 - r^n)}{(1 - r)}$** which is valid when $r < 1$

Subtracting equation (1) from equation (2) gives:

$$S_n = \frac{a(r^n - 1)}{(r - 1)} \qquad \text{which is valid when } r > 1$$

For example, the sum of the first 8 terms of the GP $1, 2, 4, 8, 16, \ldots$ is given by:

$$S_8 = \frac{1(2^8 - 1)}{(2 - 1)} \qquad \text{since '}a\text{'} = 1 \quad \text{and} \quad r = 2$$

i.e. $\quad S_8 = \dfrac{1(256 - 1)}{1} = \textbf{255}$

Sum to infinity of a GP

When the common ratio r of a GP is less than unity, the sum of n terms,

$$S_n = \frac{a(1 - r^n)}{(1 - r)}, \text{ which may be written as:}$$

$$S_n = \frac{a}{(1 - r)} - \frac{ar^n}{(1 - r)}$$

Since $r < 1, r^n$ becomes less as n increases, i.e. $r^n \to 0$ as $n \to \infty$

Hence, $\qquad \dfrac{ar^n}{(1 - r)} \to 0 \quad \text{as} \quad n \to \infty$

Thus, $\qquad S_n \to \dfrac{a}{(1 - r)} \quad \text{as} \quad n \to \infty$

The quantity $\dfrac{a}{(1 - r)}$ is called the **sum to infinity, S_∞**, and is the limiting value of the sum of an infinite number of terms,

i.e. $\quad S_\infty = \dfrac{a}{(1 - r)}$ which is valid when $-1 < r < 1$

For example, the sum to infinity of the GP: $1, \dfrac{1}{2}, \dfrac{1}{4}, \ldots$ is:

$$S_\infty = \frac{1}{1 - \frac{1}{2}}, \quad \text{since} \quad a = 1 \quad \text{and} \quad r = \frac{1}{2}$$

i.e. $\qquad S_\infty = \textbf{2}$

Here are some worked problems to help understanding of geometric progressions.

Problem 18. Determine the tenth term of the series $3, 6, 12, 24, \ldots$

$3, 6, 12, 24, \ldots$ is a geometric progression with a common ratio r of 2.

The nth term of a GP is ar^{n-1}, where 'a' is the first term.

Hence, **the 10th term is**:

$(3)(2)^{10-1} = (3)(2)^9 = 3(512) = \textbf{1536}$

Problem 19. Find the sum of the first 7 terms of the series, $0.5, 1.5, 4.5, 13.5, \ldots$

$0.5, 1.5, 4.5, 13.5, \ldots$ is a GP with a common ratio $r = 3$

The sum of n terms, $S_n = \dfrac{a(r^n - 1)}{(r-1)}$

Hence, **the sum of the first 7 terms**,

$$S_7 = \frac{0.5(3^7 - 1)}{(3-1)} = \frac{0.5(2187 - 1)}{2} = \textbf{546.5}$$

Problem 20. The first term of a geometric progression is 12 and the fifth term is 55. Determine the 8th term and the 11th term.

The 5th term is given by: $ar^4 = 55$, where the first term, $a = 12$

Hence, $r^4 = \dfrac{55}{a} = \dfrac{55}{12}$ and $r = \sqrt[4]{\left(\dfrac{55}{12}\right)}$

$= 1.4631719\ldots$

The 8th term is: $ar^7 = (12)(1.4631719\ldots)^7$

$= \textbf{172.3}$

The 11th term is: $ar^{10} = (12)(1.4631719\ldots)^{10}$

$= \textbf{539.7}$

Problem 21. Which term of the series $2187, 729, 243, \ldots$ is $\dfrac{1}{9}$?

$2187, 729, 243, \ldots$ is a GP with a common ratio, $r = \dfrac{1}{3}$ and first term, $a = 2187$

The nth term of a GP is given by: ar^{n-1}

Hence, $\dfrac{1}{9} = (2187)\left(\dfrac{1}{3}\right)^{n-1}$ from which,

$$\left(\frac{1}{3}\right)^{n-1} = \frac{1}{(9)(2187)} = \frac{1}{3^2 3^7} = \frac{1}{3^9} = \left(\frac{1}{3}\right)^9$$

Thus, $(n-1) = 9$ from which, $\boldsymbol{n = 9 + 1 = 10}$

i.e. $\dfrac{1}{9}$ **is the 10th term of the GP**

Problem 22. Find the sum of the first 9 terms of the series $72.0, 57.6, 46.08, \ldots$

The common ratio, $r = \dfrac{ar}{a} = \dfrac{57.6}{72.0} = 0.8$

(also $\dfrac{ar^2}{ar} = \dfrac{46.08}{57.6} = 0.8$)

The sum of 9 terms, $S_9 = \dfrac{a(1 - r^n)}{(1-r)}$

$$= \frac{72.0(1 - 0.8^9)}{(1 - 0.8)}$$

$$= \frac{72.0(1 - 0.1342)}{0.2}$$

$$= \textbf{311.7}$$

Problem 23. Find the sum to infinity of the series $3, 1, \dfrac{1}{3}, \ldots$

$3, 1, \dfrac{1}{3}, \ldots$ is a GP of common ratio, $r = \dfrac{1}{3}$

The sum to infinity, $S_\infty = \dfrac{a}{1-r} = \dfrac{3}{1 - \frac{1}{3}} = \dfrac{3}{\frac{2}{3}}$

$$= \frac{9}{2} = \textbf{4.5}$$

Now try the following Practice Exercise

Practice Exercise 192   Geometric progressions (answers on page 461)

1. Find the 10th term of the series $5, 10, 20, 40, \ldots$

2. Determine the sum of the first 7 terms of the series $0.25, 0.75, 2.25, 6.75, \ldots$

3. The 1st term of a geometric progression is 4 and the 6th term is 128. Determine the 8th and 11th terms.

4. Find the sum of the first 7 terms of the series $2, 5, 12.5, \ldots$ (correct to 4 significant figures).

5. Determine the sum to infinity of the series $4, 2, 1, \ldots$

6. Find the sum to infinity of the series $2\frac{1}{2}, -1\frac{1}{4}, \frac{5}{8}, \ldots$

Here are some further worked problems on geometric progressions.

Problem 24. In a geometric progression the sixth term is 8 times the third term and the sum of the seventh and eighth terms is 192. Determine (a) the common ratio, (b) the first term and (c) the sum of the fifth to eleventh terms, inclusive.

(a) Let the GP be $a, ar, ar^2, ar^3, \ldots, ar^{n-1}$

The 3rd term $= ar^2$ and the sixth term $= ar^5$

The 6th term is 8 times the 3rd

Hence, $ar^5 = 8ar^2$ from which,

$r^3 = 8$ and $r = \sqrt[3]{8}$

i.e. **the common ratio $r = 2$**

(b) The sum of the 7th and 8th terms is 192. Hence $ar^6 + ar^7 = 192$

Since $r = 2$, then $64a + 128a = 192$

$192a = 192$ from which, **the first term, $a = 1$**

(c) **The sum of the 5th to 11th terms** (inclusive) **is given by**:

$$S_{11} - S_4 = \frac{a(r^{11} - 1)}{(r - 1)} - \frac{a(r^4 - 1)}{(r - 1)}$$

$$= \frac{1(2^{11} - 1)}{(2 - 1)} - \frac{1(2^4 - 1)}{(2 - 1)}$$

$$= (2^{11} - 1) - (2^4 - 1) = 2^{11} - 2^4$$

$$= 2048 - 16 = \mathbf{2032}$$

Problem 25. A tool hire firm finds that their net return from hiring tools is decreasing by 10% per annum. If their net gain on a certain tool this year is £400, find the possible total of all future profits from this tool (assuming the tool lasts for ever).

The net gain forms a series: £400 + £400 × 0.9 + £400 × 0.9² + ⋯, which is a GP with $a = 400$ and $r = 0.9$

The sum to infinity, $S_\infty = \dfrac{a}{(1-r)} = \dfrac{400}{(1-0.9)}$

$$= \mathbf{£4000}$$

$$= \textbf{total future profits}$$

Problem 26. If £100 is invested at compound interest of 3% per annum, determine (a) the value after 10 years, (b) the time, correct to the nearest year, it takes to reach more than £150.

(a) Let the GP be $a, ar, ar^2, \ldots ar^n$

The first term, $a = £100$ and the common ratio, $r = 1.03$

Hence, the second term is: $ar = (100)(1.03) = £103$, which is the value after 1 year, the third term is: $ar^2 = (100)(1.03)^2 = £106.09$, which is the value after 2 years, and so on.

Thus, **the value after 10 years**

$$= ar^{10} = (100)(1.03)^{10} = \mathbf{£134.39}$$

(b) When £150 has been reached, $150 = ar^n$

i.e. $150 = 100(1.03)^n$

and $1.5 = (1.03)^n$

Taking logarithms to base 10 of both sides gives:

$$\lg 1.5 = \lg(1.03)^n = n\lg(1.03)$$

by the laws of logarithms

from which, $n = \dfrac{\lg 1.5}{\lg 1.03} = 13.7$

Hence, it will take 14 years to reach more than £150

Problem 27. A drilling machine is to have six speeds ranging from 50 rev/min to 750 rev/min. If the speeds form a geometric progression determine their values, each correct to the nearest whole number.

Let the GP of n terms be given by: $a, ar, ar^2, \ldots, ar^{n-1}$

The first term $a = 50$ rev/min

The 6th term is given by ar^{6-1}, which is 750 rev/min, i.e. $ar^5 = 750$

from which, $r^5 = \dfrac{750}{a} = \dfrac{750}{50} = 15$

Thus, the common ratio, $r = \sqrt[5]{15} = 1.7188$

The first term is: $\quad\quad a = 50\,\text{rev/min}$

The second term is: $\quad ar = (50)(1.7188) = 85.94,$

The third term is: $\quad ar^2 = (50)(1.7188)^2 = 147.71,$

The fourth term is: $\quad ar^3 = (50)(1.7188)^3 = 253.89,$

The fifth term is: $\quad ar^4 = (50)(1.7188)^4 = 436.39,$

The sixth term is: $\quad ar^5 = (50)(1.7188)^5 = 750.06$

Hence, correct to the nearest whole number, **the 6 speeds of the drilling machine are**:

50, 86, 148, 254, 436 and 750 rev/min

Now try the following Practice Exercise

Practice Exercise 193   Geometric progressions (answers on page 461)

1. In a geometric progression the 5th term is 9 times the 3rd term and the sum of the 6th and 7th terms is 1944. Determine (a) the common ratio, (b) the first term and (c) the sum of the 4th to 10th terms inclusive.

2. Which term of the series $3, 9, 27, \ldots$ is 59049?

3. The value of a lathe originally valued at £3000 depreciates 15% per annum. Calculate its value after 4 years. The machine is sold when its value is less than £550. After how many years is the lathe sold?

4. If the population of Great Britain is 64 million and is decreasing at 2.4% per annum, what will be the population in 5 years' time?

5. 100 g of a radioactive substance disintegrates at a rate of 3% per annum. How much of the substance is left after 11 years?

6. If £250 is invested at compound interest of 2.5% per annum, determine (a) the value after 15 years, (b) the time, correct to the nearest year, it takes to reach £425.

7. A drilling machine is to have 8 speeds ranging from 100 rev/min to 1000 rev/min. If the speeds form a geometric progression determine their values, each correct to the nearest whole number.

Practice Exercise 194   Multiple-choice questions on number sequences (Answers on page 461)

Each question has only one correct answer

1. The fifth term of an arithmetic progression is 18 and the twelfth term is 46. The eighteenth term is:
   (a) 72   (b) 74   (c) 68   (d) 70

2. The first term of a geometric progression is 9 and the fourth term is 45. The eighth term is:
   (a) 384.7   (b) 150.5   (c) 225   (d) 657.9

3. The 5th term of an arithmetic progression is 17 and the 9th term is 37. The 1st term is:
   (a) 0   (b) −3   (c) −6   (d) 3

4. The first and last term of an arithmetic progression are 1 and 13. If the sum of its terms is 42, then the number of terms is:
   (a) 5   (b) 6   (c) 7   (d) 8

5. The sum of three consecutive terms of an increasing arithmetic progression is 51 and the product of their terms is 4641. The third term is:
   (a) 9   (b) 13   (c) 17   (d) 21

6. The 7th and 12th terms of an arithmetic progression are 34 and 59, respectively. Its 18th term is:
   (a) 87   (b) 88   (c) 89   (d) 90

7. An arithmetic progression is given by 3, 7, 11, 15,.... If the sum of the series is 406, the number of terms is:
   (a) 5   (b) 10   (c) 12   (d) 14

8. The sum of the first 9 terms of the series 1, 3, 9, 27, … is:
   (a) 3280   (b) 9841   (c) 19682   (d) 6560

9. The sum to infinity of the series 4, 2, 1, … is:
   (a) 2   (b) 4   (c) 8   (d) 16

10. The 12th term of the series 4, 8, 16, 32, … is:
    (a) 8192   (b) 4096   (c) 16384   (d) 2048

For fully worked solutions to each of the problems in Exercises 188 to 193 in this chapter, go to the website:

www.routledge.com/cw/bird

Chapter 38

Binary, octal and hexadecimal numbers

Why it is important to understand: **Binary, octal and hexadecimal number**

There are infinite ways to represent a number. The four commonly associated with modern computers and digital electronics are decimal, binary, octal and hexadecimal. All four number systems are equally capable of representing any number. Furthermore, a number can be perfectly converted between the various number systems without any loss of numeric value. At a first look, it seems like using any number system other than decimal is complicated and unnecessary. However, since the job of electrical and software engineers is to work with digital circuits, engineers require number systems that can best transfer information between the human world and the digital circuit world. Thus the way in which a number is represented can make it easier for the engineer to perceive the meaning of the number as it applies to a digital circuit, i.e. the appropriate number system can actually make things less complicated.

At the end of this chapter you should be able to:

- recognise a binary number
- convert binary to decimal and vice versa
- add binary numbers
- recognise an octal number
- convert decimal to binary via octal and vice versa
- recognise a hexadecimal number
- convert from hexadecimal to decimal and vice versa
- convert from binary to hexadecimal and vice versa

38.1 Introduction

Man's earliest number or counting system was probably developed to help determine how many possessions a person had. As daily activities became more complex, numbers became more important in trade, time, distance and all other phases of human life. Ever since people discovered that it was necessary to count objects, they have been looking for easier ways to do so. The **abacus**, developed by the Chinese, is one of the earliest known calculators; it is still in use in some parts of the world. **Blaise Pascal**[*] invented the first adding machine in 1642. Twenty years later, an Englishman, **Sir Samuel Morland**[*], developed a more compact device that could multiply, add and subtract. About 1672, **Gottfried Wilhelm Leibniz**[*] perfected a machine that could perform all the basic operations (add, subtract, multiply, divide), as well as extract the square root. Modern electronic digital computers still use Leibniz's principles.

Computers are now employed wherever repeated calculations or the processing of huge amounts of data is needed. The greatest applications are found in the military, scientific and commercial fields. They have applications that range from mail sorting, and engineering design, to the identification and destruction of enemy targets. The advantages of digital computers include speed, accuracy and man-power savings. Often computers are able to take over routine jobs and release personnel for more important work that cannot be handled by a computer. People and computers do not normally speak the same language. Methods of translating information into forms that are understandable and usable to both are necessary. Humans generally speak in words and numbers expressed in the decimal number system, while computers only understand coded electronic pulses that represent digital information.

All data in modern computers is stored as series of **bits**, a bit being a **bi**nary dig**it**, and can have one of two values, the numbers 0 and 1. The most basic form of representing computer data is to represent a piece of data as a string of '1's and '0's, one for each bit. This is called a **binary** or base-2 number.

Because binary notation requires so many bits to represent relatively small numbers, two further compact notations are often used, called **octal** and **hexadecimal**. Computer programmers who design sequences of number codes instructing a computer what to do, would have a very difficult task if they were forced to work with nothing but long strings of '1's and '0's, the 'native language' of any digital circuit.

Octal notation represents data as base-8 numbers with each digit in an octal number representing three bits. Similarly, hexadecimal notation uses base-16 numbers, representing four bits with each digit. Octal numbers use only the digits 0–7, while hexadecimal numbers use all ten base-10 digits (0–9) and the letters A–F (representing the numbers 10–15).

This chapter explains how to convert between the decimal, binary, octal and hexadecimal systems.

*Who was Pascal? **Blaise Pascal** (19 June 1623–19 August 1662), was a French polymath. A child prodigy educated by his father, Pascal's earliest work was in the natural and applied sciences where he made important contributions to the study of fluids, and clarified the concepts of pressure and vacuum. To find out more go to www.routledge.com/cw/bird

*Who was Morland? **Sir Samuel Morland, 1st Baronet** (1625–1695), was an English academic, diplomat, spy, inventor and mathematician of the 17th century, a polymath credited with early developments in relation to computing, hydraulics and steam power. To find out more go to www.routledge.com/cw/bird

*Who was Leibniz? See page 388. To find out more go to www.routledge.com/cw/bird

38.2 Binary numbers

The system of numbers in everyday use is the **denary** or **decimal** system of numbers, using the digits 0 to 9. It has ten different digits (0, 1, 2, 3, 4, 5, 6, 7, 8 and 9) and is said to have a **radix** or **base** of 10.

The **binary** system of numbers has a radix of 2 and uses only the digits 0 and 1.

(a) Conversion of binary to decimal

The decimal number 234.5 is equivalent to
$$2 \times 10^2 + 3 \times 10^1 + 4 \times 10^0 + 5 \times 10^{-1}$$

In the binary system of numbers, the base is 2, so 1101.1 is equivalent to:
$$1 \times 2^3 + 1 \times 2^2 + 0 \times 2^1 + 1 \times 2^0 + 1 \times 2^{-1}$$

Thus, the decimal number equivalent to the binary number 1101.1 is:

$$8 + 4 + 0 + 1 + \frac{1}{2} \text{ that is } 13.5$$

i.e. $\mathbf{1101.1_2 = 13.5_{10}}$, the suffixes 2 and 10 denoting binary and decimal systems of numbers, respectively.

Problem 1. Convert 1010_2 to a decimal number.

From above:

$$1010_2 = 1 \times 2^3 + 0 \times 2^2 + 1 \times 2^1 + 0 \times 2^0$$
$$= 8 + 0 + 2 + 0$$
$$= \mathbf{10_{10}}$$

Problem 2. Convert 11011_2 to a decimal number.

$$11011_2 = 1 \times 2^4 + 1 \times 2^3 + 0 \times 2^2 + 1 \times 2^1 + 1 \times 2^0$$
$$= 16 + 8 + 0 + 2 + 1$$
$$= \mathbf{27_{10}}$$

Problem 3. Convert 0.1011_2 to a decimal fraction.

$$0.1011_2 = 1 \times 2^{-1} + 0 \times 2^{-2} + 1 \times 2^{-3} + 1 \times 2^{-4}$$
$$= 1 \times \frac{1}{2} + 0 \times \frac{1}{2^2} + 1 \times \frac{1}{2^3} + 1 \times \frac{1}{2^4}$$
$$= \frac{1}{2} + \frac{1}{8} + \frac{1}{16}$$
$$= 0.5 + 0.125 + 0.0625$$
$$= \mathbf{0.6875_{10}}$$

Problem 4. Convert 101.0101_2 to a decimal number.

$$101.0101_2 = 1 \times 2^2 + 0 \times 2^1 + 1 \times 2^0 + 0 \times 2^{-1}$$
$$+ 1 \times 2^{-2} + 0 \times 2^{-3} + 1 \times 2^{-4}$$
$$= 4 + 0 + 1 + 0 + 0.25 + 0 + 0.0625$$
$$= \mathbf{5.3125_{10}}$$

Now try the following Practice Exercise

Practice Exercise 195 Conversion of binary to decimal numbers (answers on page 461)

In Problems 1 to 4, convert the binary numbers given to decimal numbers.

1. (a) 110 (b) 1011 (c) 1110 (d) 1001

2. (a) 10101 (b) 11001 (c) 101101 (d) 110011

3. (a) 101010 (b) 111000
   (c) 1000001 (d) 10111000

4. (a) 0.1101 (b) 0.11001
   (c) 0.00111 (d) 0.01011

5. (a) 11010.11 (b) 10111.011
   (c) 110101.0111 (d) 11010101.10111

(b) Conversion of decimal to binary

An integer decimal number can be converted to a corresponding binary number by repeatedly dividing by 2 and noting the remainder at each stage, as shown below for 39_{10}

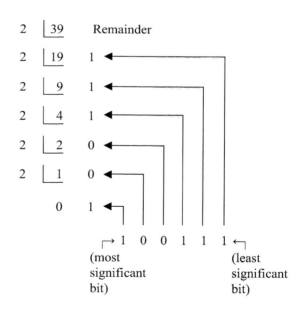

The result is obtained by writing the top digit of the remainder as the least significant bit, (the least significant bit is the one on the right). The bottom bit of the remainder is the most significant bit, i.e. the bit on the left.

Thus, $39_{10} = 100111_2$

The fractional part of a decimal number can be converted to a binary number by repeatedly multiplying by 2, as shown below for the fraction 0.625

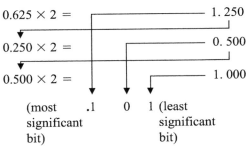

For fractions, the most significant bit of the result is the top bit obtained from the integer part of multiplication by 2. The least significant bit of the result is the bottom bit obtained from the integer part of multiplication by 2.

Thus, $0.625_{10} = 0.101_2$

Problem 5.   Convert 49_{10} to a binary number.

From above, repeatedly dividing by 2 and noting the remainder gives:

2	49	Remainder
2	24	1
2	12	0
2	6	0
2	3	0
2	1	1
	0	1

1 1 0 0 0 1
(most significant bit)   (least significant bit)

Thus, $49_{10} = 110001_2$

Problem 6.   Convert 56_{10} to a binary number.

The integer part is repeatedly divided by 2, giving:

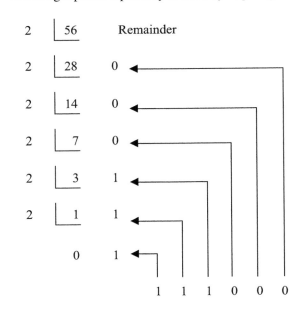

Thus, $56_{10} = 111000_2$

Problem 7.   Convert 0.40625_{10} to a binary number.

From above, repeatedly multiplying by 2 gives:

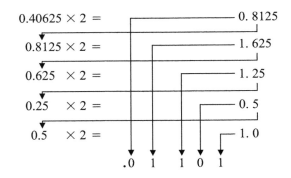

i.e. $0.40625_{10} = 0.01101_2$

Problem 8.   Convert 58.3125_{10} to a binary number.

The integer part is repeatedly divided by 2, giving:

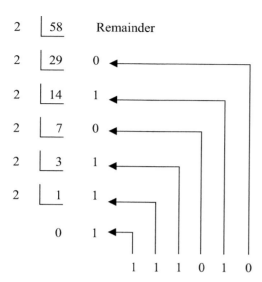

The fractional part is repeatedly multiplied by 2, giving:

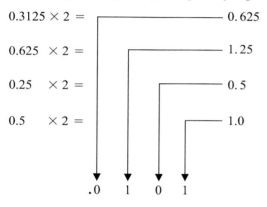

Thus, $58.3125_{10} = 111010.0101_2$

Now try the following Practice Exercise

Practice Exercise 196   Conversion of decimal to binary numbers (answers on page 462)

In Problems 1 to 5, convert the decimal numbers given to binary numbers.

1.   (a) 5   (b) 15   (c) 19   (d) 29

2.   (a) 31   (b) 42   (c) 57   (d) 63

3.   (a) 47   (b) 60   (c) 73   (d) 84

4.   (a) 0.25   (b) 0.21875
      (c) 0.28125   (d) 0.59375

5.   (a) 47.40625   (b) 30.8125
      (c) 53.90625   (d) 61.65625

(c) Binary addition

Binary addition of two/three bits is achieved according to the following rules:

	sum	carry		sum	carry
$0+0=0$		0	$0+0+0=0$		0
$0+1=1$		0	$0+0+1=1$		0
$1+0=1$		0	$0+1+0=1$		0
$1+1=0$		1	$0+1+1=0$		1
			$1+0+0=1$		0
			$1+0+1=0$		1
			$1+1+0=0$		1
			$1+1+1=1$		1

These rules are demonstrated in the following worked problems.

Problem 9.   Perform the binary addition: $1001 + 10110$

$$\begin{array}{r} 1001 \\ + \underline{10110} \\ \mathbf{\underline{11111}} \end{array}$$

Problem 10.   Perform the binary addition: $11111 + 10101$

$$\begin{array}{r} 11111 \\ + \underline{10101} \\ \text{sum } \mathbf{\underline{110100}} \\ \text{carry } 11111 \end{array}$$

Problem 11.   Perform the binary addition: $1101001 + 1110101$

$$\begin{array}{r} 1101001 \\ + \underline{1110101} \\ \text{sum } \mathbf{\underline{11011110}} \\ \text{carry } 11 \quad 1 \end{array}$$

Problem 12.   Perform the binary addition: $1011101 + 1100001 + 110101$

$$1011101$$
$$1100001$$
$$+ \quad 110101$$
sum **11110011**
carry 1 1 1 1 1 1

Now try the following Practice Exercise

Perform the following binary additions:

1. $10 + 11$
2. $101 + 110$
3. $1101 + 111$
4. $1111 + 11101$
5. $110111 + 10001$
6. $10000101 + 10000101$
7. $11101100 + 111001011$
8. $110011010 + 11100011$
9. $10110 + 1011 + 11011$
10. $111 + 10101 + 11011$
11. $1101 + 1001 + 11101$
12. $100011 + 11101 + 101110$

38.3   Octal numbers

For decimal integers containing several digits, repeatedly dividing by 2 can be a lengthy process. In this case, it is usually easier to convert a decimal number to a binary number via the octal system of numbers. This system has a radix of 8, using the digits 0, 1, 2, 3, 4, 5, 6 and 7. The decimal number equivalent to the octal number **4317$_8$** is:

$$4 \times 8^3 + 3 \times 8^2 + 1 \times 8^1 + 7 \times 8^0$$

i.e. $4 \times 512 + 3 \times 64 + 1 \times 8 + 7 \times 1 = \mathbf{2255_{10}}$

An integer decimal number can be converted to a corresponding octal number by repeatedly dividing by 8 and noting the remainder at each stage, as shown below for 493_{10}

Thus, $\mathbf{493_{10} = 755_8}$

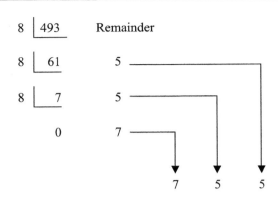

The fractional part of a decimal number can be converted to an octal number by repeatedly multiplying by 8, as shown below for the fraction 0.4375_{10}

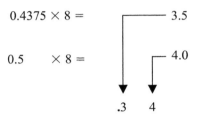

For fractions, the most significant bit is the top integer obtained by multiplication of the decimal fraction by 8, thus,

$$\mathbf{0.4375_{10} = 0.34_8}$$

The natural binary code for digits 0 to 7 is shown in Table 38.1, and an octal number can be converted to a binary number by writing down the three bits corresponding to the octal digit.

Table 38.1

Octal digit	Natural binary number
0	000
1	001
2	010
3	011
4	100
5	101
6	110
7	111

Thus, $437_8 = 100\ 011\ 111_2$

and $26.35_8 = 010\ 110.011\ 101_2$

The '0' on the extreme left does not signify anything, thus

$$26.35_8 = 10\ 110.011\ 101_2$$

Conversion of decimal to binary via octal is demonstrated in the following worked problems.

Problem 13. Convert 3714_{10} to a binary number via octal.

Dividing repeatedly by 8, and noting the remainder gives:

```
8 | 3714    Remainder

8 | 464      2
8 |  58      0
8 |   7      2
  |   0      7

            7  2  0  2
```

From Table 38.1, $7202_8 = 111\ 010\ 000\ 010_2$

i.e. $3714_{10} = 111\ 010\ 000\ 010_2$

Problem 14. Convert 0.59375_{10} to a binary number via octal.

Multiplying repeatedly by 8, and noting the integer values, gives:

$0.59375 \times 8 = 4.75$

$0.75 \times 8 = 6.00$

$.4 6$

Thus, $0.59375_{10} = 0.46_8$

From Table 38.1, $0.46_8 = 0.100\ 110_2$

i.e. $\mathbf{0.59375_{10} = 0.100\ 11_2}$

Problem 15. Convert 5613.90625_{10} to a binary number via octal.

The integer part is repeatedly divided by 8, noting the remainder, giving:

```
8 | 5613    Remainder

8 | 701      5
8 |  87      5
8 |  10      7
8 |   1      2
  |   0      1

            1  2  7  5  5
```

This octal number is converted to a binary number (see Table 38.1)

$$12755_8 = 001\ 010\ 111\ 101\ 101_2$$

i.e. $5613_{10} = 1\ 010\ 111\ 101\ 101_2$

The fractional part is repeatedly multiplied by 8, and noting the integer part, giving:

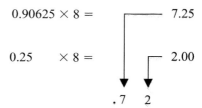

$0.90625 \times 8 = 7.25$

$0.25 \times 8 = 2.00$

$.7 2$

This octal fraction is converted to a binary number (see Table 38.1)

$$0.72_8 = 0.111\ 010_2$$

i.e. $0.90625_{10} = 0.111\ 01_2$

Thus, $\mathbf{5613.90625_{10} = 1\ 010\ 111\ 101\ 101.111\ 01_2}$

▶ **Problem 16.** Convert 11 110 011.100 01$_2$ to a decimal number via octal.

Grouping the binary number in three's from the binary point gives:

$$011\ 110\ 011.100\ 010_2$$

Using Table 38.1 to convert this binary number to an octal number gives:

$$363.42_8$$

and $363.42_8 = 3 \times 8^2 + 6 \times 8^1 + 3 \times 8^0$
$$+ 4 \times 8^{-1} + 2 \times 8^{-2}$$
$$= 192 + 48 + 3 + 0.5 + 0.03125_{10}$$
$$= 243.53125_{10}$$

i.e. **11 110 011.100 01$_2$ = 363.42$_8$ = 243.53125$_{10}$**

Now try the following Practice Exercise

Practice Exercise 198   Conversion between decimal and binary numbers via octal (answers on page 462)

In Problems 1 to 3, convert the decimal numbers given to binary numbers, via octal.

▶ 1.   (a) 343   (b) 572   (c) 1265

▶ 2.   (a) 0.46875   (b) 0.6875   (c) 0.71875

▶ 3.   (a) 247.09375 (b) 514.4375 (c) 1716.78125

▶ 4.   Convert the following binary numbers to decimal numbers via octal:

   (a) 111.011 1   (b) 101 001.01
   (c) 1 110 011 011 010.001 1

38.4 Hexadecimal numbers

The hexadecimal system is particularly important in computer programming, since four bits (each consisting of a one or zero) can be succinctly expressed using a single hexadecimal digit. Two hexadecimal digits represent numbers from 0 to 255, a common range used, for example, to specify colours. Thus, in the HTML language of the web, colours are specified using three pairs of hexadecimal digits RRGGBB, where RR is the amount of red, GG the amount of green, and BB the amount of blue.

A **hexadecimal numbering system** has a radix of 16 and uses the following 16 distinct digits:

   0, 1, 2, 3, 4, 5, 6, 7, 8, 9, A, B, C, D, E and F

'A' corresponds to 10 in the denary system, B to 11, C to 12, and so on.

(a) Converting from hexadecimal to decimal

For example, $1A_{16} = 1 \times 16^1 + A \times 16^0$
$$= 1 \times 16^1 + 10 \times 1$$
$$= 16 + 10 = 26$$

i.e. **$1A_{16} = 26_{10}$**

Similarly, $2E_{16} = 2 \times 16^1 + E \times 16^0$
$$= 2 \times 16^1 + 14 \times 16^0$$
$$= 32 + 14 = 46_{10}$$

and $1BF_{16} = 1 \times 16^2 + B \times 16^1 + F \times 16^0$
$$= 1 \times 16^2 + 11 \times 16^1 + 15 \times 16^0$$
$$= 256 + 176 + 15 = 447_{10}$$

Table 38.2 compares decimal, binary, octal and hexadecimal numbers and shows, for example, that

$$23_{10} = 10111_2 = 27_8 = 17_{16}$$

▶ **Problem 17.** Convert the following hexadecimal numbers into their decimal equivalents:
(a) $7A_{16}$ (b) $3F_{16}$

(a) $7A_{16} = 7 \times 16^1 + A \times 16^0 = 7 \times 16 + 10 \times 1$
$$= 112 + 10 = 122$$
Thus, **$7A_{16} = 122_{10}$**

(b) $3F_{16} = 3 \times 16^1 + F \times 16^0 = 3 \times 16 + 15 \times 1$
$$= 48 + 15 = 63$$
Thus, **$3F_{16} = 63_{10}$**

▶ **Problem 18.** Convert the following hexadecimal numbers into their decimal equivalents:
(a) $C9_{16}$ (b) BD_{16}

(a) $C9_{16} = C \times 16^1 + 9 \times 16^0 = 12 \times 16 + 9 \times 1$
$$= 192 + 9 = 201$$
Thus, **$C9_{16} = 201_{10}$**

Table 38.2

Decimal	Binary	Octal	Hexadecimal
0	0000	0	0
1	0001	1	1
2	0010	2	2
3	0011	3	3
4	0100	4	4
5	0101	5	5
6	0110	6	6
7	0111	7	7
8	1000	10	8
9	1001	11	9
10	1010	12	A
11	1011	13	B
12	1100	14	C
13	1101	15	D
14	1110	16	E
15	1111	17	F
16	10000	20	10
17	10001	21	11
18	10010	22	12
19	10011	23	13
20	10100	24	14
21	10101	25	15
22	10110	26	16
23	10111	27	17
24	11000	30	18
25	11001	31	19
26	11010	32	1A
27	11011	33	1B
28	11100	34	1C
29	11101	35	1D
30	11110	36	1E
31	11111	37	1F
32	100000	40	20

(b) $BD_{16} = B \times 16^1 + D \times 16^0 = 11 \times 16 + 13 \times 1$
$$= 176 + 13 = 189$$

Thus, $\mathbf{BD_{16} = 189_{10}}$

Problem 19. Convert $1A4E_{16}$ into a decimal number.

$1A4E_{16} = 1 \times 16^3 + A \times 16^2 + 4 \times 16^1 + E \times 16^0$
$$= 1 \times 16^3 + 10 \times 16^2 + 4 \times 16^1 + 14 \times 16^0$$
$$= 1 \times 4096 + 10 \times 256 + 4 \times 16 + 14 \times 1$$
$$= 4096 + 2560 + 64 + 14 = 6734$$

Thus, $\mathbf{1A4E_{16} = 6734_{10}}$

(b) Converting from decimal to hexadecimal

This is achieved by repeatedly dividing by 16 and noting the remainder at each stage; as shown below for 26_{10}

Hence, $\mathbf{26_{10} = 1A_{16}}$

Similarly, for 446_{10}

Thus, $\mathbf{446_{10} = 1BE_{16}}$

▶ Problem 20. Convert the following decimal numbers into their hexadecimal equivalents:
(a) 39_{10} (b) 107_{10}

(a)   16 ⌐ 39   Remainder

   16 ⌐ 2 $7 \equiv 7_{16}$ ────┐

      0      $2 \equiv 2_{16}$ ──┐

                       ┌→ 2   7 ←┐
                     (most   (least
                     significant significant
                     bit)   bit)

Hence, $39_{10} = 27_{16}$

(b)   16 ⌐ 107   Remainder

   16 ⌐ 6 $11 \equiv B_{16}$ ────┐

      0      $6 \equiv 6_{16}$ ──┐

                              6   B

Hence, $107_{10} = 6B_{16}$

▶ Problem 21. Convert the following decimal numbers into their hexadecimal equivalents:
(a) 164_{10} (b) 239_{10}

(a)   16 ⌐ 164   Remainder

   16 ⌐ 10 $4 \equiv 4_{16}$ ────┐

      0      $10 \equiv A_{16}$ ──┐

                            A   4

Hence, $164_{10} = A4_{16}$

(b)

   16 ⌐ 239   Remainder

   16 ⌐ 14 $15 \equiv F_{16}$ ──────┐

      0      $14 \equiv E_{16}$ ──┐

                              E   F

Hence, $239_{10} = EF_{16}$

Now try the following Practice Exercise

In Problems 1 to 4, convert the given hexadecimal numbers into their decimal equivalents.

▶1. $E7_{16}$ ▶2. $2C_{16}$ ▶3. 98_{16} ▶4. $2F1_{16}$

In Problems 5 to 8, convert the given decimal numbers into their hexadecimal equivalents.

▶5. 54_{10} ▶6. 200_{10} ▶7. 91_{10} ▶8. 238_{10}

(c) Converting from binary to hexadecimal

The binary bits are arranged in groups of four, starting from right to left, and a hexadecimal symbol is assigned to each group. For example, the binary number 1110011110101001

is initially grouped in
          fours as: 1110   0111   1010   1001

and a hexadecimal symbol
assigned to each group as: E   7   A   9
                                          from Table 38.2

Hence, $1110011110101001_2 = E7A9_{16}$

▶ Problem 22. Convert the following binary numbers into their hexadecimal equivalents:
(a) 11010110_2 (b) 1100111_2

(a)   Grouping bits in fours from the
                right gives: 1101   0110
      and assigning hexadecimal symbols
                to each group gives: D   6
                                     from Table 38.2

Thus, $11010110_2 = D6_{16}$

(b) Grouping bits in fours from the
right gives: 0110   0111
and assigning hexadecimal symbols
to each group gives: 6   7
from Table 38.2

Thus, $1100111_2 = 67_{16}$

Problem 23. Convert the following binary numbers into their hexadecimal equivalents:
(a) 11001111_2 (b) 110011110_2

(a) Grouping bits in fours from the
right gives: 1100   1111
and assigning hexadecimal symbols
to each group gives: C   F
from Table 38.2

Thus, $11001111_2 = CF_{16}$

(b) Grouping bits in fours
from the right gives: 0001   1001   1110
and assigning hexadecimal
symbols to each group gives: 1   9   E
from Table 38.2

Thus, $110011110_2 = 19E_{16}$

(d) Converting from hexadecimal to binary

The above procedure is reversed, thus, for example,

$6CF3_{16} = 0110\ 1100\ 1111\ 0011$ from Table 38.2

i.e. $6CF3_{16} = 110110011110011_2$

Problem 24. Convert the following hexadecimal numbers into their binary equivalents:
(a) $3F_{16}$ (b) $A6_{16}$

(a) Spacing out hexadecimal
digits gives: 3   F
and converting each into
binary gives: 0011   1111
from Table 38.2

Thus, $3F_{16} = 111111_2$

(b) Spacing out hexadecimal
digits gives: A   6
and converting each into
binary gives: 1010   0110
from Table 38.2

Thus, $A6_{16} = 10100110_2$

Problem 25. Convert the following hexadecimal numbers into their binary equivalents:
(a) $7B_{16}$ (b) $17D_{16}$

(a) Spacing out hexadecimal
digits gives: 7   B
and converting each into
binary gives: 0111   1011
from Table 38.2

Thus, $7B_{16} = 1111011_2$

(b) Spacing out hexadecimal
digits gives: 1   7   D
and converting each into
binary gives: 0001   0111   1101
from Table 38.2

Thus, $17D_{16} = 101111101_2$

Now try the following Practice Exercise

Practice Exercise 200   Hexadecimal numbers (answers on page 462)

In Problems 1 to 4, convert the given binary numbers into their hexadecimal equivalents.

1. 11010111_2 2. 11101010_2
3. 10001011_2 4. 10100101_2

In Problems 5 to 8, convert the given hexadecimal numbers into their binary equivalents.

5. 37_{16} 6. ED_{16}
7. $9F_{16}$ 8. $A21_{16}$

Practice Exercise 201   Multiple-choice questions on binary, octal and hexadecimal numbers (answers on page 462)

Each question has only one correct answer

1. The base of a hexadecimal number system is:
   (a) 6 (b) 8 (c) 16 (d) 10

2. The number of digits in a binary number system is:
   (a) 10 (b) 2 (c) 4 (d) 6

3. The number system which uses the alphabet as well as numbers is called:
   (a) binary   (b) octal
   (c) decimal   (d) hexadecimal

4. The value of the base of a decimal number system is:
   (a) 8   (b) 2   (c) 10   (d) 16

5. The hexadecimal number equivalent of 15 is:
   (a) A   (b) F   (c) D   (d) E

6. The symbol D in the hexadecimal system corresponds to:
   (a) 8   (b) 16   (c) 13   (d) 14

7. Converting the binary number 1000 to decimal gives:
   (a) 2   (b) 4   (c) 6   (d) 8

8. The number of digits required to represent a decimal number 31 in the equivalent binary form is:
   (a) 2   (b) 4   (c) 5   (d) 6

9. In decimal, the binary number 1101101 is equal to:
   (a) 218   (b) 31   (c) 127   (d) 109

10. In hexadecimal, the decimal number 123 is:
    (a) 7B   (b) 123   (c) 173   (d) 1111011

11. Converting the binary number 11001 to decimal gives:
    (a) 3   (b) 50   (c) 25   (d) 13

12. Converting the hexadecimal number B2 to binary gives:
    (a) 10110010   (b) 101101
    (c) 11100   (d) 11000010

13. Converting the binary number 11011 to hexadecimal gives:
    (a) 1A   (b) B1   (c) A1   (d) 1B

14. Converting the decimal number 20 to hexadecimal gives:
    (a) 11   (b) 14   (c) 1B   (d) 1A

15. Converting the hexadecimal number 2C to decimal gives:
    (a) 3A   (b) 34   (c) 44   (d) 54

16. The binary addition $1111 + 1111$ gives:
    (a) 0000   (b) 11111   (c) 11110   (d) 2222

17. Converting the decimal number 21 to binary gives:
    (a) 10101   (b) 10001
    (c) 10000   (d) 11111

18. The hexadecimal equivalent of the binary number 1110 is:
    (a) E   (b) 0111   (c) 15   (d) 14

19. The maximum value of a single digit in an octal system is:
    (a) 7   (b) 9   (c) 6   (d) 5

20. The binary number 111 in octal is:
    (a) 6   (b) 7   (c) 8   (d) 5

21. Converting the octal number 22_8 into its corresponding decimal number is:
    (a) 18   (b) 26   (c) 81   (d) 82

22. Which of the following binary numbers is equivalent to the decimal number 24?
    (a) 1101111   (b) 11000
    (c) 111111   (d) 11001

23. The binary number corresponding to the hexadecimal number B2 is:
    (a) 100011   (b) 11011
    (c) 10110010   (d) 10010010

24. In binary, the decimal number 45 is:
    (a) 11100   (b) 101101
    (c) 10100   (d) 110101

25. In decimal, the hexadecimal number 52_{16} is:
    (a) 28   (b) 83   (c) 80   (d) 82

For fully worked solutions to each of the problems in Practice Exercises 195 to 200 in this chapter, go to the website:
www.routledge.com/cw/bird

This assignment covers the material contained in Chapters 37 and 38. *The marks for each question are shown in brackets at the end of each question.*

1. Determine the 20th term of the series 15.6, 15, 14.4, 13.8, . . . (3)

2. The sum of 13 terms of an arithmetic progression is 286 and the common difference is 3. Determine the first term of the series. (5)

3. An engineer earns £21 000 per annum and receives annual increments of £600. Determine the salary in the 9th year and calculate the total earnings in the first 11 years. (5)

4. Determine the 11th term of the series 1.5, 3, 6, 12, . . . (2)

5. Find the sum of the first eight terms of the series $1, 2\frac{1}{2}, 6\frac{1}{4}, \ldots$, correct to 1 decimal place. (4)

6. Determine the sum to infinity of the series 5, $1, \frac{1}{5}, \ldots$ (3)

7. A machine is to have seven speeds ranging from 25 rev/min to 500 rev/min. If the speeds form a geometric progression, determine their value, each correct to the nearest whole number. (9)

8. Convert the following to decimal numbers:
   (a) 11010_2 (b) 101110_2 (6)

9. Convert the following decimal numbers into binary numbers:
   (a) 53   (b) 29 (8)

10. Determine the binary addition: $1011 + 11011$ (3)

11. Convert the hexadecimal number 3B into its binary equivalent. (2)

12. Convert 173_{10} into hexadecimal. (4)

13. Convert 1011011_2 into hexadecimal. (3)

14. Convert DF_{16} into its binary equivalent. (3)

List of formulae

Length in metric units:

$$1 \text{ m} = 100 \text{ cm} = 1000 \text{ mm}$$

Areas in metric units:

$$1 \text{ m}^2 = 10^4 \text{ cm}^2 \qquad 1 \text{ cm}^2 = 10^{-4} \text{m}^2$$

$$1 \text{ m}^2 = 10^6 \text{ mm}^2 \qquad 1 \text{ mm}^2 = 10^{-6} \text{ m}^2$$

$$1 \text{ cm}^2 = 10^2 \text{ mm}^2 \qquad 1 \text{ mm}^2 = 10^{-2} \text{ cm}^2$$

Volumes in metric units:

$$1 \text{ m}^3 = 10^6 \text{ cm}^2 \qquad 1 \text{ cm}^3 = 10^{-6} \text{m}^3$$

$$1 \text{ litre} = 1000 \text{ cm}^3$$

$$1 \text{ m}^3 = 10^9 \text{ mm}^3 \qquad 1 \text{ mm}^3 = 10^{-9} \text{ m}^3$$

$$1 \text{ cm}^3 = 10^3 \text{ mm}^3 \qquad 1 \text{ mm}^3 = 10^{-3} \text{ cm}^3$$

Laws of indices:

$$a^m \times a^n = a^{m+n} \qquad \frac{a^m}{a^n} = a^{m-n} \qquad \left(a^m\right)^n = a^{mn}$$

$$a^{m/n} = \sqrt[n]{a^m} \qquad a^{-n} = \frac{1}{a^n} \qquad a^0 = 1$$

Quadratic formula:

$$\text{If } ax^2 + bx + c = 0 \quad \text{then} \quad x = \frac{-b \pm \sqrt{b^2 - 4ac}}{2a}$$

Equation of a straight line:

$$y = mx + c$$

Definition of a logarithm:

$$\text{If } y = a^x \quad \text{then} \quad x = \log_a y$$

Laws of logarithms:

$$\log(A \times B) = \log A + \log B$$

$$\log\left(\frac{A}{B}\right) = \log A - \log B$$

$$\log A^n = n \times \log A$$

Exponential series:

$$e^x = 1 + x + \frac{x^2}{2!} + \frac{x^3}{3!} + \cdots \quad \text{(valid for all values of } x\text{)}$$

Theorem of Pythagoras:

$$b^2 = a^2 + c^2$$

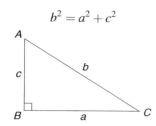

Areas of plane figures:

(i) **Rectangle**   Area $= l \times b$

(ii) **Parallelogram**   Area $= b \times h$

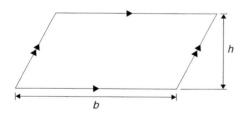

(iii) **Trapezium**   Area $= \dfrac{1}{2}(a+b)h$

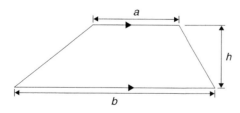

(iv) **Triangle**   Area $= \dfrac{1}{2} \times b \times h$

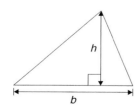

(v) **Circle**   Area $= \pi r^2$ Circumference $= 2\pi r$

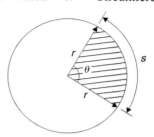

Radian measure: 2π radians $= 360$ degrees

For a sector of circle:

arc length, $s = \dfrac{\theta^\circ}{360}(2\pi r) = r\theta$ (θ in rad)

shaded area $= \dfrac{\theta^\circ}{360}(\pi r^2) = \dfrac{1}{2}r^2\theta$ (θ in rad)

Equation of a circle, centre at origin, radius r:

$$x^2 + y^2 = r^2$$

Equation of a circle, centre at (a, b), radius r:

$$(x-a)^2 + (y-b)^2 = r^2$$

Volumes and surface areas of regular solids:

(i) **Rectangular prism (or cuboid)**

Volume $= l \times b \times h$

Surface area $= 2(bh + hl + lb)$

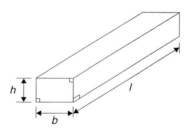

(ii) **Cylinder**

Volume $= \pi r^2 h$

Total surface area $= 2\pi rh + 2\pi r^2$

(iii) **Pyramid**

If area of base $= A$ and

perpendicular height $= h$ then:

$$\text{Volume} = \frac{1}{3} \times A \times h$$

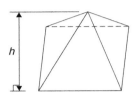

Total surface area = sum of areas of triangles forming sides + area of base

(iv) **Cone**

$$\text{Volume} = \frac{1}{3}\pi r^2 h$$

$$\text{Curved surface area} = \pi rl$$

$$\text{Total surface area} = \pi rl + \pi r^2$$

(v) **Sphere**

$$\text{Volume} = \frac{4}{3}\pi r^3$$

$$\text{Surface area} = 4\pi r^2$$

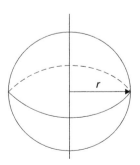

Areas of irregular figures by approximate methods:

Trapezoidal rule

$$\text{Area} \approx \begin{pmatrix} \text{width of} \\ \text{interval} \end{pmatrix} \left[\frac{1}{2} \begin{pmatrix} \text{first } + \text{ last} \\ \text{ordinate} \end{pmatrix} \right. $$
$$\left. + \text{ sum of remaining ordinates} \right]$$

Mid-ordinate rule

$$\text{Area} \approx (\text{width of interval})(\text{sum of mid-ordinates})$$

Simpson's rule

$$\text{Area} \approx \frac{1}{3} \begin{pmatrix} \text{width of} \\ \text{interval} \end{pmatrix} \left[\begin{pmatrix} \text{first } + \text{ last} \\ \text{ordinate} \end{pmatrix} \right.$$
$$\left. + 4 \begin{pmatrix} \text{sum of even} \\ \text{ordinates} \end{pmatrix} + 2 \begin{pmatrix} \text{sum of remaining} \\ \text{odd ordinates} \end{pmatrix} \right]$$

Mean or average value of a waveform:

$$\text{mean value, y} = \frac{\text{area under curve}}{\text{length of base}}$$

$$\approx \frac{\text{sum of mid-ordinates}}{\text{number of mid-ordinates}}$$

Triangle formulae:

Sine rule: $\quad \dfrac{a}{\sin A} = \dfrac{b}{\sin B} = \dfrac{c}{\sin C}$

Cosine rule: $\quad a^2 = b^2 + c^2 - 2bc\cos A$

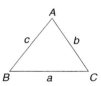

Area of any triangle

$$= \frac{1}{2} \times \text{base} \times \text{perpendicular height}$$

$$= \frac{1}{2}ab\sin C \quad \text{or} \quad \frac{1}{2}ac\sin B \quad \text{or} \quad \frac{1}{2}bc\sin A$$

$$= \sqrt{[s(s-a)(s-b)(s-c)]} \quad \text{where} \quad s = \frac{a+b+c}{2}$$

For a **general sinusoidal function** $y = A \sin(\omega t \pm \alpha)$, then

$$A = \text{amplitude}$$

$$\omega = \text{angular velocity} = 2\pi f \,\text{rad/s}$$

$$\frac{\omega}{2\pi} = \text{frequency}, f \text{ hertz}$$

$$\frac{2\pi}{\omega} = \text{periodic time } T \text{ seconds}$$

$$\alpha = \text{angle of lead or lag (compared with}$$
$$y = A \sin \omega t)$$

Cartesian and polar co-ordinates:

If co-ordinate $(x, y) = (r, \theta)$ then

$$r = \sqrt{x^2 + y^2} \quad \text{and} \quad \theta = \tan^{-1} \frac{y}{x}$$

If co-ordinate $(r, \theta) = (x, y)$ then

$$x = r \cos \theta \quad \text{and} \quad y = r \sin \theta$$

Arithmetic progression:

If $a = $ first term and $d = $ common difference, then the arithmetic progression is: $a, a + d, a + 2d, \ldots$

The nth term is: $a + (n - 1)d$

Sum of n terms, $S_n = \dfrac{n}{2}[2a + (n - 1)d]$

Geometric progression:

If $a = $ first term and $r = $ common ratio, then the geometric progression is: $a, ar, ar^2, \ldots$

The nth term is: ar^{n-1}

Sum of n terms, $S_n = \dfrac{a(1 - r^n)}{(1 - r)}$ or $\dfrac{a(r^n - 1)}{(r - 1)}$

If $-1 < r < 1$, $S_\infty = \dfrac{a}{(1 - r)}$

Statistics:

Discrete data:

$$\text{mean}, \bar{x} = \frac{\sum x}{n}$$

$$\text{standard deviation}, \sigma = \sqrt{\left[\frac{\sum(x - \bar{x})^2}{n}\right]}$$

Grouped data:

$$\text{mean}, \bar{x} = \frac{\sum fx}{\sum f}$$

$$\text{standard deviation}, \sigma = \sqrt{\left[\frac{\sum\{f(x - \bar{x})^2\}}{\sum f}\right]}$$

Standard derivatives

y or $f(x)$	$\dfrac{dy}{dx} = $ or $f'(x)$
ax^n	anx^{n-1}
$\sin ax$	$a \cos ax$
$\cos ax$	$-a \sin ax$
e^{ax}	ae^{ax}
$\ln ax$	$\dfrac{1}{x}$

Standard integrals

y	$\int y\, dx$
ax^n	$a\dfrac{x^{n+1}}{n+1} + c$ (except when $n = -1$)
$\cos ax$	$\dfrac{1}{a}\sin ax + c$
$\sin ax$	$-\dfrac{1}{a}\cos ax + c$
e^{ax}	$\dfrac{1}{a}e^{ax} + c$
$\dfrac{1}{x}$	$\ln x + c$

For a copy of the list of formulae, go to:
www.routledge.com/cw/bird

Answers

Answers to Practice Exercises

Chapter 1

Exercise 1 (page 3)

1. 19 kg
2. 16 m
3. 479 mm
4. −66
5. £565
6. −225
7. −2136
8. −36 121
9. £107 701
10. −4
11. 1487
12. 5914
13. 189 g
14. −70872
15. $15 333
16. $d = 64$ mm, $A = 136$ mm, $B = 10$ mm

Exercise 2 (page 5)

1. (a) 468 (b) 868
2. (a) £1827 (b) £4158
3. (a) 8613 kg (b) 584 kg
4. (a) 351 mm (b) 924 mm
5. (a) 10 304 (b) −4433
6. (a) 48 m (b) 89 m
7. (a) 259 (b) 56
8. (a) 1648 (b) −1060
9. (a) 8067 (b) 3347
10. 18 kg
11. 29
12. 14 000

Exercise 3 (page 6)

1. (a) 4 (b) 24
2. (a) 12 (b) 360
3. (a) 10 (b) 350
4. (a) 90 (b) 2700
5. (a) 2 (b) 210
6. (a) 3 (b) 180
7. (a) 5 (b) 210
8. (a) 15 (b) 6300
9. (a) 14 (b) 420 420
10. (a) 14 (b) 53 900

Exercise 4 (page 8)

1. 59
2. 14
3. 88
4. 5
5. 33
6. 22
7. 68
8. 5
9. 2
10. 5
11. −1

Exercise 5 (page 8)

1. (c)
2. (c)
3. (b)
4. (a)
5. (a)
6. (b)
7. (a)
8. (d)
9. (c)
10. (b)
11. (d)
12. (d)
13. (c)
14. (d)
15. (b)

Chapter 2

Exercise 6 (page 13)

1. $2\frac{1}{7}$
2. $7\frac{2}{5}$
3. $\frac{22}{9}$
4. $\frac{71}{8}$
5. $\frac{4}{11}$
6. $\frac{3}{7}, \frac{4}{9}, \frac{1}{2}, \frac{3}{5}, \frac{5}{8}$
7. $\frac{8}{25}$
8. $\frac{11}{15}$
9. $\frac{17}{30}$
10. $\frac{9}{10}$
11. $\frac{3}{16}$
12. $\frac{43}{77}$
13. $\frac{47}{63}$
14. $1\frac{1}{15}$
15. $\frac{4}{27}$
16. $8\frac{51}{52}$
17. $1\frac{9}{40}$
18. $1\frac{16}{21}$
19. $\frac{17}{60}$
20. $\frac{17}{20}$
21. 14

Exercise 7 (page 14)

1. $\dfrac{8}{35}$ **2.** $2\dfrac{2}{9}$ **3.** $\dfrac{6}{11}$ **4.** $\dfrac{5}{12}$ **5.** $\dfrac{3}{28}$

6. $\dfrac{3}{5}$ **7.** 11 **8.** $\dfrac{1}{13}$ **9.** $1\dfrac{1}{2}$ **10.** $\dfrac{8}{15}$

11. $2\dfrac{2}{5}$ **12.** $\dfrac{5}{12}$ **13.** $3\dfrac{3}{4}$ **14.** $\dfrac{12}{23}$ **15.** 4

16. $\dfrac{3}{4}$ **17.** $\dfrac{1}{9}$ **18.** 13 **19.** 15 **20.** 400 litres

21. (a) £60, (b) P£36, Q£16   **22.** 2880 litres

Exercise 8 (page 16)

1. $2\dfrac{1}{18}$ **2.** $-\dfrac{1}{9}$ **3.** $1\dfrac{1}{6}$ **4.** $4\dfrac{3}{4}$ **5.** $\dfrac{13}{20}$

6. $\dfrac{7}{15}$ **7.** $4\dfrac{19}{20}$ **8.** 2 **9.** $7\dfrac{1}{3}$ **10.** $\dfrac{1}{15}$

11. 4 **12.** $2\dfrac{17}{20}$

Exercise 9 (page 16)

1. (c) **2.** (b) **3.** (c) **4.** (a) **5.** (d)
6. (a) **7.** (c) **8.** (a) **9.** (b) **10.** (a)
11. (c) **12.** (b) **13.** (d) **14.** (b) **15.** (d)

Chapter 3

Exercise 10 (page 21)

1. $\dfrac{13}{20}$ **2.** $\dfrac{9}{250}$ **3.** $\dfrac{7}{40}$ **4.** $\dfrac{6}{125}$

5. (a) $\dfrac{33}{50}$ (b) $\dfrac{21}{25}$ (c) $\dfrac{1}{80}$ (d) $\dfrac{141}{500}$ (e) $\dfrac{3}{125}$

6. $4\dfrac{21}{40}$ **7.** $23\dfrac{11}{25}$ **8.** $10\dfrac{3}{200}$ **9.** $6\dfrac{7}{16}$

10. (a) $1\dfrac{41}{50}$ (b) $4\dfrac{11}{40}$ (c) $14\dfrac{1}{8}$ (d) $15\dfrac{7}{20}$ (e) $16\dfrac{17}{80}$

11. 0.625 **12.** 6.6875 **13.** 0.21875 **14.** 11.1875

15. 0.28125

Exercise 11 (page 22)

1. 14.18 **2.** 2.785 **3.** 65.38
4. 43.27 **5.** 1.297 **6.** 0.000528

Exercise 12 (page 23)

1. 80.3 **2.** 329.3 **3.** 72.54 **4.** -124.83
5. 295.3 **6.** 18.3 **7.** $12.52\,\text{mm}$

Exercise 13 (page 24)

1. 4.998 **2.** 47.544 **3.** 385.02 **4.** 582.42
5. 456.9 **6.** 434.82 **7.** 626.1 **8.** 1591.6
9. 0.444 **10.** 0.62963 **11.** 1.563 **12.** 53.455
13. 13.84 **14.** 8.69 **15.** (a) 24.81 (b) 24.812
16. (a) 0.00639 (b) 0.0064 **17.** (a) $8.\dot{4}$ (b) $62.\dot{6}$
18. 2400 **19.** 89.25 cm

Exercise 14 (page 25)

1. (d) **2.** (a) **3.** (b) **4.** (c) **5.** (c)
6. (b) **7.** (b) **8.** (d) **9.** (d) **10.** (a)

Chapter 4

Exercise 15 (page 27)

1. 30.797 **2.** 11927 **3.** 13.62 **4.** 53.832
5. 84.42 **6.** 1.0944 **7.** 50.330 **8.** 36.45
9. 10.59 **10.** 12.325

Exercise 16 (page 28)

1. 12.25 **2.** 0.0361 **3.** 46.923 **4.** 1.296×10^{-3}
5. 2.4430 **6.** 2.197 **7.** 30.96 **8.** 0.0549
9. 219.26 **10.** 5.832×10^{-6}

Exercise 17 (page 29)

1. 0.571 **2.** 40 **3.** 0.13459 **4.** 137.9
5. 14.96 **6.** 19.4481 **7.** 515.36×10^{-6}
8. 1.0871 **9.** 15.625×10^{-9} **10.** 52.70

Exercise 18 (page 30)

1. 2.182 **2.** 11.122 **3.** 185.82
4. 0.8307 **5.** 0.1581 **6.** 2.571
7. 5.273 **8.** 1.2555 **9.** 0.30366
10. 1.068 **11.** 3.5×10^{6} **12.** 37.5×10^{3}
13. 4.2×10^{-6} **14.** 202.767×10^{-3} **15.** 18.32×10^{6}

Exercise 19 (page 31)

1. 0.4667 2. $\dfrac{13}{14}$ 3. 4.458 4. 2.732

5. $\dfrac{1}{21}$ 6. 0.7083 7. $-\dfrac{9}{10}$ 8. $3\dfrac{1}{3}$

9. 2.567 10. 0.0776

Exercise 20 (page 32)

1. 0.9205 2. 0.7314 3. 2.9042 4. 0.2719
5. 0.4424 6. 0.0321 7. 0.4232 8. 0.1502
9. −0.6992 10. 5.8452

Exercise 21 (page 32)

1. 4.995 2. 5.782 3. 25.72 4. 69.42
5. 0.6977 6. 52.92 7. 591.0 8. 17.90
9. 3.520 10. 0.3770

Exercise 22 (page 34)

1. $A = 66.59\,\text{cm}^2$ 2. $C = 52.78\,\text{mm}$ 3. $R = 37.5$
4. 159 m/s 5. 0.407 A 6. 5.02 mm
7. 0.144 J 8. $628.8\,\text{m}^2$ 9. 224.5
10. $14\,230\,\text{kg}/\text{m}^3$ 11. 281.1 m/s 12. $2.526\,\Omega$

Exercise 23 (page 35)

1. £589.27 2. 508.1 W 3. $V = 2.61\,\text{V}$
4. $F = 854.5$ 5. $I = 3.81\,\text{A}$ 6. $t = 14.79\,\text{s}$
7. $E = 3.96\,\text{J}$ 8. $I = 12.77\,\text{A}$ 9. $s = 17.25\,\text{m}$
10. $A = 7.184\,\text{cm}^2$ 11. $v = 7.327$
12. (a) 12.53 h (b) 1 h 40 min, 33 m.p.h.
   (c) 13.02 h (d) 13.15 h

Exercise 24 (page 36)

1. (b) 2. (d) 3. (a) 4. (c) 5. (b)
6. (c) 7. (b) 8. (a) 9. (d) 10. (d)

Chapter 5

Exercise 25 (page 40)

1. 0.32% 2. 173.4% 3. 5.7% 4. 37.4%
5. 128.5% 6. 0.20 7. 0.0125 8. 68.75%
9. 38.462% 10. (a) 21.2% (b) 79.2% (c) 169%

11. (b), (d), (c), (a) 12. $\dfrac{13}{20}$ 13. $\dfrac{5}{16}$ 14. $\dfrac{9}{16}$

15. $A = \dfrac{1}{2}$, $B = 50\%$, $C = 0.25$, $D = 25\%$, $E = 0.30$,
   $F = \dfrac{3}{10}$, $G = 0.60$, $H = 60\%$, $I = 0.85$, $J = \dfrac{17}{20}$

Exercise 26 (page 42)

1. 21.8 kg 2. 9.72 m
3. (a) 496.4 t (b) 8.657 g (c) 20.73 s 4. 2.25%
5. (a) 14% (b) 15.67% (c) 5.36% 6. 37.8 g
7. 14 minutes 57 seconds 8. 76 g 9. £611
10. 37.49% 11. 39.2% 12. 17% 13. 38.7%
14. 2.7% 15. 5.60 m 16. 3.5%
17. (a) (i) $544\,\Omega$ (ii) $816\,\Omega$
   (b) (i) $44.65\,\text{k}\Omega$ (ii) $49.35\,\text{k}\Omega$
18. 2592 rev/min 19. 126 20. 95.9238%

Exercise 27 (page 44)

1. 2.5% 2. 18% 3. £310 4. £175 000
5. £260 6. £20 000 7. £8241.80 8. £50.25
9. £39.60 10. £917.70 11. £185 000 12. 7.2%
13. A 0.6 kg, B 0.9 kg, C 0.5 kg
14. 54%, 31%, 15%, 0.3 t
15. 20 000 kg (or 20 tonnes)
16. 13.5 mm, 11.5 mm 17. 600 kW

Exercise 28 (page 45)

1. (d) 2. (a) 3. (c) 4. (d) 5. (c)
6. (b) 7. (c) 8. (b) 9. (a) 10. (a)
11. (d) 12. (b) 13. (a) 14. (b) 15. (c)

Chapter 6

Exercise 29 (page 49)

1. 36 : 1 2. 3.5 : 1 or 7 : 2 3. 47 : 3
4. 96 cm, 240 cm 5. $5\dfrac{1}{4}$ hours or 5 hours 15 minutes
6. £3680, £1840, £920 7. 12 cm 8. £2172
9. P 4, Q 32 10. 12 kg, 20 kg, 24 kg

Exercise 30 (page 50)

1. 1 : 15 2. 76 ml 3. 25% 4. 12.6 kg
5. 14.3 kg 6. 25 000 kg

Exercise 31 (page 52)

1. £556 2. £66 3. 264 kg 4. 450 N 5. 14.56 kg
6. (a) 0.00025 (b) 48 MPa 7. (a) 440 K (b) 5.76 litre
8. 8960 bolts

Exercise 32 (page 54)

1. (a) 2 mA (b) 25 V 2. 372 fr   3. 685.8 mm
4. 83 lb 10 oz   5. (a) 159.1 litres (b) 16.5 gallons
6. 29.4 MPa   7. 584.2 mm 8. $1012

Exercise 33 (page 55)

1. 3.5 weeks   2. 20 days
3. (a) 9.18 (b) 6.12 (c) 0.3375   4. 50 minutes
5. (a) 300×10^3 (b) 0.375 m^3 (c) $240 \times 10^3 \text{ Pa}$
6. (a) 32 J (b) 0.5 m

Exercise 34 (page 56)

1. (c)   2. (d)   3. (c)   4. (a)   5. (b)
6. (c)   7. (d)   8. (b)   9. (a)   10. (a)

Chapter 7

Exercise 35 (page 59)

1. 27   2. 128   3. 100 000   4. 96   5. 2^4
6. ± 5 7. ± 8 8. 100   9. 1   10. 64

Exercise 36 (page 61)

1. 128   2. 3^9 3. 16   4. $\frac{1}{9}$ 5. 1 6. 8

7. 100   8. 1000   9. $\frac{1}{100}$ or 0.01 10. 5 11. 7^6

12. 3^6 13. 3^6 or 729 14. 3^4 15. 1   16. 25

17. $\frac{1}{3^5}$ or $\frac{1}{243}$ 18. 49 19. $\frac{1}{2}$ or 0.5   20. 1

Exercise 37 (page 62)

1. $\frac{1}{3 \times 5^2}$ 2. $\frac{1}{7^3 \times 3^7}$ 3. $\frac{3^2}{2^5}$ 4. $\frac{1}{2^{10} \times 5^2}$

5. 9   6. 3   7. $\frac{1}{2}$ 8. $\pm\frac{2}{3}$

9. $\frac{147}{148}$ 10. $-1\frac{19}{56}$ 11. $-3\frac{13}{45}$ 12. $\frac{1}{9}$

13. $-\frac{17}{18}$ 14. 64   15. $4\frac{1}{2}$

Exercise 38 (page 63)

1. (d)   2. (b)   3. (a)   4. (c)   5. (c)
6. (a)   7. (d)   8. (b)   9. (c)   10. (c)
11. (b)   12. (b)   13. (b)   14. (a)   15. (d)
16. (b)   17. (c)   18. (a)   19. (c)   20. (d)
21. (c)   22. (a)   23. (d)   24. (a)   25. (d)

Chapter 8

Exercise 39 (page 68)

1. cubic metres, m^3 2. farad
3. square metres, m^2 4. metres per second, m/s
5. kilogram per cubic metre, kg/m^3
6. joule   7. coulomb   8. watt
9. radian or degree   10. volt   11. mass
12. electrical resistance 13. frequency
14. acceleration   15. electric current
16. inductance   17. length
18. temperature   19. pressure
20. angular velocity   21. $\times 10^9$ 22. m, $\times 10^{-3}$
23. $\times 10^{-12}$ 24. M, $\times 10^6$

Exercise 40 (page 69)

1. (a) 7.39×10 (b) 2.84×10 (c) 1.9762×10^2
2. (a) 2.748×10^3 (b) 3.317×10^4 (c) 2.74218×10^5
3. (a) 2.401×10^{-1} (b) 1.74×10^{-2} (c) 9.23×10^{-3}
4. (a) 1.7023×10^3 (b) 1.004×10 (c) 1.09×10^{-2}
5. (a) 5×10^{-1} (b) 1.1875×10
   (c) 3.125×10^{-2} (d) 1.306×10^2
6. (a) 1010 (b) 932.7 (c) 54100 (d) 7
7. (a) 0.0389 (b) 0.6741 (c) 0.008
8. (a) 1.35×10^2 (b) 1.1×10^5
9. (a) 2×10^2 (b) 1.5×10^{-3}
10. (a) $2.71 \times 10^3 \text{ kg m}^{-3}$ (b) 4.4×10^{-1}
    (c) $3.7673 \times 10^2 \Omega$ (d) $5.11 \times 10^{-1} \text{ MeV}$
    (e) $9.57897 \times 10^7 \text{ C kg}^{-1}$
    (f) $2.241 \times 10^{-2} \text{ m}^3 \text{ mol}^{-1}$

Exercise 41 (page 71)

1. 60 kPa	**2.** 150 μW or 0.15 mW
3. 50 MV	**4.** 55 nF
5. 100 kW	**6.** 0.54 mA or 540 μA
7. 1.5 MΩ	**8.** 22.5 mV
9. 35 GHz	**10.** 15 pF
11. 17 μA	**12.** 46.2 kΩ
13. 3 μA	**14.** 2.025 MHz
15. 50 kN	**16.** 0.3 nF
17. 62.50 m	**18.** 0.0346 kg
19. 13.5×10^{-3}	**20.** 4×10^3
21. 3.8×10^5 km	**22.** 0.053 nm
23. 5.6 MPa	**24.** 4.3×10^{-3} m or 4.3mm
25. 62.83 kN	**26.** 720 g
27. 76.92 MN	

Exercise 42 (page 72)

1. 2450 mm   **2.** 167.5 cm   **3.** 658 mm

4. 25.4 m   **5.** 56.32 m   **6.** 4356 mm

7. 87.5 cm   **8.** (a) 4650 mm (b) 4.65 m

9. (a) 504 cm (b) 5.04 m

10. 148.5 mm to 151.5 mm

Exercise 43 (page 74)

1. 8×10^4 cm^2 **2.** 240×10^{-4} m^2 or 0.024 m^2

3. 3.6×10^6 mm^2 **4.** 350×10^{-6} m^2 or 0.35×10^{-3} m^2

5. 5000 mm^2 **6.** 2.5 cm^2

7. (a) 288000 mm^2 (b) 2880 cm^2 (c) 0.288 m^2

Exercise 44 (page 76)

1. 2.5×10^6 cm^3 **2.** 400×10^{-6} m^3

3. 0.87×10^9 mm^3 or 870×10^6 mm^3 **4.** 2.4 m^3

5. 1500×10^{-9} m^3 or 1.5×10^{-6} m^3

6. 400×10^{-3} cm^3 or 0.4 cm^3 **7.** 6.4×10^3 mm^3

8. 7500×10^{-3} cm^3 or 7.5 cm^3

9. (a) 0.18 m^3 (b) 180000 cm^3 (c) 180 litres

Exercise 45 (page 79)

1. 18.74 in	**2.** 82.28 in
3. 37.95 yd	**4.** 18.36 mile
5. 41.66 cm	**6.** 1.98 m
7. 14.36 m	**8.** 5.91 km

9. (a) 22,745 yd (b) 20.81 km	**10.** 9.688 in^2
11. 18.18 yd^2	**12.** 30.89 acres
13. 21.89 mile^2	**14.** 41 cm^2
15. 28.67 m^2	**16.** 10117 m^2
17. 55.17 km^2	**18.** 12.24 in^3
19. 7.572 ft^3	**20.** 17.59 yd^3
21. 31.7 fluid pints	**22.** 35.23 cm^3
23. 4.956 m^3	**24.** 3.667 litre
25. 47.32 litre	**26.** 34.59 oz
27. 121.3 lb	**28.** 4.409 short tons
29. 220.6 g	**30.** 1.630 kg
31. (a) 280 lb (b) 127.2 kg	**32.** 131°F
33. 75° C	

Exercise 46 (page 80)

1. (d)	**2.** (b)	**3.** (c)	**4.** (c)	**5.** (c)
6. (a)	**7.** (d)	**8.** (a)	**9.** (a)	**10.** (b)
11. (c)	**12.** (d)	**13.** (c)	**14.** (a)	**15.** (d)
16. (b)	**17.** (d)	**18.** (a)	**19.** (b)	**20.** (b)

Chapter 9

Exercise 47 (page 85)

1. $-3a$	**2.** $a + 2b + 4c$
3. $3x - 3x^2 - 3y - 2y^2$	**4.** $6ab - 3a$
5. $6x - 5y + 8z$	**6.** $1 + 2x$
7. $4x + 2y + 2$	**8.** $3a + 5b$
9. $-2a - b + 2c$	**10.** $3x^2 - y^2$

Exercise 48 (page 87)

1. $p^2 q^3 r$	**2.** $8a^2$	**3.** $6q^2$
4. 46	**5.** $5\frac{1}{3}$	**6.** $-\frac{1}{2}$
7. 6	**8.** $\frac{1}{7y}$	**9.** $5xz^2$

10. $3a^2 + 2ab - b^2$

11. $6a^2 - 13ab + 3ac - 5b^2 + bc$

12. $\frac{1}{3b}$	**13.** $2ab$	**14.** $3x$
15. $2x - y$	**16.** $3p + 2q$	**17.** $2a^2 + 2b^2$

Exercise 49 (page 88)

1. z^8 **2.** a^8 **3.** n^3
4. b^{11} **5.** b^{-3} **6.** c^4
7. m^4 **8.** x^{-3} or $\dfrac{1}{x^3}$ **9.** x^{12}
10. y^{-6} or $\dfrac{1}{y^6}$ **11.** t^8 **12.** c^{14}
13. a^{-9} or $\dfrac{1}{a^9}$ **14.** b^{-12} or $\dfrac{1}{b^{12}}$ **15.** b^{10} **16.** s^{-9}
17. $p^6q^7r^5$ **18.** $x^{-2}yz^{-2}$ or $\dfrac{y}{x^2z^2}$
19. $x^5y^4z^3, 13\dfrac{1}{2}$ **20.** $a^3b^{-2}c$ or $\dfrac{a^3c}{b^2}, 9$

Exercise 50 (page 89)

1. $a^2b^{1/2}c^{-2}, \pm4\dfrac{1}{2}$ **2.** $\dfrac{1+a}{b}$ **3.** $a\,b^6\,c^{3/2}$
4. $a^{-4}b^5c^{11}$ **5.** $\dfrac{p^2q}{q-p}$ **6.** $xy^3\sqrt[6]{z^{13}}$
7. $\dfrac{1}{ef^2}$
8. $a^{11/6}b^{1/3}c^{-3/2}$ or $\dfrac{\sqrt[6]{a^{11}}\sqrt[3]{b}}{\sqrt{c^3}}$

Exercise 51 (page 89)

1. (c) **2.** (b) **3.** (b) **4.** (d) **5.** (a)
6. (b) **7.** (a) **8.** (a) **9.** (d) **10.** (d)
11. (c) **12.** (b) **13.** (b) **14.** (c) **15.** (c)

Chapter 10

Exercise 52 (page 92)

1. x^2+5x+6 **2.** $2x^2+9x+4$
3. $4x^2+12x+9$ **4.** $2j^2+2j-12$
5. $4x^2+22x+30$ **6.** $2pqr+p^2q^2+r^2$
7. $a^2+2ab+b^2$ **8.** $x^2+12x+36$
9. $a^2-2ac+c^2$ **10.** $25x^2+30x+9$
11. $4x^2-24x+36$ **12.** $4x^2-9$
13. $64x^2+64x+16$ **14.** $r^2s^2+2rst+t^2$
15. $3ab-6a^2$ **16.** $2x^2-2xy$
17. $2a^2-3ab-5b^2$ **18.** $13p-7q$
19. $7x-y-4z$ **20.** $4a^2-25b^2$
21. $x^2-4xy+4y^2$ **22.** $9a^2-6ab+b^2$
23. 0 **24.** $4-a$

25. $4ab-8a^2$ **26.** $3xy+9x^2y-15x^2$
27. $2+5b^2$ **28.** $11q-2p$

Exercise 53 (page 94)

1. $2(x+2)$ **2.** $2x(y-4z)$
3. $p(b+2c)$ **4.** $2x(1+2y)$
5. $4d(d-3f^5)$ **6.** $4x(1+2x)$
7. $2q(q+4n)$ **8.** $r(s+p+t)$
9. $x(1+3x+5x^2)$ **10.** $bc(a+b^2)$
11. $3xy(xy^3-5y+6)$ **12.** $2pq^2(2p^2-5q)$
13. $7ab(3ab-4)$ **14.** $2xy(y+3x+4x^2)$
15. $2xy(x-2y^2+4x^2y^3)$ **16.** $7y(4+y+2x)$
17. $\dfrac{3x}{y}$ **18.** 0 **19.** $\dfrac{2r}{t}$
20. $(a+b)(y+1)$ **21.** $(p+q)(x+y)$
22. $(x-y)(a+b)$ **23.** $(a-2b)(2x+3y)$
24. $\dfrac{A^2}{pg}\left(\dfrac{A}{pg^2}-\dfrac{1}{g}+A^3\right)$

Exercise 54 (page 96)

1. $2x+8x^2$ **2.** $12y^2-3y$
3. $4b-15b^2$ **4.** $4+3a$
5. $\dfrac{3}{2}-4x$ **6.** 1
7. $10y^2-3y+\dfrac{1}{4}$ **8.** $9x^2+\dfrac{1}{3}-4x$
9. $6a^2+5a-\dfrac{1}{7}$ **10.** $-15t$
11. $\dfrac{1}{5}-x-x^2$ **12.** $10a^2-3a+2$

Exercise 55 (page 96)

1. (a) **2.** (a) **3.** (b) **4.** (b) **5.** (b)
6. (d) **7.** (d) **8.** (c) **9.** (d) **10.** (c)

Chapter 11

Exercise 56 (page 99)

1. 1 **2.** 2 **3.** 6 **4.** -4 **5.** 2
6. 1 **7.** 2 **8.** $\dfrac{1}{2}$ **9.** 0 **10.** 3
11. 2 **12.** -10 **13.** 6 **14.** -2 **15.** 2.5
16. 2 **17.** 6 **18.** -3

Exercise 57 (page 101)

1. 5   **2.** −2   **3.** $-4\frac{1}{2}$ **4.** 2   **5.** 12

6. 15   **7.** −4   **8.** $5\frac{1}{3}$ **9.** 2   **10.** 13

11. −10   **12.** 2   **13.** 3   **14.** −11   **15.** −6

16. 9   **17.** $6\frac{1}{4}$ **18.** 3   **19.** 4   **20.** 10

21. ±12   **22.** $-3\frac{1}{3}$ **23.** ±3   **24.** ±4

Exercise 58 (page 103)

1. 10^{-7} **2.** $8\,\text{m/s}^2$ **3.** 3.472
4. (a) $1.8\,\Omega$ (b) $30\,\Omega$
5. digital camera battery £9, camcorder battery £14
6. $800\,\Omega$ **7.** $30\,\text{m/s}^2$
8. 176 MPa

Exercise 59 (page 105)

1. $12\,\text{cm}, 240\,\text{cm}^2$ **2.** 0.004   **3.** 30
4. 45°C   **5.** 50   **6.** £312, £240
7. 30 kg   **8.** 12 m, 8 m   **9.** 3.5 N

Exercise 60 (page 105)

1. (c)   **2.** (b)   **3.** (a)   **4.** (c)   **5.** (b)
6. (a)   **7.** (d)   **8.** (c)   **9.** (d)   **10.** (a)

Chapter 12

Exercise 61 (page 110)

1. $d = c - e - a - b$ **2.** $x = \dfrac{y}{7}$

3. $v = \dfrac{c}{p}$ **4.** $a = \dfrac{v - u}{t}$

5. $R = \dfrac{V}{I}$ **6.** $y = \dfrac{1}{3}(t - x)$

7. $r = \dfrac{c}{2\pi}$ **8.** $x = \dfrac{y - c}{m}$

9. $T = \dfrac{I}{PR}$ **10.** $c = \dfrac{Q}{m\Delta T}$

11. $L = \dfrac{X_L}{2\pi f}$ **12.** $R = \dfrac{E}{I}$

13. $x = a(y - 3)$ **14.** $C = \dfrac{5}{9}(F - 32)$

15. $f = \dfrac{1}{2\pi\,CX_C}$ **16.** $R = \dfrac{pV}{mT}$

Exercise 62 (page 112)

1. $r = \dfrac{S - a}{S}$ or $1 - \dfrac{a}{S}$

2. $x = \dfrac{d}{\lambda}(y + \lambda)$ or $d + \dfrac{yd}{\lambda}$

3. $f = \dfrac{3F - AL}{3}$ or $f = F - \dfrac{AL}{3}$

4. $D = \dfrac{AB^2}{5Cy}$ **5.** $t = \dfrac{R - R_0}{R_0\alpha}$ **6.** $R_2 = \dfrac{RR_1}{R_1 - R}$

7. $R = \dfrac{E - e - Ir}{I}$ or $R = \dfrac{E - e}{I} - r$

8. $V_2 = \dfrac{P_1 V_1 T_2}{P_2 T_1}$ **9.** $b = \sqrt{\left(\dfrac{y}{4ac^2}\right)}$

10. $x = \dfrac{ay}{\sqrt{(y^2 - b^2)}}$ **11.** $L = \dfrac{gt^2}{4\pi^2}$

12. $u = \sqrt{v^2 - 2as}$ **13.** $R = \sqrt{\left(\dfrac{360A}{\pi\theta}\right)}$

14. $T_2 = \dfrac{P_2 V_2 T_1}{P_1 V_1}$ **15.** $a = N^2 y - x$

16. $L = \dfrac{\sqrt{Z^2 - R^2}}{2\pi f}, 0.080$ **17.** $v = \sqrt{\dfrac{2L}{\rho ac}}$

18. $V = \dfrac{k^2 H^2 L^2}{\theta^2}$ **19.** 750 GHz

20. $x = \sqrt{\left(a^2 - \left(\dfrac{V}{W}\right)^2\right)}$

Exercise 63 (page 115)

1. $a = \sqrt{\left(\dfrac{xy}{m - n}\right)}$ **2.** $R = \sqrt[4]{\left(\dfrac{M}{\pi} + r^4\right)}$

3. $r = \dfrac{3(x + y)}{(1 - x - y)}$ **4.** $L = \dfrac{mrCR}{\mu - m}$

5. $b = \dfrac{c}{\sqrt{1 - a^2}}$ **6.** $r = \sqrt{\left(\dfrac{x - y}{x + y}\right)}$

7. $b = \dfrac{a(p^2 - q^2)}{2(p^2 + q^2)}$ **8.** $v = \dfrac{uf}{u - f}, 30$

9. $t_2 = t_1 + \dfrac{Q}{mc}, 55$

10. $v = \sqrt{\left(\dfrac{2dgh}{0.03L}\right)}, 0.965$

11. $l = \dfrac{8S^2}{3d} + d, 2.725$

12. $C = \dfrac{1}{\omega\left\{\omega L - \sqrt{Z^2 - R^2}\right\}}, 63.1 \times 10^{-6}$

13. 64 mm

14. $\lambda = \sqrt[5]{\left(\dfrac{a\mu}{\rho C Z^4 n}\right)^2}$

15. $w = \dfrac{2R - F}{L}$; 3 kN/m   **16.** $t_2 = t_1 - \dfrac{Qd}{kA}$

17. $r = \dfrac{v}{\omega}\left(1 - \dfrac{s}{100}\right)$

18. $F = EI\left(\dfrac{n\pi}{L}\right)^2$; 13.61 MN   **19.** $r = \sqrt[4]{\left(\dfrac{8\eta\ell V}{\pi p}\right)}$

20. $\ell = \sqrt[3]{\left(\dfrac{20gH^2}{I\rho^4 D^2}\right)^2}$ **21.** 500 MN/m^2

22. 37.0

Exercise 64 (page 116)

1. (c)   **2.** (c)   **3.** (d)   **4.** (b)   **5.** (a)

6. (b)   **7.** (c)   **8.** (b)   **9.** (d)   **10.** (c)

11. (d)   **12.** (d)   **13.** (a)   **14.** (a)   **15.** (b)

Chapter 13

Exercise 65 (page 120)

1. $x = 4, y = 2$ **2.** $x = 3, y = 4$
3. $x = 2, y = 1.5$ **4.** $x = 4, y = 1$
5. $p = 2, q = -1$ **6.** $x = 1, y = 2$
7. $x = 3, y = 2$ **8.** $a = 2, b = 3$
9. $a = 5, b = 2$ **10.** $x = 1, y = 1$
11. $s = 2, t = 3$ **12.** $x = 3, y = -2$
13. $m = 2.5, n = 0.5$ **14.** $a = 6, b = -1$
15. $x = 2, y = 5$ **16.** $c = 2, d = -3$

Exercise 66 (page 122)

1. $p = -1, q = -2$ **2.** $x = 4, y = 6$
3. $a = 2, b = 3$ **4.** $s = 4, t = -1$
5. $x = 3, y = 4$ **6.** $u = 12, v = 2$
7. $x = 10, y = 15$ **8.** $a = 0.30, b = 0.40$

Exercise 67 (page 124)

1. $x = \dfrac{1}{2}, y = \dfrac{1}{4}$ **2.** $a = \dfrac{1}{3}, b = -\dfrac{1}{2}$

3. $p = \dfrac{1}{4}, q = \dfrac{1}{5}$ **4.** $x = 10, y = 5$

5. $c = 3, d = 4$ **6.** $r = 3, s = \dfrac{1}{2}$

7. $x = 5, y = 1\dfrac{3}{4}$ **8.** 1

Exercise 68 (page 127)

1. $a = 0.2, b = 4$
2. $I_1 = 6.47, I_2 = 4.62$
3. $u = 12, a = 4, v = 26$ **4.** £15 500, £12 800
5. $m = -0.5, c = 3$
6. $\alpha = 0.00426, R_0 = 22.56\,\Omega$ **7.** $a = 15, b = 0.50$
8. $a = 4, b = 10$
9. $F_1 = 1.5, F_2 = -4.5$
10. $R_1 = 5.7$ kN, $R_2 = 6.3$ kN   **11.** $r = 0.258,$
                                     $\omega = 32.3$

Exercise 69 (page 129)

1. $x = 2, y = 1, z = 3$ **2.** $x = 2, y = -2, z = 2$
3. $x = 5, y = -1, z = -2$ **4.** $x = 4, y = 0, z = 3$
5. $x = 2, y = 4, z = 5$ **6.** $x = 1, y = 6, z = 7$
7. $x = 5, y = 4, z = 2$ **8.** $x = -4, y = 3, z = 2$
9. $x = 1.5, y = 2.5, z = 4.5$
10. $i_1 = -5, i_2 = -4, i_3 = 2$
11. $F_1 = 2, F_2 = -3 \; F_3 = 4$

Exercise 70 (page 129)

1. (c)   **2.** (d)   **3.** (c)   **4.** (d)   **5.** (b)
6. (a)   **7.** (b)   **8.** (a)   **9.** (b)   **10.** (a)

Chapter 14

Exercise 71 (page 135)

1. 4 or -4 **2.** 4 or -8 **3.** 2 or -6

4. -1.5 or 1.5 **5.** 0 or $-\dfrac{4}{3}$ **6.** 2 or -2

7. 4   **8.** -5 **9.** 1

10. -2 or -3 **11.** -3 or -7 **12.** 2 or -1

13. 4 or -3 **14.** 2 or 7   **15.** -4

16. 2   **17.** -3 **18.** 3 or -3

19. -2 or $-\dfrac{2}{3}$ **20.** -1.5 **21.** $\dfrac{1}{8}$ or $-\dfrac{1}{8}$

22. 4 or -7 **23.** -1 or 1.5   **24.** $\dfrac{1}{2}$ or $\dfrac{1}{3}$

25. $\dfrac{1}{2}$ or $-\dfrac{4}{5}$ **26.** $1\dfrac{1}{3}$ or $-\dfrac{1}{7}$ **27.** $\dfrac{3}{8}$ or -2

28. $\dfrac{2}{5}$ or -3 **29.** $\dfrac{4}{3}$ or $-\dfrac{1}{2}$ **30.** $\dfrac{5}{4}$ or $-\dfrac{3}{2}$

31. $x^2 - 4x + 3 = 0$ **32.** $x^2 + 3x - 10 = 0$
33. $x^2 + 5x + 4 = 0$ **34.** $4x^2 - 8x - 5 = 0$
35. $x^2 - 36 = 0$ **36.** $x^2 - 1.7x - 1.68 = 0$

Exercise 72 (page 137)

1. -3.732 or -0.268 **2.** -3.137 or 0.637
3. 1.468 or -1.135 **4.** 1.290 or 0.310
5. 2.443 or 0.307 **6.** -2.851 or 0.351

Exercise 73 (page 138)

1. 0.637 or -3.137 **2.** 0.296 or -0.792
3. 2.781 or 0.719 **4.** 0.443 or -1.693
5. 3.608 or -1.108 **6.** 1.434 or 0.232
7. 0.851 or -2.351 **8.** 2.086 or -0.086
9. 1.481 or -1.081 **10.** 4.176 or -1.676
11. 4 or 2.167 **12.** 7.141 or -3.641
13. 4.562 or 0.438

Exercise 74 (page 140)

1. $1.191\,\text{s}$ **2.** $0.345\,\text{A}$ or $0.905\,\text{A}$ **3.** $7.84\,\text{cm}$
4. $0.619\,\text{m}$ or $19.38\,\text{m}$ **5.** 0.0133
6. $1.066\,\text{m}$ **7.** $86.78\,\text{cm}$
8. $1.835\,\text{m}$ or $18.165\,\text{m}$ **9.** $7\,\text{m}$
10. 12 ohms, 28 ohms
11. $0.52\,\text{s}$ and $5.73\,\text{s}$ **12.** $400\,\text{rad/s}$
13. $2.602\,\text{m}$ **14.** $1229\,\text{m}$ or $238.9\,\text{m}$ **15.** 9.67

Exercise 75 (page 141)

1. $x = 1, y = 3$ and $x = -3, y = 7$

2. $x = \dfrac{2}{5}, y = -\dfrac{1}{5}$ and $x = -1\dfrac{2}{3}, y = -4\dfrac{1}{3}$

3. $x = 0, y = 4$ and $x = 3, y = 1$

Exercise 76 (Page 142)

1. (d)   **2.** (d)   **3.** (a)   **4.** (c)   **5.** (a)
6. (c)   **7.** (b)   **8.** (d)   **9.** (b)   **10.** (c)
11. (b)   **12.** (a)   **13.** (a)   **14.** (d)   **15.** (c)

Chapter 15

Exercise 77 (page 145)

1. 4   **2.** 4   **3.** 3   **4.** -3 **5.** $\dfrac{1}{3}$

6. 3   **7.** 2   **8.** -2 **9.** $1\dfrac{1}{2}$ **10.** $\dfrac{1}{3}$

11. 2   **12.** 10000   **13.** 100000   **14.** 9   **15.** $\dfrac{1}{32}$

16. 0.01   **17.** $\dfrac{1}{16}$ **18.** e^3

Exercise 78 (page 147)

1. $\log 6$ **2.** $\log 15$ **3.** $\log 2$ **4.** $\log 3$
5. $\log 12$ **6.** $\log 500$ **7.** $\log 100$ **8.** $\log 6$
9. $\log 10$ **10.** $\log 1 = 0$ **11.** $\log 2$
12. $\log 243$ or $\log 3^5$ or $5 \log 3$
13. $\log 16$ or $\log 2^4$ or $4 \log 2$
14. $\log 64$ or $\log 2^6$ or $6 \log 2$
15. 0.5   **16.** 1.5   **17.** $x = 2.5$ **18.** $t = 8$
19. $b = 2$ **20.** $x = 2$ **21.** $a = 6$ **22.** $x = 5$

Exercise 79 (page 149)

1. 1.690   **2.** 3.170   **3.** 0.2696   **4.** 6.058
5. 2.251   **6.** 3.959   **7.** 2.542   **8.** -0.3272
9. 316.2   **10.** $0.057\,\text{m}^3$

Exercise 80 (Page 150)

1. (c)   **2.** (c)   **3.** (d)   **4.** (b)   **5.** (a)
6. (b)   **7.** (c)   **8.** (b)   **9.** (d)   **10.** (c)
11. (d)   **12.** (d)   **13.** (a)   **14.** (a)   **15.** (b)

Chapter 16

Exercise 81 (page 152)

1. (a) 0.1653   (b) 0.4584   (c) 22030
2. (a) 5.0988   (b) 0.064037   (c) 40.446

3. (a) 4.55848 (b) 2.40444 (c) 8.05124
4. (a) 48.04106 (b) 4.07482 (c) −0.08286
5. 2.739   **6.** 120.7 m

Exercise 82 (page 154)

1. 2.0601   **2.** (a) 7.389 (b) 0.7408
3. $1 - 2x^2 - \dfrac{8}{3}x^3 - 2x^4$
4. $2x^{1/2} + 2x^{5/2} + x^{9/2} + \dfrac{1}{3}x^{13/2} + \dfrac{1}{12}x^{17/2} + \dfrac{1}{60}x^{21/2}$

Exercise 83 (page 155)

1. 3.95, 2.05   **2.** 1.65, −1.30
3. (a) 28 cm^3 (b) 116 min   **4.** (a) 70°C (b) 5 minutes

Exercise 84 (page 158)

1. (a) 0.55547   (b) 0.91374   (c) 8.8941
2. (a) 2.2293   (b) −0.33154   (c) 0.13087
3. −0.4904 **4.** −0.5822 **5.** 2.197   **6.** 816.2
7. 0.8274 **8.** 11.02 **9.** 1.522 **10.** 1.485
11. 1.962 **12.** 3   **13.** 4
14. 147.9 **15.** 4.901   **16.** 3.095
17. $t = e^{b+a\ln D} = e^b e^{a\ln D} = e^b e^{\ln D^a}$ i.e. $t = e^b D^a$
18. 500   **19.** $W = PV\ln\left(\dfrac{U_2}{U_1}\right)$
20. 992 m/s **21.** 348.5 Pa **22.** 25 minutes

Exercise 85 (page 161)

1. (a) 150°C (b) 100.5°C   **2.** 99.21 kPa
3. (a) 29.32 volts (b) 71.31 × 10^{-6} s
4. (a) 1.993 m (b) 2.293 m
5. (a) 50°C (b) 55.45 s
6. 30.37 N
7. (a) 3.04 A (b) 1.46 s
8. 2.45 mol/cm^3
9. (a) 7.07 A (b) 0.966 s
10. £1732
11. (a) 100% (b) 67.03% (c) 1.83%
12. 2.45 mA   **13.** 142 ms
14. 99.752%   **15.** 20 min 38 s

Exercise 86 (Page 162)

1. (b) **2.** (b) **3.** (a) **4.** (c) **5.** (c)
6. (a) **7.** (b) **8.** (d) **9.** (d) **10.** (c)

Chapter 17

Exercise 87 (page 170)

1. (a) Horizontal axis: 1 cm = 4 V (or 1 cm = 5 V), vertical axis: 1 cm = 10 Ω
 (b) Horizontal axis: 1 cm = 5 m, vertical axis: 1 cm = 0.1 V
 (c) Horizontal axis: 1 cm = 10 N, vertical axis: 1 cm = 0.2 mm
2. (a) −1 (b) −8 (c) −1.5 (d) 4   **3.** 14.5
4. (a) −1.1 (b) −1.4
5. The 1010 rev/min reading should be 1070 rev/min; (a) 1000 rev/min (b) 167 V

Exercise 88 (page 176)

1. Missing values: −0.75, 0.25, 0.75, 1.75, 2.25, 2.75; Gradient = $\dfrac{1}{2}$
2. (a) 4, −2 (b) −1, 0 (c) −3, −4 (d) 0, 4
3. (a) 2, $\dfrac{1}{2}$ (b) 3, −2$\dfrac{1}{2}$ (c) $\dfrac{1}{24}$, $\dfrac{1}{2}$
4. (a) 6, −3 (b) −2, 4 (c) 3, 0 (d) 0, 7
5. (a) 2, −$\dfrac{1}{2}$ (b) −$\dfrac{2}{3}$, −1$\dfrac{2}{3}$ (c) $\dfrac{1}{18}$, 2 (d) 10, −4$\dfrac{2}{3}$
6. (a) $\dfrac{3}{5}$ (b) −4 (c) −1$\dfrac{5}{6}$
7. (a) and (c), (b) and (e)
8. (2, 1)   **9.** (1.5, 6)   **10.** (1, 2)
11. (a) 89 cm (b) 11 N (c) 2.4 (d) $l = 2.4W + 48$
12. $P = 0.15W + 3.5$ **13.** $a = -20, b = 412$

Exercise 89 (page 181)

1. (a) 40°C (b) 128 Ω
2. (a) 850 rev/min (b) 77.5 V
3. (a) 0.25 (b) 12 (c) $F = 0.25L + 12$
 (d) 89.5 N (e) 592 N (f) 212 N

4. $-0.003, 8.73 \, \text{N}/\text{cm}^2$

5. (a) 22.5 m/s (b) 6.5 s (c) $v = 0.7t + 15.5$

6. $m = 26.8L$

7. (a) $1.25t$ (b) 21.6% (c) $F = -0.095w + 2.2$

8. (a) 96×10^9 Pa (b) 0.00022 (c) 29×10^6 Pa

9. (a) $\dfrac{1}{5}$ (b) 6 (c) $E = \dfrac{1}{5}L + 6$ (d) 12 N (e) 65 N

10. $a = 0.85, b = 12, 254.3$ kPa, 275.5 kPa, 280 K

Exercise 90 (Page 183)

1. (d)	**2.** (a)	**3.** (c)	**4.** (d)	**5.** (b)
6. (d)	**7.** (b)	**8.** (a)	**9.** (c)	**10.** (c)
11. (d)	**12.** (b)	**13.** (a)	**14.** (b)	**15.** (a)

Chapter 18

Exercise 91 (page 188)

1. (a) y (b) x^2 (c) c (d) d **2.** (a) y (b) $\sqrt{x}$ (c) b (d) a

3. (a) y (b) $\dfrac{1}{x}$ (c) f (d) e **4.** (a) $\dfrac{y}{x}$ (b) x (c) b (d) c

5. (a) $\dfrac{y}{x}$ (b) $\dfrac{1}{x^2}$ (c) a (d) b

6. $a = 1.5, b = 0.4, 11.78 \, \text{mm}^2$ **7.** $y = 2x^2 + 7, 5.15$

8. (a) 950 (b) 317 kN

9. $a = 0.4, b = 8.6$ (a) 94.4 (b) 11.2

Exercise 92 (page 192)

1. (a) $\lg y$ (b) x (c) $\lg a$ (d) $\lg b$

2. (a) $\lg y$ (b) $\lg x$ (c) L (d) $\lg k$

3. (a) $\ln y$ (b) x (c) n (d) $\ln m$

4. $I = 0.0012 \, \text{V}^2, 6.75$ candelas

5. $a = 3.0, b = 0.5$

6. $a = 5.6, b = 2.6, 37.86, 3.0$

7. $R_0 = 25.1, c = 1.42$ **8.** $y = 0.08e^{0.24x}$

9. $T_0 = 35.3 \, \text{N}, \mu = 0.27, 64.8 \, \text{N}, 1.29$ radians

Exercise 93 (Page 193)

1. (a)	**2.** (c)	**3.** (d)	**4.** (b)	**5.** (c)
6. (b)	**7.** (d)	**8.** (a)		

Chapter 19

Exercise 94 (page 195)

1. $x = 2, y = 4$	**2.** $x = 1, y = 1$
3. $x = 3.5, y = 1.5$	**4.** $x = -1, y = 2$
5. $x = 2.3, y = -1.2$	**6.** $x = -2, y = -3$
7. $a = 0.4, b = 1.6$	

Exercise 95 (page 199)

1. (a) Minimum $(0, 0)$ (b) Minimum $(0, -1)$
   (c) Maximum $(0, 3)$ (d) Maximum $(0, -1)$

2. -0.4 or 0.6 **3.** -3.9 or 6.9

4. -1.1 or 4.1 **5.** -1.8 or 2.2

6. $x = -1.5$ or -2, Minimum at $(-1.75, -0.1)$

7. $x = -0.7$ or 1.6 **8.** (a) ± 1.63 (b) 1 or -0.3

9. $(-2.6, 13.2), (0.6, 0.8); x = -2.6$ or 0.6

10. $x = -1.2$ or 2.5 (a) -30 (b) 2.75 and -1.50
   (c) 2.3 or -0.8

Exercise 96 (page 200)

1. $x = 4, y = 8$ and $x = -0.5, y = -5.5$

2. (a) $x = -1.5$ or 3.5 (b) $x = -1.24$ or 3.24
   (c) $x = -1.5$ or 3.0

Exercise 97 (page 201)

1. $x = -2.0, -0.5$ or 1.5

2. $x = -2, 1$ or 3, Minimum at $(2.1, -4.1)$,
   Maximum at $(-0.8, 8.2)$

3. $x = 1$ **4.** $x = -2.0, 0.4$ or 2.6

5. $x = -1.2, 0.70$ or 2.5

6. $x = -2.3, 1.0$ or 1.8 **7.** $x = -1.5$

Exercise 98 (Page 202)

1. (b)	**2.** (b)	**3.** (c)	**4.** (d)	**5.** (a)

Chapter 20

Exercise 99 (page 206)

1. $a = 12, n = 1.8, 451, 28.5$

2. $k = 1.5, n = -1$ **3.** $m = 3, n = 2.5$

Exercise 100 (page 208)

1. (i) $a = -8$, $b = 5.3$, $p = -8(5.3)^q$ (ii) -224.7
(iii) 3.31

Exercise 101 (page 210)

1. $a = 76$, $k = -7 \times 10^{-5}$, $p = 76\, e^{-7 \times 10^{-5} h}$,
37.74 kPa

2. $\theta_0 = 152$, $k = -0.05$

Exercise 102 (Page 210)

1. (c)   **2.** (a)   **3.** (d)   **4.** (a)   **5.** (c)
6. (d)   **7.** (b)   **8.** (b)   **9.** (d)   **10.** (b)

Chapter 21

Exercise 103 (page 216)

1. $82°27'$ **2.** $27°54'$ **3.** $51°11'$ **4.** $100°6'52''$
5. $15°44'17''$ **6.** $86°49'1''$ **7.** $72.55°$ **8.** $27.754°$
9. $37°57'$ **10.** $58°22'52''$

Exercise 104 (page 218)

1. reflex   **2.** obtuse   **3.** acute   **4.** right angle

5. (a) $21°$ (b) $62°23'$ (c) $48°56'17''$

6. (a) $102°$ (b) $165°$ (c) $10°18'49''$

7. (a) $60°$ (b) $110°$ (c) $75°$ (d) $143°$ (e) $140°$
(f) $20°$ (g) $129.3°$ (h) $79°$ (i) $54°$

8. Transversal (a) 1 & 3, 2 & 4, 5 & 7, 6 & 8
(b) 1 & 2, 2 & 3, 3 & 4, 4 & 1, 5 & 6, 6 & 7,
7 & 8, 8 & 5, 3 & 8, 1 & 6, 4 & 7 or 2 & 5
(c) 1 & 5, 2 & 6, 4 & 8, 3 & 7 (d) 3 & 5 or 2 & 8

9. $59°20'$ **10.** $a = 69°, b = 21°, c = 82°$ **11.** $51°$

12. 1.326 rad   **13.** 0.605 rad   **14.** $40°55'$

Exercise 105 (page 222)

1. (a) acute-angled scalene triangle
(b) isosceles triangle (c) right-angled isosceles triangle
(d) obtuse-angled scalene triangle
(e) equilateral triangle (f) right-angled triangle

2. $a = 40°, b = 82°, c = 66°$,
$d = 75°, e = 30°, f = 75°$

3. DF, DE   **4.** $52°$ **5.** $122.5°$

6. $\phi = 51°, x = 161°$

7. $40°, 70°, 70°, 125°$, isosceles

8. $a = 18°50', b = 71°10', c = 68°, d = 90°$,
$e = 22°, f = 49°, g = 41°$

9. $a = 103°, b = 55°, c = 77°, d = 125°$,
$e = 55°, f = 22°, g = 103°, h = 77°$,
$i = 103°, j = 77°, k = 81°$

10. $17°$ **11.** $A = 37°, B = 60°, E = 83°$

Exercise 106 (page 224)

1. (a) congruent BAC, DAC (SAS)
(b) congruent FGE, JHI (SSS)
(c) not necessarily congruent
(d) congruent QRT, SRT (RHS)
(e) congruent UVW, XZY (ASA)

2. proof

Exercise 107 (page 227)

1. $x = 16.54$ mm, $y = 4.18$ mm   **2.** 9 cm, 7.79 cm

3. (a) 2.25 cm (b) 4 cm   **4.** 3 m

Exercise 108 (page 229)

1–5. Constructions – see similar constructions in
worked Problems 30 to 33 on pages 227–229.

Exercise 109 (Page 229)

1. (a)   **2.** (b)   **3.** (b)   **4.** (d)   **5.** (a)
6. (c)   **7.** (c)   **8.** (b)   **9.** (a)   **10.** (d)

Chapter 22

Exercise 110 (page 232)

1. 9 cm   **2.** 24 m   **3.** 9.54 mm
4. 20.81 cm   **5.** 7.21 m   **6.** 11.18 cm
7. 24.11 mm   **8.** $8^2 + 15^2 = 17^2$
9. (a) 27.20 cm each   (b) $45°$ **10.** 20.81 km
11. 3.35 m, 10 cm   **12.** 132.7 nautical miles
13. 2.94 mm   **14.** 24 mm

Exercise 111 (page 234)

1. $\sin Z = \dfrac{9}{41}, \cos Z = \dfrac{40}{41}, \tan X = \dfrac{40}{9}, \cos X = \dfrac{9}{41}$

2. $\sin A = \dfrac{3}{5}, \cos A = \dfrac{4}{5}, \tan A = \dfrac{3}{4}, \sin B = \dfrac{4}{5}$,
$\cos B = \dfrac{3}{5}, \tan B = \dfrac{4}{3}$

3. $\sin A = \dfrac{8}{17}, \tan A = \dfrac{8}{15}$

4. $\sin X = \dfrac{15}{113}, \cos X = \dfrac{112}{113}$

5. (a) $\dfrac{15}{17}$ (b) $\dfrac{15}{17}$ (c) $\dfrac{8}{15}$

6. (a) $\sin\theta = \dfrac{7}{25}$ (b) $\cos\theta = \dfrac{24}{25}$

7. (a) 9.434 (b) -0.625

Exercise 112 (page 237)

1. 2.7550 **2.** 4.846 **3.** 36.52
4. (a) 0.8660 (b) -0.1010 (c) 0.5865
5. 42.33° **6.** 15.25° **7.** 73.78° **8.** 7°56′
9. 31°22′ **10.** 41°54′ **11.** 29.05° **12.** 20°21′
13. 0.3586 **14.** 1.803
15. (a) 40° (b) 6.79 m **16.** $-68.37°$

Exercise 113 (page 239)

1. (a) 12.22 (b) 5.619 (c) 14.87 (d) 8.349
   (e) 5.595 (f) 5.275
2. (a) $AC = 5.831$ cm, $\angle A = 59.04°$, $\angle C = 30.96°$
   (b) $DE = 6.928$ cm, $\angle D = 30°$, $\angle F = 60°$
   (c) $\angle J = 62°$, $HJ = 5.634$ cm, $GH = 10.60$ cm
   (d) $\angle L = 63°$, $LM = 6.810$ cm, $KM = 13.37$ cm
   (e) $\angle N = 26°$, $ON = 9.124$ cm, $NP = 8.201$ cm
   (f) $\angle S = 49°$, $RS = 4.346$ cm, $QS = 6.625$ cm
3. 6.54 m **4.** 9.40 mm **5.** 5.63 m

Exercise 114 (page 242)

1. 36.15 m **2.** 48 m **3.** 249.5 m **4.** 110.1 m
5. 53.0 m **6.** 9.50 m **7.** 107.8 m
8. 9.43 m, 10.56 m **9.** 60 m

Exercise 115 (Page 243)

1. (d) **2.** (a) **3.** (b) **4.** (b) **5.** (c)
6. (c) **7.** (a) **8.** (b) **9.** (c) **10.** (d)
11. (d) **12.** (b) **13.** (c) **14.** (a) **15.** (d)

Chapter 23

Exercise 116 (page 250)

1. (a) 42.78° and 137.22° (b) 188.53° and 351.47°
2. (a) 29.08° and 330.92° (b) 123.86° and 236.14°
3. (a) 44.21° and 224.21° (b) 113.12° and 293.12°
4. $t = 122°7′$ and 237°53′
5. $\alpha = 218°41′$ and 321°19′
6. $\theta = 39°44′$ and 219°44′

Exercise 117 (page 254)

1. 5 **2.** 180° **3.** 30 **4.** 120°
5. 1, 120° **6.** 2, 144° **7.** 3, 90° **8.** 5, 720°
9. 3.5, 960° **10.** 6, 360° **11.** 4, 180° **12.** 5 ms
13. 40 Hz **14.** 100 μs or 0.1 ms
15. 625 Hz **16.** leading **17.** leading

Exercise 118 (page 256)

1. (a) 40 mA (b) 25 Hz (c) 0.04 s or 40 ms
   (d) 0.29 rad (or 16.62°) leading $40\sin 50\pi t$
2. (a) 75 cm (b) 6.37 Hz (c) 0.157 s
   (d) 0.54 rad (or 30.94°) lagging $75\sin 40t$
3. (a) 300 V (b) 100 Hz (c) 0.01 s or 10 ms
   (d) 0.412 rad (or 23.61°) lagging $300\sin 200\pi t$
4. (a) $v = 120\sin 100\pi t$ volts
   (b) $v = 120\sin\left(100\pi t + 0.43\right)$ volts
5. $i = 20\sin\left(80\pi t - \dfrac{\pi}{6}\right)$ A or
   $i = 20\sin(80\pi t - 0.524)$ A
6. $3.2\sin(100\pi t + 0.488)$ m
7. (a) 5 A, 50 Hz, 20 ms, 24.75° lagging
   (b) -2.093 A (c) 4.363 A (d) 6.375 ms (e) 3.423 ms

Exercise 119 (Page 257)

1. (b) **2.** (a) **3.** (a) **4.** (c) **5.** (d)

Chapter 24

Exercise 120 (page 260)

1. $C = 83°, a = 14.1$ mm, $c = 28.9$ mm,
   area $= 189$ mm^2
2. $A = 52°2′, c = 7.568$ cm, $a = 7.152$ cm,
   area $= 25.65$ cm^2

3. $D = 19°48', E = 134°12', e = 36.0\,\text{cm}$,
   area $= 134\,\text{cm}^2$
4. $E = 49°0', F = 26°38', f = 15.09\,\text{mm}$,
   area $= 185.6\,\text{mm}^2$
5. $J = 44°29', L = 99°31', l = 5.420\,\text{cm}$,
   area $= 6.133\,\text{cm}^2$, or, $J = 135°31', L = 8°29'$,
   $l = 0.811\,\text{cm}$, area $= 0.917\,\text{cm}^2$
6. $K = 47°8', J = 97°52', j = 62.2\,\text{mm}$,
   area $= 820.2\,\text{mm}^2$ or $K = 132°52', J = 12°8'$,
   $j = 13.19\,\text{mm}$, area $= 174.0\,\text{mm}^2$

Exercise 121 (page 262)

1. $p = 13.2\,\text{cm}, Q = 47.34°, R = 78.66°$,
   area $= 77.7\,\text{cm}^2$
2. $p = 6.127\,\text{m}, Q = 30.83°, R = 44.17°$,
   area $= 6.938\,\text{m}^2$
3. $X = 83.33°, Y = 52.62°, Z = 44.05°$,
   area $= 27.8\,\text{cm}^2$
4. $X = 29.77°, Y = 53.50°, Z = 96.73°$,
   area $= 355\,\text{mm}^2$

Exercise 122 (page 264)

1. $193\,\text{km}$ 2. (a) $122.6\,\text{m}$ (b) $94.80°, 40.66°, 44.54°$
3. (a) $11.4\,\text{m}$ (b) $17.55°$ 4. $163.4\,\text{m}$
5. $BF = 3.9\,\text{m}, EB = 4.0\,\text{m}$ 6. $6.35\,\text{m}, 5.37\,\text{m}$
7. $32.48\,\text{A}, 14.31°$

Exercise 123 (page 266)

1. $80.42°, 59.38°, 40.20°$ 2. (a) $15.23\,\text{m}$ (b) $50.07°$
3. $40.25\,\text{cm}, 126.05°$ 4. $19.8\,\text{cm}$ 5. $36.2\,\text{m}$
6. $x = 69.3\,\text{mm}, y = 142\,\text{mm}$ 7. $130°$ 8. $13.66\,\text{mm}$

Exercise 124 (Page 267)

1. (d) 2. (d) 3. (a) 4. (a) 5. (c)
6. (c) 7. (b) 8. (c) 9. (a) 10. (b)

Chapter 25

Exercise 125 (page 270)

1. $(5.83, 59.04°)$ or $(5.83, 1.03\,\text{rad})$
2. $(6.61, 20.82°)$ or $(6.61, 0.36\,\text{rad})$
3. $(4.47, 116.57°)$ or $(4.47, 2.03\,\text{rad})$
4. $(6.55, 145.58°)$ or $(6.55, 2.54\,\text{rad})$
5. $(7.62, 203.20°)$ or $(7.62, 3.55\,\text{rad})$

6. $(4.33, 236.31°)$ or $(4.33, 4.12\,\text{rad})$
7. $(5.83, 329.04°)$ or $(5.83, 5.74\,\text{rad})$
8. $(15.68, 307.75°)$ or $(15.68, 5.37\,\text{rad})$

Exercise 126 (page 271)

1. $(1.294, 4.830)$ 2. $(1.917, 3.960)$
3. $(-5.362, 4.500)$ 4. $(-2.884, 2.154)$
5. $(-9.353, -5.400)$ 6. $(-2.615, -3.027)$
7. $(0.750, -1.299)$ 8. $(4.252, -4.233)$
9. (a) $40\angle 18°, 40\angle 90°, 40\angle 162°, 40\angle 234°, 40\angle 306°$
   (b) $(38.04, 12.36), (0, 40), (-38.04, 12.36)$,
   $(-23.51, -32.36), (23.51, -32.36)$
10. $47.0\,\text{mm}$

Exercise 127 (Page 272)

1. (a) 2. (d) 3. (b) 4. (a) 5. (c)

Chapter 26

Exercise 128 (page 276)

1. $p = 105°, q = 35°$ 2. $r = 142°, s = 95°$
3. $t = 146°$

Exercise 129 (page 280)

1. (i) rhombus (a) $14\,\text{cm}^2$ (b) $16\,\text{cm}$ (ii) parallelogram
   (a) $180\,\text{mm}^2$ (b) $80\,\text{mm}$ (iii) rectangle (a) $3600\,\text{mm}^2$
   (b) $300\,\text{mm}$ (iv) trapezium (a) $190\,\text{cm}^2$ (b) $62.91\,\text{cm}$
2. $35.7\,\text{cm}^2$ 3. (a) $80\,\text{m}$ (b) $170\,\text{m}$ 4. $27.2\,\text{cm}^2$
5. $18\,\text{cm}$ 6. $1200\,\text{mm}$
7. (a) $29\,\text{cm}^2$ (b) $650\,\text{mm}^2$ 8. $560\,\text{m}^2$
9. $3.4\,\text{cm}$ 10. $6750\,\text{mm}^2$ 11. $43.30\,\text{cm}^2$
12. 32 13. $230, 400$

Exercise 130 (page 282)

1. $482\,\text{m}^2$
2. (a) $50.27\,\text{cm}^2$ (b) $706.9\,\text{mm}^2$ (c) $3183\,\text{mm}^2$
3. $2513\,\text{mm}^2$ 4. (a) $20.19\,\text{mm}$ (b) $63.41\,\text{mm}$
5. (a) $53.01\,\text{cm}^2$ (b) $129.9\,\text{mm}^2$ 6. $5773\,\text{mm}^2$
7. $1.89\,\text{m}^2$ 8. $0.196\,\text{m}^2, 1.785\,\text{m}$ 9. $586.9\,\text{cm}^2$

Exercise 131 (page 284)

1. $1932\,\text{mm}^2$ 2. $1624\,\text{mm}^2$ 3. (a) $0.918\,\text{ha}$ (b) $456\,\text{m}$
4. $306\,\text{turns}$ 5. $0.416\,\text{m}^2$

Exercise 132 (page 285)

1. 80 ha   **2.** 80 m^2 **3.** 3.14 ha

Exercise 133 (Page 285)

1. (b)   **2.** (a)   **3.** (b)   **4.** (d)   **5.** (b)
6. (c)   **7.** (a)   **8.** (c)   **9.** (c)   **10.** (d)

Chapter 27

Exercise 134 (page 289)

1. 45.24 cm   **2.** 259.5 mm   **3.** 2.629 cm   **4.** 47.68 cm
5. 38.73 cm   **6.** 12730 km   **7.** 97.13 mm

Exercise 135 (page 290)

1. (a) $\dfrac{\pi}{6}$ (b) $\dfrac{5\pi}{12}$ (c) $\dfrac{5\pi}{4}$

2. (a) 0.838 (b) 1.481 (c) 4.054
3. (a) 210° (b) 80° (c) 105°
4. (a) 0°43′ (b) 154°8′ (c) 414°53′   **5.** 104.7 rad/s

Exercise 136 (page 292)

1. 113 cm^2 **2.** 2376 mm^2 **3.** 1790 mm^2
4. 802 mm^2 **5.** 1709 mm^2 **6.** 1269 m^2
7. 1548 m^2 **8.** (a) 106.0 cm (b) 783.9 cm^2
9. 21.46 m^2 **10.** 17.80 cm, 74.07 cm^2
11. (a) 59.86 mm (b) 197.8 mm   **12.** 26.2 cm
13. 202 mm^2 **14.** 8.67 cm, 54.48 cm   **15.** 80°30′
16. 748   **17.** (a) 0.698 rad (b) 804.2 m^2
18. 10.47 m^2 **19.** 19.63 m^2 **20.** 2107 mm^2
21. 2880 mm^2 **22.** 4.49 m^2 **23.** 8.48 m
24. 9.55 m   **25.** 60 mm
26. (a) 396 mm^2 (b) 42.24%   **27.** 701.8 mm
28. 7.74 mm

Exercise 137 (page 296)

1. (a) 2 (b) (3, −4)   **2.** Centre at (3, −2), radius 4
3. Circle, centre (0, 1), radius 5
4. Circle, centre (0, 0), radius 6

Exercise 138 (Page 296)

1. (d)   **2.** (b)   **3.** (c)   **4.** (c)   **5.** (c)
6. (d)   **7.** (a)   **8.** (b)   **9.** (a)   **10.** (b)

Chapter 28

Exercise 139 (page 302)

1. 1.2 m^3 **2.** 5 cm^3 **3.** 8 cm^3
4. (a) 3840 mm^3 (b) 1792 mm^2
5. 972 litres   **6.** 15 cm^3, 135 g   **7.** 500 litres
8. 1.44 m^3 **9.** (a) 35.3 cm^3 (b) 61.3 cm^2
10. (a) 2400 cm^3 (b) 2460 cm^2 **11.** 37.04 m
12. 1.63 cm   **13.** 8796 cm^3
14. 4.709 cm, 153.9 cm^2
15. 2.99 cm   **16.** 28060 cm^3, 1.099 m^2
17. 8.22 m by 8.22 m   **18.** 62.5 min
19. 4 cm   **20.** 4.08 m^3
21. (a) 17.36 m^2 (b) 4.778 m^3, 4778 litre   **22.** 272

Exercise 140 (page 306)

1. 201.1 cm^3, 159.0 cm^2 **2.** 7.68 cm^3, 25.81 cm^2
3. 113.1 cm^3, 113.1 cm^2 **4.** 5.131 cm   **5.** 3 cm
6. 2681 mm^3 **7.** (a) 268 083 mm^3
or 268.083 cm^3 (b) 20 106 mm^2 or 201.06 cm^2
8. 8.53 cm
9. (a) 512×10^6 km^2 (b) 1.09×10^{12} km^3 **10.** 664
11. 92 m^3, 92,000 litres   **12.** 8.48 m^3

Exercise 141 (page 311)

1. 5890 mm^2 or 58.90 cm^2
2. (a) 56.55 cm^3 (b) 84.82 cm^2 **3.** 13.57 kg
4. 29.32 cm^3 **5.** 393.4 m^2
6. (i) (a) 670 cm^3 (b) 523 cm^2 (ii) (a) 180 cm^3
(b)154 cm^2 (iii) (a) 56.5 cm^3 (b) 84.8 cm^2
(iv) (a) 10.4 cm^3 (b) 32.0 cm^2 (v) (a) 96.0 cm^3
(b) 146 cm^2 (vi) (a) 86.5 cm^3 (b) 142 cm^2
(vii) (a) 805 cm^3 (b) 539 cm^2
7. (a) 17.9 cm (b) 38.0 cm   **8.** 125 cm^3
9. @110.3 m^3, 25.5 m^2 **10.** 6560 litres
11. 657.1 cm^3, 1027 cm^2 **12.** 220.7 cm^3
13. (a) 1458 litres (b) 9.77 m^2 (c) £140.45
14. 7.5 m^3 **15.** (a) 66429 m^2
(b) 202074 m^3
16. 31.13 kg

Exercise 142 (page 316)

1. 147 cm^3, 164 cm^2 **2.** 403 cm^3, 337 cm^2
3. 10480 m^3, 1852 m^2 **4.** 1707 cm^2
5. 10.69 cm   **6.** 55910 cm^3, 6051 cm^2
7. 5.14 m   **8.** 72790 cm^3, 0.07279 m^3

Exercise 143 (page 317)

1. $8:125$ **2.** $137.2\,\text{g}$

Exercise 144 (Page 317)

1. (d)   **2.** (c)   **3.** (a)   **4.** (b)   **5.** (c)
6. (a)   **7.** (b)   **8.** (c)   **9.** (d)   **10.** (a)

Chapter 29

Exercise 145 (page 320)

1. 4.5 square units   **2.** 54.7 square units   **3.** 63.33 m
4. 4.70 ha   **5.** $143\,\text{m}^2$

Exercise 146 (page 321)

1. $42.59\,\text{m}^3$ **2.** $147\,\text{m}^3$ **3.** $20.42\,\text{m}^3$

Exercise 147 (page 325)

1. (a) 2 A (b) 50 V (c) 2.5 A   **2.** (a) 2.5 mV (b) 3 A
3. 0.093 As, 3.1 A   **4.** (a) 31.83 V (b) 0
5. $49.13\,\text{cm}^2$, 368.5 kPa

Exercise 148 (Page 325)

1. (c)   **2.** (d)   **3.** (b)   **4.** (c)   **5.** (a)

Chapter 30

Exercise 149 (page 330)

1. A scalar quantity has magnitude only; a vector
quantity has both magnitude and direction.
2. scalar   **3.** scalar   **4.** vector   **5.** scalar
6. scalar   **7.** vector   **8.** scalar   **9.** vector

Exercise 150 (page 336)

1. 17.35 N at $18.00°$ to the 12 N force
2. 13 m/s at $22.62°$ to the 12 m/s velocity
3. 16.40 N at $37.57°$ to the 13 N force
4. 28.43 N at $129.29°$ to the horizontal
5. 32.31 m at $21.80°$ to the 30 m displacement
6. 14.72 N at $-14.72°$ to the 5 N force
7. 29.15 m/s at $29.04°$ to the horizontal
8. 9.28 N at $16.70°$ **9.** 6.89 m/s at $159.56°$
10. 15.62 N at $26.33°$ to the 10 N force
11. 21.07 knots, E $9.22°$ S

Exercise 151 (page 339)

1. (a) 54.0 N at $78.16°$ (b) 45.64 N at $4.66°$
2. (a) 31.71 m/s at $121.81°$ (b) 19.55 m/s at $8.63°$

Exercise 152 (page 340)

1. 83.5 km/h at $71.6°$ to the vertical
2. 4 minutes 55 seconds, $60°$
3. 22.79 km/h, E $9.78°$ N

Exercise 153 (page 341)

1. $i-j-4k$ **2.** $4i+j-6k$
3. $-i+7j-k$ **4.** $5i-10k$
5. $-3i+27j-8k$ **6.** $-5i+10k$
7. $i+7.5j-4k$ **8.** $20.5j-10k$
9. $3.6i+4.4j-6.9k$ **10.** $2i+40j-43k$

Exercise 154 (Page 341)

1. (c)   **2.** (c)   **3.** (a)   **4.** (a)   **5.** (d)
6. (b)   **7.** (d)   **8.** (a)   **9.** (b)   **10.** (c)

Chapter 31

Exercise 155 (page 345)

1. $4.5\sin(A+63.5°)$
2. (a) $20.9\sin(\omega t+0.63)$ volts
      (b) $12.5\sin(\omega t-1.36)$ volts
3. $13\sin(\omega t+0.393)$ volts

Exercise 156 (page 346)

1. $4.5\sin(A+63.5°)$
2. (a) $20.9\sin(\omega t+0.62)$ volts
      (b) $12.5\sin(\omega t-1.33)$ volts   **3.** $13\sin(\omega t+0.40)$

Exercise 157 (page 348)

1. $4.472\sin(A+63.44°)$
2. (a) $20.88\sin(\omega t+0.62)$ volts
      (b) $12.50\sin(\omega t-1.33)$ volts
3. $13\sin(\omega t+0.395)$ **4.** $11.11\sin(\omega t+0.324)$
5. $8.73\sin(\omega t-0.173)$ **6.** $1.01\sin(\omega t-0.698)$A
7. $s=85\sin(\omega t+0.490)$ mm

Exercise 158 (page 350)

1. $11.11\sin(\omega t+0.324)$A
2. $8.73\sin(\omega t-0.173)$V

3. $i = 21.79\sin(\omega t - 0.639)$A

4. $v = 5.695\sin(\omega t + 0.695)$V

5. $x = 14.38\sin(\omega t + 1.444)$m

6. (a) $305.3\sin(314.2t - 0.233)$ V (b) 50 Hz

7. (a) $10.21\sin(628.3t + 0.818)$ V (b) 100 Hz

(c) 10 ms

8. (a) $79.83\sin(300\pi t + 0.352)$V (b) 150 Hz

(c) 6.667 ms

9. $150.6\sin(\omega t - 0.247)$ volts

Exercise 159 (Page 350)

1. (b) **2.** (d) **3.** (c) **4.** (b) **5.** (a)

Chapter 32

Exercise 160 (page 355)

1. (a) continuous (b) continuous (c) discrete
(d) continuous

2. (a) discrete (b) continuous (c) discrete (d) discrete

Exercise 161 (page 358)

1. If one symbol is used to represent 10 vehicles, working correct to the nearest 5 vehicles, gives 3.5, 4.5, 6, 7, 5 and 4 symbols respectively.

2. If one symbol represents 200 components, working correct to the nearest 100 components gives: Mon 8, Tues 11, Wed 9, Thurs 12 and Fri 6.5.

3. Six equally spaced horizontal rectangles, whose lengths are proportional to 35, 44, 62, 68, 49 and 41, respectively.

4. Five equally spaced horizontal rectangles, whose lengths are proportional to 1580, 2190, 1840, 2385 and 1280 units, respectively.

5. Six equally spaced vertical rectangles, whose heights are proportional to 35, 44, 62, 68, 49 and 41 units, respectively.

6. Five equally spaced vertical rectangles, whose heights are proportional to 1580, 2190, 1840, 2385 and 1280 units, respectively.

7. Three rectangles of equal height, subdivided in the percentages shown in the columns of the question. P increases by 20% at the expense of Q and R.

8. Four rectangles of equal height, subdivided as follows: week 1: 18%, 7%, 35%, 12%, 28%; week 2: 20%, 8%, 32%, 13%, 27%; week 3: 22%, 10%, 29%, 14%, 25%; week 4: 20%, 9%, 27%, 19%, 25%. Little change in centres A and B, a reduction of about 8% in C, an increase of about 7% in D and a reduction of about 3% in E.

9. A circle of any radius, subdivided into sectors having angles of $7.5°, 22.5°, 52.5°, 167.5°$ and $110°$, respectively.

10. A circle of any radius, subdivided into sectors having angles of $107°, 156°, 29°$ and $68°$, respectively.

11. (a) £495 (b) 88 **12.** (a) £16 450 (b) 138

Exercise 162 (page 364)

1. There is no unique solution, but one solution is:
39.3–39.4 1; 39.5–39.6 5; 39.7–39.8 9; 39.9–40.0 17; 40.1–40.2 15; 40.3–40.4 7; 40.5–40.6 4; 40.7–40.8 2.

2. Rectangles, touching one another, having midpoints of $39.35, 39.55, 39.75, 39.95, \ldots$ and heights of $1, 5, 9, 17, \ldots$

3. There is no unique solution, but one solution is:
20.5–20.9 3; 21.0–21.4 10; 21.5–21.9 11; 22.0–22.4 13; 22.5–22.9 9; 23.0–23.4 2.

4. There is no unique solution, but one solution is:
1–10 3; 11–19 7; 20–22 12; 23–25 11; 26–28 10; 29–38 5; 39–48 2.

5. 20.95 3; 21.45 13; 21.95 24; 22.45 37; 22.95 46; 23.45 48

6. Rectangles, touching one another, having midpoints of 5.5, 15, 21, 24, 27, 33.5 and 43.5. The heights of the rectangles (frequency per unit class range) are 0.3, 0.78, 4, 4.67, 2.33, 0.5 and 0.2.

7. (10.95 2), (11.45 9), (11.95 19), (12.45 31), (12.95 42), (13.45 50)

8. A graph of cumulative frequency against upper class boundary having co-ordinates given in the answer to Problem 7.

9. (a) There is no unique solution, but one solution is:
2.05–2.09 3; 2.10–2.14 10; 2.15–2.19 11; 2.20–2.24 13; 2.25–2.29 9; 2.30–2.34 2.

(b) Rectangles, touching one another, having midpoints of $2.07, 2.12, \ldots$ and heights of $3, 10, \ldots$

(c) Using the frequency distribution given in the solution to part (a) gives 2.095 3; 2.145 13; 2.195 24; 2.245 37; 2.295 46; 2.345 48

(d) A graph of cumulative frequency against upper class boundary having the co-ordinates given in part (c).

Exercise 163 (Page 365)

1. (a) **2.** (d) **3.** (c) **4.** (c) **5.** (b)

Chapter 33

Exercise 164 (page 369)

1. Mean 7.33, median 8, mode 8
2. Mean 27.25, median 27, mode 26
3. Mean 4.7225, median 4.72, mode 4.72
4. Mean 115.2, median 126.4, no mode

Exercise 165 (page 370)

1. 23.85 kg   **2.** 171.7 cm
3. Mean 89.5, median 89, mode 88.2
4. Mean 2.02158 cm, median 2.02152 cm, mode 2.02167 cm

Exercise 166 (page 372)

1. 4.60   **2.** 2.83 μF
3. Mean 34.53 MPa, standard deviation 0.07474 MPa
4. 0.296 kg   **5.** 9.394 cm   **6.** 0.00544 cm

Exercise 167 (page 373)

1. 30, 27.5, 33.5 days   **2.** 27, 26, 33 faults
3. $Q_1 = 164.5$ cm, $Q_2 = 172.5$ cm, $Q_3 = 179$ cm, 7.25 cm
4. 37 and 38; 40 and 41   **5.** 40, 40, 41; 50, 51, 51

Exercise 168 (Page 373)

1. (d) **2.** (a) **3.** (b) **4.** (c) **5.** (d)
6. (b) **7.** (b) **8.** (c) **9.** (a) **10.** (d)

Chapter 34

Exercise 169 (page 379)

1. (a) $\frac{2}{9}$ or 0.2222 (b) $\frac{7}{9}$ or 0.7778

2. (a) $\frac{23}{139}$ or 0.1655 (b) $\frac{47}{139}$ or 0.3381
(c) $\frac{69}{139}$ or 0.4964

3. (a) $\frac{1}{6}$ (b) $\frac{1}{6}$ (c) $\frac{1}{36}$ **4.** $\frac{7}{10}$ or 0.7 or 70%

5. $\frac{5}{36}$ **6.** (a) $\frac{2}{5}$ (b) $\frac{1}{5}$ (c) $\frac{4}{15}$ (d) $\frac{13}{15}$

7. (a) $\frac{1}{250}$ (b) $\frac{1}{200}$ (c) $\frac{9}{1000}$ (d) $\frac{1}{50000}$

Exercise 170 (page 381)

1. (a) 0.6 (b) 0.2 (c) 0.15   **2.** (a) 0.64 (b) 0.32
3. 0.0768   **4.** (a) 0.4912 (b) 0.4211
5. (a) 89.38% (b) 10.25%
6. (a) 0.0227 (b) 0.0234 (c) 0.0169

Exercise 171 (Page 382)

1. (c) **2.** (d) **3.** (c) **4.** (b) **5.** (b)
6. (d) **7.** (a) **8.** (a) **9.** (b) **10.** (c)

Chapter 35

Exercise 172 (page 386)

1. 1, 5, 21, 9, 61   **2.** 0, 11, −10, 21   **3.** proof

Exercise 173 (page 387)

1. 16, 8

Exercise 174 (page 390)

1. $28x^3$ **2.** 2   **3.** $2x-1$
4. $6x^2-5$ **5.** $-\frac{1}{x^2}$ **6.** 0
7. $1+\frac{2}{x^3}$ **8.** $15x^4-8x^3+15x^2+2x$
9. $-\frac{6}{x^4}$ **10.** $4-8x$ **11.** $\frac{1}{2\sqrt{x}}$
12. $\frac{3}{2}\sqrt{t}$ **13.** $-\frac{3}{x^4}$
14. $3+\frac{1}{2\sqrt{x^3}}-\frac{1}{x^2}$ **15.** $2x+2$
16. $1+\frac{3}{2\sqrt{x}}$ **17.** $2x-2$
18. $-\frac{10}{x^3}+\frac{7}{2\sqrt{x^9}}$ **19.** $6t-12$ **20.** $1-\frac{4}{x^2}$

21. (a) 6 (b) $\dfrac{1}{6}$ (c) 3 (d) $-\dfrac{1}{16}$ (e) $-\dfrac{1}{4}$ (f) -7

22. $12x - 3$ (a) -15 (b) 21

23. $6x^2 + 6x - 4, 32$ **24.** $-6x^2 + 4, -9.5$

Exercise 175 (page 392)

1. (a) $12\cos 3x$ (b) $-12\sin 6x$
2. $6\cos 3\theta + 10\sin 2\theta$ **3.** -0.707 **4.** -3
5. $270.2\,\text{A/s}$ **6.** $1393.4\,\text{V/s}$
7. $12\cos(4t + 0.12) + 6\sin(3t - 0.72)$

Exercise 176 (page 394)

1. (a) $15e^{3x}$ (b) $-\dfrac{4}{7e^{2x}}$ **2.** $\dfrac{5}{\theta} - \dfrac{4}{\theta} = \dfrac{1}{\theta}$ **3.** 16
4. 2.80   **5.** 664

Exercise 177 (page 394)

1. (a) -1 (b) 16
2. $-\dfrac{4}{x^3} + \dfrac{2}{x} + 10\sin 5x - 12\cos 2x + \dfrac{6}{e^{3x}}$

Exercise 178 (page 395)

1. (a) $36x^2 + 12x$ (b) $72x + 12$ **2.** $8 + \dfrac{2}{x^3}$
3. (a) $\dfrac{4}{5} - \dfrac{12}{t^5} + \dfrac{6}{t^3} + \dfrac{1}{4\sqrt{t^3}}$ (b) -4.95
4. $-12\sin 2t - \cos t$ **5.** Proof

Exercise 179 (page 396)

1. $-2542\,\text{A/s}$ **2.** (a) $0.16\,\text{cd/V}$ (b) $312.5\,\text{V}$
3. (a) $-1000\,\text{V/s}$ (b) $-367.9\,\text{V/s}$
4. $-1.635\,\text{Pa/m}$ **5.** $1.947\,\text{m/s}$

Exercise 180 (page 397)

1. $x\cos x + \sin x$ **2.** $x(2\cos 3x - 3x\sin 3x)$
3. $4(1 + \ln 3x)$ **4.** $e^{3t}\left(4\cos 4t + 3\sin 4t\right)$
5. $e^{4\theta}\left(\dfrac{1}{\theta} + 4\ln 3\theta\right)$ **6.** 3.947
7. 1.256

Exercise 181 (page 399)

1. $\dfrac{x\cos x - \sin x}{x^2}$
2. $-\dfrac{6}{x^4}(x\sin 3x + \cos 3x)$
3. $\dfrac{3(1 - x^2)}{(x^2 + 1)^2}$ **4.** $\dfrac{2(1 - \ln 2x)}{4x^2}$
5. $\dfrac{e^x}{\cos^2 x}(x\cos x + \cos x + x\sin x)$
6. -18 **7.** -1.99

Exercise 182 (page 400)

1. $8(2x - 1)^3$ **2.** $5(2x^3 - 5x)^4(6x^2 - 5)$
3. $-20\sin(4x + 3)$ **4.** $6\cos(3\theta - 2)$
5. $-10\cos^4 x\sin x$ **6.** $\dfrac{5(2 - 3x^2)}{(x^3 - 2x + 1)^6}$
7. $12e^{3t+1}$ **8.** 1.86

Exercise 183 (Page 400)

1. (b)	**2.** (d)	**3.** (b)	**4.** (c)	**5.** (c)
6. (a)	**7.** (b)	**8.** (a)	**9.** (d)	**10.** (c)
11. (a)	**12.** (c)	**13.** (d)	**14.** (a)	**15.** (b)

Chapter 36

Exercise 184 (page 405)

1. (a) $4x + c$ (b) $\dfrac{7x^2}{2} + c$

2. (a) $\dfrac{5}{4}x^4 + c$ (b) $\dfrac{3}{8}t^8 + c$

3. (a) $\dfrac{2}{15}x^3 + c$ (b) $\dfrac{5}{24}x^4 + c$

4. (a) $\dfrac{2}{5}x^5 - \dfrac{3}{2}x^2 + c$ (b) $2t - \dfrac{3}{4}t^4 + c$

5. (a) $\dfrac{3x^2}{2} - 5x + c$

     (b) $4\theta + 2\theta^2 + \dfrac{\theta^3}{3} + c$

6. (a) $\dfrac{5}{2}\theta^2 - 2\theta + \theta^3 + c$

     (b) $\dfrac{3}{4}x^4 - \dfrac{2}{3}x^3 + \dfrac{3}{2}x^2 - 2x + c$

7. (a) $-\dfrac{4}{3x} + c$ (b) $-\dfrac{1}{4x^3} + c$

8. (a) $\frac{4}{5}\sqrt{x^5}+c$ (b) $\frac{1}{9}\sqrt[4]{x^9}+c$

9. (a) $\frac{10}{\sqrt{t}}+c$ (b) $\frac{15}{7}\sqrt[5]{x}+c$

10. (a) $\frac{3}{2}\sin 2x+c$ (b) $-\frac{7}{3}\cos 3\theta+c$

11. (a) $-6\cos\frac{1}{2}x+c$ (b) $18\sin\frac{1}{3}x+c$

12. (a) $\frac{3}{8}e^{2x}+c$ (b) $\frac{-2}{15e^{5x}}+c$

13. (a) $\frac{2}{3}\ln x+c$ (b) $\frac{u^2}{2}-\ln u+c$

14. (a) $8\sqrt{x}+8\sqrt{x^3}+\frac{18}{5}\sqrt{x^5}+c$

    (b) $-\frac{1}{t}+4t+\frac{4t^3}{3}+c$

Exercise 185 (page 407)

1. (a) 1.5 (b) 0.5   **2.** (a) 105 (b) -0.5

3. (a) 6 (b) -1.333 **4.** (a) -0.75 (b) 0.8333

5. (a) 10.67 (b) 0.1667   **6.** (a) 0 (b) 4

7. (a) 1 (b) 4.248   **8.** (a) 0.2352 (b) 2.638

9. (a) 19.09 (b) 2.457   **10.** (a) 0.2703 (b) 9.099

11. $77.7\,\text{m}^3$

Exercise 186 (page 411)

1. proof   **2.** proof   **3.** 32   **4.** 29.33 N m

5. 37.5   **6.** 7.5   **7.** 1

8. 1.67   **9.** 2.67   **10.** 140 m

Exercise 187 (Page 412)

1. (d)   **2.** (a)   **3.** (b)   **4.** (a)   **5.** (c)
6. (a)   **7.** (b)   **8.** (d)   **9.** (d)   **10.** (c)
11. (b)   **12.** (b)   **13.** (d)   **14.** (c)   **15.** (c)

Chapter 37

Exercise 188 (page 417)

1. 21, 25   **2.** 48, 96   **3.** 14, 7
4. $-3, -8$ **5.** 50, 65   **6.** 0.001, 0.0001
7. 54,79

Exercise 189 (page 418)

1. 1, 3, 5, 7, ..   **2.** 7, 10, 13, 16, 19, ..
3. 6, 11, 16, 21, ..   **4.** $5n$
5. $6n-2$ **6.** $2n+1$
7. $4n-2$ **8.** $3n+6$
9. $6^3(=216)$, $7^3(=343)$

Exercise 190 (page 419)

1. 68   **2.** 6.2   **3.** 85.25
4. 23.5   **5.** 11th   **6.** 209
7. 346.5

Exercise 191 (page 420)

1. -0.5 **2.** 1.5, 3, 4.5
3. 7808   **4.** 25
5. 8.5, 12, 15.5, 19
6. (a) 120 (b) 26070 (c) 250.5
7. £19,840, £223,680   **8.** £8720

Exercise 192 (page 422)

1. 2560   **2.** 273.25   **3.** 512, 4096
4. 812.5   **5.** 8   **6.** $1\frac{2}{3}$

Exercise 193 (page 424)

1. (a) 3 (b) 2 (c) 59022
2. 10 th
3. £1566, 11 years
4. 56.68 M
5. 71.53 g
6. (a) £362.07 (b) 22 years
7. 100, 139, 193, 268, 373, 518, 720, 1000 rev/min.

Exercise 194 (Page 424)

1. (d)   **2.** (a)   **3.** (b)   **4.** (b)   **5.** (d)
6. (c)   **7.** (d)   **8.** (b)   **9.** (c)   **10.** (a)

Chapter 38

Exercise 195 (page 427)

1. (a) 6_{10} (b) 11_{10} (c)14_{10} (d) 9_{10}
2. (a) 21_{10} (b) 25_{10} (c) 45_{10} (d) 51_{10}

3. (a) 42_{10} (b) 56_{10} (c) 65_{10} (d) 184_{10}
4. (a) 0.8125_{10} (b) 0.78125_{10} (c) 0.21875_{10}
   (d) 0.34375_{10}
5. (a) 26.75_{10} (b) 23.375_{10} (c) 53.4375_{10}
   (d) 213.71875_{10}

Exercise 196 (page 429)

1. (a) 101_2 (b) 1111_2 (c) 10011_2 (d) 11101_2
2. (a) 11111_2 (b) 101010_2 (c) 111001_2 (d) 111111_2
3. (a) 101111_2 (b) 111100_2 (c) 1001001_2
   (d) 1010100_2
4. (a) 0.01_2 (b) 0.001111_2 (c) 0.01001_2 (d) 0.10011_2
5. (a) 101111.01101_2 (b) 11110.1101_2
   (c) 110101.11101_2 (d) 111101.10101_2

Exercise 197 (page 430)

1. 101
2. 1011
3. 10100
4. 101100
5. 1001000
6. 100001010
7. 1010110111
8. 1001111101
9. 111100
10. 110111
11. 110011
12. 1101110

Exercise 198 (page 432)

1. (a) 101010111_2 (b) 1000111100_2 (c) 10011110001_2
2. (a) 0.01111_2 (b) 0.1011_2 (c) 0.10111_2
3. (a) 11110111.00011_2 (b) 1000000010.0111_2
   (c) 11010110100.11001_2
4. (a) 7.4375_{10} (b) 41.25_{10} (c) 7386.1875_{10}

Exercise 199 (page 434)

1. 231_{10} 2. 44_{10} 3. 152_{10}
4. 753_{10} 5. 36_{16} 6. $C8_{16}$
7. $5B_{16}$ 8. EE_{16}

Exercise 200 (page 435)

1. $D7_{16}$ 2. EA_{16} 3. $8B_{16}$
4. $A5_{16}$ 5. 110111_2 6. 11101101_2
7. 10011111_2 8. 101000100001_2

Exercise 201 (Page 435)

1. (c) 2. (b) 3. (d) 4. (c) 5. (b)
6. (c) 7. (d) 8. (c) 9. (d) 10. (a)
11. (c) 12. (a) 13. (d) 14. (b) 15. (c)
16. (c) 17. (a) 18. (a) 19. (a) 20. (b)
21. (a) 22. (b) 23. (c) 24. (b) 25. (d)

Index